THE FAUNA OF BRITISH INDIA

INCLUDING

CEYLON AND BURMA

THE FAUNA OF BRITISH INDIA

INCLUDING

CEYLON AND BURMA.

ODONATA.

IN THREE VOLUMES

VOL. II

BY

LT.-COL. F. C. FRASER, I.M.S.

GYAN

Published by

Gyan Publishing House
5, Ansari Road
Daryaganj, New Delhi-110002
Phone: 011-47034999, 9811692060
E-mail: books@gyanbooks.com

Distribution Network
gyanbooks.com
India, USA, Canada, UK, Australia

ISBN: 978-81-212-0211-4 (Set)
978-81-212-0418-7 (HB)
First Published, 1934

2nd Impression 2024

Printed at: Gyan Press, Delhi.

THE FAUNA OF BRITISH INDIA INCLUDING CEYLON AND BURMA (VOL. II)

Author: F. C. FRASER

THE FAUNA OF BRITISH INDIA,

INCLUDING

CEYLON AND BURMA.

Published under the authority of the Secretary of State for India in Council.

EDITED BY LT.-COL. R. B. S. SEWELL, C.I.E., Sc.D., F.R.S., I.M.S.
ASSISTED BY F. W. EDWARDS, M.A., Sc.D.

ODONATA.

VOL. II.

BY

LT.-COL. F. C. FRASER, I.M.S.

LONDON:

TAYLOR AND FRANCIS, RED LION COURT, FLEET STREET.

October, 1934.

CONTENTS.

PREFACE.

THE first volume of 'The Fauna of British India' dealing with the Dragonflies (Order ODONATA) was published in 1933, and, in addition to furnishing an introduction to the study of these insects, dealt with the whole of the CŒNAGRIIDÆ or first family of the suborder ZYGOPTERA.

In this second volume the ZYGOPTERA are completed and the second suborder, ANISOPTERA, is introduced. Volume II thus deals with the families AGRIIDÆ and GOMPHIDÆ, the former forming a natural transition to the latter, which family contains the most archaic species of the ANISOPTERA, and is related to the AGRIIDÆ by the wide separation of the eyes.

A third suborder (ANISOZYGOPTERA) is recognized by most authors, but only one species is known from India, and that species only from a single, rather doubtful larva; I have included a brief reference to it for the sake of completeness (p. 151).

As in the case of the first volume, much of the text and several of the text-figures represent a mere revision of the serial "Indian Dragonflies" which appeared in the pages of the 'Journal of the Bombay Natural History Society.' The whole of the bibliography has been revised and brought up to date, species not included in that serial have here been dealt with, and most of the text-figures have been redrawn, many of those depicting wings being based on actual photographs, as giving more and truer details. Moreover, since the serial was written more material has been collected, the previously unknown opposite sexes of several species have been discovered, and the identity of other species has been more satisfactorily determined. Several notes published elsewhere have also been incorporated, and other species, new to our fauna, have been introduced.

My thanks are again due to many willing helpers; some have assisted by supplying material, whilst others have given valuable advice where needed. Among the former may be mentioned Mr. T. Bainbrigge Fletcher, former Imperial Entomologist to the Government of India, who sent me from time to time large consignments of dragonflies from Assam, Burma, and Bihar; Mr. Charles M. Inglis and Mr. H. V. O'Donel, who collected for me in Bihar, Bengal, and Sikkim; Mr. Charles Antram, for collections from Eastern Assam; Col. F. Wall, for collections from Burma and Ceylon; Lt.-Col. E. Frere, for collections from the Palni and Nilgiri Hills; and lastly, Mr. Charles Souter, for collections from the Agency Tracts and South Kanara.

In addition to this material, I have had access to the collections of several institutions, among which may be mentioned those of the Indian Museum, Calcutta; the Agriculture College, Pusa, Bihar; the Forest Research Institute, Dehra Dun; the Darjeeling Museum; the British Museum; and the Brussels Natural History Museum, these last two containing many of the types of the Indian fauna. More recently I have been enabled to examine the Indian fauna contained in the Williamson collection at Michigan University, during a brief stay at Ann Arbor.

For advice I have to thank Mr. Kenneth Morton of Edinburgh and Dr. F. F. Laidlaw of Uffculme, Devon, also Mr. D. E. Kimmins of the British Museum, who very kindly read through the proofs and undertook the laborious task of checking the Bibliography; and lastly, Dr. F. W. Edwards, who acted as Editor in the absence of Lt.-Col. R. B. S. Sewell. The work of preparing the volume was much lightened by the permission given to use the subject matter and text-figures of the serial "Indian Dragonflies" by the Committee of the Bombay Natural History Society, to whom my thanks are due.

Since the publication of the first volume we have to mourn the loss of Mr. E. B. Williamson of Indiana, one of the most painstaking students of the Order ODONATA of our time. His enthusiasm infected many other entomologists, and directed their attention to the insects of which he had made a life-long study; his inspiration remains to quicken others, and the magnificent collection of ODONATA, which is now housed at Ann Arbor, Michigan, remains a monument to his zeal as a field-collector.

The publication of Volume I of this series was beset with many difficulties owing to the death of our late Editor, Lt.-Col. J. Stephenson—a great loss to science. His death, coming shortly before the publication of Volume I, led to some of the final corrections not being included; these are given at the commencement of this volume, and it has also been thought advisable to republish the Glossary in a somewhat extended form.

F. C. FRASER.

London,
September, 1934.

GLOSSARY.

Allotype.—The specimen of a species of the opposite sex to the type first described, either at the same time as the type or later, and either by the original describer of the species or another author.

Anal appendages.—Short processes at the end of the abdomen of the adult insect which, in the male, are employed for seizing the female during copulation. There are either two pairs (ZYGOPTERA) or one pair and a single inferior appendage (ANISOPTERA) in the male, but only one pair in the female of both suborders.

Anal bridge (*Ab*) (fig. 5, B & C, vol. i).—A short longitudinal nervure running from *Ac* parallel to the under side of the discoidal cell and continued as the nervure *IA* in the ZYGOPTERA. It ends distally by meeting the short transverse nervure descending from the distal end of the discoidal cell. (The nervure is absent in some genera of the CŒNAGRIIDÆ.)

Anal crossing (*Ac*) (fig. 5, B & C, vol. i).—A short transverse nervure at the base of the wing, traversing the cubital space. It is the vestigial crossing-over of the anal nervure, to be continued as the nervure *IA*. In many genera of the AGRIIDÆ the cubital space is traversed by many accessory nervures, so that, in these, the nervure *Ac* becomes indistinguishable.

Anal loop (*Al*) (fig. 5, A, vol. i, and fig. 45, vol. ii).—An area of cells situated at the base of the hind-wing in many genera of the ANISOPTERA. It lies adjacent to and posterior to the discoidal cell. It may be altogether absent in some species and rudimentary in others, in which latter it consists of but 2 or 3 cells. In the ÆSCHNIDÆ, CORDULEGASTERIDÆ, and CORDULIINÆ it is oval or quadrate in shape, but in the LIBELLULINÆ is more or less elongated and stocking-shaped.

Anal triangle (*At*) (fig. 45, vol. ii).—A triangular area at the extreme base of the hind-wing in the ANISOPTERA, made up of one or more cells, occurring in the male only and occasionally absent even in this sex.

Antealar sinus.—A small triangular area on the dorsum of the thorax lying just in front of the attachments of the wings.

Anteclypeus.—The anterior or lower part of the clypeus to which the labrum is attached.

Antehumeral stripe.—A coloured stripe on the dorsum of the thorax situated internal to the humeral suture. It is to be contrasted with a *humeral stripe*, which lies on or borders the humeral suture.

Antennæ.—A pair of short, jointed, filamentous organs situated just in front of and to the inner side of the eyes. They are made up of four to seven segments in both the larva and imago, but some segments may be rudimentary in the former.

Antenodal nervures (fig. 5, A, B, & C, vol. i).—Short transverse nervures running from the costal or anterior border of the wings to the radius (*Ri*), proximal to the node. In the CŒNAGRIIDÆ there are only two of these, which are known as the "primary antenodals," and are found in many of the higher forms of the ANISOPTERA as especially thickened nervures, quite easily distinguishable from all the others. In other families they are numerous, ranging from six in *Libellago* to an enormous number in others. The anterior half of each nervure, traversing the costal space, is usually continuous with the posterior half, traversing the subcostal space; the two halves are invariably continuous in the ZYGOPTERA, and the same applies to the LIBELLULIDÆ, but in other Anisopterous genera they are separated.

Apical.—Parts of the wing or of segments of the abdomen or legs which are furthest from the thorax.

Arc (fig. 5, A & B, vol. i).—A short, oblique, transverse nervure near the base of the wings, which forms the outer boundary of the basal (or median) space.

Archaic.—Generalized or ancient. Applied to species in which the predominant characters are believed to be of ancient origin.

Auricle.—A small ear-shaped process situated on each side of the second abdominal segment. Variable in size; in the female always very small and rudimentary, and often entirely absent. Most commonly present in the GOMPHIDÆ and ÆSCHNIDÆ.

Basal.—Parts of the wing or of segments of the abdomen or legs which are nearest to the thorax.

Basal antenodal nervure.—An incomplete antenodal nervure situated at the extreme base of the wings in a few species of ODONATA It runs from the subcostal nervure to the radius (*Ri*), its anterior half in the costal space being absent. An important generic or specific character when present.

Basal postcostal nervure.—An accessory transverse nervure traversing the cubital space proximal to the nervure *Ac* and situated at the extreme base of the wings. It is present only in the most primitive genera of the CŒNAGRIIDÆ.

Basal space.—An elongated space at the extreme base of the wings bounded distally by the *arc*, the radius anteriorly, and the nervure *Cuii* posteriorly. In some genera it is traversed by one or more nervures.

Bicolorous.—Of two colours.

Bifurcation of Rs.—See "Forking of *Rs*."

Bridge (fig. 5, A, *Br*, vol. i).—A triangular space bounded outwardly by the oblique nervure descending from the node, anteriorly by the nervure *Ri* and posteriorly by *IRii*. Present only in the ANISOPTERA.

Carina.—A chitinous ridge on the dorsum of the thorax or of the abdominal segments.

Caudal gills.—Accessory respiratory organs found at the tail end of Zygopterous larvæ. They are three in number except for the LIBELLAGINÆ, which have only two, and are of very variable shape, thin and acuminate, lamellate and obtuse, or triquetral or bladder-shaped.

Cells of wings.—The reticulation of areolets forming the network of the wings. An area of the wings bounded by nervures.

Clypeus.—The lower or anterior part of the face, which is divided into the ante- and the postclypeus, the labrum being attached to the former.

Compressed.—Flattened from side to side.

Costa (fig. 5, A, B, & C, vol. i).—The anterior border of wing, which is also known as the *costal nervure* or costal border of wing.

Cotype.—Strictly, in cases where the author of a species does not select a single type (holotype), any typical specimen of the original series on which the description is based is a cotype. The term is sometimes more loosely used as equivalent to paratype.

Coxa.—The basal segment of the legs.

Crest of frons.—The anterior ridge of the frons which divides the lower anterior part from the upper posterior part of the frons. It is usually prominent in all the ANISOPTERA, but ill-defined in the ZYGOPTERA.

Cubital nervure (*Ac*).—A name commonly used for the *anal crossing*, which see.

Cubital space.—A space at the extreme base of the wings situated immediately posterior to the basal space and extending out as far as the base of the discoidal cell. It is traversed by the nervure *Ac*, and often by many cubital nervures.

Depressed.—Flattened from above downwards.

Discoidal cell.—A well-defined triangular or quadrilateral space situated near the base of the wings, immediately distal to the cubital space. It is invariably four-sided in the ZYGOPTERA, although the sides are not equal. It is triangular in the ANISOPTERA, and only occasionally four-sided in archaic species. The Anisopterous cell is actually divided by a longitudinal nervure into an upper or superior triangle known as the "hypertrigone," and an inferior triangle, which latter is here regarded as the discoidal cell proper. Conversely the discoidal cell of the ZYGOPTERA may be regarded as a fusion of the hypertrigone with the inferior triangle.

Discoidal field.—The space lying distal to the discoidal cell and bounded outwardly by the border of the wing. The nervures *MA* and *Cuii* constitute its anterior and posterior borders respectively.

Distal.—Equivalent to *apical*, which see. "Distalia" is a collective term sometimes applied to several distal segments of the abdomen.

Divaricate.—Diverging from the base or origin. A term usually applied to the anal appendages when they splay out from their attachments.

Dorsal.—Adjective applied to the upper surface of any part, especially the superior surface of the thorax or any abdominal segment.

Dorsum.—The back or superior surface of any part, such as the thorax or any abdominal segment.

Enfumed.—Smoky or brownish in tint.

Entire.—Adjective applied to any space of the wings when such are not traversed by any nervures.

Epimeron.—The posterior division of the side of the thorax.

Epistome.—The postclypeus or upper part of clypeus, which see.

Exuviæ.—The cast skin of a larva after each moult or final metamorphosis. (The word only exists in the plural, although employed to designate single objects.)

Femur (pl. *femora*).—The thigh or 3rd segment of the legs.

Foramen.—An opening or enclosed space. The superior anal appendages sometimes curve and finally meet at their apices so as to enclose a foramen.

Forcipate.—Shaped like a pair of forceps.

Forking of Rs.—The bifurcation of the superior sector of the arc into *Rii* and *Riv+v*. This forking may be equal or unequal in the GOMPHIDÆ, and is an important character employed in classification.

Frons.—The forehead.

Fusiform.—Spindle-shaped.

Genitalia.—The sexual organs. In the male dragonfly these organs are situated on the 2nd and 9th and 10th segments, at the junction of the latter two of which is found a gonopore from which the spermatozoa are passed into the lobe on segment 2. This gonopore is very similar in all dragonflies, but the structures on the 2nd segment vary very widely, and so have been employed largely for purposes of classification. In this volume, following the practice of most students of dragonflies, the word "genitalia," is used to designate the structures on the 2nd abdominal segment only.

The genitalia of the female are situated on the ventral surface of the 8th and 9th abdominal segments.

Genotype.—The type-species of a genus, to which reference should be made in the event of any question arising as to the definition of the genus.

Hamules.—Two pairs of minute hooks found in the genitalia on the 2nd segment of the male, and employed in copulation.

Hirsute.—Coated with long hair.

Holotype.—Equivalent to type in the strict sense, *q. v.*

Humeral—The shoulders of the thorax or outer borders of the dorsum of thorax.

Humeral stripe.—A coloured stripe situated on the humeral suture of thorax or bordering it inwardly.

Humeral suture.—A false joint of the thorax situated along the outer border of the dorsum. It is indicated by a fine, shallowly sunken line.

Hyaline.—Transparent or colourless.

Hypertrigone.—A narrow triangular cell situated longitudinally above the discoidal cell in the Anisopterous wing. See "Discoidal cell."

Imago.—The adult insect.

Inferior anal appendages.—See "Anal appendages."

Jugal suture.—Transverse sutures found on the basal abdominal segments of many species of Anisopterous dragonflies.

Labium.—The lower lip.

Labrum.—The upper lip.

Lamella.—The flattened and broadened segment of the antennæ found in larvæ belonging to the genera *Lamelligomphus* and *Sieboldius.*

Lamina.—An overlapping plate found at the anterior end of the genital sac on the second abdominal segment.

Larva.—A term applied to any stage preceding the imago or final stage of an insect, which differs so markedly from the imago that a definite metamorphosis must take place from one to the other. The term "nymph," has been applied by many students to the larval stages of dragonflies on account of the metamorphosis being incomplete, in that there is no true resting stage comparable to the pupal stage of the LEPIDOPTERA.

Lilaceous.—Pale purplish, the colour of lilac blossom.

MA.—The anterior median nervure or the continuation of the lower sector of the arc.

Median space.—An alternative term for the "basal space," which see.

Membrane.—A narrow membrane bordering the base of the wing in Anisopterous dragonflies. It is sometimes obsolete. In this volume the term is used to denote this special membrane, and not the whole of the wing between the nervures as understood by most entomologists.

Mesepimeron.—The middle division of the sides of the thorax.

Mesothorax.—The 2nd segment of the thorax, which, in dragonflies, is fused to the metathorax to form the "synthorax."

Metathorax.—The posterior or 3rd segment of the thorax. See "Synthorax."

Metepimeron.—The 2nd or posterior part of the sides of the thorax.

Mid-dorsal carina.—See "Carina."

Nervure.—One of the veins or skeletal structures of the wing.

Neuration.—The network of veins, nervures or thickened ridges on the wings.

Nodal index.—The number of ante- and postnodal transverse nervures present in the fore- and hind-wings. The numbers nearest the vertical line in the formula indicate the number of antenodal nervures of each of the four wings of one specimen.

Node.—A thickening situated at an indentation of the costal margin of the wings. In the ZYGOPTERA the node usually lies nearer the base of the wing than the apex, but in the ANISOPTERA it may be at the middle of the wing or actually nearer the apex than the base of wing.

Nymph.—See "Larva."

Oblique nervure.—An oblique nervure situated between the nervures *Riii* and *IRiii* slightly distal to the outer end of the bridge. It is duplicated in some archaic species.

Obsolete.—Degenerating or becoming invisible.

Obtuse.—Blunt.

Occiput.—The back of the head.

Ocellus.—A simple eye as contrasted with the compound eyes. All dragonflies possess three such simple eyes, which are situated in a line in front of the vesicle or arranged in a triangle around it on the vertex of the head.

Ovipositor.—The female's apparatus for depositing her eggs either in water or in the tissues of plants. It is situated on the ventral surfaces of the 8th and 9th abdominal segments and, in the ZYGOPTERA and many Æschnines, is enclosed between two strong vulvar scales which lie on either side of it. The ovipositor is poorly developed in those females which drop their eggs in water, but is a conspicuous organ in those which insert their eggs in plants.

Paratype.—Any typical specimen, except the actual type (holotype), belonging to the original series before an author at the time when he describes a species, whether the specimens were all collected at the same time and place or not. In the first of these volumes on ODONATA the term was used to denote specimens lodged in accredited public or reference collections for the use of students, and diagnosed by approved specialists as similar in all respects to the type.

Pectinate.—Comb-like. A term applied to the nervure *IA* when it is multi-branched.

Petiolate.—Stalked. A term applied to the wings when the base is markedly constricted like a stalk. The length of the stalk of a wing is measured from its base to where the posterior border meets the nervure *IA*. All the wings of the CŒNAGRIIDÆ, and most of those of the AGRIIDÆ, are stalked, but it is rare to find even a vestige of such in the ANISOPTERA.

Postanal cell.—Cells lying between the cubital nervure (*Cuii*) and the posterior border of the wings proximal to the level of the discoidal cell. Those postanal cells adjacent to the discoidal cell in the hind-wings of the ANISOPTERA form the "anal loop."

Postclypeus.—The posterior or upper part of the clypeus. See "Clypeus."

Posterior lobe of prothorax.—The prothorax is composed of three lobes fused together, of which the posterior is the most variable in shape. For this reason its shape is extensively employed for differentiating genera and species of dragonflies (fig. 121, *c*, vol. i).

Postnodal nervures (fig. 5, A, B, & C, *pn*, vol. i, and fig. 1, *Pns*, vol. ii).—Short transverse nervures running between the costal border of wing and the radius (*Ri*), distal to the node.

Primary antenodals.—The two thickened antenodal nervures found in the wings of many species of ANISOPTERA, which represent the two primitive antenodal nervures of the CŒNAGRIIDÆ. One of these is always the antenodal nearest the base of the wing, whilst the other is about the 4th to the 7th from the base.

Prothorax.—The anterior segment of the thorax, which in the Odonata is separated from the two posterior segments. See "Posterior lobe of prothorax."

Pruinescence.—A white powdery efflorescence which appears on the bodies of adult dragonflies, and on the males especially. It is often confined to the abdomen, but may be found occasionally on most of the body. In other species it is strictly confined to the vertex of the head, dorsum of thorax, and dorsum of the terminal segments of the abdomen.

Pruinosed.—Coated with pruinescence.

Pterostigma (fig. 5, A, B, & C, *Pt*, vol. i.).—A small, thickened area of the wing situated on the costal border near its apex, usually dark or differing conspicuously in colour from the remainder of the wing, but extremely variable in shape and colour. It may be absent in one pair or all of the wings, or in one or both sexes. When traversed by nervures it is said to be a false or "pseudo-stigma."

Pubescent.—Coated with short hairs.

Pulverulent.—Pruinosed, which see.

Radius (fig. 5, A, B, & C, *Ri*, vol. i, and fig. 1, *Ri*, vol. ii).—The principal nervure of the wing situated between the subcosta and the media.

Riv+v (fig. 5, A, B, & C, *Riv+v*, vol. i, and fig. 1, *Riv+v*, vol. ii).—One of the branches of the radius in prolongation of the superior sector of the arc.

Reticulation.—The minute network or nervures on the wings.

Sagittate.—Shaped like the head of an arrow.

Sectors of arc.—The origins of the nervures *Riv+v* and *MA*. These may have a common origin, as is usual, or they may be slightly separated at their origins from the arc.

Serrate.—Saw-like edge.

Spiracle.—Tracheal openings or breathing pores. There are ten pairs of these in the dragonfly imago, two pairs being situated on the thorax and the others on each of the first eight abdominal segments.

Sternite.—The ventral plate of any segment of the thorax or abdomen.

Subcosta (fig. 5, A, B, & C, *Sc*, vol. i).—A nervure running parallel to and next to the costa, extending from the base of the wing to as far as the node in dragonflies.

Subnode.—An oblique nervure extending obliquely outward into the body of the wing and traversing the space lying between *Ri* and *Rii*.

Subtrigone (fig. 1, *St*, vol. ii).—A triangular area of the wings of ANISOPTERA lying adjacent and proximal to the discoidal cell. It may consist of a single cell or be split up by traversing nervures into three or more cells. In many cases it is merged with and is indistinguishable from the distal end of the cubital space.

Superior anal appendages.—See "Anal appendages."

Suture.—A joint between two segments or subdivisions of segments.

Synthorax.—The thorax of the dragonfly is composed of two segments, the prothorax bearing the fore-legs and the synthorax bearing the wings and middle and hind legs, the synthorax being formed by a fusion of the meso- and metathorax.

Tarsus.—The distal or terminal part of the legs, which is made up of three segments in the adult dragonfly. In the larvæ of the GOMPHIDÆ the tarsi of the fore and middle legs possess only two joints, but the hind tarsus is similar to that of the imago. In all other dragonfly larvæ the tarsi of each leg possess three segments.

Teneral.—Adjective applied to the freshly emerged dragonfly before it has hardened and assumed its mature colouring.

Tergum.—The upper part of the thorax lying between the roots of the wings.

Termen.—The posterior border of the wings.

Thorax.—The middle part of the body, bearing the wings and the legs; in these volumes the word is generally employed as synonymous with synthorax (*i. e.*, excluding the prothorax).

Tibia.—The 4th segment of the legs, situated between the femur and the tarsus.

Tornus.—The angle formed by the meeting of the basal and posterior borders of the wings, which is only pronounced in the ANISOPTERA.

Trochanter.—The small 2nd segment of the legs, between the coxa and femur.

Type.—In the strict sense (equivalent to holotype), as used in 'The Fauna of British India,' the type is a specimen selected by the author of a description of a species as the single original of that description, and it must be the ultimate criterion for reference as to the identity of the species. (In a wider and older sense, not used in these volumes, the term is sometimes equivalent to *cotype*, i.e., it covers all the specimens of the original series before the author when a species was described *.)

Ventral.—Adjective applicable to the under surface of any part, especially the underside of the thorax or abdomen.

Vertex.—The upper surface of the head, which bears the vesicle and ocelli.

Vesicle.—A small eminence situated on the vertex of the head and which either overhangs the ocelli or is situated between them. In some species, such as those of the genus *Idionyx*, the vesicle is highly specialized in the males. In the GOMPHIDÆ it is usually deeply bifid or produced into two points.

Vulvar scales.—The protective sheath of the ovipositor, formed of two plate-like structures and attached to the sides of the ventral plates of the 8th and 9th abdominal segments of the female imago. They are small and inconspicuous in most of the ANISOPTERA, but prominent structures in most of the ZYGOPTERA and archaic species of ANISOPTERA.

---

* Various lists defining the terms *type*, *paratype*, etc., have been published. A list, including definitions of the several senses in which they are used, and of other cognate terms not used in these volumes, will be found in an article by Dr. Walther Horn, "Ueber den Musealen Missbrauch mit Insekten-Typen," X. Congrès International de Zoologie (Budapest, 1927), pt. 2, pp. 1021–1042 1929).

CORRIGENDA TO VOLUME I.

Page xii, line 14, for "*eximis*" read "*eximia*."

„ xii, line 37, *for* "Burmeister" *read* "(Burmeister)."

„ 30. After the last line of couplet 8 *add* :—

"Sides of thorax greenish-blue, with two thick black stripes enclosing a narrow stripe of the ground-colour .... *L. nigriceps* [Fras., p. 20."

„ 54, line 27, *for* "Baghwonie" *read* "Baghownie."

„ 78, line 13, *for* "3" *read* "2 to nearly 4."

„ 78, line 36, *after* "resembles" *add* "Genotype, *Indolestes indica* Fras."

„ 78, after line 12 from bottom of page *insert* the following key :—

Key to the Species of Indolestes.

1. { The mid-dorsal coloured stripe with superior and medial processes running out as far as the humeral suture .... *indica* Fras., [p. 78.
   The mid-dorsal thoracic colour-stripe without processes on its borders .... 2.

2. { The pterostigma four times as long as broad .... *assamica* Fras., [p. 83.
   The pterostigma only twice as long as broad .... *bilineata* (Selys). [p. 82.

„ 88, line 5 from bottom of page, *for* "Selys" *read* "(Selys)."

„ 92. **Rhinagrion mima**. *Delete* whole of line 4 of the synonymy.

Page 96. **Mesopodagrion tibetanum.** *Delete* lines 4 and 5 of the synonymy.

„ 121, line 7 from bottom of page, *after* "yellow" *add* "apices tipped with black in full adults only."

„ 126, line 27, *after* "India" *add* "Since found quite commonly on many streams in Travancore during September."

„ 141, line 5 from bottom of page, *for* "those" *read* "that."

„ 212. Key to Species of *Caconeura*, couplet 5, after "*verticalis verticalis*" add the following lines :—

"A red stripe traversing head from eye to eye; thorax entirely black beneath	*verticalis burmanensis*, p. 216."

„ 278, *for* couplet 6 at top of page, *substitute* the following :—

"Segment 9 with the black dorsal marking bifid; no apical black ring to this same segment; ground-colour differing from that of the male	*rubriceps* Selys, p. 298.
Segment 9 with the black dorsal marking bifid and with a black apical ring; ground-colour similar to that of the male	*ceylanicum* (Kirby), p. 299.
Segment 9 with the black dorsal marking not bifid	7."

„ 294, line 3, *for* "Kajibthatkhana" *read* "Rajabhatkhana."

„ 305, line 1, *for* "Burmeister" *read* "(Burmeister)."

„ 305, line 2, *for* "Burm." *read* "(Burm.)."

„ 310, line 34, *after* "Stockholm" *insert* "Museum."

„ 318, line 14, for "*risi*" read "*prætermissum.*"

„ 339, line 21, for "*occidentalis*" read "*occidentale.*"

„ 383, line 42, *for* "," *read* "and."

„ 383, line 43, *after* "Jalpaiguri" *insert* "District."

„ 395, line 22, *for* "Duars" *read* "Jalpaiguri District."

„ 406, lines 17, 21, and 25, for "*Agriocnemis*" read "*Argiocnemis*."

Page 406, line 18, *for* " Odom." *read* " Odon."

,, 408, line 23, *for* " Duars " *read* " Jalpaiguri District."

,, 419, line 10 from bottom of page, for " *camponi* " read " *campioni*."

,, 419, line 8 from bottom of page, for " *antelopsoides* " read " *antelopoides*."

,, 421, line 8, middle column, for "*Agriocnemis* " read "*Argiocnemis*."

,, 421, line 33, right column, for "*Agriocnemis* " read "*Argiocnemis*."

,, 421, line 40, right column, for " *nursi* " read " *nursei*."

,, 422, line 11 from bottom of page, middle column, for "*Agriocnemis* " read "*Argiocnemis*."

SYSTEMATIC INDEX

Suborder **ZYGOPTERA** (*continued*).

Family AGRIIDÆ. (Fig. 1.)

Dragonflies of medium or large size, with the characters of the suborder ZYGOPTERA, of which they comprise, roughly, one-third of the species.

Wings long and usually much broader than in CŒNAGRIIDÆ, of equal size and shape, but hind-wings not uncommonly broader than fore-wings, petiolate or not, and, in the former case, petiole always very short; venation intricate, very close; antenodal nervures usually numerous and always more than two in number; discoidal cells of same shape in

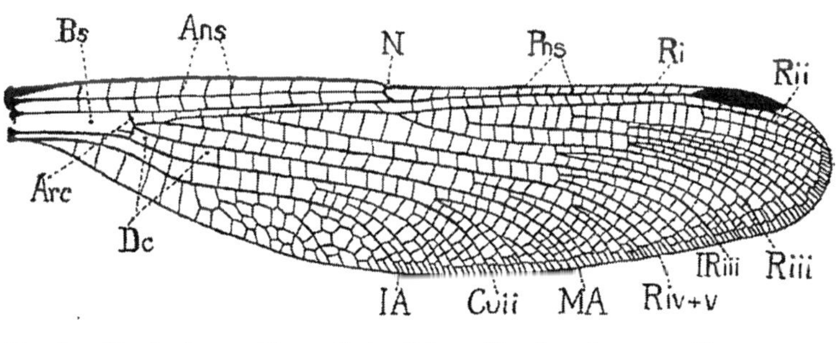

Fig. 1.—Typical venation of Agriidæ. *Bs*, basal or median space; *Ans*, *Pns*, antenodal and postnodal nervures; *N*, node; *Arc*, arc or arculus; *Dc*, discoidal cell; *Ri*, *Rii*, *Riii*, *IRiii*, *Riv+v*, *MA*, *Cuii* and *IA*, principal longitudinal nervures.

fore- and hind-wings, nearly always traversed, elongate, and rhomboidal; arc situated nearer base of wing than node; node situated about middle of wing. Abdomen variable in length, rarely longer than wings. Larvæ very different in the several subfamilies, possessing two or three caudal, and occasionally paired abdominal, gills; always breeding in running waters.

Distribution.—Cosmopolitan. The Indian fauna includes representatives of five subfamilies.

Key to Subfamilies of the Agriidæ.

1.	Petiolation of wings ending opposite to, or slightly proximal to, level of arc	2.
	Petiolation absent or ending far proximal to level of arc	3.
2.	Pterostigma short; discoidal cell traversed by a nervure, convex costalwards and of only half length of median space; cubital nervures numerous	CALIPHÆINÆ, [p. 148.
	Pterostigma long; discoidal cell entire, costal and hinder borders straight, not more than one-fifth length of median space; only a single cubital nervure	PHILOGANGINÆ, [p. 112.
3.	Sectors of arc arising from lower third of arc; discoidal cell convex costalwards and as long as median space	AGRIINÆ, p. 118.
	Sectors of arc arising from middle, or above middle, of arc; discoidal cell straight and shorter than median space	4.
4.	Epistome tumid and projecting markedly, like a nose, in front of face; abdomen shorter than wings	LIBELLAGINÆ, p. 2.
	Epistome normal, face depressed and sloping; abdomen always longer than wings........	EPALLAGINÆ, p. 71.

Subfamily LIBELLAGINÆ. (Fig. 2.)

Libellaginæ Laidlaw, Rec. Ind. Mus. vol. xiii, p. 24 (1917).

Head robust, triangular; eyes oval, large, projecting posteriorly, moderately separated from one another; frons horizontal, rather longer than broad; occiput linear, very narrow; face projecting, as long as the head is broad, ante- and post-clypeus very tumid and projecting, rounded and vesiculated; labium slightly longer than broad, split nearly to its base into two narrow triangular parts; labrum oval; antennæ with first segment rudimentary, second segment cylindrical, the most robust of all, third segment longer and more slender, with its apex slightly tumid, terminal or apical segment longer than third.

Prothorax elongate, narrower in front than behind, anterior border raised, hind border with a large lobe, tumid and oval in shape.

Thorax moderately robust, compressed, forming about one-quarter total length of body, elongate, dorsum flattened, mid-thoracic carina often bifid so as to enclose a triangular space variable in length, extending from one-fourth to whole length of carina (when present, known as the "*mesothoracic triangle*," and important for differentiating species).

Legs long and slender, extending to apical end of segment 4 in male, to end of segment 7 in female. Femora and tibiæ with long hair-like spines; tibiæ sometimes dilated, often densely pulverulent on the flexor surface; tarsal claws with a small spine at end, barely visible in some species (*Rhinocypha*).

Wings hyaline or opaque, metallic or iridescent in male, usually uncoloured in female and all long and narrow, of same length, but in some genera hind-wings of male distinctly broadened, always longer than abdomen, especially in female, markedly petiolated; reticulation close, cells tetragonal in shape. *Rii* and *IRii* not fused with *Riv+v*: node nearer base of wing than apex; basal space entire, slightly longer than half cubital space; discoidal cell straight, very narrow, traversed by one or several nervures, about one-third shorter than basal space, its ends squared or outer end oblique; arc

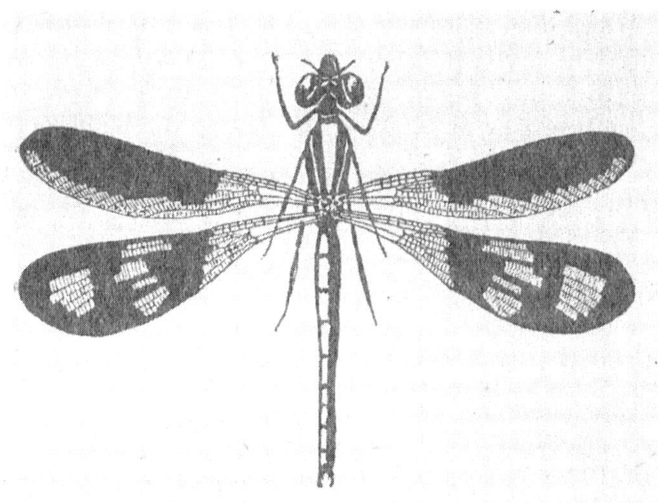

Fig. 2.—*Rhinocypha spuria* Selys, male.

nearly straight or markedly bent; its sectors arising from near middle, from a single point or moderately separated; *Cuii* nearly straight or making a costalwards curve after leaving discoidal cell; *IA* without an inferior branch, straight or curved posteriorwards or undulated; no intercalated sectors between *Cuii* and *IA*; most sectors running straight, with but a little curve towards termen of wing; many intercalated sectors; 4 to 20 antenodal nervures, first and second or third being the primaries, all others not coinciding in the costal and subcostal series; pterostigma present in all wings, except in fore-wings of male of *Libellago*, long and narrow.

Abdomen broad, depressed, especially in male, short, segments 1 and 2 very short, remainder nearly equal in length, squared. Anal appendages all very similar, superiors twice as long as segment 10, cylindrical, slightly broader at base, curved pincer-like towards one another at apices; inferiors very short, more robust, cylindrical, blunt at apices. Genitalia different in the several genera.

Larvæ.—Only three larvæ of the LIBELLAGINÆ, two of *Rhinocypha* and one of *Libellago*, are known. All three agree in everything except minor details, so that those of other genera of the subfamily are doubtless similar.

These larvæ are very sluggish in habits, cryptic in colouring, and therefore difficult to find. They cling to roots or broken-up débris in slow-running streams or at the bottoms of pools in streams with swifter current. I have found the easiest way to take them is to dredge out a quantity of such débris and spread it on the foreshore of the stream, exposed to the sun. As soon as it has drained and the creatures have begun to feel the heat of their new environment they will begin to stir from their lethargy, and as a consequence are at once easily detected.

They possess three spine-like caudal gills, which are undeveloped in the earlier instars. All appear to have banded limbs and most of them have cryptic brown markings on the head, thorax, and abdomen. The mask is deeply fissured, the edges of the fissures overlapping; the lateral lobes are cheliform and furnished with a long movable hook. The whole mask is of remarkable length, extending back when in repose to the hinder legs.

The gizzard has numerous folds, usually 16 in number, each bearing a row of 4 to 5 teeth. The antennæ long and 8-jointed, the basal one very short, the penultimate of great length.

Autotomy of the caudal gills is accomplished with the greatest ease, so that it is difficult to obtain a specimen with both gills intact; probably the gills serve more for defence than for respiration, as the larvæ certainly live quite well without them.

For details of the respective larvæ the reader is referred to J. Bombay Nat. Hist. Soc. vol. xxxii, pl. iii, figs. 1–3, p. 690 (1928).

Distribution.—Africa, Southern Asia, Philippines, and Australia.

Key to Genera of the Libellaginæ.

1.	Sectors of arc separated at origin; pterostigma present in all four wings of both sexes	2.
	Sectors of arc arising from a common point; fore-wing of male without a pterostigma	LIBELLAGO Selys, p. 58.

2.	Wings stalked (petiolated) to far distal of nervure *Ac* (cubital nervure) ……	[p. 52. CALOCYPHA Fraser,
	Wings stalked to slightly proximal of nervure *Ac* ……………………	3.
3.	Wings (at least hind-wings, except in *Rhinocypha immaculata*) coloured in male; *IA* not zigzagged at its origin ..	[p. 5. RHINOCYPHA Rambur,
	Wings uncoloured in both sexes; *IA* zigzagged from its origin ……………	[p. 55. CHLOROCYPHA Fraser,

Genus RHINOCYPHA Rambur.

Agrion Guérin, Mag. Zool., Ins. vol. i, t. 15 (1831); Percheron, Gen. Ins., Neur. t. 2 (1835).
Calopteryx Burmeister, Handb. Ent. vol. ii, p. 826 (1839).
Libellago (pars) Selys, Mon. Lib. Europ. p. 200 (1840).
Rhinocypha Rambur, Ins. Névrop. p. 232 (1842); Selys, Syn. Cal. p. 59 (1853); Walker, List Neur. Ins. Brit. Mus. vol. iv, p. 645 (1853); Selys, Mon. Cal. p. 198 (1854); Kirby, Cat. Odon. p. 112 (1890); Williamson, Proc. U.S. Nat. Mus. vol. xxviii, pp. 173–175 (1905); Ris, Ent. Mitteil. vol. v, pp. 310–311 (1916); Laidlaw, Rec. Ind. Mus. vol. xiii, pp. 33–39 (1917); Fraser, J. Bombay Nat. Hist. Soc. vol. xxix, p. 478 (1923); id., ibid. vol. xxxii, pp. 185–189 (1927).

Characters as for the subfamily; wings of male, less rarely of female also, coloured vividly with metallic blues, violet, green or fiery coppery, at least in hind-wings, often with clear vitreous spots and stripes on hind-wings, which glow with opalescent rainbow tints; antenodal nervures numerous, always more than six; *IA* not undulated at its origin; petiolation always ending proximal to *ac*; shape of wings very variable, either all four very long and narrow, or hind-wing considerably dilated at its middle. Mesothoracic triangle nearly always present and brightly coloured in most species, of variable length, sometimes extending up to alar sinus, but often much shorter than this. Legs long and slim, hind pairs of tibiæ nearly always pruinosed white or yellow on the flexor surface. Anal appendages as for subfamily.

Larva.—Body short and robust, head squarish, eyes representing outer angles, antennæ seven-jointed, segments becoming progressively smaller from base to apex; prothorax small, furnished with four small protuberances on dorsum; wing-cases lying parallel along dorsum; abdomen short and cylindrical; legs of great length in earlier instars, of moderate length in adult, hind tibiæ extending to a short distance beyond anal end of abdomen, minutely spined, striped with pigment; mask deeply bifid in final instars, edges of fissure usually overlapping to enclose a small foramen, very long and flattened, extending as far as base of middle pair of legs in adult, furnished with two lateral setæ and three hooks which may be bifid at apex, movable hook of great length; caudal gills two in number, triquetral in shape, heavily spined, 11th abdominal tergite present as an *appendix dorsalis* unmodified as a gill. Gizzard with sixteen folds, each bearing four to five teeth. Found in running water, usually montane or submontane streams, clinging to weed or submerged twigs, etc.

Distribution.—Oriental and Australasian regions.

Genotype, *R. tincta* Rambur.

Rhinocypha forms a large genus of medium-sized dragonflies, and includes some of the most beautiful species of the order. Not only do their wings display an inimitable play of scintillating colours, ranging through blues, greens, violet, pink to gorgeous fiery coppery red *, but the bodies in most cases are also gaily decorated with red and blue or yellow of many shades. The picture afforded by a couple of males circling round one another, their wings glittering with multitudinous rainbow hues, as they compete for the affections of an admiring female, is not readily forgotten. They are the living gems of tropic streams, and delight the eye of even a casual observer.

When mating, the males perform a kind of nuptial dance before the female, during which they make a great display of the white pulverulent flexor surface of the hinder pairs of tibiæ. The legs are trailed and show up dazzlingly white in the strong sunshine. Meanwhile the fore-wings flutter rapidly to support the insect, whilst the hinder pair are held flat to display their wealth of colour. The female, perched on a prominent twig beside the stream, appears to be totally unconcerned by her mate's efforts to attract her. One very rarely sees a pair *in copula*, although vast numbers of both sexes may be present on the banks of the stream. Less rarely a female may be seen ovipositing on a floating twig or broken reed.

Because of the large number of species, it is extremely difficult to establish relationships, and their distribution presents many apparently insoluble problems. Roughly the whole genus may be split up into two large divisions :—

(*a*) Those with narrow wings, the hind scarcely broader than the fore, and in which *Riii* is widely separated and distal to the subnode : e. g., *R. biforata*, *perforata*, *bisignata*, etc.

(*b*) Those with broader wings, the hind usually being much broader than the fore, and in which *Riii* comes off from the subnode : e. g. *R. cuneata*, *spuria*, *quadrimaculata*, *ignipennis*, etc.

* The colours of the wings of species of *Rhinocypha* are entirely due to iridescence. By transmitted light the wings merely show varying shades of brown.

R. ignipennis, *trimaculata*, and *unimaculata* are clearly fairly closely related, not only from the colour of the wings, coppery in all, but also from the fact that all have an aborted mesothoracic triangle.

R. iridea, which shares this latter character, I am inclined to view as closely related, as the fore-wings are often coppery.

A quite natural series, which may indicate a possible line of evolution, is as follows :—

R. immaculata — *hilaryæ* — *bifasciata* — *trifasciata* — *bifenestrata*—*spuria*—and so on to the *fenestrella* group.

So far as the seventeen Indian species are concerned, it is possible to place them in four more or less natural groups, of which group *perforata* appears to be the most primitive.

Group I.—*Fenestrella.*

Third branch of radius (*Riii*) arising at subnode ; wings opaque black, with iridescent vitreous spots of blue, violet, or emerald-green ; hind-wings considerably broader than fore-wings ; mesothoracic triangle very large, extending to antealar sinus, coloured : *cuneata*—*spuria*—*fenestrella*—*quadrimaculata.*

Group II.—*Bifasciata.*

Third branch of radius arising at subnode or very slightly distal to it ; wings hyaline, or hind-wings marked with one or more opaque black bands running from costa to posterior border ; hind-wings only slightly broader than fore-wings ; mesothoracic triangle very large, extending to antealar sinus, coloured : *immaculata* — *hilaryæ* — *bifasciata* — *trifasciata* — *bifenestrata.*

Group III.—*Unimaculata.*

Third branch of radius arising from subnode or slightly distal to it ; wings fiery metallic coppery, with iridescent vitreous spots of pink, violet, or green (*iridea* has hind-wing and costal border of fore-wing opaque black, with golden or violaceous vitreous spots and stripe) ; hind-wings usually broader than fore-wings ; mesothoracic triangle obsolete : *unimaculata*—*ignipennis*—*trimaculata*—*iridea.*

Group IV.—*Perforata.*

Third branch of radius arising well distal to subnode ; hind-wings not broader than fore-wings, both pairs very narrow ; hind-wings opaque only in outer third or half, this area marked with one or two rows of iridescent vitreous spots ; mesothoracic triangle short, not extending more than half-way towards antealar sinus : *perforata*—*biforata*—*bisignata*—*whiteheadi.*

Key to Indian Species of Rhinocypha.

1.	Wings uncoloured in both sexes	*immaculata* Selys, [p. 35.
	Wings coloured, at least partly, in male.	2.
2.	Dorsal mesothoracic triangle extending as far as root of wings	3.
	Dorsal mesothoracic triangle either absent, uncoloured, or extending not more than half-way up to root of wings....	8.
3.	Hind-wings of male with opaque bands .	4.
	Hind-wings of male opaque, with vitreous spots	6.
4.	Hind-wings of male with apex narrowly black and an incomplete (more rarely complete) black band just proximal to pterostigma	*hilaryæ* Fras., p. 36.
	Hind-wing with apex narrowly black and a black band lying about mid-way between pterostigma and node	*bifasciata* Selys, p. 29.
	Similar to last, but with a third, incomplete, black band at level of node	5.
5.	Apical, medial, and nodal bands separated.	*trifasciata* Selys, p. 31.
	All three bands connected narrowly along costal and hinder borders of wing, thus enclosing two large spaces.	*bifenestrata* Fras., p. 33.
6.	The large apical vitreous spot separated from the costal border by not more than one or two rows of cells	*cuneata* Selys, p. 9.
	Apical vitreous spot separated from costal border by at least three or four rows of cells.....................	7.
7.	Apical vitreous spot moderately small, lying exactly under pterostigma; medium-sized species	*fenestrella* Ramb., [p. 17.
	Apical vitreous spot large, lying almost entirely proximal to line of pterostigma; mid-row of vitreous spots more or less confluent; large species, hind-wing 27–28 mm.	*spuria* Selys, p. 12.
	Apical vitreous spot moderately large, lying partly proximal to line of pterostigma; mid-row of vitreous spots always well separated, costal spot extending much nearer to node than middle spot; small species, hind-wing 20–24 mm.	*quadrimaculata* Selys, p. 14, and race *hemihyalina* Fraser, p. 17.
8.	Mesothoracic triangle uncoloured or entirely obsolete; hind-wing distinctly broader than fore-wing	9.
	Mesothoracic triangle coloured, extending from one-third to half-way up dorsum; hind-wing nearly same as fore-wing	12.

9.	Wings fiery burnished coppery metallic; no opaque areas in fore-wings........	10.
	Wings opaque black, hind-wing with a sickle-shaped vitreous spot at distal end which partially encircles outer row of vitreous spots; costal border of fore-wing narrowly black nearly as far as node	*iridea* Selys, p. 20.
10.	Hind-wing with a large subquadrate violet-green vitreous spot at its centre; very large species	*unimaculata* Selys, [p. 27.
	Hind-wing with a medial row of vitreous spots and an inner linear spot	11.
11.	Small species; hind-wing 23 mm. or less; inner vitreous spot very elongate	*trimaculata* Selys, p. 25.
	Larger species; hind-wing 26 mm. or more; inner vitreous spot much shorter	*ignipennis* Selys, p. 23.
12.	Only apical third of hind-wing opaque; a single row of vitreous spots	13.
	Apical half of hind-wing opaque; two rows of vitreous spots	*perforata* Perch. and [subsp., pp. 41–44.
13.	Mesothoracic triangle very small, pink; vitreous spots elongate; only extreme apex of fore-wing tipped with black..	*biforata* subsp. *biforata* Selys and *abbreviata* Fras., pp. 45 & 48.
	Mesothoracic triangle larger, extending half-way along dorsum, pink; vitreous spots shorter; apex of fore-wing tipped with black as far as inner end of pterostigma	*biforata* subsp. *delimbata* Selys and *beesoni* Fras., pp. 45 & 48.
	Similar to the last, but nearly the apical third of fore-wing black.............	*bisignata* Selys, p. 49.
	Mesothoracic triangle small, blue; an isolated spot distal to the row of vitreous spots; opaque area of fore-wing equal to that of hind-wing, intensely black, the vitreous spots peacock-blue; wings markedly rounded at apices	*whiteheadi* Kirby, p. 39.

169. **Rhinocypha cuneata** Selys. (Fig. 3 and Pl. I, fig. 1.)

Rhinocypha cuneata Selys, Syn. Cal. p. 60 (1853); id., Mon. Cal. p. 206 (1854); Kirby, Cat. Odon. p. 113 (1890); Williamson, Proc. U.S. Nat. Mus. vol. xxviii, pp. 173, 174 (1905); Laidlaw, Rec. Ind. Mus. vol. xiii, p. 36, pl. ii (1917); Fraser, J. Bombay Nat. Hist. Soc. vol. xxix, p. 478 (1923); id., ibid. vol. xxxii, pp. 189, 190, pl. ii, fig. 2 (1927).

Libellago cuneata Walker, List Neur. Ins. Brit. Mus. iv, p. 650 (1853).

Rhinocypha adamantina Förster, Ann. Mus. Nat. Hung. vol. i, p. 547 (1903).

Male.—Abdomen 24 mm. Hind-wing 27–28 mm.

Head: ground-colour velvety black; labium black; labrum black, with a pale blue spot on either side; rest of head unmarked save for a tiny spot of rust-red on outer side of ocelli, and another smaller, bluish spot on each side of occiput; eyes brown. *Prothorax* black, with a pale blue longitudinal streak on the mid-dorsum of posterior lobe. *Thorax* black; mesothoracic triangle very large, extending as far as antealar sinus, pointed at apex, palest blue in colour, a tiny humeral point just behind upper end of humeral suture, a linear streak along anterior border of second lateral suture, not extending up as far as wing-roots, and dividing into two small spots below; lastly, an upper short streak on metepimeron, all reddish-yellow. *Legs* black, the two hinder pairs of tibiæ pure white from dense pruinescence on flexor surfaces, distal portion of hind femora less so. *Wings* opaque from apices to about 5 to 7 cells proximal to node, bases hyaline, palely saffronated; opaque area in fore-wings occupying roughly about costal half of wing, hinder border of this area serrate; in hind-wings this area begins 4 to 5 cells proximal to node and runs obliquely back and out from this point in a very

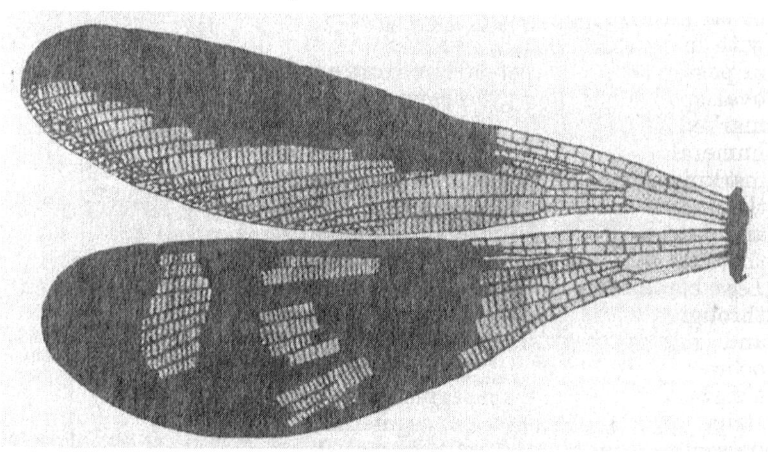

Fig. 3.—Wings of *Rhinocypha cuneata* Selys, male.

irregular and serrate manner, leaving a vitreous streak of pale violaceous anterior to discoidal cell, which deeply indents the opaque area and runs inward to a little beyond inner end of discoidal cell; inner border of the opaque area, posterior to the discoidal cell, also violaceous vitreous. The opaque area presents a medial row of spots, all more or less confluent, and presenting an infinite variety of patterns. The anterior or costal spot lies between *Ri* and *Riii* (first and third branches of radius), and is usually separated from the hinder two; it extends inward well beyond the others. Posterior spots usually confluent, extending to within one cell-row of hind border of wing. In addition to this medial row, a very large preapical spot extends from *Ri* nearly to hind border of wing, oval in outline, its inner and outer borders serrate, its outer border in line with inner end of pterostigma. The latter black, with an elongate spot of pale blue at its centre. Iridescent spots and vitreous hind area of fore-wings peacock-blue or pale violaceous, according to angle from which they are viewed. *Abdomen* black, unmarked. Superior *anal appendages* slender, moderately separated, slightly curved in at apices.

Female.—Hind-wing 28–30 mm. Abdomen 21 mm.

The following description is made from specimens from Turzum, near Darjeeling :—

Head marked as in male, but with the following additional spots, all bright ochreous :—A small spot on each side of the anterior ocellus and just anterior to its level ; a large spot at base of antennæ, a pyriform spot on each side of frons, cheeks narrowly, bases of mandibles, and, finally, a small spot or stripe on each side of rhinarium. *Prothorax* black, marked with bright ochreous as follows :—A median streak on dorsum of posterior lobe as in male, a short streak on sides and a large oval spot laterally, just below posterior lobe. *Thorax* black, marked with bright ochreous as follows :—Fine lines on humeral and median lateral sutures, incomplete below, same markings as seen in male but considerably broader, especially the metepimeral marking, which is broad and roughly triangular ; finally, a minute antehumeral streak on lower part of thorax. Mesothoracic triangle mapped out in lines of yellow. *Legs* black, not pruinosed. *Wings* hyaline, evenly enfumed throughout, so that they appear brown when overlapping and folded over dorsum. *Abdomen* black, marked with ochreous as follows :—Dorsal carina finely as far as segment 7, a zigzag lateral stripe on segment 1, a lateral basal longitudinal stripe and an apical spot on segments 2 to 4, and the spot only on segments 5 to 7 ; ventro-lateral stripe on segments 2 to 6. In some specimens many of these markings are more or less obsolete, younger specimens being as a rule better marked. *Anal appendages* black, pointed, cylindrical, twice as long as segment 10 ; vulvar scale very robust, extending to end of abdomen ; pterostigma brownish-black, clouded with enfumed white, oblique, and rather squared distally, pointed within.

Distribution.—NORTHERN BENGAL and ASSAM. I possess specimens from Gopaldhara, Assam, and from Mungpoo and Turzum, Darjeeling District, collected by Messrs. H. Stevens, Oscar Lindgren, and Chas. M. Inglis. The *type* is said to have come from Tibet and may well have done so ; it is a male in the Selysian collection.

R. adamantina Först. is obviously merely a variety of this species; the type, from Sikkim, is in the Michigan University collection.

The species is easily distinguished by the apical vitreous spot encroaching to within one or two cells of the costal margin of the wing.

170. **Rhinocypha spuria** Selys. (Fig. 2).

Rhinocypha spuria Selys, Bull. Acad. Belg. (2) vol. xlvii, p. 388 (1879); Kirby, Cat. Odon. p. 113 (1890); Selys, Ann. Mus. Civ. Genov. (2) vol. x (xxx), p. 491 (1891); Williamson, Proc. U.S. Nat. Mus. vol. xxviii, pp. 173, 174, & 177 (1905); Laidlaw, Rec. Ind. Mus. vol. xiii, pp. 36, 37 (1917); Fraser, J. Bombay Nat. Hist. Soc. vol. xxix, p. 478 (1923); id., ibid. vol. xxxii, pp. 190–192 (1927).

Male.—Hind-wing 27–28 mm. Abdomen 24 mm.

Head: black, marked with citron-yellow as follows:—A small oval spot on outer side of each posterior ocellus and a tiny occipital spot on each side, these markings being similar to those of *R. cuneata*; labium black, labrum unmarked. *Prothorax* black, marked with citron-yellow as follows:—A linear streak on mid-dorsum of posterior lobe and a small spot at each of its outer angles, another larger spot below this and a large triangular spot on each side of middle lobe. *Thorax* black, marked with yellow as follows:—An upper linear antehumeral spot, an upper humeral stripe broken above, close to hind border of suture, a broad irregular stripe on sides on hind part of mesepimeron, and an elongate triangular streak on hind part of metepimeron, which is often broken in two; lastly, and occasionally, a small spot on upper part of mesepimeron and a tiny streak along upper part of first lateral suture. All these markings, both prothoracic and thoracic, subject to variation, the prothoracic and smaller markings on thorax often obsolete. Mesothoracic triangle very long and broad, lilaceous. *Legs* black, the two posterior pairs of tibiæ snowy white from pruinescence on flexor surface. *Wings* opaque black, with a steely blue or green reflex in outer three-fourths, marked with vitreous spots and areas of lilaceous or purple; in some lights these spots have a mother-of-pearl reflex. In the fore-wing the opaque area covers costal three-fourths, the vitreous area posterior to it having a beautiful violaceous reflex; border of opaque area markedly serrate; inwardly it begins four cells proximal to node and has a markedly jagged border; in the hind-wings the same area begins about 2 to 3 cells proximal to the node, and is indented by two vitreous areas, the anterior of which extends in for a distance of 4 to 5 cells, the posterior, which is on the wing-border, for a distance of 3 to 4 cells. The vitreous areas in this wing consist of a middle row of spots and an apical, the latter lying with its outer two-thirds under the pterostigma, and nearly quadrate in shape, being bounded by *IRii* and *IRiii*. Middle row of spots rather variable, consisting of four linear spots, the anterior or costal one made up of 2 rows of cells, the next of 2 or 3 at its outer end, the third of some isolated cells and the hinder of 2 or 3 rows at its outer end. The inner border of these four rows forms a slightly convex outline, the second being a little more distal than the others. About 20 antenodal nervures in both wings; 4 to 5 traversing nervures in discoidal cell. Pterostigma blackish-brown, marked with a very large, elongate, lilaceous spot in the hind-wings. *Abdomen* black, with dark metallic reflex, segment 1 with a small lateral yellow spot, 2 and 3 with a linear streak along ventral border. *Anal appendages* similar to those of *R. cuneata*.

Female.—Abdomen 22 mm. Hind-wing 30 mm.

Head velvety black, marked with bright ochreous spots; labium black, its lateral lobes white; labrum black, the whole of its central area yellow, deeply notched by a tongue-like prolongation of the black at its base. In addition to the spots found in the male, bases of mandibles broadly, cheeks and basal segment of antennæ, a small triangular spot on each side of rhinarium, a large triangular spot on each side of upper surface of frons, a small spot on each side of the anterior ocellus, and, lastly, a small medial oval spot on occiput, yellow. *Prothorax* marked as in male, but rather more extensively. *Thorax* with similar yellowish markings as in male, but these are greenish-yellow in hue and much more extensive. The upper and anterior lines bordering the humeral and first lateral sutures are generally complete, and the antehumeral marking becomes a well-marked stripe extending in a curved manner inwards from the upper end of the humeral suture towards the antealar sinus and then straight down the whole length of the dorsum parallel to the mesothoracic triangle, which is finely outlined in yellow. *Wings* evenly enfumed and palely saffronated throughout, hyaline, unmarked; pterostigma black, its central area ochreous; 14 to 16 antenodal nervures in hind- and fore-wings. *Legs* black, the two posterior pairs of tibiæ pruinosed. *Abdomen* metallic-black, with the following yellow markings:—Segments 2 to 7 with the mid-dorsal carina narrowly yellow; segment 1 with its apical border and a large curved or angulated lateral spot; segments 2 to 5 with a longitudinal basal stripe and an apical spot, as well as a ventro-lateral stripe. On segment 6, and sometimes 7, the apical spot only present. Segments 2 to 6 with apical border laterally yellowish. *Anal appendages* and vulvar scale as in *cuneata*.

Distribution.—ASSAM: Shillong. BURMA: Kalaw, Southern Shan States; Chin Hills.

R. spuria is one of the largest and most beautiful species of the genus. It is specifically distinct from *R. quadrimaculata*, which is one of the smallest species of the genus. The difference in size is so striking that when placed side by side there is never any doubt as to which is which. Both Williamson and Laidlaw, however, express the opinion that, judging by the descriptions, the two are conspecific, so that it seems advisable to settle this point once and for all by pointing out the differences. In a large number of specimens I find:—

R. spuria.	*R. quadrimaculata.*
Reticulation of wings rather open.	Reticulation very close.
Vitreous stripes in hind-wing invading opaque black area for a distance of 3 to 5 cells.	Vitreous stripes extending distalwards into opaque area for a distance of 7 to 8 cells.
Apical vitreous spot nearly square, limited posteriorly by nervure *IRiii*.	Apical vitreous spot transversely elongate, much narrower than long, extending one cell-row posterior to *IRiii*.
Middle row of spots always four in number; second spot from costal border equal in length to hinder spot, inset but slightly; inner border of spots shallowly concave.	Middle row of spots almost always three in number; middle spot much shorter than hinder spot and inset markedly to the others, so that the inner border of spots is deeply concave.
Abdomen 24 mm. Hind-wing 28 mm.	Abdomen 19 mm. Hind-wing 21 mm.

The extent and shape of the apical spot is the best guide.

Type in the Brussels Museum.

171. **Rhinocypha quadrimaculata** Selys. (Fig. 4.)

Rhinocypha quadrimaculata Selys, Syn. Cal. p. 60 (1853); id., Mon. Cal. p. 202 (1854); Kirby, Cat. Odon. p. 112 (1890); Williamson, Proc. U.S. Nat. Mus. vol. xxviii, pp. 173, 174, 175, 176–7 (1905); Laidlaw, Rec. Ind. Mus. vol. xiii, pp. 36, 37 (1917); Fraser, Mem. Dept. Agric. India (Ent.), vol. vii, p. 64 (1921); id., J. Bombay Nat. Hist. Soc. vol. xxxii, pp. 192, 193 (1927).

Libellago quadrimaculata Walker, List Neur. Ins. Brit. Mus. iv, p. 651 (1853).

Male.—Abdomen 19 mm. Hind-wing 21 mm.

Head black, with a small spot on outer side of each posterior ocellus and a smaller spot on each side of occiput; labium and labrum black, unmarked. *Prothorax* black, unmarked. *Thorax* black, marked with yellow as follows:—A fine humeral line interrupted below, and a moderately broad lateral stripe on anterior border of hind lateral suture. Occasionally a small elongate spot on metepimeron, but this, as well as the humeral marking, usually absent. (Teneral specimens have markings very similar to those of *R. spuria*, both on prothorax and thorax, but, although retained in adults of this species, they are invariably lost in adults of *R. quadrimaculata*, excepting the broader lateral stripe.) Mesothoracic triangle narrow, long, pink. *Legs* black, distal halves of two posterior pairs of femora, and tibiæ, snowy-white on the flexor surfaces from pruinescence. *Wings* shaped as in *R. cuneata* and *R. spuria*, hind-wings considerably broader than fore-wings, and dilated at middle, opaque black, with dark plum-coloured steely reflex, and marked with vitreous areas and spots as follows :—An apical spot, much broader than long, bounded by *Rii* costalwards, overlapping by one row of cells *IRiii* posteriorly, its inner and outer borders serrate ; a medial row of spots, usually 3 in number, but occasionally 4, the

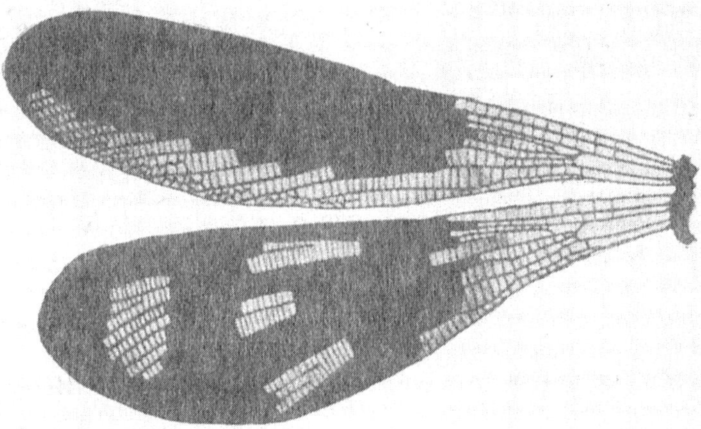

Fig. 4.—Wings of *Rhinocypha quadrimaculata* Selys, male.

anterior or costal spot made up of 2 rows of cells, and extending in towards the node much nearer than the other two spots of the series, the second or middle spot much shorter than the other two, and situated more distal than either ; third spot, when it exists, made up of 2 or 3 cells only ; lastly, posterior spot situated obliquely, so that its outer posterior corner may be in apposition to hind border of wing, made up of 3 rows of cells. At the inner border of the opaque area, which is very irregular, are seen two deep indentations due to invasion of hyaline vitreous areas ; these vary in extent, but usually invade the opaque area for a distance of 6 to 8 cells. In the fore-wing the opaque area covers the costal three-fourths, its hind border being markedly serrate, and its inner border, which extends 4 to 5 cells proximal to the node, markedly dentate. The area posterior to the black area vitreous and peacock-blue or purple, according to the angle of view. Vitreous spots and areas in hind-wing purple or emerald-green, according to angle from which viewed. Antenodal nervures 16 to 18 in number; discoidal cell traversed 4 to 5 times; pterostigma black. *Abdomen* black, with a vestigial lateral spot on sides of first segment and ventro-lateral stripes on third and fourth. *Anal appendages* as for *R. spuria*.

Female.—Abdomen 20 mm. Hind-wing 25 mm.

Head velvety black, marked with ochreous spots as follows:—A reniform spot on outer side of each posterior ocellus, a largish rounded spot on each side of occiput, and a medial oval or linear spot between and behind them on hind border of occiput. An oval spot in front of, and on each side of, anterior ocellus, a large triangular spot on each side of frons, a large spot on each side of rhinarium and a larger one on the summit of that structure. Cheeks, bases of mandibles broadly, second joint of antennæ, and, lastly, two larger triangular spots on labrum, bright ochreous; labium black, inner borders of lateral lobes yellowish-white. *Prothorax* black, with a longitudinal medial stripe on posterior lobe, and another, rounded or triangular, at each of its outer corners; a large oval medio-lateral spot. *Thorax* black, marked with ochreous as follows:—Mesothoracic triangle finely mapped out in yellow, an antehumeral fine stripe, which begins at upper end of humeral suture, curves rapidly in and then down to anterior border of thorax, running parallel to mesothoracic triangle; a fine post-humeral line, complete, running close to and behind humeral suture, another finer line running close to first lateral suture, and separated narrowly from an upper spot on mesepimeron; a broad irregular stripe bordering anterior aspect of second lateral suture; and, finally, an elongate spot on metepimeron. *Legs* black, two posterior pairs of tibiæ and femora sparsely pruinosed. *Wings* moderately, deeply, and evenly enfumed with brown; pterostigma blackish, with a diffuse yellowish centre more conspicuous distally; antenodal nervures 14 to 15 in number. *Abdomen* black, marked with yellow as follows:—A very large spot on each side of segment 1, a broad elongate spot at base and an apical spot on sides of 2; these are repeated on sides of 3 to 7, but the elongate longitudinal basal spot linear and gradually shortening from segment to segment; segments 8 and 9 with only a lateral spot, largest on the latter segment; dorsal carina narrowly yellow from segment 2 to 7; segment 10 unmarked. *Anal appendages* and vulvar scale similar to those of *R. cuneata*.

Distribution.—Rare in BURMA: its zoo-centre seems to be about NEPAL. I have seen specimens from Almora in Kumaon (UNITED PROVINCES) and Falodhi, and there is a specimen in the Vienna Museum from KASHMIR. Mr. Chas. Inglis has taken it in SIKKIM and around Darjeeling, but sparingly, whilst Col. F. Wall, has sent me specimens from Maymyo, UPPER BURMA. It has also been collected round Dehra Dun (UNITED PROVINCES).

It would appear that *R. quadrimaculata* is gradually replaced eastwards by *R. spuria* and *R. fenestrella*, but that it is specifically distinct from both. The differences between the former are set forth above under the description of *R. spuria*.

Type probably in the Selysian collection, Brussels Museum.

Race *hemihyalina* Fraser, Mem. Dept. Agr. India (Ent.), vol. vii, p. 64 (1921).

Male.—Abdomen 23 mm. Hind-wing 23 mm.

Differs from typical *quadrimaculata* by opaque area of fore-wing covering only costal half and by apical vitreous spot in hind-wing extending back to 3 rows of cells posterior to *IRiii*, instead of only one. Mesothoracic triangle lilaceous, as in *R. spuria*. The opaque area of the hind-wing is sharply bevelled off proximally, so that the vitreous spots indenting it are merged in the adjacent vitreous area, and the border appears regularly indented or serrate. There is a fourth row of cells, vestigial in nature, in the medial row of spots.

Female.—Abdomen 21 mm. Hind-wing 28 mm. Not differing from the type.

A pair from Shillong, ASSAM, collected by Mr. T. Bainbrigge Fletcher, now deposited in the British Museum.

In a male from the Naga Hills, in my own collection, the apical spot is similar to that of race *hemihyalina*, but the fore-wing is as broadly opaque as in typical *R. quadrimaculata*.

172. **Rhinocypha fenestrella** Rambur. (Fig. 5.)

Rhinocypha fenestrella Rambur, Ins. Névrop. p. 236 (1842); Selys, Syn. Cal. p. 60 (1853); id., Mon. Cal. p. 204 (1854); id., Bull. Acad. Belg. (2) vol. xlvii, p. 387 (1879); Kirby, Cat. Odon. p. 113 (1890); Williamson, Proc. U.S. Nat. Mus. vol. xxviii, pp. 173–179 (1905); Laidlaw, Rec. Ind. Mus. vol. xiii, pp. 35–37 (1917); Fraser, J. Bombay Nat. Hist. Soc. vol. xxxii, pp. 193, 194, pl. iii, fig. 2 (1927).

Libellago fenestrella Walker, List Neur. Ins. Brit. Mus. iv, p. 650 (1853).

Male.—Abdomen 19 mm. Hind-wing 21 mm.

Head velvety black, with an oval ochreous spot on outer side of each posterior ocellus and another rounded spot on each side of occiput. In the type the former of these spots is vestigial and the latter obsolete, but in specimens from Borneo and Burma which I have examined they are clearly visible. *Prothorax* black, with a longitudinal mid-dorsal stripe on posterior lobe white (absent in the type). *Thorax* black, with a bright citron-yellow irregular stripe bordering second lateral suture anteriorly, and an elongate spot on metepimeron. The type has also a fine incomplete line on the humeral suture and another on the upper part of the first lateral suture. Mesothoracic triangle extending to antealar sinus, palest pink or lilaceous. Four yellow spots beneath thorax. *Legs* black, the two posterior pairs of tibiæ pulverulent-white on flexor surface. *Abdomen* glossy black with steely reflex, first segment with a small bright citron-yellow spot on each side. *Wings* hyaline at base, opaque black for the distal three-fourths, hyaline area tinted yellow, opaque area steely green or blue, marked with vitreous spots which

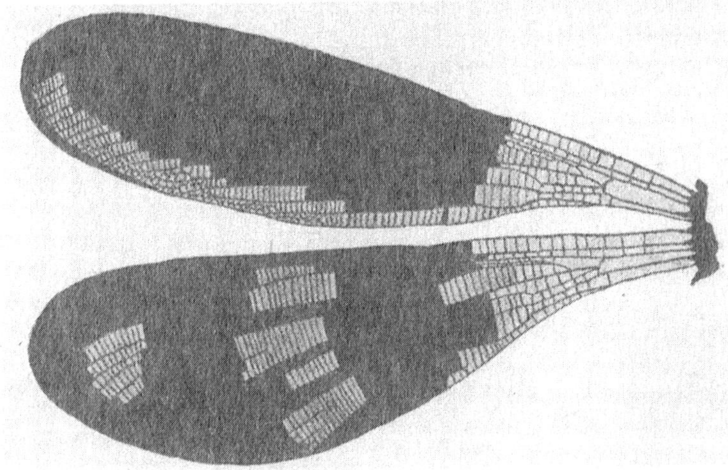

Fig. 5.—Wings of *Rhinocypha fenestrella* Rambur, male.

glow emerald-green, lilaceous, mother-of-pearl or a beautiful rose-pink, according to angle from which viewed. The opaque area in the fore-wing occupies variably the anterior three-fourths or half of wing, and in specimens from Sintang, Borneo, the vitreous area extends up at the apex as far as the pterostigma. The hyaline area posterior to it in this wing, and for some distance proximally at the base, is a beautiful vitreous peacock-blue. The dark area has a very irregular border basally and is markedly serrate behind. In the hind-wing the opaque area begins about 5 cells proximal to the node and extends back in a very ragged manner as in *R. quadrimaculata*, leaving two deep indentations made by the bordering vitreous area exactly as in *R. quadrimaculata*. In the Bornean examples the opaque area has a very oblique border, as seen in *R. quadrimaculata* race *hemihyalina*, so that these two indentations are barely distinguishable from the other irregular serrations, and the bordering vitreous area is correspondingly broadened. At the apex of the wing, *and situated exactly beneath the pterostigma*, is a variably sized spot, but usually much smaller than the corresponding apical spot found in *R. cuneata*, *R. spuria*, and *R. quadrimaculata*, being rarely more than 4 rows of cells in depth. Costalwards it is bounded by *IRii*, and posteriorly it extends about one row of cells beyond *Riii*, but is subject to slight variation from this. The middle row of spots shows even greater variability, consisting of 3 or 4 spots, the two middle of which are often confluent. The spots do not vary markedly in length as a rule, as in *R. quadrimaculata*, and their inner and outer borders form a regular curve. Pterostigma black, broadly lilaceous beneath, in the hind-wing. Antenodal nervures 13 to 16 in number; discoidal cell traversed 3 to 4 times. *Anal appendages* similar to those of *R. quadrimaculata*.

Female.—Abdomen 17 mm. Hind-wing 24 mm. (Rambur's type.)
Abdomen 19 mm. Hind-wing 28 mm. (Specimen from Upper Burma.)

Head black, marked exactly the same as in *R. quadrimaculata*. *Prothorax* marked similarly to that of *R. quadrimaculata*, but the two lateral spots much larger. *Thorax* with an antehumeral fine line broadening below, and curved out above along border of antealar sinus as far as upper end of humeral suture; a fine line which starts above and behind the humeral suture crosses its middle, and ends shortly before the anterior border of the thorax. Laterally the markings are similar to those of *R. quadrimaculata*. Below are found two long yellow spots shaped like the broad head of a stabbing spear. (These two spots serve to separate this species from the female of *R. quadrimaculata*.) *Legs* black, the two posterior pairs of femora distinctly pruinosed white. *Wings* faintly and evenly enfumed, tinted with yellow at base; antenodal nervures 16 to 17 in number; discoidal space traversed four times; pterostigma black, dark yellow for the greater part of its centre. *Abdomen* black, marked with ochreous spots as follows:—A largish lateral spot on segment 1, a smaller apical spot on segments 2 to 7, and a ventro-lateral stripe on segments 2 to 4. Mid-dorsal carina very finely ochreous from segments 2 to 7. *Anal appendages* and vulvar scale similar to those of *R. cuneata*.

Distribution.—The *type*, a male taken in 1825, is supposed to have come from Malacca, and is in the Paris Museum; a second male was taken at Pulo-Penang I. A third male, which was in the Dale collection, now in the British Museum, is labelled "India." The co-type female described by Rambur is in the Paris Museum, and is labelled as from China. I have males from Sintang, Borneo, and a pair from Gokteik, just above Maymyo, UPPER BURMA, collected by Col. Wall. It is apparently found scattered throughout Southern Asia in submontane areas.

The female described by Rambur is teneral and incomplete ; the description given above was made from Col. Wall's specimen.

R. fenestrella is easily determined by the position of the apical spot on the hind-wing, situated exactly under the pterostigma, whilst the female may be recognized by the markings on the under side of the thorax.

173. **Rhinocypha iridea** Selys. (Fig. 6 and Pl. I, fig. 3.)

Rhinocypha iridea Selys, Ann. Mus. Civ. Genova, (2) vol. x (xxx), pp. 492–494 (1891); Williamson, Proc. U.S. Nat. Mus. vol. xxviii, pp. 173, 181 (1905); Laidlaw, Rec. Ind. Mus. vol. xiii, p. 37 (1917); Fraser, J. Bombay Nat. Hist. Soc. vol. xxxii, pp. 195, 196 (1927).

Male.—Abdomen 20–22 mm. Hind-wing 25–26 mm.

Head velvety black, marked with yellow as follows :—A small spot on outer side of each posterior ocellus, a rounded spot on each side of occiput, and an oval spot on posterior border at its centre ; labium black, basal halves of middle and lateral lobes yellowish-white ; bases of mandibles bluish ; labrum with a broad transverse stripe pale blue, indented by a small prolongation of the basal black in the middle line. *Prothorax* black, marked with azure-blue as follows :—A medial longitudinal stripe on posterior lobe, a large lateral spot on each side of middle lobe, and a small spot at outer angles of posterior lobe. In teneral specimens these markings are yellowish, the hind margin of posterior lobe being narrowly yellow. *Thorax* black, marked with azure-blue as follows :—A fine antehumeral stripe, curved outwardly above, expanding slightly below, an equally narrow humeral line lying along hind border of humeral suture, a similar line adjacent to first lateral suture, a broad oblique stripe on posterior half of mesepimeron not extending to roots of wings, and broadening abruptly and markedly on lower half of mesepimeron ; finally, the whole central part of the metepimeron is azure-blue. *Legs* black, the two posterior pairs of tibiæ and the distal two-thirds of the two posterior pairs of femora densely pruinosed white. *Wings* partly opaque, hind-wings distinctly broader than fore-wings ; fore-wings hyaline at base, vitreous towards apex, costa for a depth of two cell-rows opaque black as far as 10 to 12 cells distal to node, behind which the wings glow with an iridescence of many colours, according to the angle from which they are viewed. Extreme apex dull violaceous, greater part of wing golden-green as far proximal as node. Opaque area along costa often extending nearer node in space between *Ri* and *Rii* than between costa and *Ri*. In hind-wing opaque area extending irregularly basally as far as middle of discoidal cell at that level, and for about 3 cells proximal to node along costa; this area marked by vitreous spots and stripes as follows :—An elongate sickle-shaped stripe, made up of a single row of cells, margined costalwards by *Rii* and beginning about 15 cells distal to the node, extends distal to level of middle of pterostigma, at which point it turns down in a regular sickle-shaped curve, this end of the stripe made up of 12 rows of cells, broad at

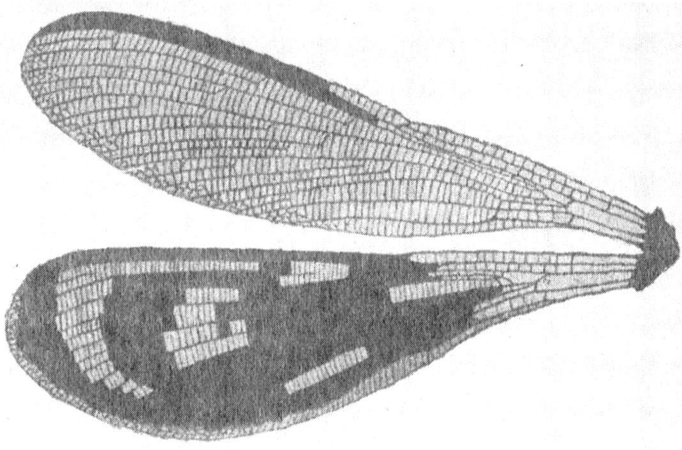

Fig. 6.—Wings of *Rhinocypha iridea* Selys, male.

costal end, tapering towards hind border of wing, which it nearly meets; costal end of this curved stripe nearly continuous with an elongate spot which lies in the angle formed by *Rii* and *Riii* at the subnode, and which is about 18 cells long. Lying within the angle or curve formed by the curved band is another vitreous spot, which lies between *IRii* and *Riv+v*, is irregularly quadrate, and very broken and jagged on its inner and outer borders. Immediately posterior to the first spot is a third, separated from it by 4 rows of cells and bordered behind by *Cuii*, equal in length to subnodal spot. The whole posterior border of the wings from nervure *ac* to outer end of *Ri* narrowly hyaline. Finally, an elongate spot which begins adjacent to the arc, bordered behind by $Riv+v$, and extending distally into the opaque area for a distance of about 18 cells. Sickle-shaped band and border of wing glowing golden-green; spots rose-pink or purple or emerald-green, according to angle from which viewed. Pterostigma black, yellow outwardly. *Abdomen* black, marked with azure-blue as follows:—A triangular spot on each side of segment 1 and its apical border, an elongate spot and an apical point on each side of segments 2 and 3, the long spot abbreviated on 2, long and tapering to a point on 3; segments 4 to 8 with the elongate spot only. *Anal appendages* similar to those of *R. cuneata*.

Female.—Abdomen 20 mm. Hind-wing 28 mm.

Head black, marked with yellow spots as follows:—Similar spots as found in male, and, in addition, a large rounded spot on summit of epistome, obsolete in adults, bases of mandibles, a triangular spot on each side of epistome, two large sub-triangular spots on upper surface of frons, an oval spot on each side of anterior ocellus, and, narrowly, the cheeks. Middle and lateral lobe of labium yellowish-white, except the extreme tips. *Prothorax* and *thorax* similar to those of male, but mesothoracic triangle outlined in yellow instead of pink, and markings more greenish-yellow instead of blue. *Legs* black, tibiæ and femora pruinosed white on flexor surfaces. *Wings* entirely hyaline, tinted with greenish-yellow; pterostigma framed narrowly in black, clouded with blackish inwardly, yellow outwardly; 14 to 15 antenodal nervures; discoidal cell traversed 3 to 4 times in both sexes. *Abdomen* black, marked with azure-blue as in male, but spots confluent to form a single stripe on segments 1 to 4, and segments 8 and 9 with an apical lateral spot; segment 3 with a ventral stripe. *Anal appendages* and vulvar scale similar to those of *R. cuneata*.

Distribution.—UPPER BURMA only. It occurs from May to August at Maymyo, Southern Shan States, Upper Burma, where Col. F. Wall found it comparatively common.

R. iridea is related to the other species of the *unimaculata* group by the primitive condition of its mesothoracic triangle, which, in the male, resembles that of the female.

This description differs in several points from the original Selysian one, which was made from a sub-adult specimen. Thus the markings of the labrum, prothorax, and thorax are given by Selys as yellow instead of azure-blue. In this species, probably better than in any other known dragonfly, one sees the gradual development of colour, especially in the wing-markings, which develop gradually in the course of several days like the details of a developing photograph. Mature specimens are comparatively rare in collections, most showing a thin greyish marking like an insufficiently exposed and under-developed photograph. The fully matured insect is probably the most beautiful of all dragonflies if one considers the combinations and harmony of colouring, together with the delicacy and artistic nature of the design in the hind-wings.

Type in the Genoa Museum, from Leito.

174. **Rhinocypha ignipennis** Selys.

Rhinocypha ignipennis Selys, Bull. Acad. Belg. (2) vol. xlvii, pp. 389, 390 (1879); Kirby, Cat. Odon. p. 113 (1890); Williamson, Proc. U.S. Nat. Mus. vol. xxviii, pp. 173, 179–181 (1905); Laidlaw, Rec. Ind. Mus. vol. xiii, p. 35 (1917); Fraser, Mem. Dept. Agric. India (Ent.), vol. viii, pp. 79, 80, pl. ix (1922); id., J. Bombay Nat. Hist. Soc. vol. xxxii, pp. 311, 312 (1927); id., ibid. p. 690, pl. iii, fig. 1 (larva) (1928).

Male.—Abdomen 22 mm. Hing-wing 22 mm.

Head: labium black; labrum black, unmarked in adults, with two oval bluish spots in tenerals; face glossy black, upper part of head velvety black, with a small oval orange spot on outer side of each posterior ocellus and a small rounded postocular spot on each side of occiput. At rear of latter, and mid-way between these two spots, a small linear orange spot. *Prothorax* velvety black, with a small median orange spot on hind part of median lobe and a long lateral spot of pale blue or yellowish on each side. *Thorax* velvety black, marked with orange as follows:—A tiny upper antehumeral spot, a fine linear posthumeral stripe broadly broken above, incomplete below, a short linear spot on upper part of first lateral suture, and an elongate, oblique, lateral, orange stripe on each side on lower part of thorax, broadly broken in two by anterior border of metepimeron. *Legs* black, the two posterior pairs of tibiæ not pulverulent on inner side as in most other species. *Wings* semi-opaque from node to apices, this area dark brown by transmitted light, brilliant fiery coppery by reflected light. Hind-wings with two rows of vitreous spots which vary in tint from mother-of-pearl to palest peach-blossom pink or pale blue; the spot nearest the base about 14 cells long, one cell-row wide, begins about 6 cells proximal to the node and ends distally about 2 to 4 postnodal cells beyond the node, lying between *MA* and *Riv+v*. Outer row consisting of three long stripes similarly coloured, the costal stripe rather the longest of the three, the posterior slightly the shortest, the costal stripe lying between *IRii* and *Riii*, about 26 cells long; middle stripe between same nervures as basal spot, about 22 cells long, one cell wide proximally, three distally; posterior stripe 15 cells long, lying between *Cuii* and *MA*. Pterostigma dark yellowish-brown framed in black, and with inner third the same colour. Inner hyaline areas of wings tinted with pale yellow; 16 to 18 antenodal nervures; discoidal cells traversed 3 to 4 times; petiolation begins nearer level of proximal antenodal than level of arc. *Abdomen* black or steely blue, marked, on sides of first two segments only, with a large orange spot on segment 1 and a tiny apical spot on segment 2. Teneral specimens are marked very nearly as in the female, described below. *Anal appendages* black, shaped as in *R. cuneata*.

Female.—Abdomen 21 mm. Hind-wing 28 mm.

Head black, spotted with yellow or orange as follows:—Base of labium and a spot on each lateral lobe, a broad oval spot on each side of base of labrum, mandibles broadly, a large spot on each cheek and basal joints of antennæ, a tiny spot on each side of epistome, a large, broadly oval, transverse spot on each side of frons, with a smaller spot just behind each, a rounded spot on each side of anterior ocellus, and, lastly, the same spots as in the male. *Prothorax* black, with bright yellow markings as follows:—A fine mid-dorsal line running from posterior lobe to anterior end, the marginal crest of the posterior lobe and an inward prolongation from this; a very large bell-shaped spot on sides of the middle lobe. *Thorax* velvety black, marked as in male, but more broadly so, the oblique lateral stripes extending the whole length of the thorax, and broken in three places by the lateral and humeral sutures. Humeral spot linear and, after a short interval, continued as a fine stripe to anterior end of thorax; post-humeral stripe complete. Beneath, two large orange spots on posterior area of thorax. Mid-dorsal carina and mid-line of antealar sinus finely yellow. *Legs* black, flexor surfaces of femora slightly but distinctly pulverulent. *Wings* hyaline, palely enfumed greenish-yellow. Pterostigma as in male; antenodal nervures about 20 in number. *Abdomen* black, marked with yellow or ochreous as follows:—Segment 1 with a large lateral spot, 2 to 4 with an apico-lateral spot and a linear spot nearly confluent with it, lengthening to a stripe on segments 3 and 4; lastly, segments 5 and 6 with a vestigial apico-lateral spot. *Anal appendages* and vulvar scale as in *R. cuneata*.

Distribution.—ASSAM and UPPER BURMA. Mr. T. Bainbrigge Fletcher has found it quite common at Shillong, Assam, from September to November. The Selysian *type* comes from this district, and is now in the Brussels Museum. The specimen described by Williamson from Burma is undoubtedly a teneral male of this species, and is the only record from Burma.

The species is quite easily distinguished from all others save *R. trimaculata*, which has the same glorious colouring, but which is smaller and has the vitreous spots much shorter than in *R. ignipennis*. *R. fulgipennis* (Siam) has similarly coloured wings, but they are broader and the thorax bears a coloured mesothoracic triangle. The colouring of the wings of these three handsome insects reminds one strongly of the gorgeous fiery Indian sunsets, and is unparalleled in nature.

175. **Rhinocypha trimaculata** Selys. (Fig. 7 and Pl. II, fig. 1.)

Rhinocypha trimaculata Selys, Syn. Cal. p. 62 (1853); id., Mon. Cal. p. 211 (1854); Kirby, Cat. Odon. p. 113 (1890); Laidlaw, Rec. Ind. Mus. vol. xiii, p. 35 (1917); Fraser, J. Bombay Nat. Hist. Soc. vol. xxxii, pp. 312, 313 (1927).

Male.—Abdomen 18 mm. Hind-wing 22 mm.

Head velvety black; labium and labrum unmarked; vertex and occiput marked as in *R. ignipennis*. *Prothorax*

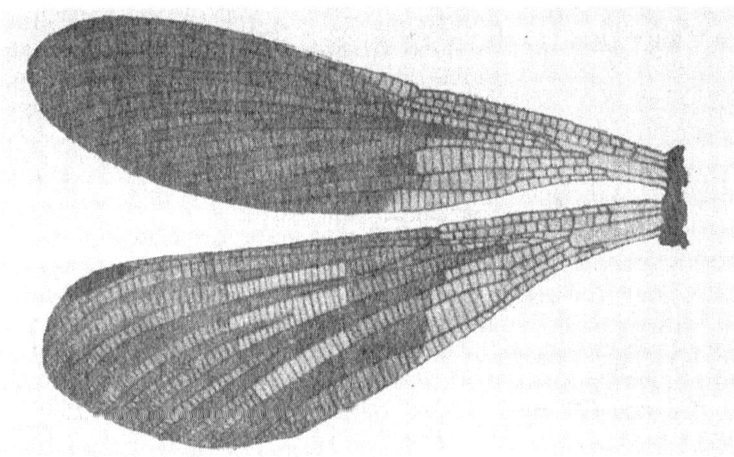

Fig. 7.—Wings of *Rhinocypha trimaculata* Selys, male.

velvety black, quite unmarked. *Thorax* glossy black, marked with bright yellow, bluish-green, or blue, according to age of specimen, as follows:—A tiny upper humeral spot, a fine linear posthumeral stripe, broken above, a short fine line on upper part of first lateral suture, an irregular oblique stripe running the full length of each side of thorax, broken into three spots, one of which occupies the central part of the metepimeron, and another, more elongate, the centre of the mesepimeron. *Legs* black; tibiæ not pulverulent. *Wings* fiery coppery red, even more brilliant than in *R. ignipennis*, semi-opaque from level of node in hind-wings, from slightly distal to that level in the fore-wings, where the costal margin of the wing is hyaline as far as beyond the pterostigma. In hind-wing, four vitreous spots of palest iridescent blue or peach-blossom pink, the palest one of which is about 17 cells long, extends from distal end of discoidal cell to level of third postnodal cell, and lies between *MA* and *Riv+v*. The outer spots consist of a series of three, all of about the same length, the costal spot lying between *IRii* and *Riii* and about 13 cells long, the medial and posterior spots each 10 cells long, the latter lying between *Cuii* and *MA*. Pterostigma blackish-brown, clouded with white in its outer half in sub-adults. Antenodal nervures 15 to 17 in number; discoidal cell traversed 2 or 3 times. *Abdomen* glossy black, with a lateral apical spot on each side of segments 1 and 2. *Anal appendages* black, shaped as for the genus.

Female.—Abdomen 17 mm. Hind-wing 23 mm.

Head velvety black, marked with pale yellow as follows :—Base of labium, which has a greenish tinge, a pair of small basal spots on labrum, bases of mandibles broadly, a large darker yellow spot on upper surface of rhinarium and two large cuneiform spots just behind it; basal segments of antennæ, a rounded spot in front and on either side of anterior ocellus, a small spot on either side of rhinarium, and, finally, the same spots as seen in the male. *Prothorax* black, with a large yellow spot on each side of mid-lobe, a short mid-dorsal line on posterior lobe and a large spot on each side, low down, near the trochanters. *Thorax* black, marked with yellow as follows :—Similar spots to those seen in the male, but larger, often with a greenish tinge, and probably bluish during life; in addition, a long linear streak on lower half of humeral region. Beneath thorax two broad longitudinal streaks of greenish-yellow. *Legs* black, not pruinescent. *Wings* deeply enfumed, burnt brown, especially the hind-wings and towards the apices. Pterostigma black, white in its outer half; 12 antenodal nervures; discoidal cells traversed twice in all wings. *Abdomen* black, marked with blue, the first seven segments with a lateral apical spot on each side, the second to fifth with a lateral stripe on each side, in line with, but not confluent with, the apical spots. *Anal appendages* and vulvar scale black, shaped as for the genus.

Distribution.—The Selysian *types*, two males in the British Museum, are from Tibet. The species, discovered in 1853, was not found again until Mr. Antram took it in numbers at Cachar, ASSAM, in September 1921. It must be remarkably restricted to particular localities, or such a handsome and brilliant insect would have been taken many times in the intervening sixty-eight years. If possible, it outshines in brilliancy its near relation *R. ignipennis*, from which it is distinguished by its smaller size and the much shorter vitreous spots of the hind-wing. The female is also distinguished by its smaller size and by its blue markings, etc.

Specimens have been placed in the British Museum, others are in my own collection.

176. **Rhinocypha unimaculata** Selys. (Fig. 8.)

Rhinocypha unimaculata Selys, Syn. Cal. p. 61 (1853); id., Mon. Cal. p. 207 (1854); Kirby, Cat. Odon. p. 113 (1890); Laidlaw, Rec. Ind. Mus. vol. xiii, p. 35 (1917); Fraser, J. Bombay Nat. Hist. Soc. vol. xxix, p. 479 (1923); id., ibid. vol. xxxii, pp. 313, 314 (1927).

Libellago unimaculata Walker, List Neur. Ins. Brit. Mus. iv, p. 649 (1853).

Male.—Abdomen 24 mm. Hind-wing 27–30 mm.

Head velvety black; labium and labrum unmarked; rounded spots of chrome-yellow on each side of occiput, one on outer side of each posterior ocellus, and a central spot on posterior border of occiput. *Prothorax* black, with a large irregular yellow spot on each side of middle lobe, a small linear mid-dorsal spot on posterior lobe, and a narrow yellow lateral margin to the same lobe. *Thorax* glossy black, marked with bright yellow lines as follows:—A short oblique, linear, upper humeral spot, a long antehumeral line, incomplete above, markedly angulate below, a fine posthumeral line broken above, a short upper line on first lateral suture, and, finally, the same broad, broken, oblique stripes on lower part of sides of thorax as those seen in *R. ignipennis* and *R. trimaculata*. The mid-dorsal carina is usually finely mapped out in yellow, and this may be continued on to the antealar sinus. *Legs* black, the two posterior pairs of femora and tibiæ densely pruinosed white, with a creamy yellow tinge. *Wings* enfumed brown from slightly distal to level of node in fore-wings, and slightly proximal to the same level in hind-wings, where this area is prolonged as a diffuse tongue in the space between *IRiii* and *Riv+v*. This area is also considerably darker in the hind-wings, and in both pairs is of a similar brilliant fiery coppery red as in *R. ignipennis* and *R. trimaculata*. In hind-wings the middle of this coppery area bears a large, roughly quadrate vitreous spot, which glows emerald-green or peacock-blue, according to the angle from which it is viewed. Its inner border is markedly indented, its outer roughly convex; its costal border is bounded by *Rii*, its posterior almost reaches the wing-margin. The spot lies considerably nearer the node than the pterostigma. Bases of wings hyaline, palely tinted with yellow. Pterostigma blackish-brown, outer half paler. Apices of wings very finely reticulated, especially between costa and *Ri*. Discoidal cells traversed from 5 to 7 times; 21 to 23 antenodal nervures. *Abdomen* black, marked with yellow as follows:—The first six segments with a ventral stripe running nearly the full length; segments 1 to 3 with an apical lateral spot on each side; intersegmental nodes narrowly yellow. *Anal appendages* black, relatively short, shaped as for the genus.

Female.—Abdomen 22 mm. Hind-wing 31–32 mm.

Markings of body very similar to those of male, but more extensive.

Head with same spots as in male, with the addition of the following:—Lateral lobes of labium, two obscure basal spots on labrum, bases of mandibles broadly, the yellow colour here confluent with a stripe which runs up the cheeks adjacent

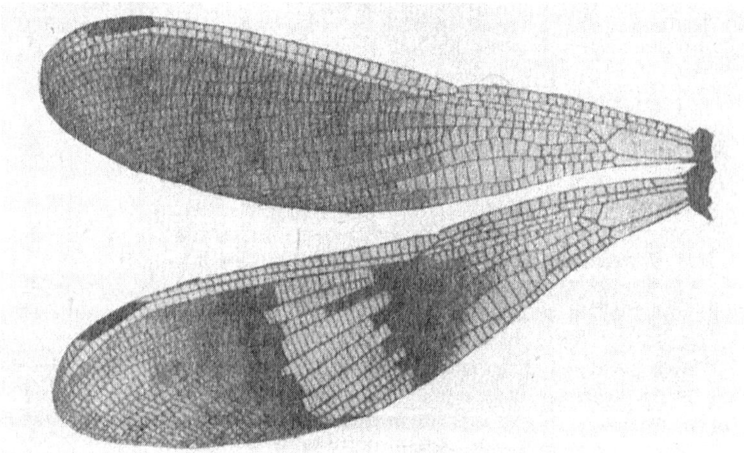

Fig. 8.—Wings of *Rhinocypha unimaculata* Selys, male.

to and bordering the eyes, also extending on to basal joints of antennæ, and confluent with two large cuneiform spots on upper surface of rhinarium. On the latter, in front above, a large rounded spot; a round spot on each side of, and in front of, the anterior ocellus, two tiny points between the posterior ocelli, and a spot on the lower part of the sides of the rhinarium. Postocular spots much larger than in male, and nearly united by a transverse spot on hind border of occiput. *Prothorax* black, marked as in male, but the mid-dorsal stripe runs nearly the whole length of the dorsum from the hind border of the posterior lobe, and the lateral markings are much better defined. *Thorax* glossy black, marked as in male, but considerably more extensively. Antehumeral line continued up and curved outwards above to become confluent with the humeral oblique spot. Six small spots beneath thorax. *Legs* black, the two posterior pairs of femora and tibiæ pruinosed yellow, as in the male. *Abdomen* glossy black, marked with parallel stripes of yellow, one ventral as in the male, extending from segments 1 to 7, and a lateral stripe on each side, on the same segments, becoming broken up into a stripe, and an apical spot on each segment from 4 to 7. Mid-dorsal carina finely yellow on segments 2 to 7. *Anal appendages* and vulvar scale as for the genus.

Distribution.—The species seems to be centred about Darjeeling, and is established in several districts in northern BENGAL and ASSAM. I have taken specimens in May and June, perching on rocks in mid-stream, at Mungpoo, 3,600 ft., British Sikkim. The *type* is in the Selysian collection at Brussels, without locality save " India."

R. unimaculata is the largest species of the genus, and is distinguished very readily by its striking markings and colours. By the absence of the mesothoracic triangle and the coppery colours of the wings it is closely related to the foregoing species. A fusion of the middle series of vitreous spots has resulted in the large median spot ; thus, in one of my own specimens the spot is practically cut in two by an invasion of the dark area above.

177. **Rhinocypha bifasciata** Selys. (Fig. 9.)

Rhinocypha bifasciata Selys, Bull. Acad. Belg. (2) vol. xlviii, p. 386 (1879); Kirby, Cat. Odon. p. 113 (1890); Laidlaw, Rec. Ind. Mus. vol. xiii, p. 34 (1917); Fraser, J. Bombay Nat. Hist. Soc. vol. xxix, p. 478 (1923); id., ibid. vol. xxxii, pp. 314, 315 (1927).

Male.—Abdomen 22 mm. Hind-wing 25 mm.

Head glossy black in front, velvety black above ; labium and labrum unmarked, a large oval spot on outer side of each posterior ocellus and a small round postocular spot on each side of occiput. *Prothorax* black, with a mid-dorsal longitudinal spot on the posterior lobe. *Thorax* glossy black, marked on sides with a broad broken yellow band, made up of a small serrate spot on metepimeron, and an elongate pyriform spot on mesepimeron capped by a U-shaped spot anteriorly. Mesothoracic triangle very large, extending from alar sinus to anterior border of thorax, palest azure-blue. *Legs* black, the two posterior pairs of femora and tibiæ heavily pruinosed white, this snow-like excrescence overlapping the margins of the tibiæ and making them appear dilated. *Wings* partially hyaline, from discoidal cell to apex greyish by transmitted light, palely yellow at base. The darkened area by reflected light is a beautiful violet-blue or emerald-green, according to the angle from which it is viewed. The wings in this area have a satiny gloss, and the hind-wings are traversed by two brown bands from costa to hind border, one at apex of hind-wings and the other a short distance on inner side of pterostigma. Both bands very narrow, and the inner very irregular and diverse in shape. The iridescent area limited anteriorly by *Rii*. Discoidal cells traversed 5 times; 15 to 18 antenodal nervures. Pterostigma black, white or yellowish outwardly. *Abdomen* glossy black, with a small apical spot of yellow on sides of segments 1 and 2. *Anal appendages* black, and similar in shape to those of *R. cuneata*.

Female.—Abdomen 18 mm. Hind-wing 28 mm.

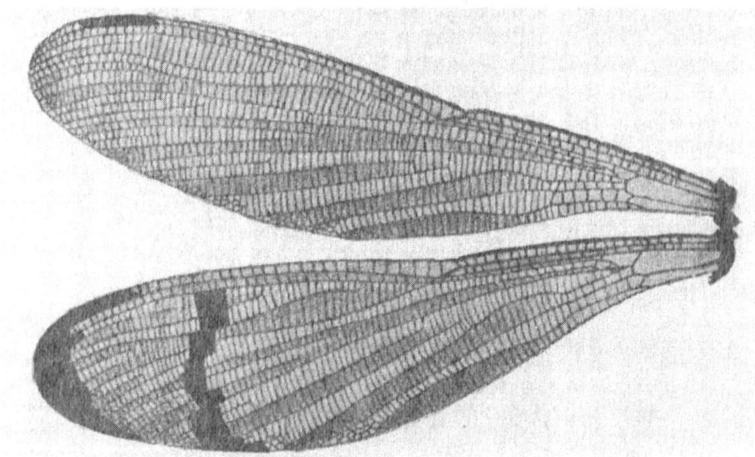

Fig. 9.—Wings of *Rhinocypha bifasciata* Selys, male.

Head similar to that of male, but with a medial spot on hind border of occiput and a moderately sized bright yellow spot on each cheek near the eye continued up as a very narrow streak against the eye. *Prothorax* with an additional large spot just below posterior lobe. *Thorax* with mesothoracic triangle present, yellow in its upper half, this colour continued narrowly along both sides of the triangle below. Lateral markings broader than in male, and there are some additional markings, as follows:—An upper antehumeral spot and a lower stripe, widely separated, a fine posthumeral stripe broadly interrupted above, where is isolated a tiny upper point, and not extending downwards as far as anterior border of thorax; there is also a fine broken line on first lateral suture and a small spot on upper part of mesepimeron. *Wings* evenly and palely enfumed throughout, with a pale greenish tint. Pterostigma black, its outer half or two-thirds yellowish. Discoidal cells with 3 to 5 traversing nervures; 15 to 18 antenodal nervures. *Abdomen* black, marked with ochreous as follows :—A large lateral spot on segment 1, an apical spot and a basal stripe on segments 2 to 4, segment 5 with a small basal spot only, while segments 2 and 3 have a vestigial ventral stripe also. *Anal appendages* and vulvar scale as usual in the genus.

Distribution.—BENGAL and ASSAM, more especially the latter. The *type* in the Selysian collection, in the Brussels Museum, is from Darjeeling. I have seen several specimens taken at Gopaldhara by Mr. H. Stevens during October and November.

There is no difficulty in distinguishing this beautiful species ; the two opaque bands at once determine it, *R. hilaryæ* having the inner band much nearer the pterostigma, and the vitreous area limited to the outer apical area of the wing.

The wings of this species and *R. trifasciata* are markedly pleated and, if held in certain positions, one set of pleats glows with a satin-like violety blue sheen, whilst the other set shines emerald-green. The restriction of the markings of the head will distinguish the female from that of other Indian species.

178. **Rhinocypha trifasciata** Selys. (Fig. 10 and Pl. II, fig. 3.)

Rhinocypha trifasciata Selys, Syn. Cal. p. 61 (1853); id., Mon. Cal. p. 207 (1854); Kirby, Cat. Odon. p. 113 (1890); Laidlaw, Rec. Ind. Mus. vol. xiii, p. 34 (1917); Fraser, J. Bombay Nat. Hist. Soc. vol. xxix, p. 478 (1923); id., ibid. vol. xxxii, pp. 315, 316 (1927).

Libellago trifasciata Walker, List Neur. Ins. Brit. Mus. vol. iv, p. 650 (1853).

Male.—Abdomen 24 mm. Hind-wing 27 mm.

This species is very similar to the last, but is invariably larger, and the male bears an additional opaque band on the hind-wings.

Head with markings exactly similar to those of *R. bifasciata*. *Prothorax* black, with a mid-dorsal longitudinal bluish spot on posterior lobe. *Thorax* black, marked very similarly to that of *R. bifasciata* ; mesothoracic triangle very large, extending from alar sinus to anterior border of thorax, pale blue. A fine yellow line on first lateral suture, broken above, a shorter line on second lateral suture, incomplete below, a small spot below root of fore-wing, and the usual irregular broken stripe along lower part of sides of thorax ; this consists of a smallish spot on the metepimeron, triangular in shape, an elongate spot on the mesepimeron, with a cuneiform angulate spot on its anterior part. *Wings* coloured like those of *R. bifasciata* vitreous area of equal extent, and bluish-violaceous or emerald-green, according to the angle from which it is viewed. Three blackish-brown bands traverse the hind-wings from costa to posterior border. An apical band extending inwards to outer end of pterostigma, its inner border concave; a medial band situate at about the junction of the outer and middle thirds of the wing, with serrate concave inner border, slightly sinuous convex outer, from 6 to 8 cells wide. An inner band with its inner border in line with, or slightly distal to, level of node, limited costalwards by *Riii*, reaching the hind border

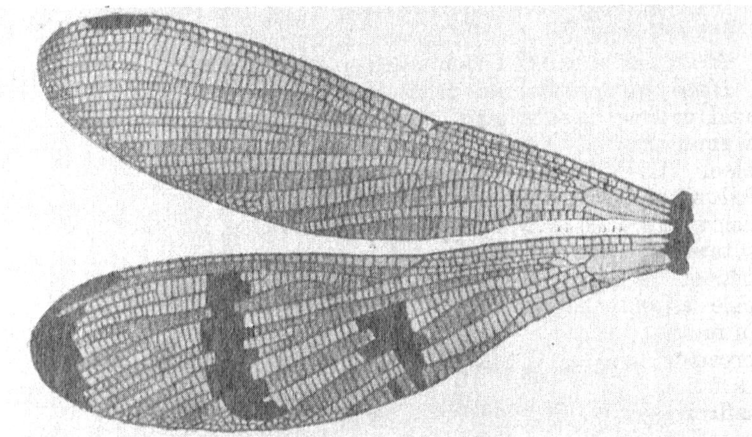

Fig. 10.—Wings of *Rhinocypha trifasciata* Selys, male.

of the wing, moderately straight on its proximal, but with three outer prolongations on its distal, border. Pterostigma blackish-brown, central part of that of hind-wings paler; 20 to 22 antenodal nervures; discoidal cells traversed 6 times. *Abdomen* black, marked on each side of segments 1 to 3 with a small apico-lateral spot of yellow. *Anal appendages* black, shaped as for the genus. *Legs* black, the two posterior pairs of femora and tibiæ heavily pruinosed white.

Female.—Abdomen 22 mm. Hind-wing 29 mm.

Very similar to the female of *R. bifasciata*, but considerably larger. Mesothoracic triangle long and broad, as in the male, dark ochreous in colour. Markings of head, thorax, and abdomen exactly as in *R. bifasciata*, with the exception of the mesothoracic triangle, which is as described above.

Distribution.—The *type* in the Selysian collection, in the Brussels Museum, is labelled "India," without further indication of its origin. Specimens in the British Museum are from Kangra Valley, PUNJAB. Three males in the Indian Museum,

Calcutta, are from Kailana, UNITED PROVINCES, so that it would appear that this species has a more northerly and westerly distribution than that of *R. bifasciata*. The two species are undoubtedly distinct, although closely related.

The opaque bands of the hind-wings are subject to considerable variation in shape and length. The description given above is from a specimen in my own collection.

179. **Rhinocypha bifenestrata** Fraser. (Fig. 11.)

Rhinocypha bifenestrata Fraser, Mem. Dept. Agric. India (Ent.) vol. vii, no. 7, p. 63 (1922); id., J. Bombay Nat. Hist. Soc. vol. xxix, p. 478 (1923); id., ibid. vol. xxxii, pp. 316, 317, pl. iii, fig. 1 (1927).

Male.—Abdomen 25 mm. Hind-wing 27 mm.

Head black; labium and labrum unmarked; a small oval yellow spot on outer side of each posterior ocellus, and a smaller rounded spot on each side of occiput. *Prothorax* black, unmarked. *Thorax* black, marked with yellow as follows:—A vestigial broken posthumeral line, incomplete below, a narrow stripe bordering front of second lateral suture, and a small spot below hind-wing. Mesothoracic triangle extending from alar sinus to anterior border of thorax, pale lilaceous. *Legs* black, the two posterior pairs of tibiæ pruinosed white. *Wings* largely opaque, by fusion of three broad bands which enclose two large vitreous spots. Base as far as outer end of discoidal cell hyaline, rather brightly saffronated. Opaque area in fore-wings variable, usually broadly serrated and limited posteriorly by *IRiii*, but sometimes more closely serrated and extending as far back as *MA* at base and *IRiv* nearer apex of wing. Hyaline area posterior to opaque area vitreous and a beautiful violet by reflected light, extending basally at one point to proximal to the outer end of the discoidal cell. In hind-wings the opaque area extends proximally as far as 3 or 4 cells proximal to node near costal margin of wing, and from thence runs obliquely out to posterior border of wing, with a very irregular border. By a confluence of the three bands, two large vitreous violaceous spots are enclosed, varying much in size and shape in individual specimens. In the type the outer two bands are not much wider than in *R. trifasciata*, being narrowly connected along the costal and hind borders of the wing, but in some specimens the apical opaque band extends as far in as slightly proximal to the inner end of the pterostigma. Inner spot usually partially bisected by a prolongation of the inner band, one cell wide, running posterior to *Riii*, and in some specimens this prolongation actually fuses with the middle band, thus completely bisecting the inner vitreous spot.

In the type the middle and inner bands are only very narrowly confluent along both costal and hind margins of the wing, so that the inner spot is of large dimensions. Pterostigma black, its outer three-fourths white, but clouded with light brown in fore-wings ; antenodal nervures 19 to 20 ; discoidal cells traversed 6 times. *Abdomen* black, quite unmarked. *Anal appendages* black, shaped as for the genus.

Female.—Abdomen 21 mm. Hind-wing 31 mm.

Head : labium and labrum black ; rest of head velvety black, marked with bright ochreous as follows :—A large spot at base of mandibles, a narrow streak against the eyes expanding on to the cheeks, a narrow longitudinal streak on lower part of epistome, a tiny point just in front and to the outer side of middle ocellus, and two linear spots on frons

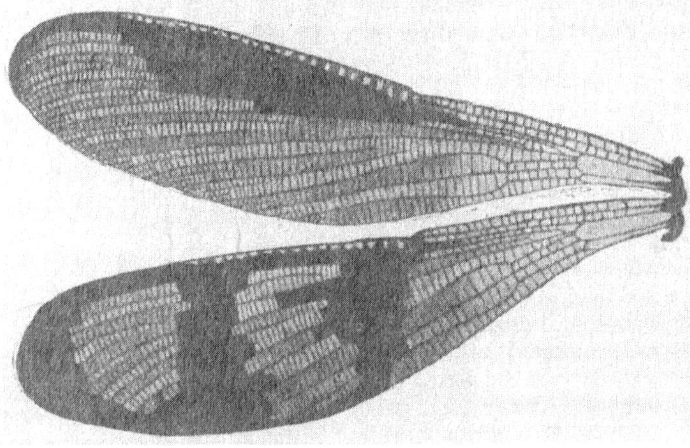

Fig. 11.—Wings of *Rhinocypha bifenestrata* Fras., male.

in front of latter, a large rounded spot on outer side of each outer ocellus and a similar postocular spot behind it ; lastly, a large triangular spot at middle of hind border of occiput. These markings liable to some variation ; thus, the spots on the frons may be very large and triangular, or entirely absent, while the occipital spot may be small and linear or large and crown-shaped. *Prothorax* black, marked with bright ochreous as follows :—A mid-dorsal longitudinal streak running from posterior lobe on to middle lobe, a small spot on lower part of posterior lobe and a spot, which may be large and triangular or reduced to a mere point, on each side of middle lobe. *Thorax* black, marked with bright ochreous as follows :—Mesothoracic triangle, except for its lower part, intersecting suture of alar sinus, a very fine antehumeral stripe, often broken into a chain of tiny points, and best defined in its lower part, a posthumeral stripe incomplete below and broadly broken in its upper part, a linear spot on upper part of mesepimeron, and a fine incomplete line bordering first lateral suture; finally, a broad irregular broken stripe on lower part of sides of thorax and two small rounded spots beneath. *Legs* black, tibiæ not pruinosed. *Wings* uniformly enfumed brown; pterostigma black, slightly clouded in its outer part with yellow; discoidal cells traversed 4 to 5 times; petiolation short, from near the first antenodal nervure or between the first and second; 20 to 21 antenodal nervures, 32 to 34 postnodals. *Anal appendages* and vulvar scale as for the genus. *Abdomen* black, marked with bright ochreous as follows:—A small cuneiform spot on each side of segment 1, a small apico-lateral spot followed by a short streak, and a ventral stripe on sides of segments 2 to 4; segments 5 and 6 similar, but upper streak obsolete and apical spot much reduced.

Distribution.—BENGAL: Mungpoo, Darjeeling District, 8,600 ft., May and June, and again in August and September. Mr. Inglis has also taken it in April, but it is rare during that month. Mr. Inglis, Mr. Shaw, and myself took upwards of forty specimens at the end of May along the beds of streams, usually settled on rocks, and never engaging in sustained flight like that of *R. bisignata*.

This species is, I think, more nearly related to *R. cuneata* than to *R. bifasciata*, but it appears to link up groups *cuneata* and *bifasciata*, and may have originated by a fusion of the opaque bands of *R. trifasciata*.

Type in the British Museum.

180. **Rhinocypha immaculata** Selys.

Rhinocypha immaculata Selys, Bull. Acad. Belg. (2) vol. xlvii, p. 385 (1879); Kirby, Cat. Odon. p. 113 (*R. unimaculata*, a *lapsus calami* for *R. immaculata*) (1890); Laidlaw, Rec. Ind. Mus. vol. xiii, pp. 34, 35 (1917); Fraser, J. Bombay Nat. Hist. Soc. vol. xxix, p. 478 (1923); id., ibid. vol. xxxii, pp. 317, 318 (1927).

Male.—Abdomen 22–25 mm. Hind-wing 27–28 mm.

Head black, marked with blue and ochreous spots as follows:—Bases of mandibles, cheeks, a stripe bordering the eyes, a reniform spot on each side of frons, and a smaller rounded spot just behind these, pale blue. An oval spot on outer side of each posterior ocellus, and a round postocular spot on each side of occiput, ochreous; labium and labrum unmarked. *Prothorax* black, marked with a mid-dorsal longitudinal blue line beginning on posterior lobe, and a large spot on each side. *Thorax* black, marked with bright yellow as follows:—A fine lower antehumeral stripe, a fine upper humeral stripe incomplete below, a small upper spot near first suture, a thick irregular broken stripe on lower part of sides, consisting of a triangular spot on metepimeron, and an irregular longitudinal stripe on mesepimeron. Four to six yellow spots on underside of thorax. Mesothoracic triangle very large, extending from anterior border of thorax to alar sinus, pale blue. *Legs* black, posterior pairs of femora yellow on flexor surface, tibiæ of the same legs pruinosed white. *Abdomen* black, marked with yellow as follows :— A small spot on each side of segment 1, a lateral stripe and an apical spot in line with the former on each side of segments 2 to 5 ; finally, a ventro-lateral stripe, not extending as far as apex of segment, on each side of segments 3 to 6. In old males the lateral stripes on segments 3 to 6 are lost, the apical spot alone remaining, or it may even be lost on segments 4 to 6. *Anal appendages* black, shaped as for the genus. *Wings* entirely hyaline (the only species of the genus in which they are so), shaped as for *R. trifasciata*, 15 to 16 antenodal nervures, discoidal cells with 5 to 6 traversing nervures ; pterostigma blackish-brown, its outer and costal part paler or yellowish.

Female.—Abdomen 22 mm. Hind-wing 28–30 mm.

Very similar to male. Markings of head better defined, and all bright yellow. *Thorax* with a large mesothoracic triangle, outlined in yellow, black at its centre ; other markings rather broader. *Legs* black, not pruinosed. *Wings* hyaline, occasionally enfumed in old adults ; pterostigma black, whitish in its outer part ; antenodal nervures 16 in number. *Abdomen* black, marked as in the male. *Anal appendages* and vulvar scale as for the genus.

Distribution.—Known only from Cherrapunji, Khasi Hills, Assam. On the wing during September and October.

This interesting species is easily distinguished from all others of the genus by the total absence of markings, or even vitreous iridescence, on the wings.

Type in the Selys collection, Brussels Museum.

181. **Rhinocypha hilaryæ** Fraser. (Fig. 12.)

Rhinocypha hilaryæ Fraser, Rec. Ind. Mus. vol. xxix, pp. 83–86, fig. 4 (1927) ; id., J. Bombay Nat. Hist. Soc. vol. xxxii, pp. 318, 319, pl. iii, fig. 2 (1927).

Male.—Abdomen 21–22 mm. Hind-wing 27 mm.

Head : labium black, lateral lobes creamy white ; labrum azure-blue, its anterior border and base narrowly black, the basal black with a median point ; epistome glossy black in front, velvety black above, with a small triangular spot of blue on each side ; frons with a large quadrangular blue spot on each side, with its outer angle pointed and prolonged outwards, the two spots very narrowly separated and confluent with two large subrotund spots of the same colour on the vertex, which lie in close apposition to the median ocellus. Laterally, bases of mandibles, penultimate segment of antennæ, and a narrow irregular stripe bordering the eyes, bluish. Lastly, a tiny point on outer side of each lateral ocellus, another on each side of occiput, and a small linear spot at its middle. (In dried specimens these markings are yellow or greenish-blue, but in life the adult markings are always blue.) *Prothorax* black, marked with blue, bluish-green, or yellow, according to age of specimen, as follows:—A narrow median linear stripe which begins on posterior lobe and tapers away anteriorly, a small triangular spot on each side of posterior lobe, and a short oval on each side of middle lobe. *Thorax* black, marked with blue and yellow as follows :—Mesothoracic

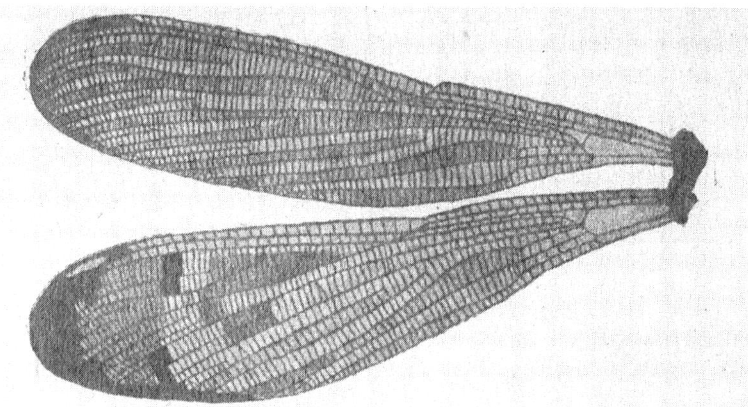

Fig. 12.—Wings of *Rhinocypha hilaryæ* Fras., male.

triangle pale turquoise-blue, elongate, extending as far up as alar sinus; a fine humeral stripe azure-blue; laterally an irregular thick interrupted stripe extending the whole length of the thorax, rather zigzagged anteriorly, and broadened into an elongate triangular spot on metepimeron. *Legs* black, tibiæ and distal parts of femora of the two posterior pairs of legs pruinosed white on flexor surface. *Wings* hyaline, palely enfumed with greenish-yellow, marked with opaque black and vitreous-violet areas as follows:—Fore-wing with only extreme apex bordered or clouded with opaque black; pterostigma black; hind-wing with apex broadly opaque black, inner border of opaque area serrated, running almost straight back from middle of pterostigma to hind border of wing, and, finally, continued as a narrow border along hind edge of the wing for about its apical fourth. At a short distance from the proximal end of the pterostigma is seen a narrow opaque black fascia, nearly always broadly broken at its middle into a small quadrate or angulate spot lying between *Rii* and *IRii*, and a markedly irregular spot confluent with the narrow black bordering of the wing; rarely the two spots are united by a narrow isthmus, but in no two specimens are they ever exactly alike. The vitreous-violet areas comprise the whole of the area included between the apical black bordering and the irregular fascia, as well as three linear spots situate between the level of the node and the narrow black fascia, nearer the latter than former. Of these, the anterior stripe is the longest, about 20 to 25 cells long, and situated between *Rii* and *Riii*; the median stripe, lying between *IRiii* and *Riv+v*, is half the length of the former, its outer end in line with the outer end of the same; finally, the third, or hinder spot, lying between *MA* and *Cuii*, is very narrow, and continues outward along the posterior border of the wing, to become confluent with the narrow opaque black border; pterostigma black, with a large oval pale blue spot in its outer two-thirds; discoidal cell traversed 4 to 5 times; petiolation variable, beginning opposite the second or third antenodal, but usually opposite the second; 17 to 19 antenodal nervures, 28 to 38 postnodal nervures (very variable). *Abdomen* black, marked with bluish-green or yellow as follows:—Segment 1 with a large lateral triangular spot on each side; segment 2 with a similar spot situate at the apex, as well as a ventral stripe nearer apex than base of segment; segment 3 similar, but apical spot smaller and ventral stripe longer, extending for nearly the whole length of the segment; segments 4 and 5 similar to 3, but the markings becoming obsolete; 6 and 7 with vestigial ventral stripes only. *Anal appendages* as usual in the genus.

Female.—Abdomen 21–22 mm. Hind-wing 30–32 mm.

Head: markings similar to those of male, but bright yellow instead of blue. Lateral spot on epistome larger, and, in addition, a lateral stripe on either side of its front. *Prothorax* similar to that of male. *Thorax* black, marked with bright yellow as follows:—Mesothoracic triangle outlined in yellow, the two lateral lines converging and fusing above, and then continued as a median line bisecting the antealar sinus; a fine complete antehumeral stripe curving slightly outwards above, and inwards below; a fine humeral stripe, incomplete below; a strongly curved stripe on upper part of mesepimeron; and, lastly, a broad irregular stripe traversing the whole length of the lower part of the thorax, as in the male. *Wings* hyaline, palely enfumed throughout, tinted slightly with yellow, especially at base; pterostigma black, its outer part bearing a broad oval yellowish spot; 15 to 19 antenodal nervures, 26 to 31 postnodals; discoidal cells traversed 2 to 4 times; petiolation begins between the first and second antenodals. *Abdomen* black, similar to that of male, but with yellow markings; segments 2 and 3 have also a lateral stripe lying above and parallel to the ventral stripe, in line with, but not confluent with, the apical spot, whilst the ventral stripe is obsolete after segment 3 and all markings are obsolete after segment 6. *Anal appendages* and vulvar scale as usual in the genus.

Distribution.—UPPER BURMA: Maymyo. Quite a number of both sexes were taken by Col. F. Wall, I.M.S., during the months of June, July, and August. The insect appears to be remarkably local.

The shape of the mesothoracic triangle and the wings show that *R. hilaryæ* belongs to group *bifasciata*. It can only be confounded with *R. bifasciata*, in which, however, the proximal fascia is complete and situated much nearer the node, the vitreous area also covering nearly the entire wings.

Type and allotype in the British Museum.

182. **Rhinocypha whiteheadi** Kirby. (Fig. 13.)

Rhinocypha whiteheadi Kirby, Ann. Mag. Nat. Hist. (7) vol. v, p. 536, pl. xii, fig. 4 (1900); Martin, Mission Pavie, Neurop. (sep.), p. 17 (1904); Laidlaw, Rec. Ind. Mus. vol. xiii, pp. 38, 39 (1917); Fraser, Mem. Dept. Agric. India (Ent.), vol. vii, pl. viii, fig. 4 (1922); id., J. Bombay Nat. Hist. Soc. vol. xxxii, pp. 450, 451 (1928).

Male.—Abdomen 17 mm. Hind-wing 21 mm.

Head: labium, labrum, and rest of head velvety black, marked with blue as follows:—On each side a reniform or oval spot on outer side of posterior ocellus, a smaller post-ocular spot behind it, and a medial linear occipital spot behind the postocular spot. *Prothorax* black, marked on each side with a rounded pale blue spot, and a large blue spot, shaped like an arrow-head, on dorsum of posterior lobe and hind part of middle lobe. *Thorax* velvety black, with a dark violet lustre, marked with azure-blue and chrome-yellow as follows:—Mesothoracic triangle pale blue, extending considerably less than half-way up dorsum of thorax, and very broad at base, far more so than in other species of the genus; an ante-humeral stripe, consisting of a small isolated spot above, and a narrow curved stripe on the lower half of the dorsum; a fine posthumeral stripe closely apposed to the humeral suture, broken above to leave a small isolated spot, behind which is another elongate larger spot, all blue. A broad interrupted stripe on lower part of sides yellow, but the portion on the metepimeron blue anteriorly; pruinosed white beneath. *Legs* black, femora and the two posterior pairs of tibiæ pruinosed white. *Wings* hyaline, palely tinted with yellow at base, opaque blackish-brown apically, hind-wing marked on this part with three rows of vitreous-violet spots. Fore-wing with apical portion opaque blackish-brown as far inwards as rather more than half-way from apex to node, extending nearer node along costal border of wing, inner border of dark area serrate and running obliquely backward and outwards; pterostigma black, cells immediately beneath it hyaline. Hind-wing with opaque area darker and extending inwards for about three-fourths the distance between apex and node, its inner border very irregular, indented and serrate. Vitreous

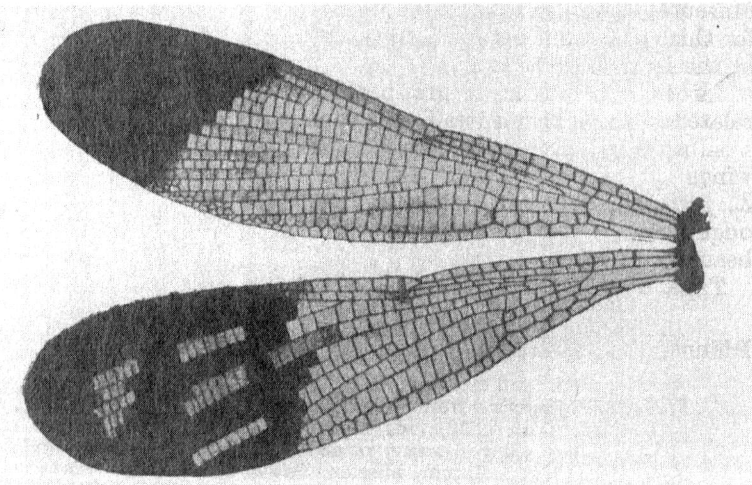

Fig. 13.—Wings of *Rhinocypha whiteheadi* Kirby, male.

spots a beautiful violet, from whatever angle they are viewed. Apical spot single, 4 cell-rows deep by 6 cells wide, irregularly shaped, situated 3 cell-rows behind pterostigma, which is black; middle row of spots 3 in number, all in alignment and of about equal length, about 10 cells wide, costal spot lying between *IRii* and *Riii*, median spot between *IRiii* and *Riv+v*, posterior spot between *Cuii* and *IA*; proximal spot single, short, only 5 cells long, lying between *IRii* and *Riv+v*, projecting slightly into opaque area. All the wings with 14 to 15 antenodal nervures; discoidal cell traversed 3 times in the fore-wing, 4 times in the hind-wing; wings petiolated to level of second antenodal; 21 to 26 postnodals. Wings of approximately the same shape and length, long and narrow, the hind-wings slightly broader than the fore-wings. *Abdomen* black, marked with azure-blue as follows:—Segments 1 to 4 with lateral rounded spot at apical end of segments, growing successively smaller from 1 to 4, segment 1 has also the apical border narrowly lined with blue, and there is a short ventro-lateral stripe on segments 2 and 3. *Anal appendages* shaped as for the genus; inferior appendages with a series of small robust spines on upper surface.

Female.—Unknown.

Distribution.—ASSAM: Cachar. Very local, but apparently not uncommon where found, to judge by the numbers taken by Mr. Antram. Martin records it from Tonkin, but I was unable to find any specimens in his collection in the Paris Museum, and infer that he has mistaken a race of *R. perforata* for this species. There is a single male from Sibsagar, Assam, in the Indian Museum.

Although belonging to group *perforata*, it is not very closely related to *R. perforata* itself, and cannot even be considered as a subspecies of that species, as has been suggested. The wings in *R. whiteheadi* are not only broader than in *R. perforata*, but are distinctly rounded at the apices. It ought not to be difficult to discover the female of this beautiful insect.

Type in the British Museum.

Rhinocypha perforata (Percheron). (Fig. 14.)

Agrion perforatus Percheron, Gen. Ins., Neur. t. 2 (1835).
Rhinocypha perforata Rambur, Ins. Névrop. p. 235 (1842); Selys, Syn. Cal. p. 63 (1853); id., Mon. Cal. p. 219 (1854); id., Bull. Acad. Belg. (2) vol. xxxv, p. 487 (1873); Kirby, Cat. Odon. p. 114 (1890); Martin, Mission Pavie, Neurop. (sep.) p. 17 (1904); Williamson, Proc. U.S. Nat. Mus. vol. xxvii, p. 174 (1905); Laidlaw, Rec. Ind. Mus. vol. xiii, pp. 38, 39 (1917); Fraser, Mem. Dept. Agric. India (Ent.), vol. vii, pl. viii, fig. 3 (1922); id., J. Bombay Nat. Hist. Soc. vol. xxix, p. 479 (1923); id., ibid. vol. xxxii, p. 451, pl. i, fig. 3 (1928).
Libellago perforata Walker, List Neur. Ins. Brit. Mus. vol. iv, p. 647, (1853).
Rhinocypha apicalis Krüger, Stett. Ent. Zeit. p. 79 (1898); Laidlaw, Fascic. Malayenses (Zool.), pt. i, p. 196 (1903).
Rhinocypha inas Laidlaw, Proc. Zool. Soc. Lond. pp. 88–90, pl. vi, fig. 6 (1902).

The typical form of *R. perforata* is not found within Indian limits, but is represented by its subspecies or races *limbata* and *beatifica*, described below. The type was taken in Cochin China, and cotypes exist in the McLachlan collection from the island of Hainan, China, collected by Swinhoe. *R. perforata perforata* differs from the form *limbata* by not having the border of the apex of the hind-wing hyaline, and by the reduction of the opaque area in both fore- and hind-wings. In the fore-wing this area occupies a little more than the apical fourth, but in the hind-wing it extends to within 4 cells of the node. In the form *limbata* this area extends

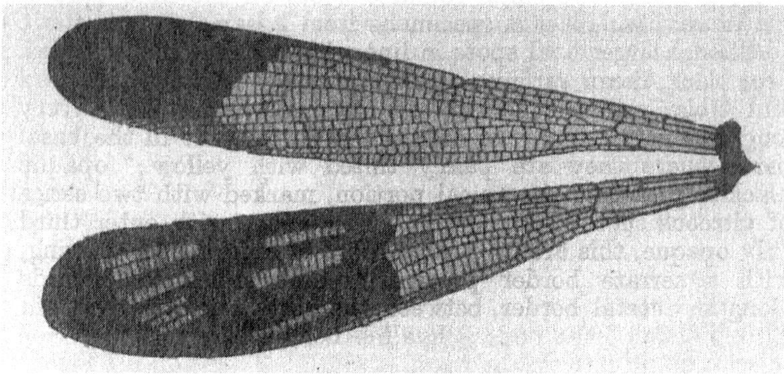

Fig. 14.—Wings of *Rhinocypha perforata perforata* Perch., male.

nearly up to the node in the fore-wings, but is variable in the hind-wings, usually extending to about the same distance from the node, whereas in the form *beatifica* it extends right up to the node.

Type in the Paris Museum.

183. **Rhinocypha perforata limbata** Selys.

Rhinocypha perforata var. *limbata* Selys, Bull. Acad. Belg. (2) vol. xlvii, p. 392 (1879); Kirby, Cat. Odon. p. 114 (1890); Laidlaw, Rec. Ind. Mus. vol. xiii, p. 38 (1917).

Rhinocypha perforata limbata Fraser, J. Bombay Nat. Hist. Soc. vol. xxxii, pp. 451, 452 (1928).

Male.—Abdomen 16–18 mm. Hind-wing 25 mm.

Head: labium dirty white, apices of lobes black; labrum and rest of head velvety black, with a small rounded spot on outer side of each posterior ocellus, a larger rounded post-ocular spot and a linear medial spot on occiput, all bluish. *Prothorax* black, with the greater part of the posterior lobe rose-pink or lilaceous, and a large blue spot on either side of the middle lobe. *Thorax* velvety black; mesothoracic triangle extending rather more than one-third up the dorsum, lilaceous. On the outer side of this triangle, and slightly longer than it, a large subtriangular azure-blue spot with its posterior border meeting the anterior part of the humeral suture. Above this spot is a small triangular upper humeral spot, also blue, whilst laterally the greater part of the sides is of the same colour, viz., nearly the whole of the mesepimeron, the upper two-thirds of the metepimeron, and a narrow prolongation along the hind border of the humeral suture, reaching as far as the root of the fore-wings. Beneath thorax 2 to 4 yellow spots which may be somewhat obscured by pruinescence. In the Siamese and Annamese forms there are only two small rounded spots on the hind part of the thorax, whereas, in specimens from Assam, there are two additional larger oval spots in line with and in front of these. *Legs* black, flexor surfaces of the two posterior pairs of femora and tibiæ pruinosed white or creamy white. *Wings* very long and very narrow, of equal breadth, hyaline in the basal part, where they are palely tinted with yellow; opaque blackish-brown in the apical portion, marked with two series of vitreous spots in hind-wing. Fore-wings with outer third only opaque, this area stopping short of hind border of wing, with a serrate border proximally and prolonged inwards along the costal border, between the costa and *Ri*, to within 2 to 4 cells of the node; hyaline border of wing, posterior to opaque area, vitreous with bluish, violaceous or green reflex, which may extend as far as the extreme apex of the wing. Pterostigma black; 15 antenodal nervures, 24 postnodals; discoidal cell traversed twice only; petiolation ending at level of second antenodal nervure. Hind-wing with opaque area extending inwards to within 1 to 4 cells of node, proximal border of this area ragged and indented deeply by a vitreous spot 8 or 9 cells long, between *IRiii* and *Riv+v*. Hind border of wing narrowly hyaline and vitreous, finely reticulated, and extending round apex of wing as far as pterostigma. Apical series of vitreous spots very variable, usually composed of 3 linear spots, which decrease in length from the costa to the hind border of the wing, and which slightly overlap the level of the inner end of the pterostigma. Costal spot 11 to 16 cells long, lying between *IRii* and *Riii*; posterior spot 3 cell-rows deep by 4 to 8 cells long, lying between *IRiii* and *Riv+v*, whilst the middle spot, which is 2 to 3 cell-rows deep by 9 to 10 cells long, lies mid-way between the two others, banded by short intercalated sectors; inner series lying rather nearer node than pterostigma, composed of 3 spots, a costal spot 1 cell-row wide by 10 cells long, lying between *IRii* and *Riii*, a medial spot 2 cell-rows wide by 10 to 15 cells long, between *IRiii* and *Riv+v*, and a posterior spot 1 cell-row wide by 9 to 12 cells long, between *MA* and *Cuii*. All these vitreous spots and areas glowing emerald-green or opalescent, according to the angle from which they are viewed; 28 postnodal nervures, 14 to 15 antenodals; discoidal cell traversed 3 to 4 times; petiolation the same as in fore-wing; pterostigma black. *Abdomen* black, marked laterally with azure-blue as follows:—Segment 1 with a large cuneiform spot on each side, 2 with a bilobed longitudinal spot extending from near base to apical border, 3 and 4 both with a long wedge-shaped spot, with base of wedge at apical border of segment, segments 5 to 9 each with a short triangular apical spot, that on 9 much reduced. *Anal appendages* black, superiors slender, cylindrical, semicircular, a little broadened at apex, which is strongly incurved, and minutely spined on outer side near apex. Inferiors half the length of the superiors, moderately separated, bluntly pointed, a little denticulated at apex.

Female.—Abdomen 18 mm. Hind-wing 26 mm.

Differs from male by having wings entirely hyaline, tinted palely with greenish-yellow; pterostigma blackish-brown, its outer half yellow at the centre, especially in hind-wings; 13 to 15 antenodal nervures in fore-wings, 26 to 30 postnodals.

Head similar to that of male, with the addition of a large spot of yellow on upper surface of epistome and two smaller ones behind it. Spots on vertex and occiput also yellow, body-markings like those of male, but yellow instead of blue, mesothoracic triangle finely outlined in yellow, its centre black. *Legs* black, tibiæ and femora not pruinosed. *Abdomen* black, with yellow markings as in male, but with additional ventro-lateral stripes on segments 2 to 5. Vulvar scale as for the genus.

Distribution.—*R. p. limbata* extends from Annam, through Siam, to BURMA and ASSAM. The *type*, in the Selysian collection, is from East Burma.

The abdominal markings, said by Selys to be more restricted than in typical *R. perforata*, are not found to be so when a series of specimens is examined. In the fore-wing, at least, the opaque area is greatly reduced as compared to *R. p. limbata* or *R. p. beatifica*.

184. **Rhinocypha perforata beatifica** Fraser.

Rhinocypha perforata beatifica Fraser, Rec. Ind. Mus. vol. xxix, pp. 86–87, fig. 5 (1927); id., J. Bombay Nat. Hist. Soc. vol. xxxii, pp. 452, 453 (1928).

Similar in size and markings to *R. p. limbata*, from which it differs by the greater extent of the opaque area in all the wings, this area reaching right up to the level of the node. This is especially noticeable in the fore-wings, where not only the costal streak, but also the body of the area, extends to the node, the serrated border running from the node and extending as far as the apex of the wing, leaving a rather broader area of hyaline border than is found in *R. p. limbata*. In the hind-wing, the hyaline border broadens out at the apex of the wing, extending right up to, or even overlapping, the outer end of the pterostigma. This latter has, in the hind-wing, a narrow bluish centre along its outer half, and that of the fore-wing is pale for the same extent. In the outer series of vitreous spots the middle spot is 2 cell-rows deep and is fused completely with the costal spot, so that in reality there are only two spots in this series.

Distribution.—ASSAM: Naga Hills, from April to June.

R. p. beatifica differs from *R. p. limbata* to about the same extent as does the latter from typical *R. perforata*, so that, given sufficient material, it might be possible to construct a series showing a gradual merging of one into the others. For the present it seems better to consider the *perforata* series as one species with a typical form and two subspecies.

Type in the Fraser collection.

Rhinocypha biforata Selys.

Rhinocypha biforata Selys, Bull. Acad. Belg. (2) vol. vii, p. 446 (1859); Kirby, Cat. Odon. p. 113 (1890); Laidlaw, Proc. Zool. Soc. Lond. (1) p. 88 (1902); Williamson, Proc. U.S. Nat. Mus. vol. xxviii, pp. 173, 179, text-fig. 12 (1905); Laidlaw, Rec. Ind. Mus. vol. xiii, pp. 37, 38 (1917); Fraser, J. Bombay Nat. Hist. Soc. vol. xxix, p. 479 (1923).
Rhinocypha biforata biforata Fraser, J. Bombay Nat. Hist. Soc. vol. xxxii, p. 453 (1928).

The species is not found in its typical form within Indian limits, but is represented by its three subspecies—*delimbata* Selys, *beesoni* Fraser, and *abbreviata* Fraser. The typical form is characterized by the greater extent of the opaque area in the hind-wings, which extends right up to the node or a little beyond it; the vitreous spots also tend to be longer and the reticulation somewhat closer. Of the subspecies, *R. b. beesoni* is closest to the type-form but, curiously enough, is separated from it geographically by the intervention of the other two subspecies.

Type in the Selysian collection, Brussels Museum.

185. **Rhinocypha biforata delimbata** Selys. (Fig. 15 and Pl. II, fig. 2.)

Rhinocypha biforata var. *delimbata* Selys, Bull. Acad. Belg. (2) vol. xlvii, p. 392 (1879); Kirby, Cat. Odon. p. 113 (1890).
Rhinocypha biforata delimbata Fraser, J. Bombay Nat. Hist. Soc. vol. xxxii, pp. 453, 454 (1928).

Male.—Abdomen 19–20 mm. Hind-wing 23–26 mm.

Head velvety black, marked with five bright ochreous spots, one rounded or oval on outer side of each posterior ocellus, one round postocular spot on each side, and one medial oval occipital spot; labium white, lobes tipped with black. *Prothorax* black, posterior lobe finely bordered with yellow, and with a large rose-pink spot in its middle; a large bluish spot on each side near trochanters, and a similar subdorsal spot on each side of middle lobe. *Thorax* velvety black, marked as follows:—Mesothoracic triangle rose-pink or lilaceous, extending less than half-way up the dorsum of thorax, but continued as a fine line of the same colour along the mid-dorsal carina as far as the antealar sinus. On each side of the mesothoracic triangle a large irregularly oval spot of the same colour, much larger than the triangle itself, a small upper antehumeral spot, a broadish stripe behind the upper two-thirds of the humeral suture, the greater part of the mesepimeron and metepimeron, azure-blue. Posterior lateral suture, lower part of metepimeron, and a quadrate area on upper anterior part of mesepimeron all black, the latter area bearing a small isolated blue spot. Beneath, 6 large yellow spots, separated by black sutures and framed narrowly in black. *Legs* black, the two posterior pairs of

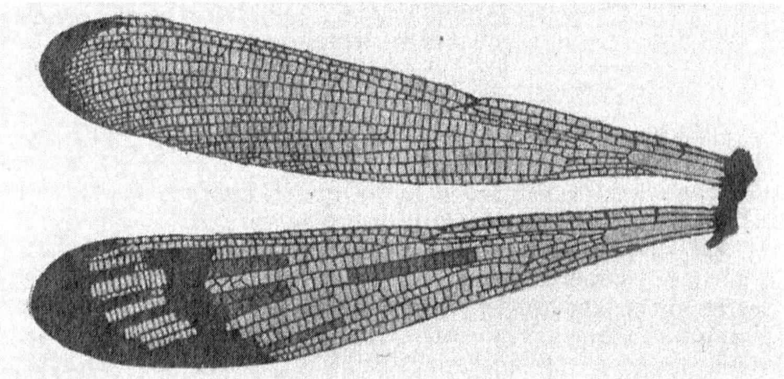

Fig. 15.—Wings of *Rhynocypha biforata delimbata* Selys, male.

tibiæ pulverulent white, the two posterior pairs of femora pulverulent yellow on the flexor surfaces. *Wings* hyaline, the greater part enfumed and tinted with yellow, which is most intense near the base. Fore-wings with apices tipped with blackish-brown, this marking gradually shading off into the hyaline area at the outer end of the pterostigma (in *R. biforata biforata* this area begins well proximal to the pterostigma, and covers about the outer fifth of the wing); discoidal cell traversed 3 times; 14 to 16 antenodal nervures, 27 to 32 postnodals; petiolation ending at level of second antenodal nervure. Hind-wings with about the apical fourth blackish-brown (the apical third or more in *R. biforata biforata*), this area marked with a series of 4 linear vitreous spots, very irregular, each varying in length, and the costal three often linked up by one or more connecting cells; inner level of spots forming a concavity towards apex of wing, as does also the inner border of the opaque area, which extends slightly nearer the base along the hind border of the wing. Just proximal to the opaque area (projecting into it in *R. biforata biforata*) is a second row of vitreous spots; the costal one of these spots, lying slightly nearer node than pterostigma, composed of one row of cells, about 12 in number, lying between *IRii* and *Riii*; second spot shorter, lying more distal, 2 cell-rows deep by 8 or 9 cells long, between *Riv+v* and *Riii*; posterior spot at same level as costal, 9 cells long by 1 row wide, lying between *MA* and *Cuii*. In the same space as the medial spot of this series, but lying behind level of node, another vitreous spot, 1 row wide by 11 to 18 cells long. Pterostigma black; 15 to 16 antenodal nervures, 26 postnodals; discoidal cell traversed 3 times. *Abdomen* black, marked with azure-blue as follows:—A large triangular spot on each side of segment 1, a mid-subdorsal and an apico-lateral spot on each side of segment 2, as well as a ventro-lateral stripe below them; similar spots and stripes, but longer, on segments 3 to 5; segments 6 and 7 with the ventro-lateral stripe only. *Anal appendages* black, shaped as for the genus.

Female.—Abdomen 15–16 mm. Hind-wing 23 mm.

Head black, with the same markings as in male, but, in addition, bases of mandibles, basal segments of antennæ, two large spots on frons, and two minute linear spots in front and to outer side of anterior ocellus, all yellow. *Prothorax* black, with the following yellow markings:—Hind border of posterior lobe outwardly finely yellow, a narrow longitudinal medial line, often broken into minute points on dorsum of same lobe, a geminate point at the anterior end of this on the medial lobe, a largish triangular spot on each side of mid-lobe, and a rounded spot on each side of anterior end. *Thorax* black, marked with blue and yellow as follows:—Meso-thoracic triangle finely mapped out in yellow, its centre black, a fine antehumeral line incomplete above and strongly hooked outwards and backwards in front, a fine incomplete humeral line, all yellow. Sides similar to those of male, blue. *Legs* black, not pruinosed, flexor surfaces of posterior pairs of femora yellow. *Abdomen* black, marked exactly as in male. *Anal appendages* and vulvar scale as usual in the genus.

Distribution.—Burma and Assam; common in parts of the former but apparently rare or local in the latter. I have specimens from Mergui, Lower Burma, collected in November, and from Maymyo, Upper Burma, collected in July, the latter specimens by Col. F. Wall. It appears to be the commonest species of *Rhinocypha* in the Maymyo district.

The measurements differ widely from those given by Selys for *R. biforata*, but it is obvious that these are incorrect, at least for the abdomen. The Maymyo specimens are very constant in their markings, especially with regard to the extent of the opaque areas of the wings, so that I think *delimbata* should at least rank as a subspecies.

Type in the Selysian collection, Brussels Museum.

186. **Rhinocypha biforata abbreviata** Fraser.

Rhinocypha biforata abbreviata Fraser, J. Bombay Nat. Hist. Soc. vol. xxxii, pp. 454, 455 (1928).

The single male specimen known to me shows the following differences from the other subspecies :—

The mesothoracic triangle is considerably smaller than in *R. biforata delimbata*, and the carina above it is unmarked; the antehumeral spot is absent; the fore-wings have only the extreme tips opaque, and the nodal index is considerably lower (12 antenodal nervures and 21 postnodals); the spots in the outer series are discrete, and all are much shorter than in the typical form (12 to 13 cells for costal spot in *R. biforata biforata*, only 6 in *R. b. abbreviata*, 18 cells in the second spot, only 14 in *R. b. abbreviata*, etc.). The spots of the middle series show the same difference, being only 4 to 5 cells in length and falling well short of the opaque area, although in the hind-wing this area is as extensive as in *R. biforata biforata*. Finally, the proximal spot is only 8 cells in length. Nodal index of hind-wing is lower, 11 antenodals and 23 postnodals. *Abdomen* with markings present on the first five segments only as follows :—A large triangular blue spot on the sides of segment 1, an apical rounded spot, a short mid-lateral stripe and a ventro-lateral stripe on segment 2; segments 3 to 5 similar to 2, but medial stripes absent and other markings much reduced.

Distribution.—ASSAM.

The small size of this insect, the abbreviated vitreous markings, and the reduced abdominal markings are, I think, sufficient to constitute it a well-defined subspecies, bearing the same relation to typical *R. biforata* as does *R. trimaculata* to *R. ignipennis*.

Type in the Fraser collection.

187. **Rhinocypha biforata beesoni** Fraser.

Rhinocypha beesoni Fraser, Mem. Dept. Agric. India (Ent.), vol. vii, pp. 61, 62; id., J. Bombay Nat. Hist. Soc. vol. xxix, p. 479 (1923).

Rhinocypha biforata beesoni Fraser, J. Bombay Nat. Hist. Soc. vol. xxxii, p. 455 (1928).

Male.—Abdomen 18 mm. Hind-wing 25 mm.

R. biforata beesoni combines the characters of typical *R. biforata* and the subspecies *delimbata*, in that the apical area of the fore-wings is as extensive as in typical *biforata*, and the same area in the hind-wings is as reduced as in *delimbata*. In the fore-wing the opaque area extends 2 to 3 cells proximal to the inner end of the pterostigma, its inner margin being somewhat irregular, bevelled outwardly from the costa posteriorwards, and prolonged slightly along the costa as far as half-way between the pterostigma and the node between the costal and radial nervures. There is also a slight prolongation along the hind margin of the wing. In the hind-wing the opaque area covers nearly the outer third of the wing, and the middle series of spots extends right up to it or even invades it for a short distance. The length of these spots is slightly greater than in subsp. *abbreviata*, but decidedly less than in typical *biforata* or subsp. *delimbata*. The vitreous spots in all these subspecies have a metallic green or coppery reflex in the opaque areas, violaceous in the hyaline area. The discoidal cells are traversed 3 to 4 times respectively, the fore-wing has 14 to 15 antenodal nervures and 22 to 25 postnodals. The body-markings are similar to those of *delimbata*.

Distribution.—The *type*, in the Forest Research Institute, Dehra Dun, is from Lachiwala, United Provinces, and was collected in the month of November. I have also seen two pairs in the Pusa collection which are recorded as having been collected at Mergui, Lower Burma, during June. There are no record of specimens having been taken in the vast stretch of country separating these two localities, and it is, therefore, more than likely that some error in labelling the supposed Mergui specimen has crept in, as Mr. C. F. Beeson collected in Lower Burma (as well as round Dehra Dun), and might have been expected to find examples, if the species occurs there.

This subspecies is easily distinguished from the two subspecies *delimbata* and *abbreviata* by the extent of the apical marking of the fore-wing, and from typical *biforata* by the much reduced opaque area in the hind-wing.

188. **Rhinocypha bisignata** Selys. (Figs. 16 & 17, and Pl. I, fig. 2.)

Rhinocypha bisignata Selys, Syn. Cal. p. 62 (1853); id., Mon. Cal. p. 214 (1855); Kirby, Cat. Odon. p. 113 (1890); Laidlaw, Rec. Ind. Mus. vol. xiii, p. 38 (1917); Fraser, Mem. Dept. Agric. India (Ent.), vol. vii, pp. 80, 81, pl. x (1922) (larva); id., J. Bombay Nat. Hist. Soc. vol. xxix, p. 478 (1923); id., Rec. Ind. Mus. vol. xxvi, p. 483 (1924); id., J. Bombay Nat. Hist. Soc. vol. xxxii, pp. 455–456 (1928); id., Rec. Ind. Mus. vol. xxxiii, p. 448 (1931).

Male.—Abdomen 20 mm. Hind-wing 24–26 mm.

Head: eyes brown; rest of head, including lips, velvety black. A small spot external to each posterior ocellus, a similar postocular spot on each side, and an occasional occipital spot, all bright ochreous. *Prothorax* black, with a large yellow spot on each side of middle lobe, a similar spot on outer ends of posterior lobe, and a large rose-pink spot covering the greater part of dorsum of the same lobe. *Thorax* black, marked as follows :—Mesothoracic triangle rose-pink, extending for more than one-third but for less than half-way up dorsum of thorax ; a large spot of the same colour, pointed at both ends, bluntly so anteriorly, lying between humeral suture and mesothoracic triangle, but well separated from the latter, and of twice its length ; a small upper yellow antehumeral spot, a narrow yellow stripe bordering the upper half of the humeral suture behind ; a broad, very broken stripe on posterior half of mesepimeron and the whole centre of metepimeron, golden-yellow. Beneath, three pairs of yellow spots, the posterior the larger, the middle pair obscured. *Legs* black, the two posterior

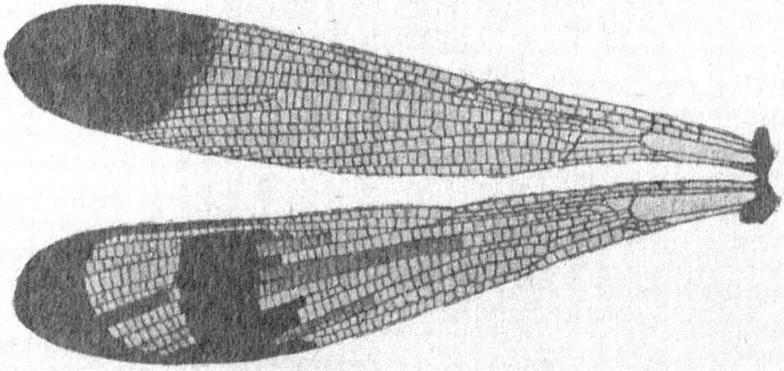

Fig. 16.—Wings of *Rhinocypha bisignata* Selys, male.

pairs of tibiæ and femora pruinosed white on flexor surfaces. *Wings* long and narrow as in *R. biforata* ; hind-wings slightly broader than fore-wings, hyaline in basal part, where they are tinted with yellow, opaque blackish-brown in apical area, the hind-wing bearing here a large vitreous fuliginous spot. Fore-wings with outer fourth or more opaque, this area being brilliant coppery or with fiery reflex as in *R. ignipennis*, with its inner border prolonged slightly basally between costa and *Ri* and running obliquely out towards hind border of the wing. Hind-wing with outer third opaque, and marked with two series of vitreous spots which glow with a coppery or violaceous reflex, the inner spot always the latter colour. Outer spot produced by a fusion of spots similar to those seen in *R. biforata*, but often completely or partially divided into two or more spots, lying between *Rii* or *MA* or *Riv+v*, and its outer border on a level with middle of pterostigma. Inner series composed of 3 vitreous spots similar to those seen in *R. biforata*, but the middle one usually prolonged inwards to fuse with the nodal spot seen in *R. biforata*. The first spot of this series one row deep by 8 cells long, lying between *IRiii* and *Riv+v*, the middle spot 20 to 22 cells long when fused with the nodal spot, otherwise only 10 cells long, lying between *IRiii* and *Riv+v*, the posterior spot 11 to 12 cells long, lying between *MA* and *Cuii*. Pterostigma black in all wings; nodal index 12 to 14 antenodals and 28 to 31 postnodals in fore-wings, 12 to 14 antenodals and 22 to 28 postnodals in hind-wings; discoidal cell traversed 2 to 3 times in fore-wings, 3 to 6 in hind-wings; petiolation begins slightly proximal to the second antenodal. *Abdomen* black, marked with yellow as follows:—Segment 1 with a subtriangular lateral spot on each side; segment 2 with a mid-lateral stripe, an

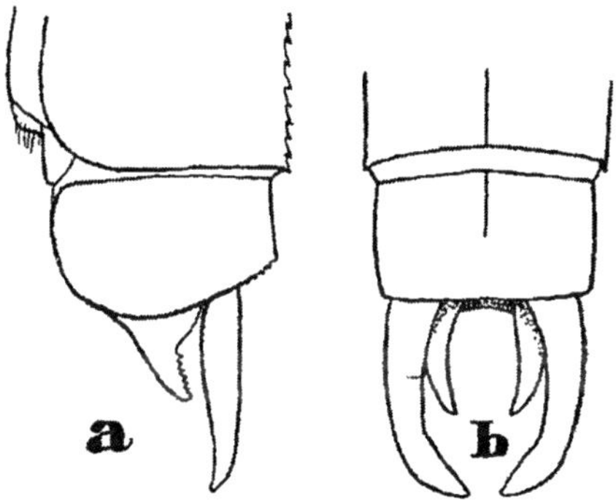

Fig. 17.—Anal appendages of *Rhinocypha bisignata* Selys, male. **a**, left lateral view; **b**, dorsal view.

apical spot and ventro-lateral stripes; 3 and 4 similar to 2, but the mid-lateral spot obsolete, whilst 5 has the apico-lateral spots only. All other segments unmarked. *Anal appendages* as for the genus, the inferiors minutely spined above.

Female.—Abdomen 16 mm. Hind-wing 22 mm.

Head: labium bluish-green, tipped with black; labrum black, with a large oval or triangular yellow spot on each side; cheeks broadly yellow, this colour continued up as a narrow bordering to the eyes as far back as the level of the posterior ocellus; basal segments of antennæ, a small triangular spot low down on sides of epistome, two large triangular spots on upper surface of frons, two transversely oval spots just behind these, lying in front and to outer side of anterior ocellus, a rounded postocular spot on each side, and a mid-occipital spot, triangular in shape, all bright ochreous. *Prothorax* black, with a large yellow spot on each side and borders of posterior lobe laterally, finely yellow. *Thorax* as in male, but the large antehumeral spot replaced by a fine antehumeral line, incomplete above, whilst the mesothoracic triangle is black, finely outlined in yellow, as also the mid-dorsal carina. *Legs* black, not pruinosed. *Wings* entirely hyaline, tinted palely with yellow, apices narrowly enfumed; pterostigma black, with pale creamy centre; discoidal cell traversed twice, or more rarely thrice, in all wings; 12 to 14 antenodal nervures, 21 to 27 postnodals in fore-wings, 19 to 24 in hind-wings. *Abdomen* black, marked with yellow as in male, but rather more extensively, thus segments 3 to 5 are marked similarly to 2, and 6 and 7 similar, the remainder unmarked. *Anal appendages* and vulvar scale as usual in the genus.

Distribution.—This species is the sole representative of the genus in Southern and Peninsular India. It is a widely distributed insect; thus I have records of its capture in localities as wide apart as Cochin (South India) and the Central Provinces. It is common throughout the Palni, Nilgiri, and Shevaroy Hills in the far south, abundant in Coorg, less common at Khandala and Igatpuri (Bombay Pres.) on the ghats near Bombay; lastly, it is widespread throughout the Agency Tracts and Jeypore (Madras Pres.) on the East Coast. Its exact limits here are yet to be worked out, but it probably stops short of Bengal. It is a submontane insect, living and breeding between 2,000 and 5,500 feet altitude. Its larva was discovered in the Nilgiris, and was one of the first of the genus to be described.

The females congregate on the bare twigs of trees near the breeding-places in great numbers, and are very rarely seen pairing. Occasionally they may be seen ovipositing on a piece of dead twig floating in mid-stream. The males frequently disport in couples before the females, circling and sparring with one another, their white pruinosed legs glistening and thrust out like the arms of a wrestler seeking for an opening; meanwhile their wings flash like living jewels of fire.

Type in the Selysian collection.

Genus **CALOCYPHA** Fraser.

Calocypha Fraser, J. Bombay Nat. Hist. Soc. vol. xxxii, p. 185 (1927); id., ibid. vol. xxxii, p. 457, pl. ii, fig. 3 (1928).

Moderately small but robust insects, with characters similar to those of *Rhinocypha*, differing mainly in details of venation and the greater petiolation of the wings. Head, thorax, abdomen, and legs similar to *Rhinocypha*. Wings equal in length, similar in shape, apices markedly rounded, reticulation more open than in *Rhinocypha*, hyaline, with opaque apices; arc very oblique, lying almost in line with costal border of the discoidal cell, latter shorter than in most species of *Rhinocypha*, and acutely pointed outwardly; petiolation beginning at or well distal to inner end of discoidal cell; nervure *ac* more basally situated than in *Rhinocypha*; anal appendages slenderer, superior pair twice the length of segment 10, narrow, cylindrical, curving in at apices which are bluntly pointed, furnished with a few hairs. Inferior pair much shorter, half the length of superior, broad at base, backwardly directed and parallel with one another, furnished with a few minute spines on upper surface.

Female with apices of hind-wings partially opaque, venation as in the male. Vulvar scale and anal appendages as in the genus *Rhinocypha*.

Genotype, *Rhinocypha laidlawi* Fraser.

189. **Calocypha laidlawi** (Fraser). (Fig. 18.)

Rhinocypha laidlawi Fraser, Rec. Ind. Mus. vol. xxvi, pp. 482, 483 (1924).

Calocypha laidlawi Fraser, J. Bombay Nat. Hist. Soc. vol. xxxii, pp. 457, 458, pl. ii, fig. 3 (1928); id., Rec. Ind. Mus. vol. xxxiii, p. 448 (1931).

Male.—Abdomen 17–18 mm. Hind-wing 20 mm.

Head black, marked with bright vermilion as follows:—A large oval spot on dorsum of anterior part of epistome; behind this spot a pair of large subtriangular spots on frons, nearly confluent with two large tongue-shaped spots on vertex, latter closely apposed to ocelli (which they partially enclose), changing from vermilion to pale ochreous at their hinder ends; a postocular spot on each side of occiput of the same colour. Eyes dark brown; labium pale yellow; labrum glossy black. *Prothorax* black, marked with creamy yellow as follows:—A narrow anterior collar, an oval spot on each side of middle lobe, a spot on each side of posterior lobe, and a large spot on trochanters. *Thorax* black, mesothoracic triangle very narrow and elongate, extending whole length of dorsal carina, expanded in its lower third, with parallel sides in its upper two-thirds, bright vermilion-red. Central part of alar sinus creamy yellow, lower parts of sides azure-blue save for a small triangular area posterior to the humeral suture and a small linear spot on upper part of second lateral suture. Dark area in front bearing a small upper posthumeral blue spot. Six large rounded yellowish spots on underside, anterior pair confluent. *Wings* hyaline, bases tinted with yellow, rather less than apical third of each wing opaque black with dark violet-metallic reflections, margin of this opaque area straight and sharply defined, without clear spots or stripes. Pterostigma black, swollen outwardly, oblique at both ends. *Legs* black, the flexor surfaces of the two hinder pairs of tibiæ pruinosed white, the same pairs of femora very slightly so. *Abdomen* black, marked with bright azure-blue on segments 1 to 8, 1 to 3 broadly blue on sides from base to apical border, 4 to 8 with a broad wedge-shaped spot tapering apically but not reaching apical border of segments, gradually diminishing in size on successive segments; 2 to 8 with narrow paired basal subdorsal blue lunules. *Anal appendages* black.

Female.—Abdomen 17 mm. Hind-wing 24 mm.

Differs from the male in having the markings bright pale ochreous instead of vermilion and azure-blue, and its fore-wings entirely hyaline. *Head* black, marked with pale ochreous as follows:—Two large oval spots covering greater part of labrum; spots on upper surface of epistome and frons as in male; bases of mandibles and a narrow stripe adjacent to

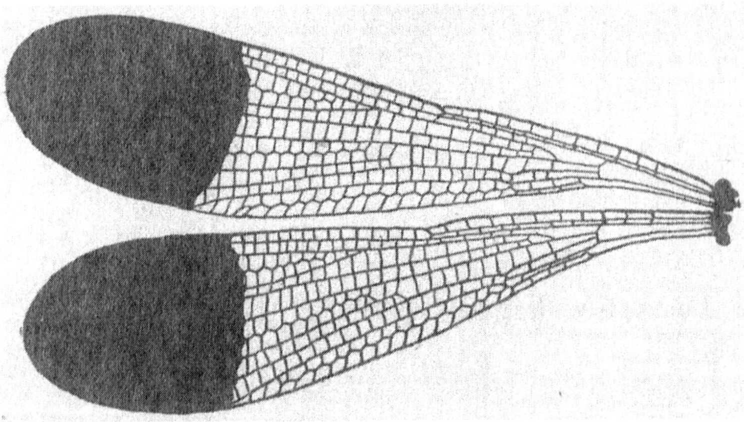

Fig. 18.—Wings of *Calocypha laidlawi* (Fras.), male.

each eye, as far as level of ocelli; basal segments of antennæ; a small spot on each side of epistome, and lastly, postocular spots as in male. *Prothorax* similar to that of male, but mid-dorsum bearing a stripe which runs from posterior lobe nearly to anterior end of prothorax. *Thorax* black, marked with greenish-yellow as follows:—Two minute spots on antealar sinus; a fine line on mid-dorsal carina; a fine antehumeral stripe, its anterior end curling abruptly outward; a short posthumeral stripe, incomplete below; a still shorter vestigial stripe just posterior to upper part of first lateral suture; a long stripe traversing whole length of sides, crossing both sutures obliquely, its upper and lower borders very irregular and serrate. *Legs* black, femora dirty yellow on inner side. *Wings* hyaline; hind pair with apices broadly brown as far proximal to inner end of pterostigma as length of latter, and of a much deeper tint in the hinder two-thirds of wing. Centres of anterior cells hyaline; brown area at apex of wing replaced by a small opaque white area. Pterostigma with inner half black, outer half whitish, margined with black. Antenodal nervures 10 to 11 in fore-wings, postnodals 18 to 19 (10 to 11, and 18 to 24 in male); discoidal cell of fore-wings traversed once only, rarely twice, in both sexes, 2 to 3 times in hind-wings. Petiolation beginning at inner end of discoidal cell in the female. *Abdomen* black, marked with bright ochreous and greenish-yellow:—Segment 1 with a large lateral spot and a minute mid-dorsal linear apical spot; 2 to 8 with the mid-dorsal carina finely yellow; 2 to 7 with a long narrow stripe followed by a large apical spot on each side, whilst beneath this marking, on all segments, is a supraventral stripe; remaining segments unmarked. *Anal appendages* black, long, acuminate. Vulvar scale as for the genus.

Distribution.—Confined in South Kanara District (BOMBAY PRES.) to the network of rivers about Sulia. A single male, however, was taken near Tamaracherri, S. Malabar (MADRAS PRES.), by the author.

The insect is a shade-lover; it is found perched on half-submerged logs in mid-stream or resting on twigs overhanging the river. The red and blue markings are very conspicuous when the insect is in flight, whilst the black tips of the wings form an almost complete black circle round it as the wings whirl with cinema effect.

Type and allotype in the British Museum; paratypes in the Fraser, Ris, Laidlaw, Morton, Pusa, and Darjeeling Museum collections.

Genus **CHLOROCYPHA** Fraser (1928).

Libellago (pars) Selys, Mon. Lib. Eur. p. 200 (1840); id., Syn. Cal. p. 57 (1853); Walker, List Neur. Ins. Brit. Mus. iv, p. 643 (1853); Selys, Mon. Cal. p. 225 (1854); Kirby, Cat. Odon. p. 112 (1890); Ris, Ent. Mitt. v, nr. 9/12, p. 304 (1916); Laidlaw, Rec. Ind. Mus. vol. xiii, p. 39 (1917); Fraser, J. Bombay Nat. Hist. Soc. vol. xxxii, p. 458 (1928).

Chlorocypha Fraser, J. Bombay Nat. Hist. Soc. vol. xxxii, p. 684 (1928).

Moderately robust dragonflies, very similar to members of the genus *Rhinocypha*. Head, thorax, abdomen, and anal appendages as in *Rhinocypha*, except for the mid-dorsal "mesothoracic triangle," which is quite undeveloped in *Chlorocypha*. Abdomen almost always brilliantly red, blue, yellow, or other colours, marked with black. Wings always entirely hyaline in both sexes, narrow, hind pair not noticeably broader than the fore pair, venation close, petiolation variable, usually extending close up to the level of *Ac*, ending a little proximal to that nervure, but in the single Indian species very short, and ending quite near the base of the wings; *MA* zigzagged almost, or quite, from or at its origin (not straight as in most species of *Rhinocypha*), so that the cells bordering it are pentagonal in shape. Legs variable, tibiæ not dilated in *C. asiatica*, but distinctly so in *C. vittata*, coloured or not on the inner surfaces.

Distribution.—Only two species are found in Asia, the genus being confined almost entirely to the African continent,

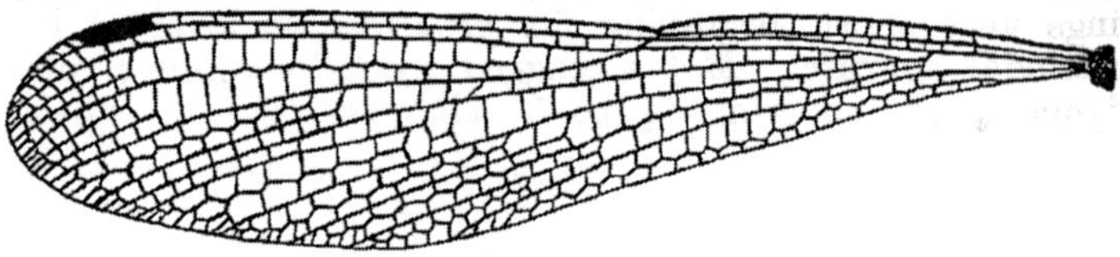

Fig. 19.—Wing of *Chlorocypha vittata* (Selys), male.

which is rich in species. It is extremely doubtful if the two Asiatic species are congeneric with those of Africa, and they are certainly not closely related, as is evidenced by the striking differences in the legs and in the petiolation of the wings. Of the two Asiatic species, *C. asiatica* is probably confined to the Philippines and *C. vittata* to the mainland of Asia (Assam, Burma, and Indo-China), although forms from the latter area may perhaps be subspecies or distinct species.

Genotype, *Agrion dispar* Beauv.

In discussing the genus *Libellago* (p. 59) I have set forth the reasons for introducing the new name *Chlorocypha*.

190. **Chlorocypha vittata** (Selys).

Libellago asiatica subsp. *vittata* Selys, Ann. Mus. Civ. Genova, (2) x (xxx), pp. 490, 491 (1891); Martin, Mission Pavie, Néurop. (sep.), p. 17 (1904); Williamson, Proc. U.S. Nat. Mus. vol. xxviii, p. 173 (1905); Laidlaw, Rec. Ind. Mus. vol. xiii, pp. 39, 40 (1917).

Male.—Abdomen 23 mm. Hind-wing 25 mm.

Head: labium and labrum black; rest of head black, marked with yellow as follows:—Extreme bases of mandibles; a narrow streak on genæ against eyes; a small round spot on each side of ante-clypeus; three pairs of spots on frons in front of ocellar space; basal joints of antennæ; a small spot in front of each posterior ocellus; and two large triangular spots on occiput behind ocellar space. Eyes blackish-brown. *Prothorax* black, with three irregular longitudinal stripes, a mid-dorsal and two lateral, made up of two or three spots;

anterior lobe also greenish-yellow on dorsum. *Thorax* black, marked with yellow as follows :—A fine line on mid-dorsal carina ; a narrow humeral stripe, tapering above and incomplete in its upper part ; a small linear upper humeral spot ; a narrow stripe on antero-lateral suture, which latter bisects it obliquely ; a short, strongly curved, narrow stripe on medio-lateral suture, incomplete below ; a broad irregular oblique stripe on lower part of mes- and met-epimeron, rather broadly interrupted by a black stripe on postero-lateral suture. Underside black, with a tiny median yellow point. *Wings* hyaline, pale saffron in basal third ; all discoidal cells traversed twice ; petiolation of both pairs of wings very short, beginning far proximal to level of basal antenodal nervure ; *MA* zigzagged from its origin ; pterostigma black, covering 2½ to 3 cells, but slightly dilated ; nodal index $\frac{13\text{–}18}{13\text{–}15} \Big| \frac{14\text{–}19}{13\text{–}18}$. *Legs* black, the four posterior tibiæ dilated, their flexor surfaces, as well as opposing surfaces of femora, pure white. *Abdomen* : segment 1 black, with a large lateral spot and a tiny dorsal apical triangle bluish-green ; 2 black, apical border narrowly and lateral border broadly pale blue, dorsum with a large spot shaped like an ivy-leaf, with the stalk directed apically ; 3 blue or bluish-green, ventral border broadly black, a broad black subdorsal stripe tapering from base to apical border of segment, and a pair of small transverse subapical dorsal spots ; 4 to 7 golden or greenish-yellow, with only the subapical dorsal paired spots ; 8 yellow, with two broad black triangular subdorsal stripes confluent at apical border of segment, but tapering and diverging to the base ; 9 to 10 black, 9 with a small mid-dorsal subapical crescentic yellow spot, 10 with a similar tiny point. *Anal appendages* black ; superiors nearly twice the length of segment 10, compressed, twisted on the long axis so that the apical end, which is slightly broadened, comes to look upward, the inner surface becoming the upper ; inferiors rather more than half as long as superiors, directed straight back, broad at base, tapering to a fine point ; superiors slightly curved, apices nearly meeting as viewed from dorsum.

Female unknown.

Distribution.—Lushai Hills, ASSAM ; and Bhamo, BURMA ; in June and July. *C. vittata* is easily distinguished from *C. asiatica* of the Philippine Is. by its smaller size, the very short petiolation of its wings, shorter pterostigma, and dilated tibiæ ; and from *Rhinocypha immaculata*, which has hyaline wings, by the absence of the mesothoracic triangle so conspicuous in the former, by the shorter petiolation of the wings, and by the nervure *MA* zigzagged from its origin. The abdominal markings are reminiscent of some species of

Libellago, but the more numerous antenodal nervures will prevent any error here. *C. vittata* differs from all other species of the genus by the extremely short petiole of the wings, a character which might justify raising it to generic rank.

Type in the Genoa Museum.

Genus **LIBELLAGO** Selys. (Fig. 20.)

Libellago (pars) Selys, Mon. Lib. Eur. p. 200 (1840).
Micromerus Rambur, Ins. Névrop. p. 238 (1842); Selys, Syn. Cal. p. 64 (1853); Walker, List Neur. Ins. Brit. Mus. iv, p. 652 (1853); Selys, Mon. Cal. p. 233 (1854); Kirby, Cat. Odon. p. 115 (1890); Williamson, Proc. U.S. Nat. Mus. vol. xxviii, p. 171 (1905); Ris, Ent. Mitteil. v, nr. 9/12, p. 304 (1916); Fraser, J. Bombay Nat. Hist. Soc. vol. xxxii, pp. 683, 684 (1928).

Characters as for the subfamily; epistome much more projecting than in *Rhinocypha*, so that a deep fissure intervenes between the frons and nasus. Fore- and hind-wings similarly shaped, very long and narrow, reticulation moderately open; fore-wings of male in most species tipped with black, otherwise all wings of both sexes completely hyaline except in *L. snellemani* and *bisignata* (both non-Indian species), in which the wings are marked with broad blackish-brown fasciæ as in many species of *Rhinocypha*; antenodal nervures 4 to 7 in number; sectors of arc arising from a common point, where the arc is strongly angulated at a little above its middle; *IA* undulated from its origin; petiolation ends proximal to the first antenodal nervure; intercalated nervures few in number and found only near apices of wings; *Riii* arising widely distal to node; pterostigma absent in fore-wings of male except in those species where the apices are not tipped with black, present and very elongate in hind-wings of male and in both pairs of wings of female.

Mesothoracic triangle absent or reduced, uncoloured; legs long and slim, usually all femora and tibiae pruinosed white on flexor surfaces. Abdomen short and stout, much shorter than the wings, fusiform, depressed, narrowing very gradually towards the tenth segment. Anal appendages very similar to those of *Rhinocypha*, superiors twice the length of segment 10, pincer-like, cylindrical and pointed at apices; inferiors obtuse, less than half the length of the superiors. Vulvar scale very robust, shaped as in species of *Rhinocypha*.

Larva very similar to that of *Rhinocypha*.

Distribution.—Southern Asia from Western India to Borneo and New Guinea. Most Indian species appear to have originated from *L. lineata*, which within Indian limits is only found typically in Lower Burma, though occurring also in Java and Siam; several subspecies or races occur throughout Burma, Continental India, and Ceylon. In the latter island we find a surprising development of the genus, most species found there apparently not being closely related to *L. lineata*.

The species of *Libellago* do not differ in habits from those of *Rhinocypha*, and they are found in similar habitats.

Genotype, *Calopteryx lineata* Burm.

The first mention of the genus *Libellago* was made by Selys, Mon. Lib. Eur. p. 200 (1840), who gave a short description and cited *Agrion lineata* as the genotype, but also included *fenestrata* and *fulgipennis* (now classed as *Rhinocypha*, but at that time as *Agrion*). Burmeister's *lineata* had, however, been described as a *Calopteryx*, and Selys committed an error in assigning it to *Agrion*, as is clear from the fact that he refers to the Mon. Lib. Eur. when describing the genus *Micromerus* in the Mon. Cal. p. 233. It seems perfectly clear therefore that the *A. lineata* referred to in the Mon. Lib. Eur. was

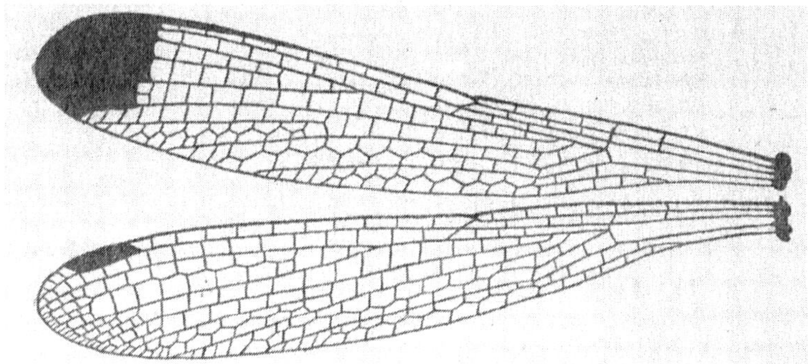

Fig. 20.—Wings of *Libellago lineata indica* (Fras.), male.

Burmeister's *Calopteryx lineata* (*Micromerus lineatus*), and equally clear that it was given as the genotype for *Libellago*. This being the case, *Micromerus* is a mere synonym of *Libellago*, and the latter name has been wrongly used.

For these reasons I have restored the original name *Libellago* to the genus *Micromerus* Ramb., 1842, and have introduced the new name of *Chlorocypha* for the genus *Libellago* as commonly understood.

Key to Indian Species of Libellago.

1.	Abdomen marked with bright orange or brick-red	2.
	Abdomen marked with bright citron-yellow or greenish	3.
2.	Greater part of abdomen bright orange; mid-dorsal carina of abdominal segments not marked with black	*aurantiaca* (Selys), [p. 70.
	Abdominal segments bearing large spots of brick-red; dorsal carina of all segments black............................	*greeni* (Laid.), p. 68.

3.	Male with six antenodal nervures to all wings	4.
	Male with five antenodal nervures	*lineata lineata* (Burm.), [p. 60.
4.	Black tip to fore-wings of male 4–5 mm. in length	5.
	Black tip to fore-wings of male only 3 mm. in length	6.
5.	Humeral stripe present at its middle; posterior lobe with a large dorsal spot; thoracic markings orange	*lineata blanda* (Selys), [p. 65.
	Humeral stripe absent; posterior lobe of prothorax unmarked; thoracic markings citron-yellow or green	*andamanensis* (Fras.), [p. 66.
6.	Small insects, hind-wing less than 18 mm. in length; linear spot on hinder border of occiput with a median anterior prolongation; abdominal markings citron-yellow	*lineata indica* (Fras.), [p. 63.
	Larger insects, with hind-wing more than 20 mm. in length; linear spot on hinder border of occiput without median prolongation; abdominal markings bluish or bluish-green	*finalis* (Selys), p. 67.

191. **Libellago lineata lineata** (Burmeister). (Fig. 21, f.)

Calopteryx lineata Burmeister, Handb. Ent. vol. ii, p. 826 (1839).
Libellago lineata Selys, Mon. Lib. Eur. p. 200 (1840).
Micromerus lineatus Rambur, Ins. Névrop. p. 238 (1842); Selys, Syn. Cal. p. 65 (1853); id., Mon. Cal. p. 237 (1854); id., Bull. Acad. Belg. (2) vol. xxvii, p. 666 (1869); Kirby, Cat. Odon. p. 115 (1890); Selys, Ann. Mus. Civ. Genova, (2) vol. x (xxx), p. 494 (1891); Laidlaw, Fascic. Malayenses (Zool.), pt. i, p. 197 (1903); Williamson, Proc. U.S. Nat. Mus. vol. xxviii, p. 171, figs. 5, 6 (1905); Laidlaw, Rec. Ind. Mus. vol. xiii, p. 39 (1917); Fraser, ibid. vol. xxvi, p. 483 (1924); id., J. Bombay Nat. Hist. Soc. vol. xxxii, pp. 684–686, pl. i, figs. 6–9 (1928).
Micromerus uxor Rambur, Ins. Névrop. p. 239 (1842).
Libellago lineata Walker, List Neur. Ins. Brit. Mus. p. 653 (1853).
Micromerus obscurus Kirby, Proc. Zool. Soc. Lond. p. 328, pl. 33, fig. 1 (1886); id., Cat. Odon. p. 115 (1890).

Male.—Abdomen 15 mm. Hind-wing 17 mm.

Head: labium dull brownish-yellow to blackish; labrum blackish-brown, anterior border finely yellow; epistome black, anterior facet blue-black metallic, above velvety black; frons and vertex velvety black, with two transverse oval citron-yellow spots on former near middle line, and a large oval spot on outer side of each posterior ocellus; occiput black, with a large postocular spot, subtriangular in shape, on each side separated more or less narrowly from a median crown-shaped spot. The latter spot rather variable, in the type transversely elongate, with a median spot in front of its middle, but in other specimens confluent with this spot, and with additional triangular points on either side. *Prothorax* black, marked with citron-yellow as follows:—A fine border to posterior lobe, expanded below into a pyriform spot;

a large transversely oval spot on dorsum of posterior lobe, this part being raised into an oval boss, coated with fine hairs; a narrow anterior collar and a large subtriangular spot on each side of middle lobe. *Thorax* black, marked with citron- or greenish-yellow as follows :—The whole of antealar sinus; mid-dorsal carina finely; a narrow ragged antehumeral stripe, incomplete above; a small upper antehumeral spot, a narrow curved post-humeral stripe, incomplete below; nearly the whole mesepimeron, which shows a narrow wedge of black at its middle, and a hook-shaped or triangular invasion of black above; the whole metepimeron except anterior sutural

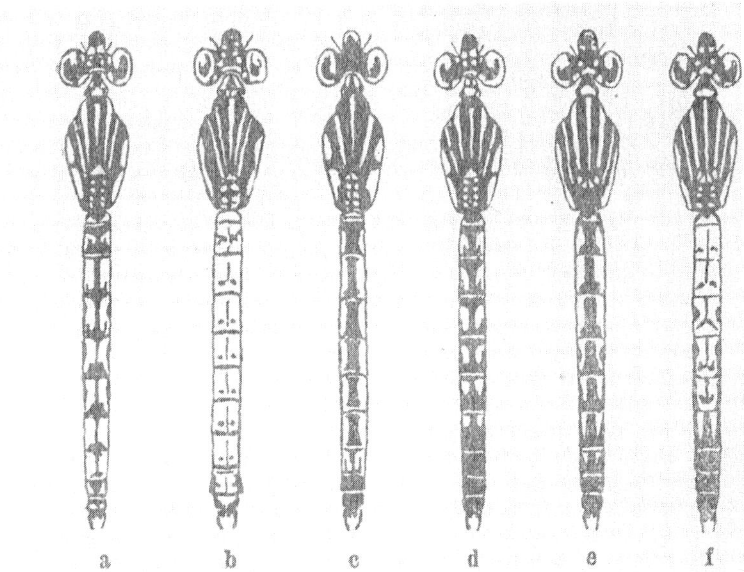

Fig. 21.—Markings of *Libellago*.
a, *L. greeni* (Laidlaw), ♂; b, *L. aurantiaca* (Selys), ♂; c, *L. andamanensis* (Fraser), ♂; d, *L. finalis* (Selys), ♂; e, *L. lineata indica* (Fraser), ♂; f, *L. lineata lineata* (Burmeister), ♂.

line and narrow ventral border. Beneath yellow. *Legs* black, flexor surfaces of tibiæ and of anterior pair of femora pruinosed white. *Wings* hyaline, the bases of all, nearly to level of node, palely tinted with amber; apices of fore-wings (except in teneral examples where no marking is visible) tipped with black as far proximal as inner end of pterostigma, but falling short of hinder border of wing. Pterostigma absent in fore-wing, black, tumid and elongate in hind-wing, where it covers from 2 to 2½ cells. Costa of fore-wing gradually thickened towards the site where pterostigma would ordinarily be; discoidal cell traversed once; nodal index, 5 antenodal nervures, 8 to 10 postnodals. *Abdomen* black and bright greenish- or golden-yellow, latter colour prevailing on anterior segments, the former on posterior segments. Segment 1 with apical border narrowly and sides broadly yellow; 2 to 5 with latero-ventral borders and intersegmental sutures narrowly black, carina also finely black on segments 2 to 4 (or in teneral specimens from 2 to 6), but stopping short of apical border, where the black expands or is connected with two wing-like projections, so that this, together with the carinal stripe, form inverted capital "T's" (in some specimens the arms of the "T" are slightly separated from the stem, but this is rare). Segments 5 and 6 with two large dorsal yellow spots tapering apically, enclosed by an expansion and confluence of the carinal T-spot, with the apical ring and latero-ventral black; segments 7 and 8 may show smaller similar spots, but older specimens have these segments unmarked, as also 9 and 10. *Anal appendages* black, superiors twice the length of segment 10, separated widely at base, converging at apices, which are turned slightly in, cylindrical and very slightly tapered. Inferior appendages only half the length of superiors, closely apposed at bases and again at apices so as to enclose a minute foramen, blunt, broad at base, narrowed thereafter.

Female.—Abdomen 13 to 16 mm. Hind-wing 18 to 22 mm.

A stouter, more robust insect than the male, marked very similarly, but the yellow more extensive.

Head with markings as in male, but with the following additional spots:—Labrum and labium, the former with a small basal median tongue of black; anterior surface of epistome (which in the male is metallic blue); bases of mandibles; cheeks; basal joints of antennæ; a large spot on either side of epistome and two large diamond-shaped spots on front of frons, narrowly separated from the smaller hinder pair. *Prothorax, thorax,* and *legs* as in male. *Abdomen* black, marked with yellow as follows:—Apical border and sides of segment 1; dorsal carina on 2 to 7; a longitudinal stripe on dorsum of 8 and 9 (that of 10 unmarked). Laterally the segments are yellow, with a large apical spot, notched in front, on 2 to 7, bifid on 8, and present only as a small yellow spot on sides of 9 and 10. (Briefly the abdomen may be said to be yellow, with two longitudinal bands of black and black articular rings at the intersegmental nodes.) *Anal appendages* black, twice as long as segment 10, pointed, widely separated throughout. *Legs* black, flexor surfaces of femora and tibiæ yellow. *Wings* entirely hyaline, basal area, especially along costa, slightly tinted with amber; pterostigma present in all wings, elongate, black, or yellow framed in black (according to age of individual), covering 2 or 3 cells; 6 antenodal nervures, 9 to 10 postnodals.

Distribution.—Throughout the Sundaic Archipelago and the adjacent mainland, including Java, Sumatra, Siam, Annam, BURMA, and NORTH INDIA. Kirby's *M. obscurus*, described from a teneral specimen from northern India, is identical with specimens which I have received from UNITED PROVINCES (Dehra Dun), ASSAM (Cachar), and UPPER BURMA. It differs only in the limitation of black markings due to immaturity, and is undoubtedly *L. lineata*.

In Peninsular and Southern India a distinct subspecies, which I have named *L. lineata indica*, replaces typical *L. lineata*. The two Ceylon forms show sufficient differences to deserve specific rank.

Type of *L. lineata* in the Halle Museum, from Java; of *M. uxor* and *M. lineatus* in the Paris Museum; of *M. obscurus* in the British Museum.

192. **Libellago lineata indica** (Fraser). (Figs. 20 & 21, e.)

Micromerus lineatus Fraser, Rec. Ind. Mus. vol. xvi, pp. 197, 198, pl. xxiii (1919) (larva); Laidlaw, Spolia Zeylanica, vol. xii, pp. 354, 355 (1924).

Micromerus lineatus indica Fraser, J. Bombay Nat. Hist. Soc. vol. xxxii, pp. 686, 687, pl. i, fig. 5, and pl. iii, fig. 2 (1928); id., Rec. Ind. Mus. vol. xxxiii, pp. 448, 463 (1931).

Male.—Abdomen 14 to 16 mm. Hind-wing 15 to 18 mm.

Head very similar to that of *L. l. lineata*, labium ochreous, labrum dark brownish-yellow, black at base; a more or less obscure spot on either side of epistome, the anterior facet of which is metallic blue as in *L. l. lineata*; eyes brown, but with a bluish metallic lustre in certain lights; bases of mandibles and cheeks narrowly yellow; oval spots on frons reniform in shape and often a second pair of smaller rounded spots behind them; a crown shaped mark on occiput with an elongate median point, but no lateral points, or, if present, then very short, usually connected to the postocular spots. *Prothorax*, *thorax*, and *legs* as in typical *L. lineata*; femora pulverulent white on basal halves only. *Wings* as in typical *L. lineata*, but apical black spot of fore-wings distinctly longer (absent in teneral specimens); nodal index differing slightly, in Kanara specimens 6 antenodals in the fore-wings, 7, less commonly 5 to 6 in the hind-wings; in specimens from the Deccan almost always 6 in both wings and 10 to 14 postnodals. The amber tinting at base usually very marked, up to or even beyond node; discoidal triangles traversed once; pterostigma as in *L. l. lineata*. *Abdomen*: as in *Chlorocypha*, so in *Libellago* specialization seems to be confined almost entirely to the colouring and markings of the abdomen. Segment 1 bright golden-yellow with the base narrowly black and a small quadrate black spot; 2 to 6 bright citron-yellow, dorsum golden-green, often very green, 2 with a thick mid-dorsal dumbbell-shaped black marking confluent with a narrow basal ring, the ends of which are prolonged along sides nearly to apical border and separated very narrowly from the ventral border; 3 to 5 with broad mid-dorsal stripes confluent with narrow basal rings and somewhat expanded apically, whilst near the apical border is a small ventro-lateral black spot; 6 and 7 with basal subdorsal wedge-shaped yellow spots enclosed by a fusion of the mid-dorsal black with an elongated lateral spot; 8 to 10 black. In some specimens the mid-dorsal marking of segment 2 is narrowly bisected with yellow, and there is nearly always an indication of the beginning of such a line apically. *Anal appendages* not differing from those of *L. l. lineata*.

Female.—Abdomen 13–17 mm. Hind-wing 17–20 mm.

Head: labium yellow; labrum yellow with a small median basal point or tongue of black; epistome entirely yellow in front, similarly coloured above but here edged with black anteriorly and laterally, its dorsum clouded with light brown and its base narrowly dark brown, with a tongue of the same colour running obliquely forwards and outwards; frons with two large quadrate citron-yellow spots, separated narrowly by a black median stripe which is confluent anteriorly and posteriorly with an equally narrow black stripe, the whole thus forming a black anchor-like mark, accentuated by the outer corners of the hinder portion angulated forwards like the flukes of an anchor; markings of rest of head similar to those of male but an additional pair of spots situated between ocelli. Eyes brown. *Prothorax* entirely similar to that of the male. *Thorax* similar, but lateral markings rather more extensive. *Legs* yellow, femora lined externally with dark brown. *Abdomen* largely yellow, with black markings as follows:—Segment 1 with a large quadrate spot on dorsum, base narrowly and apical border laterally black; 2 to 8 with the dorsum very broadly black, but bisected by a narrow mid-dorsal carinal yellow line which runs as far as segment 9, but on the last two segments is fusiform, tapering apically and basally; laterally a stripe, more or less defined, broadening abruptly subapically and not quite reaching the base of segments; on segment 9 this line expands basally and becomes confluent with the dorsal black marking, so as to enclose a large rounded subdorsal yellow spot; segment 10 black, with two transversely elongate subdorsal yellow spots. *Wings* hyaline, pale with amber tint; pterostigma present in all wings, creamy white in a black frame which expands somewhat proximally over 2 cells; nodal index, 6 to 7 antenodals, 9 to 10 postnodals; discoidal cell traversed once in all wings. *Anal appendages* blackish-brown, pointed, tapered, widely divaricate, rather longer than segment 10. Vulvar scale robust, with a large spot of yellow laterally.

Distribution.—A very common insect throughout SOUTH INDIA, especially in the Western Ghats and Deccan. It is common on the river banks at Poona and Khandala, less common in Coorg, Kanara, and Malabar.

Laidlaw reports it from CEYLON (Haragama, July), and remarks on its difference from the form occurring in Burma and Siam.

It appears to be double-brooded, as it may be found almost throughout the year. The abdominal markings will easily distinguish it from typical *L. lineata*.

Type and allotype in British Museum from Poona.

193. **Libellago lineata blanda** (Selys).

Micromerus blandus Selys, Syn. Cal. p. 64 (1853); id., Mon. Cal. p. 234 (1854); Kirby, Cat. Odon. p. 115 (1890); Laidlaw, Rec. Ind. Mus. vol. xiii, p. 39 (1917); Fraser, J. Bombay Nat. Hist. Soc. vol. xxix, p. 479 (1923).

Libellago blandus Walker, List. Neur, Ins. Brit. Mus. iv, p. 652 (1853).

Micromerus lineatus blandus Fraser, J. Bombay Nat. Hist. Soc. vol. xxxiii, p. 687 (1928).

Male.—Abdomen 17–18 mm. Hind-wing 20–21 mm.

Female.—Abdomen 12–15 mm. Hind-wing 19–21 mm.

I have not seen this form, which is said to resemble closely *L. l. lineata*, a fact which leads me to treat it as a subspecies.

It is a more robust insect and differs from *L. l. lineata* as follows :—Yellow stripe on cheeks bordering eyes absent; prothorax black, its middle lobe unmarked, but the large dorsal tumid spot on posterior lobe extending forwards as a point; thoracic markings more orange than yellow; humeral stripe incomplete above and below, present in its middle third only; the anterior of the two broad lateral fasciæ broken up into three large spots, the upper very minute; dorsal markings of abdomen broader; apical black spot of fore-wings longer than broad, about 4 mm. in length; pterostigma longer, covers about 5 cells in the hind-wings; 6 antenodal nervures in all wings.

Female.—Very similar to that of *L. l. lineata*; yellow markings more restricted, humeral stripe absent, the anterior of the two lateral fasciæ divided up into two spots; dorsal black markings of segments 3 to 6 narrower, nearly interrupted in some, ground-colour orange; nodal index higher, 14 post-nodals in hind-wings, 12 in fore-wings; pterostigma yellow, black proximally, broader, over 3 to 5 cells.

Distribution.—Confined to the NICOBARS.

Type in the Selysian collection.

194. **Libellago andamanensis** (Fraser). (Fig. 21, c.)

Micromerus andamanensis Fraser, Rec. Ind. Mus. vol. xxvi, p. 410 (1924); id., J. Bombay Nat. Hist. Soc. vol. xxxii, pp. 687, 688, pl. i, fig. 3 (1928).

Male.—Abdomen 16 mm. Hind-wing 20 mm.

Head : eyes bluish-grey ; labium and labrum black ; rest of head jet-black, with the following citron-yellow markings :— Two narrow oval spots at base of epistome above, a large oval spot on outer side of each posterior ocellus, a similar postocular spot on each side, and a linear spot on posterior border of occiput. *Prothorax* black, with a large yellow spot on each side. *Thorax* black, marked with golden greenish-yellow as follows :—Two small spots on antealar sinus ; a fine carinal dorsal line ; two fine antehumeral stripes, incomplete above ; an oblique linear spot on each side of the antealar sinus ; two irregular spots lying between the lateral sutures, the anterior golden-yellow, the posterior distinctly greenish ; and a broad stripe occupying upper anterior three-fourths of metepimeron. *Legs* black, hind tibiæ pulverulent white on the flexor surface. *Wings* hyaline, faintly amber-tinted at bases, fore-wings broadly tipped with black to 4 mm. in extent, this area metallic blue by reflected light ; hind-wings enfumed at apices ; pterostigma black, only present in hind-wings ; 5 to 6 antenodal nervures in all wings. *Abdomen* golden-yellow, marked with black as follows :—Segment 1 narrowly black at base, more broadly at sides ; 2 with a broad dorsal stripe expanded at basal and apical ends, also a broad black stripe low down on sides not extending as far as apical border of segment ; 3 to 6 with broad mid-dorsal stripe expanded only at apical border of segments, where it forms a narrow ring ; 7 similar, but apical black area extending forwards on sides nearly as far as base of segment ; 8 to 10 entirely black. *Anal appendages* black, similar to those of *L. lineata*.

Female.—Unknown.

Distribution.—ANDAMANS. The *type*, a male, is in the Indian Museum, and was taken by the late Dr. Annandale on the Western slopes of Mt. Harriet, November 30, 1923 (fluttering over the surface of a pool below a waterfall) ; altitude 500 feet.

This very beautiful insect is easily distinguished by the broad metallic blue tip to its wings as well as the extreme degree of melanism as compared with other species of the genus. The female is probably very similar to *L. l. lineata* and without a black tip to her wings.

195. **Libellago finalis** (Selys). (Fig. 21, **d.**)

Micromerus finalis Selys, Bull. Acad. Belg. (2) vol. xxvii, p. 665 (1869); id., ibid. vol. xxxvi, p. 616 (1873); Kirby, Cat. Odon. p. 115 (1890); id., J. Linn. Soc. (Zool.), vol. xxiv, p. 556 (1894); Fraser, J. Bombay Nat. Hist. Soc. vol. xxix, p. 479 (1923); Laidlaw, Spolia Zeylanica, vol. xii, pp. 351, 352 (1924); Fraser, J. Bombay Nat. Hist. Soc. vol. xxxii, pp. 688, 689, pl. i, fig. 4 (1928).

Male.—Abdomen 16 mm. Hind-wing 21 mm.

Head: labium yellow, lobes tipped with black; labrum glossy black, unmarked; epistome metallic blue or violet-blue; cheeks with a narrow yellow stripe against the eyes; frons black with a pair of large rectangular spots of citron-yellow; vertex black, with an oval citron-yellow spot on outer side of each posterior ocellus; occiput black, with a rounded postocular spot on each side, and a linear spot along posterior border, narrowly separated from the former. *Thorax* and *prothorax* black, marked exactly as in *L. l. lineata*, but the anterior of the two lateral fasciæ with a large black spot below. Underside black, marked obscurely with dark yellow areas. *Legs* black, all tibiæ pulverulent white on flexor surfaces and all femora pulverulent on same surfaces except for a short distance at the distal extremities. *Wings* hyaline, pale amber-tinted, especially the hind pair; fore-wings tipped with black for about 3 mm., pterostigma absent; hind-wing with well-formed elongate pterostigma, covering 3 cells; 6 antenodal nervures in all wings, 10 postnodals in hind-wings; discoidal cell traversed once in all wings. *Abdomen* black, marked with bright citron-yellow laterally, changing to green or bluish-green on dorsum. Segment 1 with base narrowly and sides broadly black; 2 with a large angulated subdorsal spot on either side, the black forming a large T-shaped mark on dorsum, base, and sides where the arms of the T turn back; 3 to 4 very similar, but the head of the T-shaped mark very narrow, as also its lateral prolongations backward; on sides of segment 4, the beginnings of a narrow lateral stripe apically; 5 to 7 with subdorsal elongated spots, divided, except on segment 5, by a narrow lateral black stripe; 8 to 10 unmarked. *Anal appendages* black, similar in structure to those of *L. l. lineata.*

Female.—Abdomen 16 mm. Hind-wing 25 mm.

Head black, marked with bright yellow-ochre as follows:—Labium except tips of lobes; labrum except base, which is narrowly black, and sides, which are broadly black, and also a triangular prolongation from the middle of the basal black area, which may be confluent with that of the fore-border (in very mature specimens the yellow area reduced to two minute points); a pyriform spot on dorsum of epistome, reduced in mature specimens to a tiny linear mark; anterior facet of epistome metallic blue, but more dully so than in male; bases of mandibles broadly yellow; basal segments of antennæ, cheeks, two pairs of spots on frons in front of ocelli, an oval spot on outer side of each hinder ocellus and spots on occiput as in male (in very mature specimens these spots reduced to mere points, and the median linear spot on hinder border of occiput entirely absent.) *Prothorax* black with a linear or top-shaped spot on mid-dorsum of posterior lobe, and a large spot on the sides of mid-lobe. *Thorax* black, marked very much as in the male, but the markings more restricted. The antealar sinus with two small yellow spots, the mid-dorsal carina very finely yellow, humeral stripe either absent or represented by a chain of small spots or a tiny linear upper spot; lateral fasciæ similar or the anterior one bisected below, and completely cut in two at its middle, the upper portion more or less obsolete. Beneath brownish-yellow, sutures diffusely black. *Legs* yellowish, black externally, hind femora bright yellow except at distal ends. *Wings* palely tinted and enfumed, hind pair occasionally tipped narrowly and diffusely with blackish-brown; pterostigma creamy white, framed in black and clouded with black proximally; discoidal cells traversed once or twice; 7 antenodal nervures to all wings, 11 to 13 postnodals. *Abdomen* black, marked with bright citron-yellow or greenish-yellow as follows:—Mid-dorsal carina narrowly from segment 2 to 8; segment 1 as in male; 2 to 8 with broad lateral stripes constricted subapically, markedly so on segments 2 and 3, less so on 4 to 6, tapered apically on 7, and broadening apically on 8; a large round spot replacing this stripe on segment 9, and a smaller spot on 10, absent in very mature specimens. A chain of broad yellow spots on venter. Vulvar scale robust, shaped as for *L. l. lineata*. *Anal appendages* black, conical, pointed at apices, twice the length of segment 10.

Distribution.—Confined to the montane and submontane districts of CEYLON. Col. F. Wall took a fine series at Banderawela and Nalande, the former 5,000 ft., during October, but it probably has a succession of broods throughout the monsoons. I took this species in August at Dyatalava, and Col. Wall and Mr. E. Green at Kandy, 2,000 ft., October–November.

Type in the Selysian collection.

196. **Libellago greeni** (Laidlaw). (Fig. 21, a.)

Micromerus greeni Laidlaw, Spolia Zeylanica, vol. xii, pp. 352–354 (1924); Fraser, J. Bombay Nat. Hist. Soc. vol. xxxii, pp. 689, 690, pl. i. fig. 1 (1928).

Male.—Abdomen 14 mm. Hind-wing 17–19 mm.

Head black, including entire labium and labrum. Epistome

with its anterior facet metallic bluish-violet. Bright ochre markings present as follows :—Frons with a single pair of spots, less commonly a second pair of very minute points behind anterior pair ; an oval spot on outer side of each hinder ocellus, a rounded postocular spot on each side of occiput and a linear spot along its posterior border, which in some specimens has a tiny medial point anteriorly ; cheeks with a narrow yellow bordering against eyes. *Prothorax* black, marked with citron-yellow as follows :—A narrow anterior collar, followed laterally by a small spot and then a large medio-lateral subtriangular spot. A large spot on dorsum of posterior lobe shaped like the ace of clubs, and two smaller spots low down on its sides. *Thorax* black, marked as in *L. l. lineata*, but markings more reduced as follows :—Humeral stripe present only as a tiny upper linear spot ; antealar sinus with two small rounded spots ; upper part of anterior lateral fascia tapered away above, and shaped like a sickle, the second or hinder fascia framed heavily in black. Beneath black. *Legs* black, all tibiæ pulverulent white on flexor surfaces, all femora pulverulent creamy yellow on distal halves of flexor surface. *Wings* hyaline, tinted with pale yellow, hind pair slightly enfumed, apices of fore-wings tipped with black for about 3 mm., and with a more or less marked metallic lustre at this part ; apices of hind-wings clouded with black at extreme border. Pterostigma present only in hind-wings, covering 3 to 3½ cells, black. Costa proximal to node, and the nervures behind it, bright blood-red, this colour less marked in hind-wing and darkening proximally. Discoidal cells with one traversing nervure ; 6 antenodal nervures in all wings, 10 to 12 postnodals in hind-wings. *Abdomen* black, marked with bright citron-yellow and blood-red as follows :—Segment 1 with an apical ring of citron-yellow broadening laterally ; 2 with a large subdorsal citron-yellow spot, not reaching basal extremity and constricted near its apical end ; 3 with a broader spot citron-yellow tinged with reddish, extending from base to apex of segment, abruptly narrowed subapically and partially divided by a narrow longitudinal black stripe running from its apical end ; 4 to 10 with the markings blood- or brick-red, broad paired spots separated by the black dorsal carina, constricted at apical end and not quite reaching apex of segments, thus forming rather narrow inverted black capital T-like markings on each segment ; 9 and 10 with only a small round apical spot on each side. *Anal appendages* black, rather slenderer than in other species, twice as long as segment 10.

Female.—Abdomen 12–13 mm. Hind-wing 19–20 mm.

Head marked very similarly to the male, differing as follows :—Labrum with a pair of large triangular yellow spots ; epistome with metallic colouring poorly developed; two small linear spots on dorsum of epistome, obsolete in fully mature specimens; bordering medial spot on occiput absent. *Prothorax* black, with a narrow linear medial stripe on dorsum of posterior lobe, a large spot low down on its sides, a large spot on each side of medial lobe, and, rarely, a fine anterior collar. *Thorax* black, marked exactly as in male, but in fully mature specimens with the antero-lateral fascia cut into two spots at its middle. *Legs* black, some slight pruinescence on flexor surface of femora, and the same surface on anterior pair of tibiæ pulverulent creamy white except at distal ends. *Abdomen* black, marked with citron-yellow, with a greenish tinge on the anterior segments as follows:—Segment 1 with a lateral quadrate spot; 2 to 8 with the mid-dorsal carina finely yellow and with lateral stripes, constricted apically on 2 to 5, tapered on 6 and 7, reduced to a spot on the others, or in fully mature specimens absent on segments 9 and 10, less often on segment 8 and rarely present on segment 10. *Anal appendages* and vulvar scale as in *L. l. lineata*. *Wings* hyaline, not enfumed or tinted; 5 to 6 antenodal nervures, 10 to 12 postnodals in all wings; discoidal cell traversed once; pterostigma white, framed in black and clouded at its proximal end.

Distribution.—Col. F. Wall took a number of these insects at Nalande, CEYLON, during September and October, whilst Mr. E. Green took specimens at Haragama, Peradeniya, and Kandy from June to December, so that it appears to have a long season, like most species of the genus.

It closely resembles *L. aurantiaca*, but the black markings are more extensive; it is easily distinguished from all other species of the genus by its red abdomen, strongly reminiscent of the African species of *Calocypha*.

Type in the British Museum.

197. **Libellago aurantiaca** (Selys). (Fig. 21, b.)

Micromerus aurantiacus Selys, Bull. Acad. Belg. (2) vol. vii, p. 448 (1859); id., ibid. (2) vol. xlvii, p. 398 (1879); Kirby, Cat. Odon. p. 115 (1890); Williamson, Proc. U.S. Nat. Mus. vol. xxviii, p. 172, text-fig. 7 (1905); Laidlaw, Proc. Zool. Soc. Lond. p. 331, (1920); Fraser, J. Bombay Nat. Hist. Soc. vol. xxxii, pp. 690, 691 (1928).

Male.—Abdomen 12–15 mm. Hind-wing 15–16 mm.

Head black, spotted with citron-yellow as follows:—A lateral spot on each side of epistome, which latter is metallic blue at its anterior extremity; two pairs of squarish spots on frons, a rounded spot on outer side of each hinder ocellus, a postocular rounded spot on each side of occiput and a linear spot on its hinder border. In fully mature specimens the latter spot and the hinder pair of frontal spots absent. *Prothorax* black, with a small geminate spot on mid-dorsum of posterior lobe and a large spot on each side of mid-lobe yellow. *Thorax* black, marked with citron-yellow as follows :—A fine mid-dorsal carinal line ; an antehumeral stripe broadening below ; a fine humeral stripe incomplete below or almost obsolete ; two small spots on antealar sinus ; and two broad fasciæ on the sides of thorax, the anterior of which is bifid below and narrowed and sickle-shaped above. *Wings* hyaline, tinted with pale amber at bases ; fore-wings tipped with black for about 3 mm. and without pterostigma. Costa reddish immediately proximal to node ; hind-wing with well-formed pterostigma, black, covering 2 to 3 cells ; discoidal cell traversed once in all wings ; 5 antenodal, 10 postnodal nervures in all wings. *Abdomen* bright orange, articulations narrowly black, also ventral border of segments. Segment 2 with an irregular black apical marking, not meeting over dorsum but prolonged obliquely forwards on sides to fuse with black of ventral border ; 3 to 8 with a pair of minute black dorsal subapical points ; 9 with the latero-ventral border extending on to dorsum, especially apically, where it indents the orange ground-colour ; 10 with its sides and apical border black. *Anal appendages* black, shaped as in *L. greeni*, twice the length of segment 10. Inferiors only half the length of superiors.

Female unknown.

Distribution.—A male from Mergui, LOWER BURMA, taken by Mr. Elton Bott, and now in the Pusa Museum, is the only specimen of *L. aurantiaca* known to have occurred in India. The species, however, is not uncommon in Malaysia and thence throughout Southern Asia. Williamson reports it from Trong, Lower Siam, Martin from Tonkin, Laidlaw from Borneo.

The markings of the abdomen are much reduced in the Burmese example. The Siamese example has the base and sides of segment 1 black, as also a lateral black stripe on segments 3 to 8, enclosing a basal spot of the ground-colour on segments 3 to 5 and incomplete basally on segments 6 to 8. There is also an interrupted black stripe on the sides of segments 2 to 8 in the type from Singapore.

Type in the Selysian collection.

Subfamily EPALLAGINÆ. (Fig. 22.)

Epallaginæ Fraser, J. Bombay Nat. Hist. Soc. vol. xxxiii, pp. 47–49 (1928).

Head robust, transversely elongate ; eyes globular, large, much more widely separated than in the LIBELLAGINÆ ; labium with mid-lobe cleft nearly to its middle, apices of lobes subacute ; labrum oval, very broadly and very shallowly notched ; occiput rounded at the middle, tumid behind eyes, broad; frons depressed; epistome ridged, not projecting markedly.

Prothorax with a large rounded boss on each side of the middle lobe, posterior lobe moderately large, transversely elongate, simple, rounded.

Thorax robust, short, shoulders broadly rounded, nearly always bearing two pairs of stripes (except in adult males where these are obscured by melanism), one pair of stripes formed by confluence of antehumeral and humeral above and below, the other by a similar confluence of a post-humeral and a lateral.

Legs short, hind femora extending to middle of second abdominal segment, furnished with very fine spines; tibial spines slightly more robust; claw-hooks inconspicuous, situated near apex of claws.

Wings hyaline or enfumed in female, hyaline or with one or both pairs coloured in male, hind pair often bearing large metallic spots or areas; fore-wings long and narrow; hind-wings similar or more or less dilated in male, shorter than abdomen in male, as long as or slightly longer than abdomen in female, shortly or not at all petiolated; reticulation very close, cells mainly tetragonal; *Rii* at its origin not in contact with $R+M$; node nearer base of wing than apex or situate at its centre; $Riv+v$ arising from *Rii* at about the level of outer end of discoidal cell; *IRiii* separating from *Rii* at about half-way from arc to node; basal space entire, about twice the length of discoidal cell, which is very short, straight, narrow, squared at the ends, with parallel sides, longer in hind-wing than in fore-wing, entire or traversed by one or more nervures; arc slightly angulated; sectors arising from about middle of arc; *IA* strongly curved especially at origin, several rows of cells between it and posterior border of wing; *Cuii* straight; many supplementary nervures between all main sectors; antenodal and postnodal nervures very numerous; primary antenodal nervures entirely absent; no incomplete basal antenodal nervures; first and second series of antenodal nervures coinciding, as in *Libellulinæ*, with occasional exceptions. Pterostigma present in all wings of both sexes, very long and narrow, tapering at both ends which are oblique.

Abdomen long, narrow, cylindrical, longer than wings in male, of the same length or shorter in female, tenth segment flat on dorsum or with a robust carinal spine or two spines.

Anal appendages very similar in all members of the subfamily, superiors forcipate, spatulate, large, at least as long as tenth abdominal segment; inferiors sometimes very small and inconspicuous, sometimes moderately large, but always shorter than superiors.

Females remarkably homogeneous in size, colouring, and general facies, so that where several genera or species exist together it is difficult to determine which are the respective pairs. Abdomen in all species marked laterally with a yellow or pale blue stripe which broadens at the base of each segment and tapers apically, becoming gradually lost as traced towards the anal end of abdomen. Anal appendages as long as segment 10, tapered, very acute. Vulvar scale robust, made up of a laminated sheath with two small stylet-like organs which act as tactile organs at end of ovipositor.

Larvæ.—We know little about the life-histories of the EPALLAGINÆ with the exception of *Anisopleura* and the *Pseudophaea* group, described and figured in J. Bombay Nat. Hist. Soc. vol. xxxiii, pl. ii, p. 56 (1928), and pl. iv, p. 300 (1929).

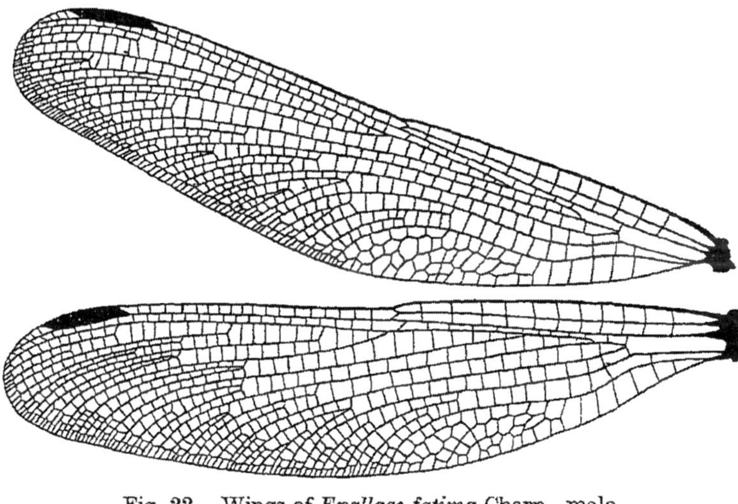

Fig. 22.—Wings of *Epallage fatima* Charp., male.

There is a striking difference between these two groups, abdominal gills being totally absent in *Anisopleura* (and probably also in *Bayadera* and *Epallage*), while seven pairs of these organs are present in *Pseudophæa* and *Indophæa*. Possibly undue importance has been placed on these structures, which have been said to function as respiratory organs. My own observations have led me to the conclusion that their real function is that of anchoring organs, for the larvæ are invariably found clinging to the flat surfaces of stones in the swiftest parts of streams, seemingly by the aid of these seven pairs of grappling hooks, the so-called abdominal gills. Were it not for these organs they would be instantly swept down stream by the rush of waters, as may easily be demonstrated by removing the gills and replacing the dismembered insect on a stone in a swift part of the current. The true legs of course are used for ordinary locomotive purposes, the pseudo-legs or gills having no powers in this respect.

The larvæ of *Pseudophæa splendens*, *Indophæa dispar*, and *Indophæa fraseri* are all very similar, the two latter being indistinguishable except by the locality which they inhabit; the former possesses a lateral beard-like group of robust spines external to and beneath the eyes. The abdominal gills are found on the first seven segments and are all S-shaped, except the distal pair, which are more or less weakly developed.

The caudal gills, as in *Anisopleura*, are triplicate bladder-like structures, thickly coated with coarse hairs, one situated medially and dorsally, the others lying on each side of it, the apposed surfaces being flattened, the under surfaces also flattened so as to lie flush with the surface on which the larva is resting. The mask is simple and without setæ.

Distribution.—Southern Asia from Western India and Ceylon to China and the Philippines and southwards to Borneo, Java, etc. One genus has spread westwards into Europe *via* Kashmir and Persia.

The subfamily is represented by nine genera, all of which save *Anisophæa* (*Mesophæa*) and *Paraphæa* have representatives within Indian limits.

Owing to the similarity of the known females of these genera our genera are founded solely on male characters; it would indeed be quite impossible to split up the subfamily effectively otherwise. The old genus *Pseudophæa* is here split up into four genera, as the differences in colouring, petiolation, and shape of wings are so great. Selys, in his classification, split up the genus into groups, and these again into subgroups; it is to the original groups that generic rank has been given.

In determining the precedence of these new as well as old genera the closeness of the similarity of the male to the female has been taken as the measure of archaism; thus, it is evident that a species like *P. ochracea*, in which the fore- and hind-wings are similarly shaped and of similar shape to those of the female, is much nearer the ancestor of the subfamily than one like *P. variegata*, where specialization in the male has led to such divergence from the female that an entomologist ignorant of the order might well be excused for determining them as different species.

Out of the twenty-four known species of the old genus *Pseudophæa* I have been able to study sixteen, viz.:—*dispar*, *cardinalis*, *fraseri*, *impar*, *inæquipar*, *tricolor*, *subnodalis*, *subcostalis*, *ornata*, *decorata*, *ochracea*, *brunnea*, *splendens*, *refulgens*, *variegata*, and *masoni*, which, together with authors' descriptions, have formed the basis of the following key to the subfamily.

Key to Genera of Subfamily Epallaginæ.

1.	Discoidal cell entire	2.
	Discoidal cell traversed	4.
2.	Costa of hind-wing of male running straight from base to node	3.
	Costa of hind-wing of male with an obtuse projecting angle between base and node. .	ANISOPLEURA Selys, [p. 84.
3.	Abdomen of male longer than wings; node situated slightly proximal to centre of wing; wings petiolated to level of first antenodal nervure	BAYADERA Selys, [p. 78.
	Abdomen of male shorter than wings; node situated at centre of wings; petiolation almost absent, ending well proximal to first antenodal nervure	EPALLAGE Charp., [p. 75.
4.	Tenth abdominal segment without a carinal spine	DYSPHÆA Selys, [p. 91.
	Tenth segment with two dorsal spines	PARAPHÆA Mart., [(not Indian).
	Tenth segment with one dorsal spine	5.
5.	Fore- and hind-wings of male similarly shaped, saffronated in part, but without any opaque areas	ALLOPHÆA Fraser, [p. 95.
	Fore- and hind-wings of male differently shaped, hind pair at least with some opaque markings	6.
6.	Both fore- and hind-wings of male more or less opaque black, hind pair markedly broader than fore pair and usually bearing a large basal metallic blue or green area. .	PSEUDOPHÆA Kirby, [p. 99.
	Only hind-wings of male partly opaque black.............................	7.
7.	Hind-wings of male with the middle part abruptly broadened, the apical portion thereafter narrowing rapidly to a falcate apex; the broadened portion bearing an opaque band	MESOPHÆA Fraser [(not Indian)*.
	Hind-wings of male usually much shorter than fore-wings and markedly rounded at apex, the apical half or less opaque black and sometimes bearing a brilliant metallic blue spot; abdomen in some species bright red, otherwise black	INDOPHÆA Fraser, [p. 104.

Genus **EPALLAGE** Charp.

Epallage Charpentier, Lib. Eur. p. 16 (1840); Selys, Rev. Odon. p. 143 (1850); Walker, List Neur. Ins. Brit. Mus. iv, p. 636 (1853); Selys, Syn. Cal. p. 49 (1853); id., Mon. Cal. p. 162 (1854); Kirby, Cat. Odon. p. 108 (1890); Fraser, J. Bombay Nat. Hist. Soc. vol. xxxiii, p. 49 (1928).

Wings of both sexes hyaline or enfumed at apices, narrow, petiolation almost absent, ceasing well proximal to the first antenodal nervure; hind-wing not broader than fore-wing, and of similar breadth in the two sexes; *Rii* not in contact

* The name *Mesophæa* being preoccupied in Coleoptera, I propose **Anisophæa**, nom. nov., as a substitute.

with $R+M$ at its origin; node situated at centre of wing; discoidal cell entire, short, less than half the length of median space; arc only slightly angulated; only 1 cubital nervure in all wings (rarely 2); not more than 4 long intercalated nervures posterior to *IA* in the hind-wing, and only 2 between *Cuii* and *IA*; *Riii* in continuation with the subnode or a shade distal; outermost antenodal nervure in all wings often incomplete; no basal incomplete antenodal nervures in subcostal space; pterostigma long and narrow.

Thorax very robust; abdomen not or only just extending beyond tips of wings; anal appendages longer than segment 10, inferiors only slightly shorter than superiors and minutely bifid at apex, superiors strongly hooked or angulated downwards at apex.

Genotype, *Epallage fatima* Charp.

Distribution.—Greece, Turkey in Europe, Asia Minor, Persia, N.W. India, and Kashmir. *E. alma* Selys, from Persia, is probably not more than a local race or variety, so that the genus contains but a single species.

Unfortunately nothing has been recorded about its habits, and noting how widely those of the allied genera *Bayadera* and *Anisopleura* differ, it is idle to speculate by comparisons. All we know is that it breeds in streams.

198. **Epallage fatima** Charpentier. (Figs. 22 & 23.)

Agrion fatima Charpentier, Lib. Eur. p. 132, t. 45, fig. 2 (1840).
Epallage fatime Schneid. Stett. Ent. Zeit. vol. vi, p. 115 (1845); Selys, Syn. Cal. p. 50 (1853); id., Mon. Cal. p. 165 (1854); id., Bull. Acad. Belg. (2) vol. xxvii, p. 659 (1869).
Euphœa fatime Selys, Rev. Odon. p. 143 (1850); Walker, List Neur. Ins. Brit. Mus. iv, p. 637 (1853).
Epallage alma Selys, Bull. Acad. Belg. (2) vol. xlvii, p. 372 (1879); Kirby, Cat. Odon. p. 108 (1890); Morton, Trans. Ent. Soc. Lond. p. 305 (1907).
Epallage fatima Kirby, Cat. Odon. p. 108 (1890); Morton, Trans. Ent. Soc. Lond. p. 305 (1907); Laidlaw, Rec. Ind. Mus. vol. xiii, p. 40 (1917); Fraser, J. Bombay Nat. Hist. Soc. vol. xxxiii, pp. 49, 50 (1928).

Male.—Abdomen 28–32 mm. Hind-wing 32 mm.

Head: labium yellow, base and apex black; labrum olivaceous, with a median basal black impression; cheeks, bases of mandibles, anteclypeus, and sides and fore-border of postclypeus pale olivaceous, rest of head black. Two rounded bosses on frons; occiput coated with rather long coarse yellow hairs. Eyes brown. *Prothorax* black, a thin pulverulent covering making it to appear dark violet, especially the small posterior lobe; laterally and subdorsally coated with long white hairs. *Thorax* black, sides and under surface markedly pulverulent, especially the latter, which may be chalky white from pruinescence. *Legs* short, hind femora not reaching middle of second abdominal segment when extended, whole leg falling short of distal end of third segment; black, femora pulverulent, the two posterior femora with a dorsal and a lateral yellow stripe throughout their length, with a row of short, widely spaced black spines on the lateral stripe; anterior pair of femora with only the dorsal stripe. *Wings* hyaline, apices narrowly bordered or tipped with dark brown; pterostigma black; 12 to 14 antenodals, 15 to 16 postnodals. *Abdomen* black, basal segments often pulverulent white. *Anal appendages* black. Superiors one-and-a-half times the length of segment 10, seen from above widely separated at bases and strongly divaricate, thick, tumid, with subacute apices, seen from the side very robust, stout at base with the true apex turned strongly downwards and inwards, like the grappling flukes of an anchor; on outer side, at point where apex is bent, a short stout hook directed downwards and outwards. Inferiors about two-thirds the length of superiors,

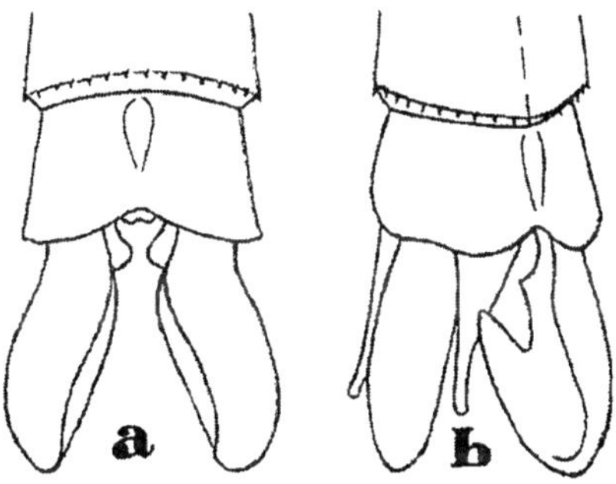

Fig. 23.—Anal appendages of *Epallage fatima* Charp., male. **a**, dorsal view; **b**, dorso-lateral view from the left side.

conical, broad at base, apex black and nipple-like, basal portion light brown; a small spine on upper side near the apex.

Juvenile males closely resemble the females in colouring. Frons yellow, black in middle, with a rounded yellow spot on each side; a small oval yellow spot on each side of anterior ocellus; occiput with a transverse yellow band. Thorax with the mid-dorsal crest, a curved humeral stripe, four oblique stripes on sides (the third broadest and the fourth shortest), some spots at roots of wings, and two small spots on antealar sinus all yellow. Abdomen with a vestigial mid-dorsal and a similar but better defined lateral stripe, both tending to become obliterated towards apical segments.

Female.—Abdomen 28–32 mm. Hind-wing 28–32 mm.

Similar to the juvenile male, but with rather more yellow markings; thoracic markings especially clearly defined and bright yellow. Hind pair of femora dark reddish-brown with the yellow stripe broadening proximally. Lateral abdominal stripe very broad, almost confluent with a narrow basal annule on each segment; segment 1 entirely yellow save for a small linear spot on each side running obliquely upwards and forwards. Wings with apices much more broadly tipped with brown; bases pale saffron, especially along costal region as far as node. Nodal index slightly higher than in the male.

Distribution.—As for genus. I possess a female from the Amanus Mts. in Asia Minor and a couple of males from the Wadi Kelt, Palestine. Mr. Bainbrigge Fletcher, after extensive collecting in Kashmir, never came across this insect, although he took all other known Kashmir species. There is a male in the Paris Museum from Baghdad and two females from Macedonia in the British Museum. Lastly, the species was taken by Col. Nurse at Quetta, BALUCHISTAN (two males in June).

Type: present location unknown; I was unable to find it in the Paris Museum.

Genus **BAYADERA** Selys (1853). (Fig. 24.)

Epallage, group *Bayadera*, Selys, Syn. Cal. p. 49 (1853).
Bayadera Selys, Mon. Cal. p. 162 (1854); id., Bull. Acad. Belg. (2) vol. xlvii, p. 373 (1879); Kirby, Cat. Odon. p. 108 (1890); Williamson, Proc. U.S. Nat. Mus. vol. xxviii, p. 169 (1905); Ris, Suppl. Ent. no. 1, pp. 48, 49 (1912); Laidlaw, Rec. Ind. Mus. vol. xiii, p. 24 (1917); Fraser, J. Bombay Nat. Hist. Soc. vol. xxix, p. 477 (1923); id., ibid. vol. xxxiii, pp. 50, 51 (1928).
Euphœa Walker, List Neur. Ins. Brit. Mus. iv, p. 636 (1853).

Wings of both sexes narrow, hyaline, apices (in two species) tipped with black; hind-wing not broader than fore-wing, and of similar breadth in both sexes; petiolation extending from about half way from base to arc and to level of first antenodal nervure; *Rii* in contact with *R+M* at its origin and for some distance; node situated at centre of wings; discoidal cell entire, short, less than half the length of median space; arc more angulated than in *Epallage*; only 1 cubital nervure in all wings (the nervure *AC*); not more than 4 long intercalated nervures between *IA* and posterior border of wing and only 2 long ones between *IA* and *Cuii*; *Riii* not in line with the subnode, either a little proximal or widely distal to it (in *B. indica* and *B. hyalina* respectively); outermost nervure in all wings complete; no basal incomplete antenodal nervures in subcostal space; pterostigma long and narrow.

Thorax very robust; abdomen longer, often much longer than wings; anal appendages considerably longer than segment 10, superiors forcipate, subcylindrical, apices obtuse or depressed, curling in and converging on one another, with or without a ventral basal spine; inferiors conical, tapering to a fine point, much less robust than superiors.

Genotype, *Epallage indica* Selys.

Distribution.—N.E. INDIA, UPPER BURMA, S. China, and Formosa. Only five species of this genus are known at present, three of which are from within Indian limits, the other two (*B. melanopteryx* and *B. brevicauda*) from S. China and Formosa respectively.

But little is known of their habits and the larva is unknown. *B. indica* is in habits strikingly similar to *Dysphæa*; the males perch on prominent rocks or twigs projecting from the water in mid-stream, and, being very shy and wary, are not easy to capture or even approach in such spots. This

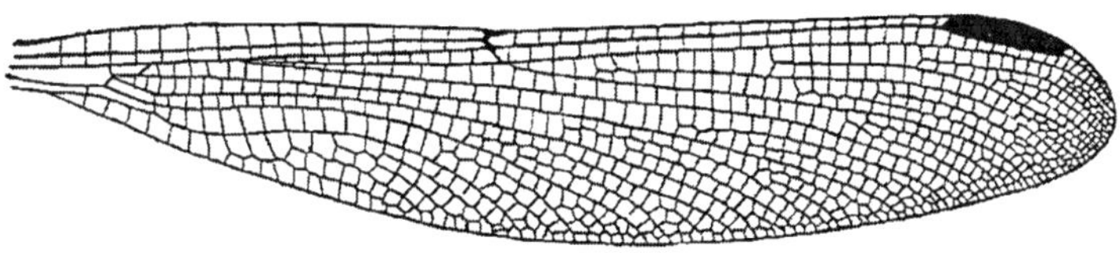

Fig. 24.—Wing of *Bayadera indica* (Selys), male.

particular species looks remarkably like *Vestalis apicalis*, which, however, is never seen in such spots. It also perches with its abdomen slanting up at a sharp angle, as does *Dysphæa*. When disturbed it moves down stream with a markedly flitting flight, soon coming to rest again. If repeatedly stalked it makes a wide circle round its pursuer and returns upstream. Many pairs may be seen *in copula*, a very unusual circumstance in the larger Zygoptera, and these pairs take long flights down stream seemingly looking for a suitable spot to deposit their eggs, though oviposition is rarely seen. From a study of the habits I am inclined to think this genus is closely related to *Dysphæa*.

199. **Bayadera indica** (Selys). (Fig. 25, a.)

Epallage indica Selys, Syn. Cal. p. 49 (1853).
Euphæa indica Walker, List Neur. Ins. Brit. Mus. iv, p. 636 (1853).
Bayadera indica Kirby, Cat. Odon. p. 108 (1890); Martin, Mission Pavie, 3, Neurop, p. 15 (1904); Ris, Suppl. Ent. no. 1, pp. 48, 49 (1912); Laidlaw, Rec. Ind. Mus. vol. xiii, p. 31 (1917); Fraser, J. Bombay Nat. Hist. Soc. vol. xxxiii, pp. 51, 52 (1928).

Male.—Abdomen 38–40 mm. Hind-wing 34 mm.

Head: labium with middle lobe and base black, lateral lobes greenish-yellow; labrum, bases of mandibles, and cheeks as far up as beyond the antennæ turquoise-blue, edges of bases of mandibles and labrum finely black, as also a small median basal virgule on labrum ; rest of head mat black ; eyes dark brown above, olivaceous below. *Prothorax* black, with a very large rounded spot on each side of mid-lobe and a smaller spot low down on each side of posterior lobe greenish-yellow ; posterior lobe large, with rounded border, lappet-shaped. *Thorax* black, marked with bright greenish-yellow as follows :—A fine antehumeral stripe curving out above and below, confluent below with a broad humeral stripe which covers the suture, and nearly confluent with it again above, so as nearly to shut in a broad oval black spot of the ground-colour ; laterally three broad irregular stripes, one on each suture, the third covering greater part of metepimeron, all three broadly confluent above, but the first and second partially separated by a short tongue of black on upper part of second lateral suture. Beneath, prothorax and thorax, as also middle lobe of labrum and basal segments of abdomen,

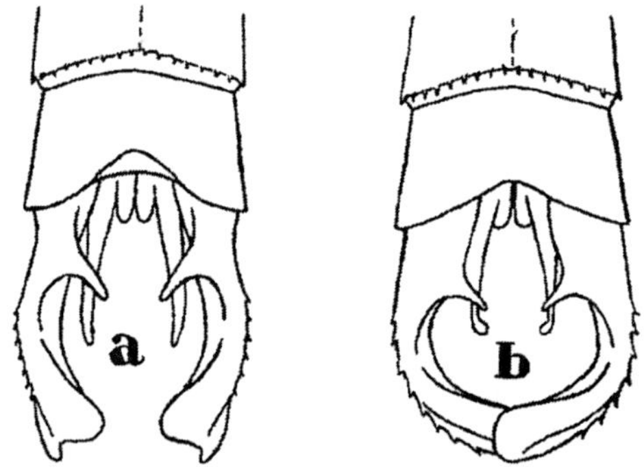

Fig. 25.—Anal appendages of (a) *Bayadera indica* (Selys), male ; (b) *Bayadera hyalina* Selys, male.

pulverulent white. *Wings* hyaline, apices of all blackish-brown to about middle of pterostigma, which is black ; extreme base saffronated ; 23 to 24 antenodal nervures in fore-wings, 21 to 22 postnodals ; 18 to 19 antenodal nervures in hind-wings, 19 to 21 postnodals ; *Riii* arises slightly proximal to subnode. *Abdomen* black, mid-dorsal carina finely yellow from segment 1 to 8 ; a narrow lateral stripe greenish-yellow, broad on sides of segments 1 and 2, narrow from 3 to 5, but broadening and confluent with an incomplete basal annule at base of each segment ; on segments 6 and 7 this line present as a mere basal vestige ; 8 to 10 unmarked. *Anal appendages* black. Superiors broad at base, narrowing slightly at middle, broadening and depressed at apices into a triangular plate which is hollowed out below ; a robust ventral spine near base, directed somewhat inwards so as to be visible from above ; a small tubercle (almost a spine, but blunt at apex) slightly basal to middle on inner side. Inferiors slightly more than half the length of superiors, broad at base, tapering rapidly to a fine acute apex. Both pairs separated at base, inferiors divaricate, superiors parallel, but apices curling in and actually overlapping. Genitalia very similar to those of *Epallage fatima*.

Female.—Abdomen 36 mm. Hind-wing 37 mm.

Much more robust than the male, with shorter stouter abdomen and relatively longer wings ; latter without black apices, but bases pale saffron ; nodal index rather higher, 24 to 25 ante- and postnodal nervures to fore-wings, 22 of each to hind ; pterostigma very long, black.

Body-markings exactly as in the male, but the lateral stripe extends to apical end of segment 7 on abdomen. Legs black, hind femora with proximal ends and a broad stripe on sides greenish-yellow. Vulvar scale robust, extending to end of abdomen. *Anal appendages* nearly twice the length of segment 10, tapering to a very fine point.

Distribution.—N.E. INDIA, N. BENGAL in the rivers about Darjeeling District. Dr. Ris stated that he had a series from Shillong, Khasi Hills, ASSAM, but I cannot help thinking that this was an error and that *B. hyalina* was meant. Mr. Bainbrigge Fletcher, who has collected almost exhaustively in this area, has taken *B. hyalina*, but never *indica* ; the type of *B. hyalina* in the Selysian collection also comes from Shillong, whereas *B. indica* comes from Bengal.

B. indica is distinguished from *B. hyalina* by its much larger size and by the apices of all wings in the male being tipped with black. From *B. melanopteryx* it is easily distinguished by the much smaller extent of black on the wings, extending in the case of the latter as far as 2 to 3 cells distal to the node, whilst from *B. brevicauda* and *longicauda* it is distinguished by having the apices of all wings black. Its habits have already been commented on above.

Type in the Selys collection, Brussels Museum.

200. **Bayadera hyalina** Selys. (Fig. 25, **b**.)

Bayadera hyalina Selys, Bull. Acad. Belg. (2) vol. xlvii, p. 373 (1879); Kirby, Cat. Odon. p. 108 (1890); Ris, Suppl. Ent. no. 1, pp. 49–52 (1912); id., Suppl. Ent. no. v, p. 3 (1916); Laidlaw, Rec. Ind. Mus. vol. xiii, p. 31 (1917); Fraser, J. Bombay Nat. Hist. Soc. vol. xxix, p. 477 (1923); id., ibid. vol. xxxiii, pp. 52, 53 (1928).

Male.—Abdomen 34 mm. Hind-wing 29 mm.

Head : labium black, lateral lobes greenish-yellow at their bases only ; labrum, bases of mandibles, and cheeks turquoise-blue, former barely margined with black and without the median virgule present in *B. indica* ; rest of head mat black. *Prothorax* black, with a very large greenish-yellow spot on each side of mid-lobe and a tiny spot on outer ends of posterior lobe, which is similarly shaped to that of *B. indica*. *Thorax* black, marked with bright citron-yellow as follows :—A fine antehumeral line not reaching upper end of dorsum and not confluent with humeral stripe either above or below, humeral stripe much narrower than in *B. indica* ; three narrow stripes on sides, of which the two latter only are confluent above and that by a mere point. A broad bar of yellow beneath thorax of which the hinder area is broadly black. *Legs* black, unmarked. *Wings* entirely hyaline ; pterostigma dark brown, covering 5 to 7 cells ; 21 to 23 antenodal nervures in fore-wings, 20 to 21 postnodals ; 17 antenodal nervures in hind-wings, 20 to 21 postnodals. *Riii* begins 1 to 2 cells distal to the subnode. *Abdomen* glossy black ; triangular spot of citron-yellow on sides of segment 1, remaining segments unmarked. *Anal appendages* black. Superiors about twice the length of segment 10, moderately broad at base, tapering slightly to beyond the middle and then dilated slightly at apices but not nearly to the same extent as in *indica* ; apices twisted and flattened, not hollowed out below, curled in towards each other ; a robust subventral spine near base, not quite as long as in *B. indica* ; no medial tubercle on inner side. Inferiors broad at base, tapering rapidly to an acute point, not quite half the length of superiors.

Female.—Abdomen and hind-wing 33 mm.

Differing very slightly from the male, but more robust in build and the abdomen of similar length to hind-wings ; antehumeral stripe well defined and reaching upper end of dorsum, but not confluent at either end with humeral stripe. Stripe at extreme end of metepimeron confluent at a point with second lateral stripe to form a hook-shaped marking. Beneath, black with a mere suggestion of yellow at centre ; this surface largely pruinosed white. Abdomen with a lateral stripe on segments 1 to 4, broad at base of 3 and 4, and tapering to a fine point which falls slightly short of apical ends of segments ; in addition a small round well-defined spot on each side of segment 10. *Anal appendages* conical, acutely pointed, slightly longer than segment 10. Wings similar to male, nodal index similar.

Distribution.—ASSAM. The *type*, in the Selys collection, Brussels Museum, is from Shillong, Khasi Hills ; I have seen several examples from this locality.

B. hyalina differs from *B. indica* and *melanopteryx* in the wings being entirely hyaline, and from *B. brevicauda* and *longicauda* in having a well-developed ventral spine to superior anal appendages, absent in the former, vestigial in the latter.

The species described by Dr. Ris from Formosa as *B. hyalina* is certainly not that species, of which I possess examples from the type-locality. The figure of the appendages given by Dr. Ris shows the superiors to be equal in length to segment 10 and without any sign of a ventral spine. In true *B. hyalina* there is a well-marked spine and the appendages are much longer than segment 10, about twice the length in fact. I have therefore renamed the Formosan species as *B. brevicauda*.

201. **Bayadera longicauda** Fraser. (Fig. 26, **a**.)

Bayadera longicauda Fraser, J. Bombay Nat. Hist. Soc. vol. xxxiii, p. 53 (1928).

Male.—Abdomen 38 mm. Hind-wing 34 mm.

Very closely allied to *B. hyalina*, but labium entirely black; labrum not margined with black; cheeks and labrum greenish-yellow, this colour passing inwards above epistome to form an incomplete band across front of frons. *Prothorax*: posterior lobe entirely yellow save for a small crenulate black mark at base. *Thorax*: antehumeral stripes complete, curving out at upper ends, remaining stripes narrow, but the two lateral ones broadly confluent at their upper ends; antealar sinus

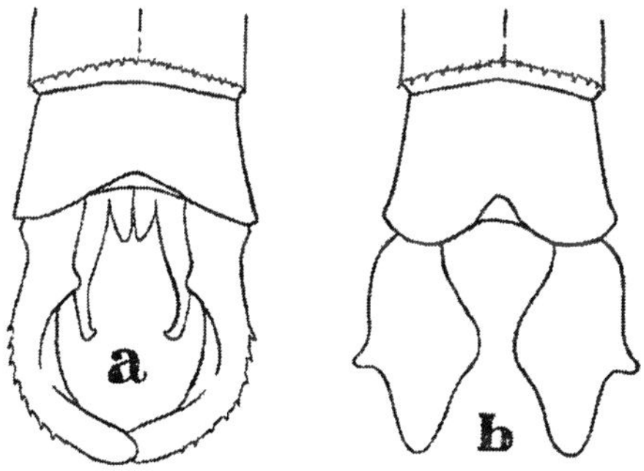

Fig. 26.—Anal appendages of (**a**) *Bayadera longicauda* Fraser, male; (**b**) *Anisopleura lestoides* Selys, male.

with two large greenish-yellow spots. *Abdomen* almost unmarked, a small lateral spot on segment 1, a tiny vestige at the base on sides of 2, and a small triangular lateral basal spot on 3–5. *Wings* with 19 to 21 antenodal nervures in fore-wings and 24 to 25 postnodals, 15 to 16 antenodals in hind-wings, 20 to 25 postnodals. Pterostigma black, covering 5 to 6 cells; *Riii* begins well distal to the subnode; all wings palely enfumed, especially the apices of hind pair. *Anal appendages* very similar to those of *B. hyalina*, but ventral spine of superiors vestigial, amounting to no more than a pointed tubercle, and, in addition, a robust obtuse tubercle at middle of upper and inner border. Inferiors very broad at base, shorter than in *B. hyalina*.

Female unknown.

Distribution.—Gangtok, 5,000 to 6,000 ft. alt., DARJEELING DISTRICT. A few specimens collected by Mr. Chas. Inglis, May 29, 1924. The shape of the appendages will readily distinguish this species from any others, as well as the relative lengths of abdomen and wing. A further striking feature is the bright yellow posterior lobe of prothorax, which offers a very useful identification mark.

Type in the Darjeeling Museum.

Genus **ANISOPLEURA** Selys. (Fig. 27.)

Anisopleura Selys, Syn. Cal. p. 48 (1853); Walker, List Neur. Ins. Brit. Mus. iv, p. 635 (1853); Selys, Mon. Cal. p. 158 (1854); Kirby, Cat. Odon. p. 108 (1890); Williamson, Proc. U.S. Nat. Mus. vol. xxviii, p. 169 (1905); Laidlaw, Rec. Ind. Mus. vol. xiii, pp. 24, 25 (1917); Fraser, J. Bombay Nat. Hist. Soc. vol. xxxiii, pp. 53, 54 (1928).

Wings of both sexes hyaline, or, at most, with apices tipped with black in the male; petiolation as in *Bayadera*, or extending a little distal to basal antenodal nervure, especially in hind-wing; all wings in both sexes narrow and of equal breadth; costal border of hind-wing in male with an abrupt angulation outwards at a point about one-third the distance from base to node, resulting widening of costal space gradually decreasing from angulation outwards; *Rii* not in contact with *R+M* at its origin; node situated at about middle of wing; discoidal cell entire, equal to about one-third or a little less than half the length of median space in fore-wing, a little more than half as long in hind-wing; arc markedly angulated; sectors of arc widely separated at origin, far more so than in *Bayadera* or *Epallage*; from 1 to 5 cubital nervures in all wings (usually only one in *A. lestoides*, several in other species); *IA* distinctly forked, or, if not, then with 3 to 4 moderately long intercalated sectors between itself and hinder margin of wing; only 2 long intercalated sectors between *IA* and *Cuii*; *Riii* widely distal to the subnode, about 1 cell more so than in *Bayadera*; outermost antenodal complete but normally not in alignment; no basal incomplete antenodal nervure in subcostal space; pterostigma long and narrow.

Thorax robust; abdomen always longer than wings, long and cylindrical; anal appendages variable, inferiors always more or less aborted, superiors subcylindrical or markedly spatulate, with or without a robust latero-ventral spine.

Genotype, *Anisopleura lestoides* Selys.

Distribution.—BENGAL, SIKKIM, ASSAM, and UPPER BURMA.

The species of *Anisopleura* breed in montane and submontane streams, and I have found them breeding in irrigation channels running through tea-plantations in Bengal. Unlike the species of *Bayadera*, those of *Anisopleura* keep to trees or bushes bordering streams, the female retiring further into the jungle and only visiting the stream when ovipositing, at which time it may be taken *in copula* with the male. Only occasionally do these insects descend to the river-bed, but immediately retreat to vegetation on being disturbed; their flight is short and flitting.

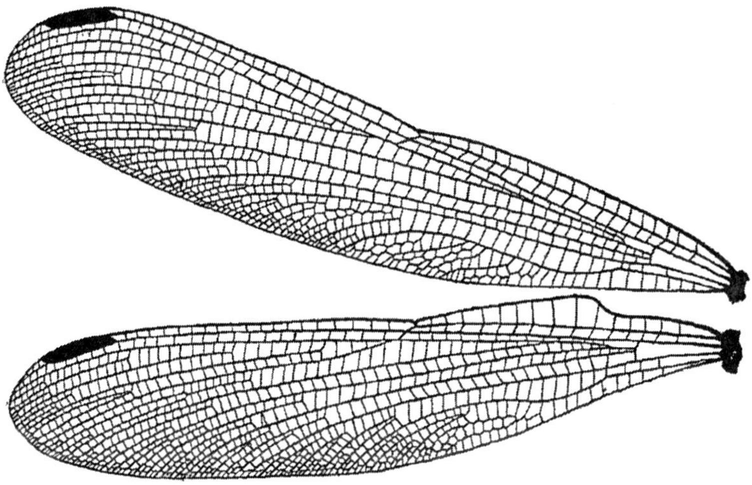

Fig. 27.—Wings of *Anisopleura comes* Selys, male.

Key to Indian Species of Anisopleura.

1.	Usually only one cubital nervure in all wings; no humeral stripe	2.
	Several cubital nervures; humeral stripe present	3.
2.	Superior anal appendages with short spine at middle	*lestoides* Selys, p. 86.
	Superior anal appendages forked	*furcata* Selys, p. 89.
3.	Superior anal appendages without spine....	*comes* Selys, p. 87.
	Superior anal appendages with short spine near base	*subplatystyla* Fraser, [p. 89.

202. **Anisopleura lestoides** Selys. (Fig. 26, **b**.)

Anisopleura lestoides Selys, Syn. Cal. p. 48 (1853); id., Mon. Cal. p. 159 (1854); Kirby, Cat. Odon. p. 108 (1890); Selys, Ann. Mus. Civ. Genova, (2) vol. x (xxx), p. 489 (1891); Laidlaw, Rec. Ind. Mus. vol. xiii, p. 31 (1917); Fraser, J. Bombay Nat. Hist. Soc. vol. xxix, p. 477 (1923); id., ibid. vol. xxxiii, pp. 54, 55 (1928).

Euphœa lestoides Walker, List Neur. Brit. Mus. iv, p. 635 (1853).

Male.—Abdomen 36–38 mm. Hind-wing 28–30 mm.

Head: labium black, pulverulent white at base; labrum greenish-yellow, narrowly bordered with black; bases of mandibles, cheeks broadly up to level of lateral ocelli, and postclypeus greenish-yellow; anteclypeus black; vertex and occiput black, with a large rounded spot of greenish-yellow on outer side of each lateral ocellus; eyes blackish-brown above, olivaceous green below. Dorsum of head sometimes pulverulent white. *Prothorax* black; outer ends of posterior lobe and a very large lateral spot on middle lobe greenish-yellow. *Thorax* black, marked with an antehumeral stripe, broad below, tapering above where it turns out as a short point; laterally, the whole central part of metepimeron and a broad irregular stripe centred over the first lateral suture greenish-yellow, dorsal black area extending well beyond humeral suture and a small tongue of black descending on upper part of first lateral suture. *Legs* short, but robust, spines fine, black, a greenish stripe on outer sides of two posterior pairs of femora, tapering towards but not extending as far as distal ends. *Wings* hyaline, fore-wings of male with extreme apex dark brown, bases pale saffron; only one cubital nervure in all wings; *IA* not clearly forked; 15 to 19 antenodal nervures in fore-wings, 13 to 16 in the hind; 20 to 22 postnodal nervures in fore-wings, 17 to 18 in the hind; pterostigma black, over 3½ to 4½ cells; petiolation shorter than in other species, often arrested short of the first or basal antenodal nervure. *Abdomen* black, marked with greenish-yellow as follows:—Sides of segment 1 broadly, a broad lateral stripe on 2, a basal ring broadly interrupted on the mid-dorsum, and a narrow lateral stripe on 3 to 6, the stripes becoming shorter and confined to middle portions of 5 and 6; mid-dorsal carina finely yellow on 3 to 6. *Anal appendages* black. Superiors subcylindrical at base, broadened and depressed near apices, which curve in towards each other and even overlap; seen from above, the upper border distinctly sinuous and curving down slightly at apex; a robust spine at about middle on ventro-lateral face, directed analwards and visible both from above and from the side. Inferiors rudimentary, not visible in profile.

Fully mature specimens show a marked pruinescence, especially beneath, on the coloured part of prothorax and thorax, the temples, and dorsum of last two abdominal segments.

Female.—Abdomen 30 mm. Hind-wing 29 mm.

Very similar to the male in markings, but these rather more extensive, especially on the abdomen, which is short and very robust.

The lateral abdominal markings consist of a long stripe and basal spot narrowly separated, and extending the whole length of segments 2 to 7, the spots being confluent on 2, where the stripe is very broad ; segment 8 has a small round apico-lateral spot and 9 a very large irregular lateral spot extending over more than half the segment ; segment 10 unmarked, very short, notched on dorsum. *Anal appendages* half as long again as segment 10, tapering to an acute point. Vulvar scale robust, not quite reaching end of abdomen.

Distribution.—BENGAL, ASSAM, and SIKKIM, at altitudes of about 3,000 to 5,000 ft. On the wing from May to September.

Type in the Selys collection, Brussels Museum.

203. **Anisopleura comes** Selys. (Figs. 27 & 28, **a**.)

Anisopleura comes Selys, C.R. Soc. Ent. Belg. xxiii, p. lxiii (1880); Kirby, Cat. Odon. p. 108 (1890); Selys, Ann. Mus. Civ. Genova, (2) vol. x (xxx), p. 489 (1891); Laidlaw, Rec. Ind. Mus. vol. xiii, p. 31 (1917); Fraser, J. Bombay Nat. Hist. Soc. vol. xxxiii, p. 55 (1928).

Male.—Abdomen 36–40 mm. Hind-wing 29–34 mm.

Head : labium black, pruinosed at base or on whole of middle lobe ; labrum citron-yellow, finely margined with black ; anteclypeus black ; postclypeus, bases of mandibles, cheeks as far up as antennæ, and a broad band across frons citron-yellow ; rest of head mat black, with a small oval spot of greenish-yellow on outer side of each lateral ocellus ; eyes black above, dark olivaceous beneath. *Prothorax* black, outer ends of posterior lobe and a very large triangular spot on each side of middle lobe citron-yellow, these marks in adult specimens pruinosed white. *Thorax* black, with a very fine, often interrupted antehumeral line more or less confluent with a broad humeral fascia, which extends up dorsum for about the lower two-thirds and broadens out markedly below ; the greater part of sides and the postero-lateral suture marked out in black ; metepimeron narrowly framed in black. On sides and beneath pruinosed white. *Legs* black, hind femora yellow on inner sides distally and often pruinosed white. *Wings* hyaline, extreme apices of fore-wings of male dark brown ; pterostigma black, covering 3½ to 4 cells ; 3 to 4 cubital nervures in fore-wings, 4 to 5 in hind-wings; *IA* often forked; petiolation begins at level of or distal to level of the basal antenodal nervure. *Abdomen* black, marked similarly to *A. lestoides*; segment 7 usually with a small latero-basal spot; 9 and 10 chalky white, with pruinescence. *Anal appendages* black. Superiors seen from above narrow at base, very broad and cupped above thereafter, usually overlapping so that the two appendages viewed from above form a triangle with a small window at the base. Seen from the side conical, tapering gradually to a subacute apex; no latero-ventral spine. Inferior appendages even more aborted than in *A. lestoides*.

Female.—Abdomen 32–33 mm. Hind-wing 33 mm.

Stouter and more robust than the male, but marked very similarly; base of labrum often narrowly black; ends of posterior lobe of prothorax more broadly yellow; antehumeral

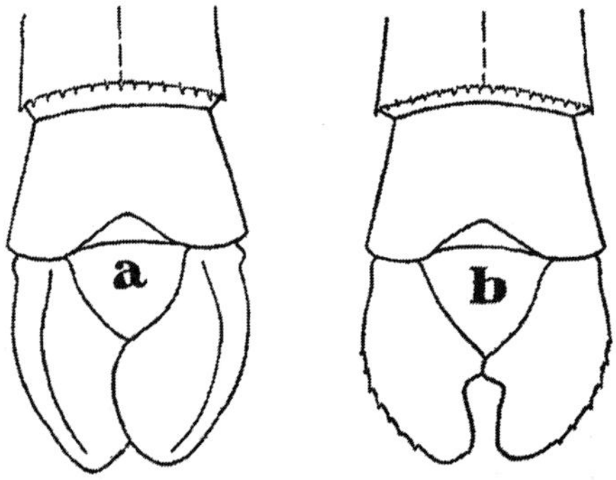

Fig. 28.—Anal appendages of (a) *Anisopleura comes* Selys, male; (b) *Anisopleura subplatystyla* Fraser, male.

stripe rarely interrupted, and usually expanded in its upper portion; abdominal markings extending on to segments 8 and 9 and exactly similar to *A. lestoides*; nodal index higher, 17 to 19 antenodal nervures in fore-wings, 20 to 23 postnodals; 17 antenodals in the hind-wings, 19 postnodals; 3 to 6 cubital nervures in fore-wings, 3 to 7 in hind-wings; pterostigma dark yellowish-brown between black nervures. Vulvar scale very robust, not extending quite to end of abdomen.

Distribution.—Bengal, Sikkim, Assam, and the Punjab Hills. It is moderately common around Darjeeling at the lower levels from about 3,000 to 4,000 ft. from April to June.

I found it breeding in irrigation channels in cinchona-plantations at Mungpoo, Darjeeling District. The imago was quite numerous on herbage beside such courses, but rose high into trees unless stalked warily.

This species is easily distinguished from the last by the characters noted in the key and by its much greater size.

Type in the Selys collection, Brussels Museum.

204. **Anisopleura furcata** Selys.

Anisopleura furcata Selys, Ann. Mus. Civ. Genova, (2) vol. x (xxx), pp. 488, 489 (1891); Williamson, Proc. U.S. Nat. Mus. vol. xxviii, p. 181, fig. 13 (1905); Fraser, J. Bombay Nat. Hist. Soc. vol. xxxiii, p. 56 (1928).

Male.—Abdomen 32 mm. Hind-wing 27 mm.

Head: labium black; labrum, postclypeus, and corners of mandibles pale blue or greenish-yellow; rest of head black save for a small rounded greenish-yellow spot on each side of ocellar space. *Prothorax* black, with a large rounded spot greenish-yellow occupying sides and subdorsum of middle lobe. *Thorax* black, marked with a greenish-yellow antehumeral stripe broad below, tapering above; laterally greenish-yellow, metepimeron finely framed in black and hinder lateral suture mapped out in black; beneath black. *Wings* hyaline, extreme bases pale saffron; extreme apices of fore-wings tipped with brownish-black; pterostigma brownish-black, covering about 4 cells; 15 to 17 antenodal nervures in fore-wings, 21 to 22 postnodals; 14 to 15 antenodals in hind-wings, 18 to 19 postnodals. *Legs* black, inner sides of hind pair of femora distally yellowish. *Abdomen* black, marked with citron-yellow exactly as in *A. lestoides*, dorsum of segments 9 and 10 pruinosed white. *Anal appendages* black. Superiors slightly longer than segment 10, subcylindrical, apices broadly obtuse and curving in slightly; on the outer side a very robust spine equal in size to apex of appendages, so that the latter appears to be forked from its middle, and seen in profile resembles a crab's claw. Inferiors rudimentary.

Female unknown.

Distribution.—UPPER BURMA. The *type* (which is in the Genoa Museum collection) is from Puepoli, taken in June.

This species closely resembles *A. lestoides* in many respects, but may be distinguished by the greater size of the spine of superior anal appendages, which gives a bifid or branched appearance to those structures.

205. **Anisopleura subplatystyla** Fraser. (Fig. 28, b.)

Anisopleura subplatystyla Fraser, Rec. Ind. Mus. vol. xxix, p. 81 (1927); id., J. Bombay Nat. Hist. Soc. vol. xxxiii, pp. 56, 57 (1928).

Male.—Abdomen 34 mm. Hind-wing 28 mm.

Head: labium black, middle lobe pruinosed white; labrum bright greenish-yellow finely margined with black;

cheeks, postclypeus, and a stripe across lower part of frons and bases of mandibles greenish-yellow; rest of head mat black, with a large reniform spot on the outer side of each lateral ocellus; eyes dark brown. *Prothorax* black, outer ends of posterior lobe and a very large oval spot on each side of mid-lobe greenish-yellow. *Thorax* black, marked with citron-yellow as follows :—Narrow antehumeral stripes running close to and parallel with the mid-dorsal carina, upper end of stripes curved outward; a narrow humeral stripe running parallel to the former and broadly confluent with it below but not above, although the two approximate; a small upper spot lying close to outer side of humeral stripe; laterally broadly yellow, metepimeron narrowly framed in black, second lateral suture mapped out in black. Beneath, and lower parts of sides, pruinosed white. *Legs* black, outer and proximal portions of femora yellow, these parts pruinosed. *Wings* hyaline, apices of fore-wings tipped with dark brown as in other species; pterostigma dark blackish-brown, covering 4 cells; 17 antenodal nervures in fore-wings, 22 postnodals; 14 antenodals in hind-wings, 20 postnodals; 3 to 4 cubital nervures in fore-wings, 4 to 5 in the hind; *Riii* well distal to subnode; *IA* usually forked. *Abdomen* black, segments 9 and 10 pruinosed white on dorsum, other segments marked as in *A. lestoides* as far as segment 6. *Anal appendages* black. Superiors very similar to those of *A. lestoides*, but the spine replaced by a much smaller one not visible from above, and situated nearer base of appendage. Inferior appendages rudimentary.

Female.—Abdomen and hind-wing 30 mm.

Closely similar to the male but more heavily built; markings differ as follows :—Labrum with anterior border more broadly black, and with its base broadly black; yellow band across frons nearly cut into two at the middle; posterior lobe more broadly yellow and an additional band of yellow forming a collar to prothorax; antehumeral and humeral bands confluent above as well as below; upper humeral spot triangular instead of linear; trefoil-like spots on coxæ and six small spots on under surface of thorax; abdomen marked similarly to the female of *A. lestoides*; wings saffronated at bases, and in subjuvenile examples brightly so as far as node in costal area, subcosta and radius also yellow for the same distance and often all nervures at base of wing to outer end of discoidal cell; nodal index slightly higher; cubital nervures 3 to 5 in all wings. Vulvar scale as in *A. lestoides*.

Distribution.—ASSAM. A number taken in Shillong by Mr. T. Bainbrigge Fletcher, who has kindly sent them to me for examination.

The species, by its interesting combination of the characters of *A. comes* and *lestoides*, is easily distinguished from all others; thus it has the combined ante-humeral and humeral stripes and numerous cubital nervures seen in *A. comes*, whilst its appendages are more similar to those of *A. lestoides*.

Type in the British Museum; paratypes in my own collection.

Genus **DYSPHÆA** Selys. (Fig. 29.)

Dysphæa Selys, Syn. Cal. p. 53 (1853); Walker, List Neur. Ins. Brit. Mus. iv, p. 641 (1853); Selys, Mon. Cal. p. 185 (1854); Kirby, Cat. Odon. p. 110 (1890); Williamson, Proc. U.S. Nat. Mus. vol. xxviii, p. 169 (1905); Fraser, J. Bombay Nat. Hist. Soc. vol. xxxiii, p. 57 (1928).

Wings of male usually marked with opaque black (except in *D. ethela*), hyaline in the female, apices rather pointed, narrow; hind-wing not noticeably broader than fore-wing and of equal breadth in the two sexes; *petiolation entirely absent*; *Rii* not in contact with *R+M*; node situated at or slightly distal to middle of wing; *discoidal cell traversed* at least once, short, about half or one-third as long again as

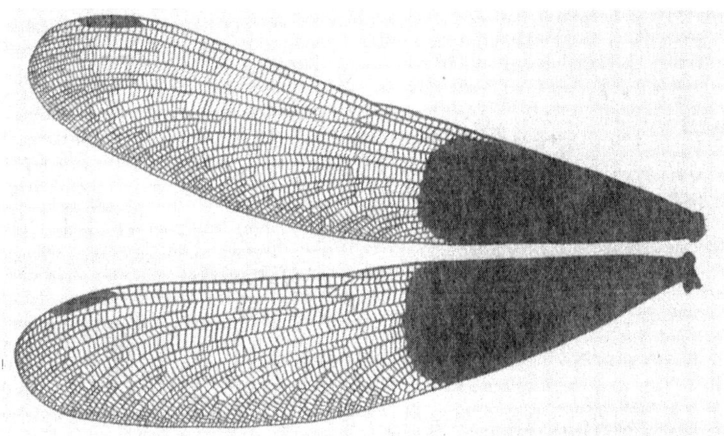

Fig. 29.—Wings of *Dysphæa walli* Fras., male.

the median space; arc almost straight; sectors of arc arising from middle of arc and well separated at origin; usually 3 cubital nervures to all wings, occasionally more in hind-wings; 4 or more intercalated nervures between *IA* and posterior border of wing, *IA* very occasionally forked; only 2 intercalated sectors between *Cuii* and *IA*, rarely more; origin of *Riii* variable even in the same species, nearly always more distal in hind-wings, usually in continuation with or proximal to subnode in fore-wing, and in continuation with or distal to subnode in hind-wing; outermost antenodal nervure complete but often not coinciding with subcostal half; no basal incomplete antenodal nervure in subcostal space; pterostigma present in all wings of both sexes, narrow, long.

Thorax robust; abdomen always extending to beyond tips of wings, often markedly so; anal appendages very homogeneous, simple, longer than segment 10, forcipate; segment 10 notched at its apical border, *its dorsum flat*, not raised into a keel-like spine as in all the following genera; vulvar scale robust, short, not reaching end of abdomen.

Genotype, *Dysphæa dimidiata* Selys.

Distribution.—From Western India and Upper Burma to Malacca, Java, Sumatra, Borneo, and New Guinea.

Species of this genus breed in swift montane and submontane streams, from 1,000 to 4,000 ft.; their habits, so far as known, are exactly similar to those of *Bayadera*. Shy, retiring insects, the females are rarely seen, spending their time feeding in the jungles not far from their parent streams, often perched high up in trees. The males will also take to such lofty roosting places during dull weather and at night—I have seen them perched on twigs at a height of 40 feet from the ground, from which vantage-point they would flit out and back again, preying on passing flies. More often the males are found in the bed of the stream resting on rocks or twigs surrounded by the boiling waters, and taking short flights, flitting in nature, out and back to the same resting place. Females have no generic characters separating them from those of the rest of the group, *Pseudophæa*, *Indophæa*, etc.

Key to Indian Species of Dysphæa.

Wings of male palely and evenly enfumed with brown throughout	*ethela* Fras., p. 92.
Wings of male with bases blackish-brown	*walli* Fras., p. 94.

206. **Dysphæa ethela** Fraser.

Dysphæa ethela Fraser, Rec. Ind. Mus. vol. xxvi, pp. 480–482 (1924); id., J. Bombay Nat. Hist. Soc. vol. xxxiii, pp. 57–59 (1928); id., Rec. Ind. Mus. vol. xxxiii, p. 448 (1931).

Male.—Abdomen 38 mm. Hind-wing 33 mm.

Head: velvety black, unmarked save for two triangular yellow marks below and hidden behind occiput; eyes dark brown above, pearly grey beneath. *Prothorax* black, with a dark ochreous rounded spot on each side of middle lobe. *Thorax* velvety black, marked (often rather obscurely) with fine antehumeral and humeral greenish-yellow stripes, confluent above, separated below, but obscured or absent in fully matured specimens. Laterally greenish-yellow shaded with olivaceous and clouded with brown, an irregular posthumeral stripe, a broad stripe on upper part of mesepimeron and the greater part of metepimeron. *Legs* black, coxæ with a spot, and femora, in subadult specimens, with a proximal inner stripe of yellow. Armature as for subfamily. *Wings* evenly enfumed with pale greenish-brown or darker brown, this varying widely according to age of specimen, whilst in some the tint is a rich golden brown throughout the whole breadth and length of wings; pterostigma very long, covering 7 to 8 cells, black. *Abdomen* black, marked with greenish-yellow as follows:—Segment 1 with a narrow apical annule and its ventral border; 2 with a broad lateral stripe constricted at its middle and then dilated at apical border, and the ventral border narrowly; 3 to 6 with baso-lateral half-rings confluent with a lateral stripe, which on 3 and 4 runs the whole length of segment, on 5 for the basal half, and on 6 for the basal third only; on 7 and 8 only the basal ring, and a ventral stripe on apical two-thirds of latter. (These markings may be largely obscured in fully matured specimens, especially on the posterior segments.) *Anal appendages* black. Superiors subcylindrical, flattened, and partially dilated towards the apical end of middle third, widely separated at base, apices curling gently in and meeting or overlapping one another, and hollowed out on their outer side. Inferiors very short, thick, closely apposed, aborted and not visible in profile.

Female.—Abdomen 32 mm. Hind-wing 33 mm.

Head: eyes dark olivaceous-brown, bluish-grey beneath; rest of head black, with a vertical stripe on cheeks bordering eyes, a transverse stripe on lower part of frons, slightly interrupted by a triangular wedge of black below, bases of mandibles and two long oblong spots on labrum creamy to citron-yellow. *Prothorax* black, with a large oval spot on each side of middle lobe, and a small spot on outer side of posterior lobe citron-yellow. *Thorax* black, marked with bright citron-yellow as follows:—Antehumeral and humeral narrow stripes, squarely confluent above and meeting at a point below, the humeral often interrupted at its middle; laterally three stripes, posterior very broad and covering whole of metepimeron except for a small elongate black area anteriorly and below, the two anterior stripes separated from each other and from the posterior yellow stripe by narrow black stripes which map out the lateral sutures. A small spot of yellow beneath, partially obscured by pruinescence. *Legs* as in male, but more yellow within. *Wings* hyaline, more or less enfumed according to age; pterostigma 4·5 mm. in length, black, covering 7 to 8 cells; venational details as in male; 30 to 36 antenodal nervures in fore-wings, 19 to 20 postnodals; 25 to 26 antenodals in hind-wing, 19 to 21 postnodals; discoidal cell traversed once or twice, usually twice in the female, once in the male; 3 cubital nervures in all wings in both sexes; *IA* not usually forked. *Abdomen* black, with markings similar to those of the male, but very vividly defined and ventral stripes continued to seventh segment. Segment 9 with a very large triangular spot on each side covering its apical half. *Anal appendages* slightly longer than segment 10, acutely pointed, black.

Distribution.—MADRAS PRES.: Coorg; Malabar; S. Kanara; Agency tracts, Jeypore. The exact distribution of this insect has yet to be worked out. In Coorg it is taken all along the course of the Cauvery and many of its tributaries, being less common on the Hatti and Harrangay Rivers. The type female was taken a quarter of a mile from the river.

Type in British Museum; paratypes in Morton, Fraser, and Laidlaw collections, etc.

207. **Dysphæa walli** Fraser.

Dysphæa walli Fraser, Rec. Ind. Mus. vol. xxix, pp. 82–83 (1927); id., J. Bombay Nat. Hist. Soc. vol. xxxiii, p. 59 (1928).

Male.—Abdomen 35 mm. Hind-wing 31 mm.

Head glossy black, with a small yellow spot on upper part of each cheek; bases of mandibles obscurely ochreous; rest of head unmarked; eyes black above, dove-grey beneath. *Prothorax* black, with a large oval citron-yellow spot on each side of middle lobe. *Thorax* black, marked with citron-yellow as follows:—Narrow antehumeral and humeral stripes confluent above by a connecting bar which runs parallel with the alar sinus, converging and confluent below; four stripes on sides and an upper vestige of a posthumeral. Of the four lateral stripes the most anterior lies on the first lateral suture, the second covers the posterior half of the mesepimeron, the third borders the metepimeron anteriorly and above, whilst the fourth covers the posterior half of that structure. Beneath black. *Legs* black, posterior pair of femora broadly yellow or obscurely so on outer side. *Wings* very palely and evenly enfumed, bases of all four blackish-brown for nearly two-thirds the distance from base to node in fore-wings, and fully three-fourths of that distance in hind-wings, outer margin of fascia slightly convex; 27 to 28 antenodal nervures in fore-wings, about 20 postnodals; 19 to 23 antenodals in hind-wings, 18 to 21 postnodals; discoidal cell traversed once or twice in all wings, more usually twice; 3 cubital nervures in all wings; pterostigma black, long, narrow, covering 6 to 8 cells. *Abdomen* black, marked with blue or citron-yellow as follows:—Segment 1 with a large triangular yellow spot on each side; 2 to 5 with a lateral stripe azure-blue, broad at base where it ascends somewhat on basal margin of dorsum, tapering to a fine point at the end of each segment. Remaining segments unmarked. *Anal appendages* black. Inferiors closely apposed, vestigial and not visible in profile. Superiors nearly twice the length of segment 10; seen laterally, broad at base, tapering to a subacute apex which is bevelled downwards; seen from above, narrow at base, broadening apically and then slightly narrowed again, markedly compressed in their apical halves; apices curling strongly in to meet or actually overlap.

Female unknown.

Distribution.—Maymyo, N. Shan States, UPPER BURMA. Four males collected by Col. F. Wall, I.M.S., May 30, 1924.

Type in British Museum, paratypes in author's collection.

Genus **ALLOPHÆA** Fraser (1929). (Fig. 30.)

Euphæa Rambur (pars), Ins. Névrop. p. 228 (1842); Selys, Syn. Cal. p. 50 (1853); Walker, List Neur. Ins. Brit. Mus. iv, p. 637 (1853); Selys, Mon. Cal. p. 167 (1854); Williamson, Proc. U.S. Nat. Mus. vol. xxviii, p. 169 (1905).
Pseudophæa Kirby (pars), Cat. Odon. p. 109 (1890).
Allophæa Fraser, J. Bombay Nat. Hist. Soc. vol. xxxiii, p. 288 (1929).

Wings hyaline, broadly saffronated in male, colourless in female; apices rather pointed; hind-wings in male not markedly broader than fore-wings, and not broader than hind-wings of female; petiolation very short, almost absent; *Rii* not in contact with *R+M*; node situated nearer base of wing than apex, about midway between base and pterostigma; discoidal

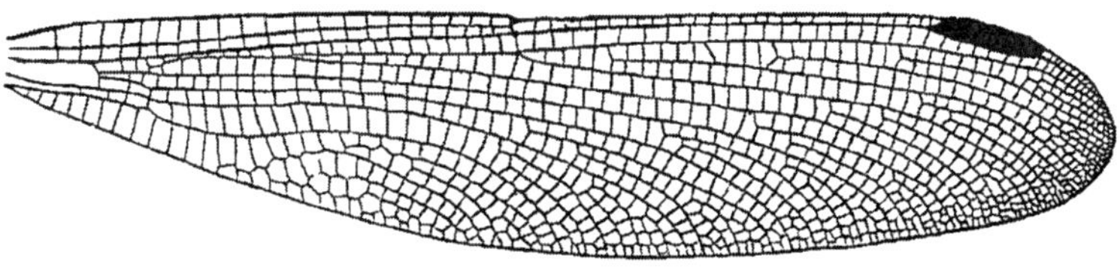

Fig. 30.—Wing of *Allophæa brunnea* (Selys).

cell traversed once or twice, short, about one-third the length of median space; arc slightly bent; sectors of arc arising from middle of arc and slightly separated at origin; usually 3 cubital nervures to all wings; about 4 intercalated sectors between *IA* and hinder border of wing, *IA* never forked; 4 or more sectors between *IA* and *Cuii*; origin of *Riii* usually very slightly distal to subnode or in continuation with it; no basal incomplete antenodal nervure in subcostal space; pterostigma present in all wings of both sexes, long, narrow.

Thorax robust but short; abdomen extending beyond tips of wings in male, sometimes markedly so, but of the same length as wings in female; superior anal appendages homogeneous, simple, forcipate, longer than segment 10; the latter rounded or arched apically and with a prominent keel or carinal spine on its mid-dorsum; vulvar scale robust, short, not extending to end of abdomen.

Genotype, *Euphæa ochracea* Selys.

Distribution.—Indo-Malay and Indo-China, ASSAM, and BURMA.

208. **Allophæa ochracea** (Selys).

Euphæa ochracea Selys, Bull. Acad. Belg. (2), vol. vii, p. 443 (1859); id., Ann. Mus. Civ. Genova, (2) vol. x (xxx), pp. 489–490 (1891); Laidlaw, Proc. Zool. Soc. Lond. (i) p. 87 (1902); Martin, Mission Pavie, Neurop. (sep.), p. 15 (1904); Williamson, Proc. U.S. Nat. Mus. vol. xxviii, pp. 181–182 (1905).

Pseudophæa ochracea Kirby, Cat. Odon. p. 109 (1890); Laidlaw, Rec. Ind. Mus. vol. xiii, pp. 32–33 (1917); Fraser, J. Bombay Nat. Hist. Soc. vol. xxix, p. 477 (1923).

Allophæa ochracea Fraser, J. Bombay Nat. Hist. Soc. vol. xxxiii, pp. 288, 289 (1929).

Male.—Abdomen 33–35 mm. Fore-wing 33–35 mm. Hind-wing 30–33 mm.

Head black, unmarked; genæ, bases of mandibles, and labrum glossy black; the rest mat black. *Prothorax* mat black, posterior border of posterior lobe narrowly, a small spot below it on each side, and a large semilunar lateral spot bright ochreous. *Thorax* mat black, marked with a series of four pairs of stripes on each side bright ochreous, first pair antehumeral, confluent above and nearly so below; second lying between humeral and first lateral sutures, confluent above, posterior member of the pair confluent with anterior member of third pair; third and fourth pairs lying between lateral sutures and on metepimeron respectively, much broader and more diffuse than the two anterior pairs, and covering greater part of mesepimeron and metepimeron; trochanters and a small spot on each side in antealar sinus ochreous. *Legs* black, unmarked. *Wings* hyaline, broadly saffron or amber-tinted, fore pair for rather more than their basal halves, hind pair as far as pterostigma, which is black, a little oblique at its proximal end and covering about 6 cells; discoidal cell of fore-wing traversed once, that of hind-wing once or twice; 3 cubital nervures in all wings; nodal index—about 30 antenodal nervures and 35 postnodals in fore-wings, about 26 antenodals and 35 postnodals in hind-wings. Hind-wing slightly broader and shorter than fore-wing ; reticulation at apices very fine. *Abdomen* black, marked with ochreous as follows :—A diffuse lateral stripe extending from segment 1, where it is very broad, to 5 or 6, brighter basal rings on 3 to 5 or 6, and an apical ring on 1 ; dorsal carina finely ochreous from 2 to 4 or 6. In subadults the general colouring of abdomen dark ochreous clouded with black ; in adults, segments 6 or 7 to the end black, unmarked. *Anal appendages* black. Superiors slightly longer than segment 10, separated at base, parallel, compressed laterally ; apices blunt, as seen from above, curling slightly in towards one another. Inferiors very short, conical, pointed. *Genitalia* : hamules projecting as two robust conical sharp teeth ; lobe black, large, scrotal-shaped.

Female.—Abdomen 29–30 mm. Hind-wing 28–29 mm.

Differs from the male by the much broader and more extensive yellow markings, more robust build, shorter abdomen, and wings only palely and diffusely saffronated at base, the colouring fading out near the node in both fore- and hind-wings. *Head* black, marked with yellow as follows :—Labium except extreme tips of lobes ; labrum except anterior border finely and a small median virgule springing from base ; mandibles ; cheeks broadly ; a broad stripe across frons and a small oblique oval spot on each side of vertex with inner ends resting on posterior ocelli. *Prothorax* and *thorax* as in male, but markings more extensive and covering mesepimeron and metepimeron except their centres, which are clouded with black. *Legs* brown, femora obscurely yellow on outer side and thinly pulverulent (as is also the underside of thorax and first two segments of abdomen). *Wings* hyaline, coloured as detailed above ; pterostigma bright ochreous between black nervures, pointed inwardly, covering 7 to 8 nervures ; nodal index lower than in male—only about 20 antenodal nervures and 28 postnodals in fore-wings, about 20 antenodals and 25 postnodals in hind-wings ; discoidal cell as in male ; 3 or 4 cubital nervures in hind-wings ; *Riii* arising a little distal to subnode in all wings. *Abdomen* black, marked with yellow as follows :—A broad lateral stripe extending from segment 1 to 7, constricted subapically and finely divided by the jugal suture near base of segments ; mid-dorsal carina finely from segment 1 to 5, obscurely so on 6 and 7 and often on 8, conspicuously so on 9 and less so on 10 ; a small rounded lateral spot on 8, a subquadrate one on each side of the apical two-thirds of 9, and apical border of 10 narrowly. *Anal appendages* black, small, conical, pointed acutely. Vulvar scale extending to end of segment 9, spotted with yellow.

Distribution.—ASSAM, BURMA, Malacca, Siam, and Annam. The *type* in the Selys collection comes from Mt. Ophir,

Malacca; other examples, Khow Sai Dow Mt., 1,000 ft., Trong, Siam, Jan.–Feb. (*Williamson*); Gokteik, Upper Burma, June (*F. Wall*); Shillong, Assam, June (*T. Bainbrigge Fletcher*), and Cachar Assam, August (*Antram*).

There is no difficulty in distinguishing this species, except from *A. brunnea*, by its hyaline wings, saffronated but without opaque areas. It has a wide distribution, extending from Assam to furthest French Indo-China, at altitudes from 1,000 to 5,000 ft. *Larva* unknown.

209. **Allophæa brunnea** (Selys).

Euphœa brunnea Selys, Bull. Acad. Belg. (2) vol. xlvii, p. 374 (1879); id., Ann. Mus. Civ. Genova, (2) vol. x (xxx), p. 490 (1891).

Pseudophœa brunnea Kirby, Cat. Odon. p. 109 (1890); Martin, Mission Pavie, Neurop. (sep.), p. 15 (1904); Laidlaw, Rec. Ind. Mus. vol. xiii, p. 33 (1917); Fraser, J. Bombay Nat. Hist. Soc. vol. xxix, p. 477 (1923).

Allophœa brunnea Fraser, J. Bombay Nat. Hist. Soc. vol. xxxiii, p. 290 (1929).

Male.—Abdomen 32–38 mm. Fore-wing 30–32 mm. Hind-wing 28–31 mm.

It is extremely doubtful whether this species is distinct from *A. ochracea*; Selys himself expresses such a doubt. Markings identical with those of *A. ochracea*, which is also taken in the same hills.

I have not seen the type of *ochracea*, but I have specimens of what probably is that insect from Siam, Burma, and Assam, and apart from size, which varies considerably even in the same locality, I can find no differences to separate them into two distinct species. In addition to indicating a larger size for *A. brunnea*, Selys states that the wings are of a darker brown. Here again we find variability.

In a single specimen which I have from Shillong, Assam, collected by Mr. Bainbrigge Fletcher, the size is greater than in other specimens that I have seen from the same and other localities—abdomen 38 mm. and hind-wing 30 mm., and the saffronated parts of the wings are clouded with dark brown along the costa in the fore-wing and in the outer and posterior part in the hind-wing. Another specimen from Kalaw, Burma, has the same dark colouring, but its size is remarkably small. It must be noted that, at the time the two species were described, they had been reported from Malacca and Assam only, two widely separated areas, but since then *A. ochracea* has been found in Burma. For the present and until more material is available this question must remain open.

Female unknown.

Distribution.—Khasi Hills, ASSAM.

Type in the Selys collection, Brussels Museum.

Genus PSEUDOPHÆA Kirby.

Euphæa Rambur, Ins. Névrop. p. 228 (1842); Selys, Syn. Cal. p. 50 (1853); Walker, List Neur. Ins. Brit. Mus. iv, p. 637 (1853); Selys, Mon. Cal. p. 167 (1854); Williamson, Proc. U.S. Nat. Mus. vol. xxviii, p. 169 (1905).
Pseudophæa Kirby, Cat. Odon. p. 109 (1890); Laidlaw, Rec. Ind. Mus. vol. xiii, p. 32 (1917); Fraser, J. Bombay Nat. Hist. Soc. vol. xxxiii, p. 290 (1929).

Wings of male marked with opaque black, often metallic blue, green, or purple; hyaline in female; apex of fore-wing pointed, that of hind-wing inclining to be rounded; hind-wing markedly broader than fore-wing in male and considerably broader than hind-wing of female; petiolation absent or nearly so; *Rii* not in contact with *R+M*; node situated nearer base of wing than apex and slightly nearer base than pterostigma; discoidal cell traversed once only, short, about one-third as long as median space; arc nearly straight; sectors of arc separated at origin and arising from centre of arc; usually only 2 cubital nervures in all wings; 4 long and numerous short intercalated sectors between *IA* and posterior border of wing; 2 long and 2 short intercalated sectors between *IA* and *Cuii*; origin of *Riii* proximal to subnode in all wings; no basal incomplete antenodal nervures in subcostal space; pterostigma present in all wings of both sexes, long and narrow.

Thorax robust, rather short. Abdomen extending well beyond apices of hind-wings in male, of the same length in female; anal appendages very homogeneous, simple, forcipate, longer than segment 10, latter with a marked mid-dorsal keel or spine, its posterior border arched or ending in the dorsal spine; vulvar scale robust, short, not extending beyond end of abdomen.

Genotype, *Euphæa variegata* Ramb.

Distribution.—CEYLON, WESTERN GHATS, BURMA, Malaya, Indo-China, Java, Sumatra, Borneo, Amboina, and Philippines. (I have cited Western India, as *splendens* has been doubtfully reported from there, although this is most certainly an error.)

Key to Indian Species of Pseudophæa.

Hind-wings of male for basal three-fourths brilliant metallic green or blue............	*splendens* (Selys), [p. 100.
Hind-wings of male opaque blackish-brown, non-metallic	*masoni* (Selys), [p. 102.

210. **Pseudophæa splendens** (Selys).

Euphæa splendens Selys, Syn. Cal. p. 52 (1853); Walker, List Neur. Ins. Brit. Mus. iv, p. 638 (1853); Selys, Mon. Cal. p. 178 (1854); id., Bull. Acad. Belg. (2) vol. xxxv, p. 485 (1873).
Pseudophæa splendens Kirby, Cat. Odon. p. 110 (1890); id., Journ. Linn. Soc. Lond., Zool. vol. xxiv, p. 559 (1893); Laidlaw, Rec. Ind. Mus. vol. xiii, p. 32 (1917); id., Spolia Zeylanica, vol. xii, pp. 356–357 (1924); Fraser, J. Bombay Nat. Hist. Soc. vol. xxxiii, pp. 291, 292 (1929).
Pseudophæa carissima Kirby, l. c. pl. xlii, fig. 4 (1893).

Male.—Abdomen 35–41 mm. Fore-wing 31–36 mm. Hind-wing 28–33 mm.

Head: labium blackish-brown; labrum, cheeks, and clypeus glossy black; rest of head deep velvety mat black. *Prothorax* and *thorax* matt black, the latter with first lateral suture and anterior border of metepimeron obscurely ochreous. In teneral specimens thoracic markings similar to those of female. *Legs* black, femora dark reddish-brown internally. *Wings* opaque black, fore-wings with base as far as midway to node hyaline, but tinted with brown, and with nervures in the outer part of this area bordered and clouded with opaque brownish-black; apices paler from level of proximal end of pterostigma; hind-wings, except for a small basal area anterior to *IA*, opaque to extreme apices and darker than fore-wings on upper surface, an area from base as far distal as half-way between node and apex of wings brilliant metallic green or peacock-blue according to angle at which viewed, outer border of this area running straight from costal to posterior border of wing, but the basal limit not including that hyaline area already mentioned above; pterostigma black, very long, acutely pointed at its proximal end, a little broadened at its centre, covering about 12–14 cells; beneath hind-wing the metallic area has a deep glossy steely blue reflex. Membrane of wings markedly pleated; about 30 antenodal nervures in fore-wings and about 40 postnodals, about 22 antenodals and 30 postnodals in the hind-wings; only 2 cubital nervures in all wings; discoidal cell traversed but once in all wings. *Abdomen* black, with an obscure ochreous lateral stripe on segments 1 and 2. *Anal appendages* black, spatulate, blunt at apex, hollowed out within, parallel, but apices curled very slightly in. Seen from above triangular in outline, with broad base and pointed apex. Inferiors very short, conical, pointed. *Genitalia* very similar to that of *Allophæa ochracea*, but hamules a little less prominent and lobe smaller and flatter.

Female.—Abdomen 31–38 mm. Hind-wing 29–37 mm.

Head: labium yellow, with tips and central portion of middle lobe black, latter area pulverulent white; rest of head black as for male, but with bases of mandibles, cheeks, a small subrotundate spot just in front of each lateral ocellus, and labrum citron-yellow, latter with its anterior border, base, and a median basal triangular tongue black. *Prothorax* with a large lateral boss on each side of middle lobe citron-yellow. *Thorax* with an antehumeral stripe not quite extending up to antealar sinus citron-yellow, a small spot of the same colour on each half of alar sinus. Laterally a thick stripe of yellow on first lateral suture and upper and anterior half of metepimeron. In old specimens these lateral yellow markings and underside of thorax pulverulent white. *Legs* black, flexor surface of femora yellow and often pulverulent. *Wings* hyaline, palely and evenly enfumed, brownish with a greenish tint; in old specimens fore-wings clear and hind-wings only enfumed brown, with the apices for a short distance proximal to inner end of pterostigma dark brown. Extreme apices of fore-wings occasionally also enfumed. Neuration similar to that of male, but occasionally only a single cubital nervure present; nodal index—about 26–30 antenodal nervures and about 30 postnodals; 20–25 antenodals and 25 to 30 postnodals to hind-wings; pterostigma dark brown between black nervures, covering about 10 cells. *Abdomen* black, marked with citron-yellow as follows:—Segment 1 broadly on the sides, segment 2 with a broadish lateral stripe sinuous in its apical half and dilated abruptly at apex, segments 2 to 4 with a narrow lateral stripe which is interrupted at base so as to leave an isolated spot, segment 5 with only the basal spot. *Anal appendages* half as long again as segment 10, conical and very acutely pointed at apex; vulvar scale robust, extending nearly to end of abdomen. (The male with a peculiar tuft of black stiff hairs springing from a small tubercle on each side of the ventral basal end of segment 9, the nature of which is unknown.)

Distribution.—Confined to CEYLON, although Selys gives "India" as one of its localities. I have specimens from Diyatalawa, 5,000 ft., August; Hatton, 4,000–5,000 ft., May; Nalande, September, and Dyraaba, September.

P. splendens, one of the most beautiful dragonflies found in Asia, is common on most of the montane streams of Ceylon, and is to be found flitting slowly up and down stream or perching upon overhanging ferns above the water. When in flight the male keeps the hind-wings fully outspread, using them as planes and the fore-wings as propellers, so that the full beauty of the hind-wing is displayed. The females are to be found in the neighbouring jungle, perched on prominent twigs, from which they launch themselves on passing prey; rarely are they seen over their parent streams, and I have never seen them *in copula* or ovipositing.

Type in the Selys collection; specimens in most European

collections and also in the Pusa, Calcutta, and Bombay Museums. Kirby's type of *P. carrissima*, which is in the British Museum, is only a teneral example of *P. splendens*.

211. **Pseudophæa masoni** (Selys).

Euphœa masoni Selys, Bull. Acad. Belg. (2) vol. xlvii, p. 377 (1879); Laidlaw, Fascic. Malayenses (Zool.), pt. i, p. 194 (1903); Martin, Mission Pavie, Neurop. (sep.), p. 15 (1904); Williamson, Proc. U.S. Nat. Mus. vol. xxviii, p. 182 (1905).

Pseudophœa masoni Kirby, Cat. Odon. p. 110 (1890); id., Ann. Mag. Nat. Hist. (6) vol. xiv, p. 113 (1894); Fraser, J. Bombay Nat. Hist. Soc. vol. xxxiii, pp. 292, 293 (1929).

Male.—Abdomen 28–35 mm. Hind-wing 24–30 mm. Fore-wing 26–31 mm.

Head: labium dark brown; labrum, cheeks, bases of mandibles, and anteclypeus glossy black; rest of head mat velvety black and, in most specimens, an obscure reniform yellowish spot running obliquely out from each posterior ocellus. *Prothorax* and *thorax* velvety black, former unmarked, latter, except in mature specimens (from Upper Burma), with the following yellowish-brown markings more or less obscure:—A narrow antehumeral line, a small spot on each half of alar sinus, a narrow humeral stripe, an equally narrow posthumeral complete in its upper half only; four parallel stripes on the sides, in two pairs, one on mesepimeron and the other on metepimeron. *Legs* black. *Wings* opaque blackish-brown; hind-wing with extreme apex hyaline or occasionally hyaline as far as distal end of pterostigma and with median space and base of subcostal space paler; fore-wings with apex hyaline for a variable distance, as far as inner end of pterostigma or for 5 or more cells proximal to proximal end of pterostigma, base also hyaline but enfumed for rather more than half-way from base to node, rarely nearly as far as node, and in all cases subcostal space opaque as far as base, so that the hyaline area is traversed by a prominent black streak; pterostigma long, narrow, black, covering 8 to 11 cells; discoidal cells traversed once (entire in one wing of a Burmese specimen); normally 2 cubital nervures, but occasionally 3; 25 to 30 antenodal nervures and 30 to 35 postnodals in fore-wing, 20 to 23 antenodals and 24 to 32 postnodals in hind-wing; opaque areas of wings beneath steely metallic blue or bronzed. *Abdomen* black, unmarked. Segment 10 with a very prominent carinal spine. *Anal appendages* black, superiors very similar to those of *P. splendens*, but more constricted at base and more expanded thereafter and with some minute spines at apex, which is curved in slightly. Inferiors very small, conical, ending in a fine point. *Genitalia* very similar to *P. splendens*.

Female.—Abdomen 33 mm. Hind-wing 31 mm.

Differing entirely, as usual, from male and very similar to female of *P. splendens*.

Head: labium dirty brown tipped with black; labrum citron-yellow narrowly encircled with black, and with a prominent black median basal tongue which nearly meets the anterior black border; ante- and postclypeus black; cheeks, bases of mandibles, a very broad band traversing frons, and an elongate spot running obliquely outward from between the ocelli citron-yellow. *Prothorax* black, marked with yellow as follows:—A small subdorsal spot on each side of anterior lobe, a large lateral boss on each side of middle lobe, posterior margin of posterior lobe narrowly and a small longitudinal median spot above this lobe, and a large boss on each side of posterior lobe. *Thorax* black, marked with moderately narrow antehumeral and humeral stripes, a fine posthumeral stripe broadly broken at middle, and lastly the whole of the sides and under surface, except for narrow black stripes outlining the lateral sutures. *Legs* black, proximal two-thirds of posterior femora and inner sides of middle femora yellow. *Wings* long and narrow, uniformly enfumed greenish-brown; pterostigma brown or dark ochreous finely framed in black, long and narrow; *Riii* arising slightly proximal to subnode, other details of venation as given for the male. *Abdomen* black, marked with greenish-yellow as follows:—Segment 1 largely yellow, with a basal dorsal patch and a lateral spot black, segment 2 with its mid-dorsal carina narrowly yellow and with a broad longitudinal lateral stripe expanded apically; segment 3 similar but with dorsal stripe tailing off and finally disappearing before apex, and with lateral stripes expanded at both ends, segments 4 to 7 similar but without dorsal marking, and with the lateral gradually narrowing, segment 8 with a round spot on each side, 9 with a large subquadrate spot on each side, 10 has a vestigial rounded spot only. *Anal appendages* small, pointed, conical, black. Vulvar scale yellow, not quite extending to end of abdomen.

Distribution.—Naga Hills, ASSAM, April; Gokteik, UPPER BURMA, May, also below Maymyo, 2,500 ft., June; Tavoy Dist., LOWER BURMA, in April; Daban, Annam, 600 ft., May; Hoa Minh, Tonkin, and near Bangkok, Siam.

P. masoni is subject to great variation in size like most other species of *Pseudophæa*, but to a less extent in its markings. It is one of the blackest dragonflies known, and must be very conspicuous on the wing. The wing-markings serve to distinguish it from others.

Pseudophæa bocki has been mentioned as occurring in Burma by the late René Martin, but this is almost certainly an error.

Type in the Selys collection from Tenasserim, Lower Burma.

Genus **INDOPHÆA** Fraser (1929). (Fig. 31.)

Euphœa (partim) Rambur, Ins. Névrop. p. 228 (1842); Selys, Syn. Cal. p. 50 (1853); id., Mon. Cal. p. 167 (1854).
Pseudophœa (partim) Kirby, Cat. Odon. p. 109 (1890); Laidlaw, Rec. Ind. Mus. vol. xiii, p. 32 (1917); Fraser, ibid. vol. xxiv, p. 9 (1922).
Indophœa Fraser, J. Bombay Nat. Hist. Soc. vol. xxxiii, pp. 293, 294 (1929).

Hind-wings of male with apices more or less broadly opaque black, nearly up to the node in one species; fore-wings hyaline as also all wings of female; fore-wings with apices pointed, hind-wings rounded, markedly so in some species and considerably shorter than fore-wings; fore- and hind-wings of equal breadth, and wings of equal breadth in the sexes; petiolation distinct, especially in the hind-wings; *Rii* not in contact with *R+M*; node situated much nearer base of wing

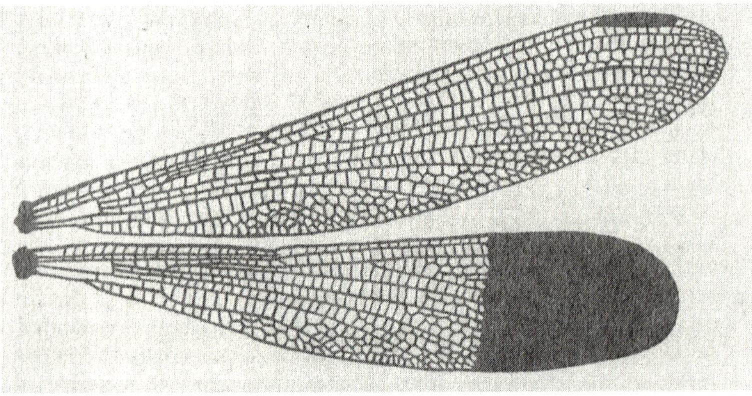

Fig. 31.—Wings of *Indophœa fraseri* (Laid.), male.

than apex, especially in fore-wings; discoidal cell traversed, usually once, but occasionally entire and occasionally three times; 2 to 5 cubital nervures, usually 3; *Riii* arising at, or slightly, or very widely distal to subnode; discoidal cell of fore-wing much shorter than that of hind-wing and less than half the length of median space; arc almost straight; sectors of arc arising from middle of arc and widely separated at origin; 3 or 4 intercalated sectors between *IA* and posterior margin of wing; *IA* never forked; several short intercalated sectors between *IA* and *Cuii*; no basal incomplete antenodal nervures in subcostal space; pterostigma present in all wings, very long and very narrow. Thorax robust. Abdomen cylindrical, very long and attenuated, usually much longer than hind-wings (markedly so in *I. fraseri*, but of nearly the same length in *I. cardinalis*), but of the same length or shorter in the female. Anal appendages very homogeneous, simple, forcipate ; segment 10 pointed apically and with a very robust dorsal keel ; vulvar scale robust, not extending to end of abdomen.

Genotype, *Euphœa dispar* Ramb.

Distribution.—WESTERN GHATS ; Malaya and Borneo.

Key to Indian Species of Indophæa.

1.	Hind-wing conspicuously shorter than fore-wing ; antehumeral stripes on thorax bright azure-blue	*fraseri* (Laid.), p. 110.
	Hind-wing only slightly shorter than fore-wing ; antehumeral stripes on thorax ochreous	2.
2.	Labrum bright ochreous ; legs dull reddish throughout ; nearly the apical half of hind-wing black	*cardinalis* (Fras.), [p. 108.
	Labrum bright azure-blue ; legs bright yellow ; only the apical third of hind-wing black	*dispar* (Ramb.), p. 105.

212. **Indophæa dispar** (Rambur).

Euphœa dispar Rambur, Ins. Névrop. p. 230 (1842); Selys, Syn. Cal. p. 51 (1853); Walker, List Neur. Ins. Brit. Mus. iv, p. 640 (1853); Selys, Mon. Cal. p. 169 (1854); id., Bull. Acad. Belg. (2) vol. xxxvi, p. 614 (1873).

Pseudophœa dispar Kirby, Cat. Odon. p. 109 (1890); Laidlaw, Rec. Ind. Mus. vol. xiii, p. 32 (1917); id., ibid. vol. xix, pp. 25–27 (1920); Fraser, ibid. vol. xxiv, p. 9 (1922); id., ibid. vol. xxvi, pp. 479–480 (1924).

Indophœa dispar Fraser, J. Bombay Nat. Hist. Soc. vol. xxxiii, pp. 294, 295 (1929); id., Rec. Ind. Mus. vol. xxxiii, pp. 448, 463 (1931).

Male.—Abdomen 39–47 mm. Hind-wing 32–40 mm. Fore-wing 35–42 mm.

Head : labium dark reddish-brown, paler at the borders of lateral lobes ; labrum turquoise-blue, finely bordered with black and with a black medio-basal tongue ; bases of mandibles with a spot of turquoise-blue ; cheeks and epistome glossy black, rest of head mat black, unmarked. *Prothorax* black, with a large boss on each side of middle lobe, a smaller reniform spot below it and posterior border of posterior lobe reddish-ochreous. *Thorax* black, marked with bright reddish-ochreous as follows :—Antehumeral and humeral stripes confluent as a broad loop above and narrowly separated below so as nearly to enclose a long oval spot of the ground-colour, the rest, posterior to the humeral suture, which is finely black, bright reddish-ochreous except for an oval spot of black between the humeral and first lateral sutures, and a small spot or beginnings of a stripe on the upper parts of the two lateral sutures. An elongate spot on each half of antealar sinus and the whole of under side of thorax bright ochreous. *Legs* bright yellow except the extensor surface of femora and tibiæ, which are dark reddish. *Wings* hyaline, palely enfumed with greenish-brown; apices of fore-wings merely tipped with blackish-brown; apices of hind-wings broadly black to as far proximal to pterostigma as nearly half-way from apex to node; apex of this wing rounded and only about 3 mm. shorter than fore-wing, but about 7 mm. shorter than abdomen; pterostigma black, covering about 12 cells; 3 cubital nervures to all wings; discoidal cell normally traversed once; *Riii* arising 1 to 2 cells distal to the subnode; about 24 ante-nodal nervures and about 40 postnodals to fore-wings, about 20 antenodals and about 38 postnodals to hind-wings (number differs widely according to size of insect). (Discoidal cell may also be traversed twice or entire; thus in one specimen the cells of the hind-wings are traversed twice, whilst that of the right fore-wing is entire and that of the left traversed once. Occasionally a specimen will be taken with the black apex of hind-wing marked by a large hyaline window.) *Abdomen* bright vermilion-red, segmental joints and entire abdomen from the apical third of segment 6 to the end black. Apex of segment 8 with a tuft of short black hairs on its ventral surface and about 8 long stiff black hairs beneath the base of segment 9; segment 10 with a very prominent dorsal keel. *Anal appendages* very similar to those of *P. masoni*, black, ungulate, laterally compressed and hollowed out within, apices blunt and furnished with a few inconspicuous spines above. Inferiors very small, conical, pointed. *Genitalia* very similar to *P. masoni*, bright ochreous, hamules finely bordered with black, lobe large, scrotal shaped, black.

Female.—Abdomen 35–38 mm. Hind-wing 34–39 mm.

Bears a remarkable likeness to females of *P. splendens*, *P. masoni*, etc.

Head: labium dirty yellow; labrum and bases of mandibles turquoise-blue, former finely bordered with black and with a medio-basal black tongue as in male; anteclypeus black; postclypeus and a broad transverse band across frons, as well as cheeks broadly, bright ochre. A rounded spot of the same colour on outer side of each posterior ocellus. *Prothorax* and *thorax* marked similarly to male, but yellow instead of bright reddish-ochreous. Dorsal stripes narrower, black posthumeral spot confluent with the black below, not entirely surrounded by brighter colour, lateral vestigial black sutural lines complete, although that on the first lateral suture is rather diffuse and often incomplete below. *Legs* as in male but more black. *Wings* hyaline, uniformly enfumed with greenish-brown, and hind-wing in old specimens with a moderately well-marked brownish-black apex extending proximally slightly beyond inner end of pterostigma ; venational details and pterostigma similar to male ; *Riii* not quite one cell distal to subnode ; cubital nervures sometimes irregular, 2 to 4 in number, and discoidal cells sometimes traversed twice ; pterostigma black, over about 12 cells ; nodal index $\frac{32\text{–}18 \quad 17\text{–}34}{30\text{–}18 \quad 19\text{–}31}$. *Abdomen* black, marked with bright yellow or ochreous as follows :—Segment 1 almost entirely greenish-yellow, 2 to 7 with a longitudinal stripe on either side, broad on 2, narrower on the rest and becoming interrupted on 6 and 7, base expanded and cut off from the rest by jugal suture ; segment 8 with a small quadrate apical lateral spot, 9 with a larger similar spot, 10 unmarked. *Anal appendages* small, conical, pointed, black. Vulvar scale robust, marked with yellowish, not extending to end of abdomen.

Variation.—In a specimen from South Kanara the markings are largely obsolete, especially on the abdomen, whilst those on the thorax are cut up into parallel lines of yellow by the black, even on the sides. In other specimens there is an additional small round spot on the inner side of each posterior ocellus, and the posterior border of the posterior lobe of prothorax is bright yellow as in the male. Unlike *P. masoni* and *P. splendens*, there is no dorsal yellow marking on any of the abdominal segments. As in all species of *Pseudophæa* there is a great disparity in size of specimens from various localities, those from lower altitudes usually being of smaller size than those from higher.

Distribution.—Confined to the WESTERN GHATS from South Kanara and Coorg to the Nilgiris (Malabar Wynaad) from 3,500 to 6,000 ft., from May to September.

Easily distinguished from other species by the extent of black on apices of hind-wings, and by its turquoise-blue labrum and mandibles. The black area of wings has steely-blue reflections in some lights, dull coppery-bronze in others. The female is distinguished from other species by its turquoise-blue labrum. Usually the males will be found perched on twigs some feet above the water, often at a great height, especially towards nightfall, when they rise to the tops of neighbouring trees.

Females are not uncommon but must be sought for in the neighbouring jungle or ridings some short distance from the streams. They are very pugnacious, and I have found them devouring newly-hatched specimens of their own species!

Rarely are they found *in copula*, but I have a pair which not even death in the cyanide bottle was able to part.

Type now, I believe, in the Selys collection. Specimens in the British Museum, Pusa, and Indian Museums, and in most private collections, to which I have been able to present specimens.

213. **Indophæa cardinalis** (Fraser). (Fig. 32.)

Pseudophæa cardinalis Fraser, Rec. Ind. Mus. vol. xxvi, pp. 512–513 (1924).

Indophæa cardinalis Fraser, J. Bombay Nat. Hist. Soc. vol. xxxiii, pp. 295, 296 (1929); id., Rec. Ind. Mus. vol. xxxiii, pp. 448, 463 (1931).

Male.—Abdomen 41–45 mm. Fore-wing 39–42 mm. Hind-wing 36–40 mm.

Head : labium dirty yellow, brown at its middle ; labrum bright ochreous, narrowly bordered with reddish-brown, and an obscure mediobasal tongue of dark brown ; anteclypeus dark blackish-brown ; postclypeus, bases of mandibles, and cheeks bright ochreous ; frons broadly reddish-ochreous, clouded with reddish-brown at its middle and with a crenulate black basal line from which spring medial and lateral short black points ; four small black points also projecting into base of postclypeus ; rest of head black save for basal joints of antennæ, and a small round point lying slightly to outer and fore side of posterior ocellus on each side bright ochreous. Eyes dark reddish-brown ; frons coated with long black hairs. *Prothorax* black, with a large boss on each side of middle lobe, posterior border of posterior lobe save at its middle, and the sides broadly bright ochreous. *Thorax* bright reddish-ochreous and black, markings very similar to those of *I. dispar*, thus—antehumeral and humeral narrow ochreous stripes confluent as a loop above and nearly confluent below ; humeral suture narrowly outlined in black ; laterally entirely ochreous save for a long black oval spot between humeral and first lateral sutures and the beginnings of narrow black lines on upper parts of lateral sutures ; beneath ochreous. *Legs* entirely reddish, tarsi dark reddish-brown, spines black. *Wings* relatively broader than in *I. dispar* and marked very similarly. Fore-wings hyaline, faintly enfumed and with a greenish tinge, tinted with yellow at extreme base ; hind-wings with apices black as far proximally as 4 to 8 mm. from node, and in some quite the outer half of wing opaque black, this part dull coppery above, dull or steely bluish-black below ; pterostigma long, covers 10–12 cells, black ; *Riii* arising from half to one and a half cells distal to subnode or even in continuation of subnode ; discoidal cell entire in all wings of some specimens, or traversed once in fore-wings or twice or thrice in hind-wings, very variable; 2 cubital nervures in fore-wings, 2 to 4 in hind-wings; 20 to 24 antenodal nervures and 38 to 47 postnodals in fore-wing, about 18 to 20 antenodals and 38 to 40 postnodals in hind-wing. Hind-wing 2 to 3 mm. shorter than fore-wing and markedly rounded. *Abdomen* bright vermilion-red as far as the basal two-thirds of segment 6, from which point it is black. Segment 10 with a very pronounced carinal spine; 8 and 9 with tufts of long ventral hairs. *Anal appendages* similar to those of *I. dispar*, as also genitalia.

Female.—Abdomen 36 mm. Hind-wing 37 mm.

Head: labium black; labrum coloured similarly to male; cheeks, bases of mandibles, a broad fascia traversing frons but slightly interrupted at its middle, and a small oval spot on outer fore side of posterior ocellus bright ochreous; rest of head black. *Prothorax* and *thorax* bright yellow, marked

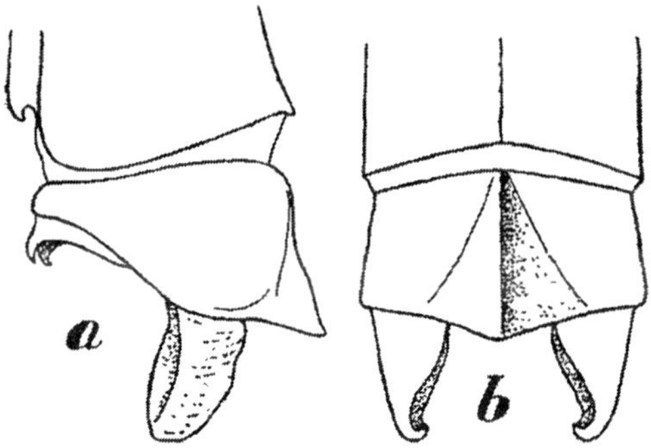

Fig. 32.—Anal appendages of *Indophœa cardinalis* (Fras.), male. *a*, left lateral view; *b*, dorsal view.

with black as in male, but lateral stripes on sutures complete. *Legs* blackish-brown, flexor surfaces of femora obscurely yellow. *Wings* hyaline, uniformly enfumed with pale greenish-brown, hind-wings more deeply than fore-wings and apices slightly clouded with darker brown; pterostigma black, narrow; 21 to 24 antenodal nervures and 31 to 33 postnodals in fore-wings, 18 antenodals and about 26 postnodals in hind-wings; discoidal cell traversed once in all wings; other venational points as in male. *Abdomen* black, marked with yellow, sides of segments 1 to 3 broadly yellow; segments 4 to 7 with a longitudinal lateral stripe expanding basally and extending up towards dorsum so as to form incomplete basal rings; 8 and 9 with subquadrate apico-lateral spots, 10 unmarked. *Anal appendages* short, conical, black. Vulvar scale very robust, yellowish, not extending to end of abdomen.

Distribution.—Confined, so far as known, to the WESTERN GHATS, South India, and to south of the Palghat Gap. Found in numerous sholas, frequenting montane streams of the Palni, Anaimalai and Mudis Hills and Travancore from April to October.

The much greater extent of the opaque area of the hind-wing, the ochreous labrum and cheeks (turquoise-blue and glossy black respectively in *dispar*), the close approximation of the length of wings to the abdomen, and the all-red legs are some of the characters distinguishing *I. cardinalis* from *I. dispar*. In addition to these, *I. cardinalis* has a large triangular tongue-like process springing from the apico-ventral border of the second abdominal segment, which is quite absent in *I. dispar*. The female is easily distinguished by its ochreous labrum.

Type in the British Museum; paratypes in the Pusa, Ris, Morton, and Fraser collections.

214. Indophæa fraseri (Laidlaw).

Pseudophæa fraseri Laidlaw, Rec. Ind. Mus. vol. xxi, pp. 23–27 (1920); Fraser, ibid. vol. xxiv, pp. 8, 9 (1922); id., ibid. vol. xxvi, p. 480 (1924).

Indophæa fraseri Fraser, J. Bombay Nat. Hist. Soc. vol. xxxiii, pp. 296–298, pl. i, figs. *c* & *e* (1929); id., Rec. Ind. Mus. vol. xxxiii, pp. 448, 463 (1931).

Male.—Abdomen 36–41 mm. Fore-wing 34–38 mm. Hind-wing 29–35 mm.

Head: labium pale yellow; labrum pale azure-blue, with its anterior border broadly black, no medio-basal black tongue; bases of mandibles azure-blue, cheeks yellowish-white; epistome glossy black; rest of head mat black, with occasionally an obscure oval yellow spot on outer side of each posterior ocellus; eyes dark brown. *Prothorax* black, with a large pale blue spot on each side of middle lobe. *Thorax* black; antehumeral and humeral stripes on dorsum, former pale sky-blue, in fine contrast to the black ground-colour, narrow in its upper half, broadening rapidly in its lower; humeral stripe yellow, turning to reddish-ochre below, very fine throughout and occasionally broken into several sections. Laterally bright ochreous with reddish tinge posteriorly and marked with a large oval black spot between humeral and first lateral suture, and with the beginning of fine lateral stripes on upper parts of the two lateral sutures; beneath bright reddish-ochreous. *Legs* as in *I. cardinalis*, but brighter red, anterior pair dark reddish-brown or almost black. *Wings* very similar to *I. dispar*, but hind-wing very markedly shorter than fore-wing and with evenly rounded apex; fore-wing hyaline, extreme apex tipped with brown; hind-wing with outer part opaque black, with coppery reflection above and bluish-violet below, this area slightly variable, usually extending from apex to about half-way to node, less extensive in specimens taken at a low altitude, more extensive in those from a higher; discoidal cell traversed once in fore-wing, twice or thrice in hind-wing; 6 cubital nervures in all wings; *Riii* arising 4 to 5 cells distal to subnode; pterostigma black, long and narrow, covers 8 to 12 cells: 18 to 20 antenodal nervures in fore-wing and 32 to 36 postnodals, 15 to 18 antenodals and 27 to 30 postnodals in hind-wing. *Abdomen* bright vermilion-red to apex of segment 6, which is clouded with black; segment 7 dark reddish-brown to black at apex; rest of abdomen black, segment 10 with a prominent carinal spine. *Anal appendages* black, very similar to those of *I. dispar*, but with a distinct bend at about their middle; inferiors as in *I. dispar*. *Genitalia* similar to *I. dispar*, but lobe smaller and bright red instead of mat black.

Female.—Abdomen 33–34 mm. Hind-wing 31–33 mm.

Head: labium, labrum, mandibles, and cheeks as for male: ante- and post-clypeus glossy black, latter with a transversely oval blue spot at its centre; frons black, the creamy white of cheeks extending on to it on either side; rest of head black, with a small oval greenish-yellow spot on outer side of each posterior ocellus. *Prothorax* black, with a large oval spot on each side of middle lobe and posterior border of posterior lobe finely greenish-yellow. *Thorax* black on dorsum, bright yellow on the sides, marked as in male but the line on posterior suture rather better defined. In some specimens the ante-humeral and humeral stripes are confluent above as a broad loop, as in other species of the genus. *Legs* yellow, femora blackish on extensor surface, tibiæ reddish. *Wings* hyaline, palely enfumed, apices in many specimens broadly dark brown to slightly proximal to outer end of pterostigma in fore-wings and for a short distance proximal to inner end in hind-wings; venational details as for male; discoidal cell traversed once in fore-wings, twice in hind-wings or less commonly once; 3 cubital nervures in all wings; pterostigma black, long, and narrow, covering 9 to 12 cells; 17 to 18 antenodal nervures and 29 to 33 postnodals in fore-wings, 15 to 16 antenodals and 25 to 27 postnodals in hind-wings. *Abdomen* black, marked with bright greenish-yellow as follows:—Segment 1 broadly so on sides, dorsal carina throughout except on segment 1 narrowly, but broadening out on 8 to 10, on 10 forming a well-defined spot; sides of 2 broadly; a longitudinal stripe on sides of 3 to 6, broad on 3, progressively finer on 4 and 5, nearly lost or interrupted on 6, apical and basal spots on sides of 7, a small apical lateral spot on 8, a very large lateral spot on 9; sides of 10 entirely. *Anal appendages* rather longer than segment 10, black, fine, tapering to a fine point. Vulvar scale robust, yellow, not extending to end of abdomen.

Distribution.—S. INDIA: North and South Kanara, Malabar, Coorg, the Nilgiris Wynaad, and Anaimalai Hills. Found on the same rivers as *I. dispar*, but at a lower elevation; thus in Malabar it occurs sparingly near sea-level. In Coorg and the Nilgiris and Malabar Wynaad it occurs up to 3,500 ft. from May to August.

The males are not uncommonly seen resting with their wings well open as in genus *Lestes*, and are usually found on low herbage along the banks of their parent streams. Females are not uncommon in the neighbouring jungle, settled on twigs at about 8 to 12 feet from the ground.

Type in the Indian Museum; other specimens in Pusa and British Museums and in several private collections.

Subfamily PHILOGANGINÆ Kennedy (1920). (Fig. 33.)

Philoganginæ Kennedy, Ohio Journ. Sci. vol. xxi, no. 1, p. 22, figs. 38–39 (1920); Fraser, J. Bombay Nat. Hist. Soc. vol. xxxiii, pp. 298, 299, pl. i, figs. *a* & *d* (1929).

Robust insects with general facies of both sexes resembling somewhat that of Epallagine females, but details of venation etc. differing widely from that subfamily.

Fore- and hind-wings of similar shape and similar in both sexes, very long and very narrow, petiolated to nearly as far as the level of arc; node at about two-fifths of the wing-length from base; discoidal cell entire, short, about one-fifth to one-sixth the length of the median space, its costal side slightly shorter than the posterior, its distal end oblique; *Ri* not in contact with *Rii*; *Riii* arising at or 1 to 2 cells distal to subnode; arc slightly bent, situated at and in line with the distal primary antenodal nervure; antenodal nervures moderately numerous, those in the subcostal space more numerous than those in the costal and, except for the two primary antenodals, not coinciding with them; 1 to 4 basal incomplete or subcostal antenodals, usually 2 or 3, and always a single subcostal antenodal between the two primaries; no cubital nervures beyond the nervure *ac* in all wings, *ac* lying much nearer the distal primary antenodal or midway between the primaries; petiolation marked, ending at a point opposite to or slightly proximal to *ac*; *IA* straight, slightly concave or markedly convex, 1 to 2 rows of cells between it and posterior margin of wing, ending on wing-margin opposite to or widely distal to node; 1 to 2 well-defined oblique nervures between *Rii* and *IRii*; intercalated nervures between *Riv+v* and *IRiii*, *IRiii* and *Riii*, *Riii* and

Rii; pterostigma present in all wings of both sexes, long and narrow.

Head robust, Gomphine-shaped; eyes rounded, tumid behind, rather widely separated from one another; labium with middle lobe deeply cleft, ends of lobes acute; labrum arched at free border.

Thorax very robust, short; legs long and slim; femora with two rows of very short, very closely set, evenly sized spines, with more robust ones set at longer but even intervals; tibial spines moderately numerous but rather short; claw-hooks situated near ends of claws. (Tibiæ in one species with a fringe of hairs in addition to the spines.)

Abdomen robust, cylindrical, slightly dilated at anal end, especially in the female, shorter than wings; segment 10 flat on dorsum.

Superior anal appendages of male considerably longer than segment 10, subcylindrical, widely separated at base, apices curving in toward one another, blunt, minutely spined at outer border; inferior appendages rudimentary. Superior anal appendages of female long and fine. Genitalia bearing

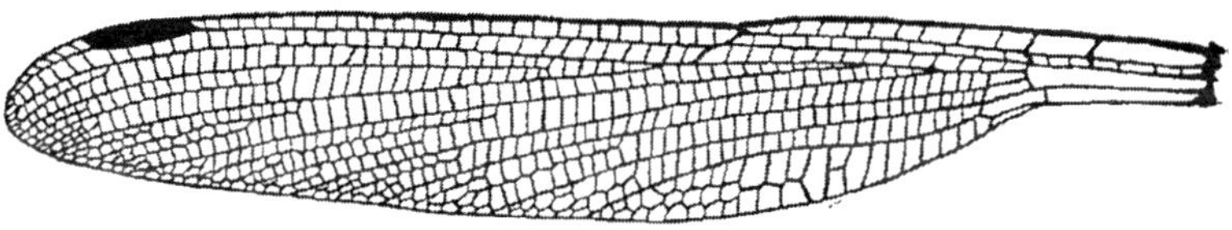

Fig. 33.—Hind-wing of *Philoganga montana* (Selys), male.

a close resemblance to that of the EPALLAGINÆ, especially the anterior hamules and lobe of penis, the former being foliate flattened quadrate processes inclined toward one another, the latter scrotal-shaped and rather longer than in the Epallagines; penis closely resembling that of *Amphipteryx*, its tentacles furnished with a fringe of spines at their apex, which is blunt. Vulvar scales very robust, extending well beyond end of abdomen and with the under border of scales coarsely serrate and evidently functioning as a saw for the insertion of ova into plant-stems.

Distribution.—ASSAM, BENGAL, BURMA, Indo-China, and South China.

I have included the three known species of *Philoganga* under a separate monogeneric subfamily as their true position is still doubtful and must remain so until the larva is discovered. From the venation, the extremely long petiolation, and the fact that they rest with the wings extended flat as in the ANISOPTERA, it is clear that they are archaic insects, probably even more so than *Epiophlebia*. I place them for convenience after the EPALLAGINÆ because there is a strong resemblance in the genitalia and the abdominal markings are typical of that subfamily. There, however, the resemblance ends, as the thoracic pattern, so characteristic of the EPALLAGINÆ, is quite different from that of *Philoganga*, whilst the coinciding costal and subcostal antenodals and the short petiolation of the wings of the EPALLAGINÆ is quite different from what is found in *Philoganga*. The long petiolation of the wings, the long legs, and the shape of the penis seems to show a relationship to *Amphipteryx*, an American genus.

Mr. Bainbrigge Fletcher has taken a number of *P. montana* along the banks of a montane stream in the Khasi Hills, so that it is evident that they breed in such spots. He has taken at least one teneral specimen here, but unfortunately failed to find its exuvia.

Genus **PHILOGANGA** Kirby (1890).

Anisoneura Selys, Bull. Acad. Belg. (2) vol. vii, p. 444 (1859); id., ibid. (2) vol. xlvii, p. 379 (1879).
Philoganga Kirby, Cat. Odon. p. 111 (1890); Needham, Proc. U.S. Nat. Mus. vol. xxvi, p. 755, fig. 44 (1903); Ris, Suppl. Ent. no. 1, pp. 44–47 (1912); Laidlaw, Rec. Ind. Mus. vol. xiii, p. 33 (1917); Fraser, J. Bombay Nat. Hist. Soc. vol. xxxiii, p. 299 (1929).

Characters and distribution as for the subfamily. Of the three known species, two occur within Indian limits.

Genotype, *Anisoneura montana* Selys.

Key to Indian Species of Philoganga.

Labrum glossy black; hind-wing not less than 45 mm. in length . . . . *montana* (Selys), p. 114.
Labrum greenish-yellow; hind-wing less than 40 mm. in length . . . . *loringæ* Fras., p. 116.

215. **Philoganga montana** (Selys). (Figs. 33 & 34.)

Anisoneura montana Selys, Bull. Acad. Belg. (2) vol. vii, p. 445 (1859); id., ibid. (2) vol. xlvii, p. 379 (1879); Kirby, Cat. Odon. p. 111 (1890).
Philoganga montana Kirby, l. c. p. 111 (1890); Laidlaw, Rec. Ind. Mus. vol. xiii, p. 33 (1917); Fraser, Mem. Dept. Agric. India (Ent.), vol. viii, no. 8, p. 87, pl. ix, figs. 2–4 (1924); id., J. Bombay Nat. Hist. Soc. vol. xxxiii, pp. 299, 300, pl. i, figs. *a* & *d* (1929).

Male.—Abdomen 48 mm. Hind-wing 45 mm.

Head: labium yellow, lobes tipped with black, bases of mandibles citron-yellow; labrum glossy black; rest of head mat black, with two transverse narrow citron-yellow stripes, one traversing cheeks and frons, the other running from eye to eye across occiput; behind eyes yellow; eyes brown.

Prothorax black, with a longitudinal mid-dorsal stripe bisected narrowly in the middle line of middle lobe and finely interrupted between middle and posterior lobes. *Thorax* black with some pruinescence beneath in adults, marked with greenish and citron-yellow as follows :—A narrow mid-dorsal carinal stripe finely bisected by the mid-dorsal black carina, a small spot on each half of antealar sinus, a narrow complete humeral stripe, and laterally two broad oblique stripes, the anterior borders of which are greenish, the first stripe on mesepimeron, the second covering entire metepimeron. *Legs* black, femora broadly yellow on outer aspect, trochanters each with a large yellow spot ; tibiæ of male with a fringe of fine short hairs in addition to the spines. *Wings* hyaline ; pterostigma blackish-brown or paler brown, surmounts $2\frac{1}{2}$ to $4\frac{1}{2}$ cells ; 2 rows of

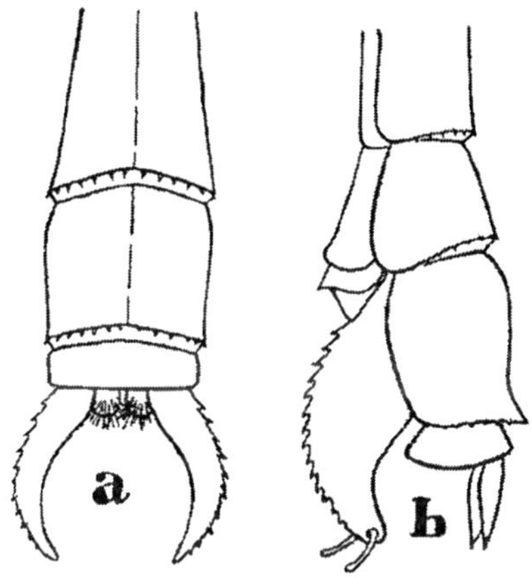

Fig. 34.—a. Anal appendages of *Philoganga montana* Selys, male.
b. The same and genitalia of the female.

cells posterior to *IA*, this nervure being flat or slightly convex ; petiolation ends at *ac* or a little proximal to it in fore-wings ; 2 to 4 incomplete basal antenodals in fore-wings, usually 2, 2 to 3 in hind-wing ; nodal index variable, $\frac{26\text{–}12/18}{24\text{–}11/17}\left|\frac{14/17\text{–}26}{14/17\text{–}24}\right.$; $\frac{21\text{–}10/15}{17\text{–}10/17}\left|\frac{10/17\text{–}19}{10/16\text{–}17}\right.$. *Abdomen* black, marked with greenish-yellow as follows :—Segment 1 broadly yellow on sides and with an apical band broadly interrupted on dorsum ; 2 with a lateral and ventral longitudinal stripe ; 3 to 7 each with a latero-basal transverse spot and a lateral stripe, latter on 3 and 4 with the apical end expanded, but on 5 to 7 tapering gradually away until much shortened on segment 7 ; 8 with a large triangular latero-apical spot, 9 with a similar but rounded spot, 10 with apical half yellow. *Anal appendages* black. Superiors nearly twice as long as segment 10, subcylindrical, slender, curving gradually and evenly towards one another, apex obtuse, the outer border near apex coarsely spined. Inferiors rudimentary, scarcely visible.

Female.—Abdomen 47 mm. Hind-wing 52 mm.

Differs only from male in size and robust build. Wings occasionally pale yellow towards base; nodal index similar but very variable; only 1 or 2 incomplete basal antenodal nervures to all wings; pterostigma as in male. *Anal appendages* long, tapering to a fine point, black. Segments 8 and 9 distinctly broadened, almost foliate laterally, 9 rather depressed. Vulvar scale very robust, prolonged well beyond end of abdomen, coarsely serrate beneath.

Distribution.—ASSAM and BENGAL. Mr. T. Bainbrigge Fletcher has taken this species in moderate numbers during May and June in two restricted localities bordering montane streams at Shillong. These were resting on bushes with their wings spread horizontally, in which position, from their shape and colouring, they looked very like Gomphines. Mr. Chas. Inglis has taken a male on the banks of a stream below Darjeeling. One male in the author's collection from Nowgong, Assam.

Type in the McLachlan collection from Assam, probably collected by Mr. Atkinson in the same localities at Shillong.

216. **Philoganga loringæ** Fraser.

Philoganga loringæ Fraser, Rec. Ind. Mus. vol. xxix, pp. 79–81 (1927); id., J. Bombay Nat. Hist. Soc. vol. xxxiii, pp. 300, 301 (1929).

Male.—Abdomen 41 mm. Hind-wing 39 mm.

Head: labium dirty yellow; labrum greenish-yellow with a small medio-basal black tongue; bases of mandibles and cheeks citron-yellow as far up as level of antennæ; rest of head mat black, pruinescent in parts; eyes dark brown. *Prothorax* black, with a mid-dorsal citron-yellow stripe broadening anteriorly and on posterior lobe. *Thorax* black, marked with citron-yellow as follows:—A moderately broad mid-dorsal stripe finely bisected by the black mid-dorsal carina, a narrow slightly sinuous antehumeral stripe; laterally entirely yellow save for the second lateral suture, which is broadly mapped out in black. *Legs* long and slim, hind femora extending to middle of segment 2; anterior pair of femora black, other two pairs dark ochreous; tibiæ and tarsi black; tibiæ not fringed with fine hairs as in *P. montana*. *Wings* hyaline, petiolated to level of *ac* or slightly proximal in fore-wing; palely and evenly enfumed; pterostigma blackish-brown, covering 3½ cells; only a single row of cells posterior to *IA*; only 1 cubital nervure to all wings; *Cuii* slightly convex; *IA* a little concave; other details of venation as in *P. montana* except that the primary antenodals are not as distinct from the others and *Riii* is more distal in its origin, arising from 1 to 2½ cells distal to the subnode; nodal index $\frac{22\text{–}13/17 \mid 13/19\text{–}21}{20\text{–}13/17 \mid 11/18\text{–}19}$. *Abdomen* dark reddish-brown; segment 1 greenish-yellow; 2 with a broad lateral bright yellow stripe narrowly bordered above with black; 3 with the black stripe continued but more diffuse and blotting out the ground-colour on dorsum of segment, ventral border dark ochreous; 4 to 10 similar but dorsum entirely black, 9 with a duplicate mid-dorsal bright ochreous spot and 10 with two similar but rounded spots on dorsum. *Anal appendages* black. Superiors nearly twice the length of segment 10, curling gradually in almost to meet at tips, which are slightly dilated and end in obtuse points; on outer side a few fine spines, much smaller than those seen in *P. montana*; in profile projecting straight back, but the apices are slightly upturned. Inferiors rudimentary, as in *P. montana*. *Genitalia* very similar to those of *P. montana*. Lobe depressed, moderately long, glossy black, resembling the flattened tumid body of a tick.

Female.—Abdomen 42 mm. Hind-wing 37 mm.

Very similar in markings to male, but a more robust insect. *Wings* petiolated distinctly proximal to *ac*; 2 basal incomplete antenodals in all wings, or occasionally only 1 (2 to 3 in the male); a well-defined oblique nervure between *Rii* and *IRii*, sometimes two such (usually not very evident in the male); pterostigma rather longer, covering 4½ cells; nodal index $\frac{22\text{–}13/17 \mid 13/19\text{–}20}{19\text{–}12/16 \mid 11/15\text{–}18}$. *Abdomen* similar to male, but sides of segments 9 and 10 broadly ochreous and dorsal spots replaced by diffuse dark ochreous. *Anal appendages* black at tips, brownish-yellow from base, short, tapering to a point. Vulvar scale exactly similar to *P. montana*.

Distribution.—Maymyo, Upper Burma. Four specimens, three males and a single female, the latter *in copula*, July 1925, collected by Col. F. Wall, I.M.S.

This very rare insect differs from *P. montana* in its much smaller size, in having the labrum yellow instead of glossy black, in the markings of abdomen, and in the point of origin of *Riii*. It and *P. montana* are at once distinguished from *P. vetusta* in having only a single row of cells posterior to *IA*. The female has the abdomen rather longer than the wings, the opposite condition being found in *P. montana*.

Type in the British Museum.

Subfamily AGRIINÆ.

Agrioninæ (pars) Kirby, Cat. Odon. p. 96 (1890).

Head very broad; eyes globular, subpetiolate in some species, and widely separated from one another; frons sloping, broader than long, or quadrate: occiput broad, depressed; ante- and postclypeus not elevated or projecting; antennæ four-jointed, segment 1 rudimentary or hidden beneath 2, which is the most robust and longest of all and usually applied flat to head, 3 and 4 much more slender and usually shorter; labium deeply fissured for its distal third or half. Prothorax robust, its posterior lobe large, tumid, triangular or crenate; middle lobe with a prominent boss on each side. Thorax slender, flattened from side to side, elongate; mesothoracic triangle absent or very small. Legs long and slender, often very long, posterior pair extending to middle of segment 3 or apex of 4 in male, or apex of 5 in female; femora and tibiæ with long, fine, closely set, bristle-like hairs; claw-hooks short, situated near end of claws. Wings hyaline, opaque or partly so, coloured or uncoloured, often of a brilliant metallic colour or iridescent in males, usually hyaline and uncoloured in females, hind-wing usually rather broader than fore-wing and considerably shorter than abdomen in both sexes, base narrow but not petiolated; reticulation very close, cells mostly rectangular in shape; node situate about middle of wing or nearer to base than to pterostigma; membrane of wing often markedly pleated; *Rii* contiguous or nearly so with radius near its origin; basal space reticulated, traversed or entire, equal to about half the length of cubital space; discoidal cell with costal border slightly convex or straight, about equal in length to basal space, very narrow, traversed by several nervures, its ends square or a little oblique; arc straight or markedly angulate, its sectors arising from a common point at its middle or below that level; *Cuii* markedly convex; *IA* markedly convex and often branching proximally; intercalated sectors between most sectors, including *Cuii* and *IA*; most sectors curving strongly towards posterior margin of wing and, in some genera, pectinated near their distal ends; ante- and postnodal nervures numerous, primaries not usually distinct from the others, and costal and subcostal antenodals usually coinciding; pterostigma present or absent, often rudimentary in one or both sexes, very variable in shape and colour, sometimes false in character, especially in females. Abdomen very long, narrow and cylindrical, end segments occasionally a little broadened and depressed; segment 10 rarely keeled and then only towards the apex. Anal appendages very similar in all genera of the subfamily, superiors rather longer than segment 10, subcylindrical, spined outwardly, curved forcipate-like towards each other, apices slightly flattened and often slightly broadened ; inferiors about two-thirds the length of superiors, broad at base, tapering to a point, straight, conical. Genitalia very homogeneous ; penile lobe flattened, elongate like the body of a tick ; lamina deeply cleft ; hamules consisting of a quadrate low-lying outer plate and an inner blunt spine.

Distribution.—Throughout the world in temperate and tropical zones. Represented within Indian limits by six genera.

Key to Indian Genera of Agriinæ.

1.	Arc angulated; basal space traversed or entire; sectors of arc separated at origin; main sectors not forked	2.
	Arc oblique, not angulated; basal space entire; sectors of arc arising from a single point; most main sectors forked; pterostigma absent	VESTALIS Selys, p. 124.
2.	Basal space traversed	3.
	Basal space entire; pterostigma present .	MNAIS Selys, p. 139.
3.	Pterostigma present....................	4.
	Pterostigma absent, or if present then abnormal and traversed by nervures ..	5.
4.	Pterostigma differing in the two sexes, node situated nearer pterostigma than base of wing	CLIMACOBASIS Laidlaw, [p. 137.
	Pterostigma similar in the two sexes, white or black in colour and broad and rather short; node situated nearer base of wing than pterostigma	ECHO Selys, p. 134.
5.	All wings of both sexes opaque black; pterostigma in female white, traversed by nervures, in male absent	MATRONA Selys, p. 144.
	Fore-wings of both sexes hyaline, hind-wings opaque or hyaline; pterostigma absent in male, false or absent in female	NEUROBASIS Selys, [p. 119.

Genus **NEUROBASIS** Selys (1853). (Fig. 35.)

Neurobasis Selys, Syn. Cal. p. 17 (1853); Walker, List Neur. Ins. Brit. Mus. iv, p. 602 (1853); Selys, Mon. Cal. p. 72 (1854); Kirby, Cat. Odon. p. 102 (1890); Williamson, Proc. U.S. Nat. Mus. vol. xxxiii. p. 170 (1905); Munz, Mem. Amer. Ent. Soc. no. 3, p. 44 (1919); Fraser, J. Bombay Nat. Hist. Soc. vol. xxxiii, pp. 577, 578 (1929).

Fore-wings of male hyaline, hind-wings opaque and coloured partly with brilliant metallic green and blue ; all wings of female hyaline but with an opaque whitish spot at node ; pterostigma absent in male, false and whitish in female ; basal space travered by several nervures ; *Rii* near its origin almost confluent with radius; *Riii* usually arising slightly proximal to subnode; sectors of arc arising from near middle of arc, which is distinctly angulated; anal area of wings reticulated, but no distinct proximal branch running from *IA*.

Larvæ very elongate and narrow, superficially resembling Phasmidæ; head rather small, eyes globular; antennæ seven-jointed, second very robust and longer than all the rest taken together; thorax long and narrow, wing-cases leaf-like, directed straight back and closely apposed to the body; abdomen cylindrical, long, narrow, tapering slightly towards the anal end; caudal gills three in number, triquetral in shape, the dorsal one shorter than the latero-ventral ones, which are very long and oar-like; legs long and spidery, slim; mask with middle lobe deeply cleft, setæ present on lateral lobes and branches of middle lobe, some robust spines on lateral lobes, one of which forms the movable hook; mask extending as far back as origin of hind legs.

Genotype, *Libellula chinensis* Linn.

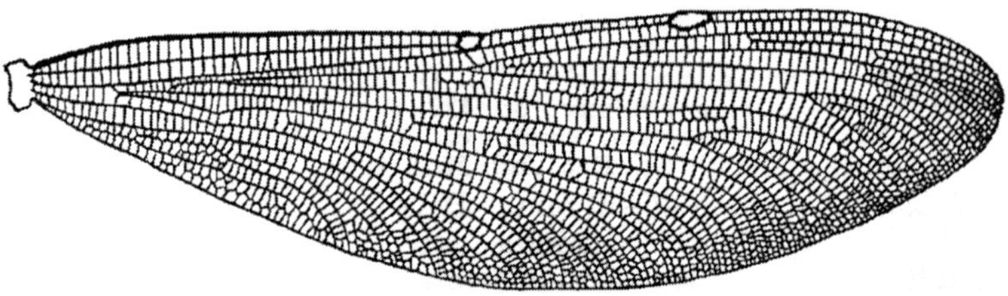

Fig. 35.—Hind-wing of *Neurobasis chinensis chinensis* (Linn.), female.

Distribution.—Southern Asia and South China, Philippines, Ceylon, Java, Borneo, New Guinea, and Australasia. The genus contains two species, of which only one, *N. chinensis*, is found within Indian limits, with several races or subspecies in the Philippines and the islands of Southern Asia. The second species, *N. kaupi*, I regard as doubtfully belonging to the genus.

Both sexes are found along the banks of streams, and rarely if ever stray into the neighbouring jungles. Males are commonly seen flitting up and down stream, hugging the surface of the water so closely that, when passing over disturbed water, such as a rapid, they are seen to rise and fall, following the undulations and ripples of the water's surface. In this act only the fore-wings of the insect are used to propel it, the hind being used as sustaining planes, outspread and motionless like the wings of an aeroplane, and flashing with a brilliant play of emerald-green and peacock-blue as the plane of the wings is altered to sustain the insect's balance. They rest on overhanging ferns and herbage beside the stream, or more commonly perch with closed wings on a rock in mid-stream. The female has somewhat similar habits, but is not given to planing over water. Mating is rarely seen; ovipositing takes place when the insects are *in copula*, and this usually in a swift current, the insects clinging to some object such as the rootlets of a tree, and descending far under water, where they are swept from side to side with every eddy of the current, the female meanwhile inserting her eggs in the substance of the root.

217. **Neurobasis chinensis chinensis** (Linn.). (Figs. 35 & 36.)

Libellula chinensis Linnæus, Syst. Nat. vol. i, p. 545, n. 15 (1758); Edwards, Nat. Hist. Birds, vol. iii, t. 112 (1750); Donovan, Ins. China, t. 46, f. i, I (1798).
Agrion nobilitata Fabricius, Gen. Ins. p. 248 (1776).
Agrion chinensis Guérin, Icon. R. Anim., Ins. p. 382, t. 60, f. 4 (1829–44).
Calopteryx chinensis Rambur, Ins. Névrop. p. 226 (1842).
Calopteryx disparilis id., ibid. p. 224 (1842).
Calopteryx sinensis Walker, List Neur. Ins. Brit. Mus. iv, p. 602 (1853).
Neurobasis chinensis Kirby, Cat. Odon. p. 102 (1890); Fraser, Rec. Ind. Mus. vol. xxvi, pp. 428, 479 (1924); id., J. Bombay Nat. Hist. Soc. vol. xxxiii, pp. 577, 578 (1929) (with full bibliography); id., Rec. Ind. Mus. vol. xxxiii. p. 448 (1931).

Male.—Abdomen 45–50 mm. Hind-wing 32–38 mm.

Head: labium with middle lobe whitish, lateral lobes metallic green; labrum turquoise-blue, with a large triangular medio-basal black spot, with its apex directed towards the anterior border and sometimes meeting it; cheeks and bases of mandibles palest blue; anteclypeus with a small median yellow spot, the remainder and postclypeus glossy metallic green; a large spot of yellow on each side of postclypeus; rest of head metallic green with a coppery reflex on occiput and vertex; antennæ with basal and second joints pale blue; upper two-thirds of eyes blackish-brown, lower third bluish-green, the two areas sharply defined. *Prothorax* bronzy-green with a coppery reflex; posterior lobe truncate, its sides bluish-white. *Thorax* brilliant metallic green; humeral and antero-lateral stripes blackish-brown, former diffusely so and clouded with bronzy brown anteriorly, latter finely, postero-lateral suture white, bordered with black, which is confluent in places, cutting up the white into spots; beneath white, barred with black; tergum brown, with metallic spots at bases of wings and two coral-white spots. *Legs* very long and slim, posterior pair extending to apical border of segment 4; femora dark bronze, white on flexor surface, changing to brown distally; tibiæ white, black on flexor surface; tarsi black. *Wings* moderately rounded at apices, especially hind-wings;

fore-wings considerably longer than hind-wings, hyaline tinted with pale yellowish-green, especially along costa and at apex, neuration brilliant emerald-green, especially the costa and main nervures, node thickened and narrowly clouded with brown; hind-wings opaque, basal two-thirds appearing brilliant metallic green or peacock-blue according to angle of view, apical third blackish-brown with violaceous reflections and green metallic nervures. The change from metallic green to black near the apex abrupt and nearly in a straight line from costa to posterior margin of wing; basal space and some adjacent cells and some of the basal cells of costal space hyaline. Beneath, fore-wings similar to above, hind-wings nearly uniformly blackish-brown with dull coppery or golden

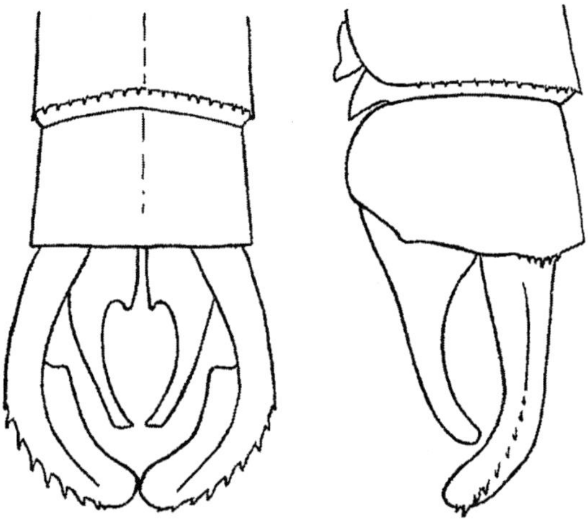

Fig. 36.—Anal appendages of *Neurobasis chinensis chinensis* (Linn.), male. Dorsal and left lateral views.

reflections, neuration finely green metallic; 6 to 9, usually 7, nervures in basal space; discoidal cell traversed 7 to 13 times in fore-wing, 11 to 13 in hind-wing, nodal index $\frac{65\text{–}44}{75\text{–}40}\Big|\frac{41\text{–}63}{40\text{–}72}$; $\frac{59\text{–}35}{70\text{–}35}\Big|\frac{42\text{–}60}{40\text{–}68}$; pterostigma absent in all wings. *Abdomen* narrow and cylindrical, much longer than wings, metallic bronzy-green above and at sides, intersegmental joints brighter emerald-green, finely bordered with black; beneath black, segments 9 and 10 whitish. *Anal appendages* black, superiors with basal half dull metallic green, inferiors white at base. Superiors slightly longer than segment 10, basal half subcylindrical, apical half broadening and flattened and furnished along outer border with 4 to 6 spines; from above they appear to curve gradually and evenly inwards, their apices almost meeting: seen from the side broad at base, tapering to a blunt apex. Inferiors one-fourth shorter than superiors, nearly straight, moderately separated, very broad at base, then narrowing abruptly at junction of basal and middle thirds, ending in a blunt apex which is furnished with a small inwardly directed spine.

Female.—Abdomen 44–50 mm. Hind-wing 36–40 mm.

Differs in many respects from male, and therefore liable to be mistaken for a different species.

Head: labium pale yellow, white at base; labrum, cheeks, bases of mandibles, lower third of eyes, and two basal segments of antennae greenish-yellow, labrum with a small medio-basal black mark; anteclypeus pale; postclypeus metallic green, with a small oval pale yellow spot on each side; upper two-thirds of eyes brownish-black; rest of head brilliant metallic green. *Prothorax* and *thorax* as in male, but humeral and lateral sutures finely white, with black borders, confluent in parts so as to cut up the white into spots; postero-lateral suture very broadly white, this colour completely framing the metepimeron enclosing an elongate triangle of metallic green which surmounts a large triangular patch of yellowish-white below. Thorax beneath greyish-white, unmarked. *Legs* as in male, but the pale parts creamy yellow. *Wings* tinted with yellow, palely enfumed with brown, especially at apices and along costa from base to node in fore-wings, and generally deeper in tint throughout the whole of hind-wings, which have occasionally a diffuse denser patch nearly traversing the wings just proximal to pterostigma. All wings with an opaque creamy yellow patch at node which usually covers one cell distal to node; a creamy white pterostigma in hind-wings, reduced or entirely absent in fore-wings, covering 7 to 14 cells in hind-wings, the nervures traversing it often incomplete and often missing in places, covering 1 to several cells in fore-wing when present. (In specimens from the Malabar Wynaad it is usually entirely absent in all wings, in those from Coorg it is very small and traversed by only 3–4 nervures and there is no widening of the space between the costa and *Ri* in the pterostigma as is usual in other species; Nilgiri specimens are variable, a vestigial pterostigma being present on one or more wings; in Siamese specimens the pterostigma is better developed and there is a definite divergence of the costa from the radius, also in some specimens the pterostigma is entire, or at the most with a single traversing nervure at each end.) Other details of neuration similar to the male. *Abdomen* dull metallic bronzy green, with golden reflections on dorsum, black beneath but thinly pruinosed. All segments with a stripe on each side, rather obscure on segments 5 to 7, broad and conspicuous on 9 and 10, expanding at apex of each segment as a large spot, which on 5 to 7 is usually the only vestige of the stripe present; the stripe bordered above and below with black. All segments with a transverse subapical black marking; 8 to 10 with the dorsal carina yellow, this stripe broadening progressively as far as the apex of segment 10, which ends in a distinct keel and a fine apical spine; laterally segment 10 has a small tubercle on each side surmounted by some minute teeth. *Anal appendages* stout, conical, pointed, rather shorter than segment 10, dark brown. Vulvar scale greenish or olivaceous yellow, extending to apex of segment 10.

Distribution.—Throughout India except in desert areas from sea-level up to about 7,500 ft., but usually found at 3,000 to 4,000 ft.

It breeds in montane and submontane streams, the larva clinging to roots and submerged water-plants. I have examined specimens from Ceylon, South Kanara, Coorg, Malabar, Palni Hills, Deccan, Bengal, Assam, Burma, Siam, and Annam, and find remarkably little variation save in size and in the pterostigma, as already commented on above; definite local races are unknown within Indian limits. Specimens from Coorg and the Nilgiri Wynaad have the pterostigma absent in the fore-wings of females, but well developed in the hind-wing. Some of the Coorg females have the hind-wings of a remarkably deep tint of burnt brown and the dark preapical fascia intensely developed. Females from the Palni Hills, 6,000 ft., have no pterostigma in fore-wings, but a single female from Ootacamund, 7,500 ft., has it well developed in all wings.

Although widely distributed the species shows a predilection for certain streams; thus its local distribution takes the form of widely scattered colonies.

The *types* of *Libellula chinensis* and *Agrion nobilitata* have apparently been lost; that of *Calopteryx disparilis* is in the Paris Museum, whilst specimens of *Neurobasis chinensis* are found in all national and most private collections.

Genus **VESTALIS** Selys (1853). (Fig. 37.)

Vestalis Selys, Syn. Cal. p. 24 (1853); Walker, List Neur. Ins. Brit. Mus. iv, p. 610 (1853); Selys, Mon. Cal. p. 79 (1854); Kirby, Cat. Odon. p. 102 (1890); Laidlaw, Rec. Ind. Mus. vol. xiii, p. 29 (1917); Munz, Mem. Amer. Ent. Soc. no. 3, p. 44 (1919); Fraser, J. Bombay Nat. Hist. Soc. vol. xxxiii, pp. 580, 581 (1929).

Vestinus Kennedy, Ohio Journ. Sci. vol. xxi, no. 2, p. 83 (1920).

Wings of both sexes rounded at apex, entirely hyaline or partly opaque or metallic; pterostigma absent in both sexes; basal space entire; *Rii* confluent near its origin and for some distance with the radius; *Riii* usually arising at or slightly proximal to subnode; sectors of arc arising from lower part of arc; discoidal cell equal in length to basal space, traversed by several nervures; arc not angulated; anal area simple, *IA* not sending off any proximal branch; *IA*, *Cuii*, *Riv+v* and *Riii* branched and pectinate at the distal ends, *MA* unbranched. Legs long and thin. Abdomen cylindrical, slim and of great length. Ground-colour of head, thorax, and abdomen metallic green or blue.

Larva very similar to that of *Neurobasis*.

Genotype, *Calopteryx luctuosa* Burm.

Distribution.—Throughout the Oriental Region, Sunda Archipelago, and Philippines.

Breeds in montane and submontane streams, gregarious. Large colonies are found inhabiting the rides of open spaces in forests, and along some shaded pathways almost every twig will be found to have its occupant. When æstivating or

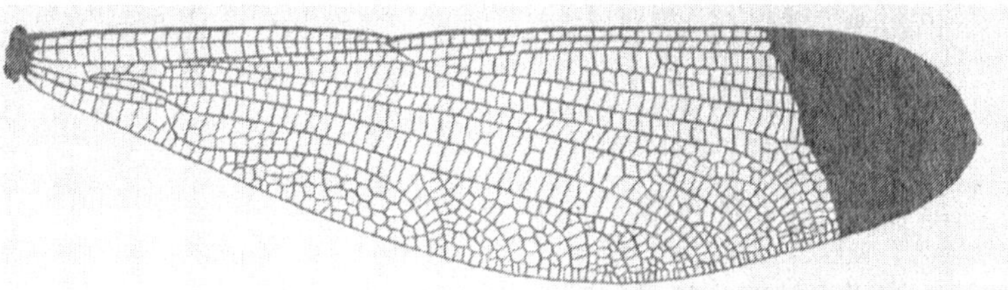

Fig. 37.—Wing of *Vestalis apicalis apicalis* Selys, male.

feeding up it will spread far inland; thus I have found it on the top of wooded hills, and it is not uncommon on Malabar Hill, Bombay. It is as much a woodland insect as *Neurobasis* is a riverine one, but appears to resort to streams for mating, as courtship is never witnessed in the jungle. Unlike *Neurobasis*, the female oviposits in blades of grass or juicy stems overhanging a stream, often several feet above the water's surface, the newly hatched larvæ dropping from thence into the water. (Similar habits have been noticed for *Lestes* and *Tetrathemis*.) By its habits and general facies the genus is closely related to the African *Phaon*, which has a similar scattered distribution throughout Africa and dominates the AGRIIDÆ there as does *Vestalis* throughout the Oriental Region.

Key to Indian Species of Vestalis.

1.	Tips of wings black	2.
	Tips of wings hyaline	3. [p. 131.
2.	Labrum entirely black. (Ceylon.)	*nigrescens* Fraser,
	Labrum yellow, marked with black. (India.)	*apicalis* Selys, p. 128.

3.	Wings tinted uniformly with golden yellow	*smaragdina* Selys, [p. 133.
	Wings untinted or but partially so	4.
4.	Two rows of cells between origins of *Cuii* and *IA*	*gracilis* (Ramb.), p. 126.
	Only a single row of cells between *Cuii* and *IA*	*amœna* Selys, p. 132.

218. **Vestalis gracilis gracilis** (Ramb.).

Calopteryx gracilis Rambur, Ins. Névrop. p. 224 (1842); Walker, List Neur. Ins. Brit. Mus. iv, p. 611 (1853).

Vestalis gracilis Selys, Syn. Cal. p. 26 (1853); id., Mon. Cal. p. 84 (1854); Kirby, Cat. Odon. p. 102 (1890); Selys, Ann. Mus. Civ. Genova, (2) vol. x (xxx), p. 487 (1891); Martin, Mission Pavie, Neurop. (sep.) p. 15 (1904); Williamson, Proc. U.S. Nat. Mus. vol. xxviii, p. 183, fig. 15 (1905); Laidlaw, Rec. Ind. Mus. vol. viii, p. 340 (1914); id., ibid. vol. xiii, p. 30 (1917); Munz, Mem. Amer. Ent. Soc. no. 3, pl. iv, fig. 18 (1919); Fraser, J. Bombay Nat. Hist. Soc. vol. xxix, p. 476 (1923); id., Rec. Ind. Mus. vol. xxvi, p. 479 (1924); id., J. Bombay Nat. Hist. Soc. vol. xxxiii, pp. 581, 582 (1929); id., Rec. Ind. Mus. vol. xxxiii, p. 448 (1931).

Male.—Abdomen 45–56 mm. Hind-wing 34–38 mm.

Head: labium yellow; labrum, bases of mandibles, cheeks, anteclypeus, two basal segments of antennæ, and an oblique oval spot on each side of frons bright yellow, labrum with a medio-basal triangle and occasionally an ill-defined basal bar black; the yellow spots on frons often obscure or even absent; basal joints of antennæ occasionally dark brown; rest of head brilliant metallic green, sometimes with golden reflections. Upper two-thirds of eyes dark brown, the rest greenish-yellow. *Prothorax* metallic emerald green on dorsum and sides, yellowish on lower parts of sides and beneath. Posterior lobe truncate, bordered finely with yellow. *Thorax* brilliant metallic emerald-green, mid-dorsal carina finely black, humeral and antero-lateral sutures finely yellow, metepimeron yellow-ochre except for a narrow elongate metallic green triangle at its centre. Under surface and ventro-lateral borders yellow-ochre, unmarked. *Legs* pale to dark brown, flexor surfaces of tibiæ and extensor surfaces of femora paler or yellow. In some specimens the legs are dark reddish or blackish-brown. *Wings* hyaline, iridescent with colours of mother-of-pearl or blue in some lights, especially the central parts of fore-wings, tinted variably with greenish-yellow as follows:—Whole of hind-wings uniformly, base of fore-wings as far as distal end of discoidal cell, whole length of costal margin between costa and nervure *Rii*, and extreme apex of wing. This tinting very variable, quite absent in teneral specimens, always more intense in mature specimens and in examples from Assam, Burma, and Siam, some of which, from the latter country, have the apices enfumed as well as tinted, although not to the same extent as in *V. apicalis*. Discoidal cell traversed 4–6 times in fore-wing, 3–5 in hind-wing, the more eastern forms seeming to be the most closely reticulated; nodal indices of three specimens chosen at random, $\frac{65–25}{53–24}\Big|\frac{26–65}{24–54}$; $\frac{66–28}{52–26}\Big|\frac{30–67}{24–57}$; $\frac{62–30}{50–24}\Big|\frac{26–54}{25–48}$. *Abdomen* metallic green or blue, usually peacock-blue in teneral specimens, emerald-green when mature, with a narrow interrupted basal yellow ring on segments 2 to 6, and sides of 1, 2 and base of 3 yellowish. Beneath black. Segment 10 with a robust keel at its apical end which ends in an apical spine, and with a more or less robust spine on the apical border on each side. The great length of the abdomen as compared with the wings, mentioned by Rambur, Selys, and other authors, is not constant, as shown below :—

	Abdomen.	Hind-wing.
Nilgiris	46–53 mm.	35–38 mm.
Burma	45 mm.	36 mm.
Siam	45 mm.	35 mm.
Bengal	56 mm.	38 mm.
Eastern Ghats....	50 mm.	34 mm.

Anal appendages black. Superiors rather longer than segment 10, widely separated, broad at base, then subcylindrical; apex flattened and bevelled, terminating in a sharp spine; a blunt dorsal conical spine at extreme base and some coarse spines along outer border, two or three longer than the rest; seen from above these appendages are curved evenly inwards almost to meet at apex. Inferiors widely separated, about two-thirds the length of superiors, conical, tapering to an obtuse rounded apex.

Female.—Abdomen 43–50 mm. Hind-wing 36–39 mm.

Exactly similar to male in colour and markings, but abdomen usually more dully metallic and segments 8–10 yellow laterally. *Anal appendages* and ovipositor dark brown, former conical, pointed, slightly shorter than segment 10, which is keeled and spined as in male.

Distribution.—That of the genus. After examining many scores of specimens from the WESTERN and EASTERN GHATS, BENGAL, ASSAM, UPPER and LOWER BURMA, Siam, Annam, and Tonkin, I am unable to find differences amounting to racial or varietal value, unless it be that specimens from the last three localities have the wings more deeply tinted. It is this tinting which serves to separate *V. gracilis* from *V. apicalis*, some specimens of the former occasionally having the apices of the wings enfumed, and so being liable to be confused with the latter.

V. gracilis and *V. apicalis* are frequently found in company, and taking pruinescence as a measure of full adulthood, we find specimens heavily pruinosed in which there is either no sign of apical darkening of the wings, or at the most *a poorly defined shadow* of such. These specimens are the true *V. gracilis*; similarly pruinosed specimens of *apicalis* found in their company have the apices of wings deep blackish-brown and *very sharply defined.*

Type in the Paris Museum.

219. **Vestalis gracilis montana,** nom. nov.

Vestalis gracilis amœna Fraser, J. Bombay Nat. Hist. Soc. vol. xxxiii, p. 583 (1929).

Male and *female.*—Measurements, colouring, and markings exactly similar to *V. gracilis gracilis*, but the venation corresponding to that of *V. amœna* in that there is only a single row of cells separating *IA* and *Cuii.* The venation is decidedly closer than in *V. amœna*, and there are 3 to 4 rows of cells between *IA* and the posterior margin of wing as in true *V. gracilis.* On the whole this subspecies appears to be rather smaller than *V. gracilis gracilis.*

Distribution.—S. INDIA: COORG and the Nilgiri Wynaad, at about 3,500 ft. altitude.

Type in the Fraser collection.

220. **Vestalis apicalis apicalis** Selys. (Figs. 37 & 38.)

Vestalis apicalis Selys, Bull. Acad. Belg. (2) vol. xxxvi, p. 612 (1873); id., ibid. (2) vol. xlvii, p. 362 (1879); Kirby, Cat. Odon. p. 102 (1890); id., J. Linn. Soc. (Zool.) vol. xxiv, pp. 558–559 (1893); Laidlaw, Rec. Ind. Mus. vol. xiii, p. 30 (1917); Fraser, J. Siam. Soc. vol. iv, p. 164 (1921); id., Rec. Ind. Mus. vol. xxvi, p. 479 (1924); Laidlaw, Spolia Zeylanica, vol. xii, parts 47 & 48, pp. 355, 356 (1924); Fraser, J. Bombay Nat. Hist. Soc. vol. xxxiii, pp. 583, 584 (1929); id., Rec. Ind. Mus. vol. xxxiii, p. 448 (1931).

Neurobasis apicalis Kirby, Proc. Zool. Soc. Lond. p. 204, pl. xx, fig. 2 (1891).

Male.—Abdomen 49–55 mm. Hind-wing 36–39 mm.

Head: labium, labrum, cheeks, bases of mandibles, and basal joints of antennæ coloured as in *V. gracilis*, rest of head metallic emerald-green; upper two-thirds of eyes brown, the rest olivaceous or yellow. Antennal basal joints often brownish-black inwardly; labrum sometimes with a broad quadrate basal spot or broad blackish-brown line. *Prothorax* and *thorax* coloured similarly to *V. gracilis*, but posterior lobe of former more conspicuously bordered with yellow. *Legs* blackish-brown, flexor surfaces of femora, extensor surfaces of tibiæ, and coxæ yellow. *Wings* hyaline, mature specimens sometimes tinted as in *V. gracilis*, but not as deeply; apices of all wings broadly tipped with blackish-brown for about the distal 5 mm.; venational details very similar to those of *V. gracilis*; discoidal cell traversed 4 to 5 times in fore-wing, 2 to 4 times in hind-wing; 10 cubital nervures in fore-wing, 8 to 9 in hind-wing; nodal index $\frac{66\text{–}30 \mid 30\text{–}65}{61\text{–}27 \mid 26\text{–}68}$; $\frac{68\text{–}24 \mid 27\text{–}69}{54\text{–}24 \mid 24\text{–}55}$. *Abdomen* metallic emerald-green, marked with yellow as in *V. gracilis* on segments 1 to 3; intersegmental nodes black; obsolete subapical black transverse spots on 1 to 4; beneath black, often more or less pruinosed. *Anal appendages* black. Superiors gradually curving inwards from base to apex, coarsely spined along outer border, broad at base, subcylindrical thereafter, but broadening

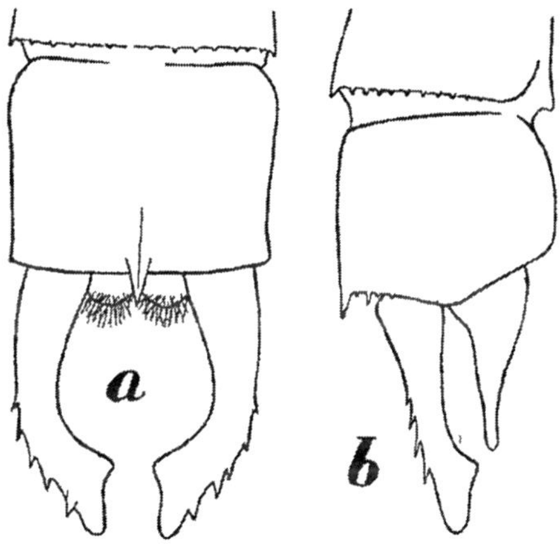

Fig. 38.—Anal appendages of *Vestalis apicalis apicalis* Selys, male. *a*, from above; *b*, from the right side.

at end and shallowly bifid at apex, which is much broadened and bevelled inwardly, the inner end bearing a sharp spine, the outer truncate and blunt. Inferior appendages about two-thirds the length of superiors, cylindrical, but broad at base, apices blunt.

Female.—Abdomen 46–50 mm. Hind-wing 38–40 mm.

Closely similar to male; labrum usually with a basal line in continuation with the medio-basal black spot; often an oblique paler area on each side of frons similar to that often seen in *V. gracilis*, and often a small pale area on outer side of each lateral ocellus. *Thorax* with humeral and lateral sutures yellow, former and antero-lateral narrowly bordered with brown. *Wings* similar to male, but the apical marking usually paler and less sharply defined. *Abdomen* more coppery and less metallic than in male; segments 2 to 5 with narrow basal interrupted yellow rings or paired sub-basal spots; ventro-lateral borders of 1 to 4 or 5 yellow; 10 with a well-marked dorsal keel ending in a sharp spine, and with a smaller spine on each side. Ovipositor as in *V. gracilis*.

Distribution.—That of *V. gracilis*.

Nearly always found in company with *V. gracilis*, and has exactly similar habits. Distinguished from other species by the black apices of wings.

Type in the Selys collection, Brussels Museum.

221. **Vestalis apicalis submontana,** nom. nov.

Vestalis apicalis amœna Fraser, J. Bombay Nat. Hist. Soc. vol. xxxiii, p. 584 (1929).

Male and female closely similar to *V. apicalis apicalis* and bearing the same relation to it as does *V. gracilis montana* to *V. gracilis gracilis*.

Differs from *V. apicalis* by the cheeks all glossy black and by the black apex of wings much restricted, occupying only about 2·5 mm. (in some specimens, usually females, this marking is not sharply defined). Venation resembling that of *V. amœna*, there being only a single row of cells between *Cuii* and *IA*; usually 3 to 4 rows of cells between *IA* and wing-border for a short distance. In some specimens the dorsal keel and spine are absent on segment 10, but the lateral spine is always present. In a specimen from the Nilgiris the labrum is bordered with black, the cheeks are entirely black; the thorax is a beautiful metallic emerald with black sutures, even the upper part of the postero-lateral suture being of this colour; the abdomen is very dull metallic, the end-segments being mat black. In a specimen from Anantagiri, Eastern Ghats, the colouring is coppery or golden bronzed green, the head being coloured and marked as in the Nilgiri form; the thorax with the sutures black except the postero-lateral one; the abdomen is more brightly metallic; the size is much smaller (abdomen 47 mm., hind-wing 34 mm.); the apices of all wings are deep black. These two forms may represent distinct races.

Distribution.—NILGIRIS and EASTERN GHATS.

Easily distinguished by the combination of apical black marking to all wings and a single row of cells between *Cuii* and *IA*, and the cheeks glossy black.

Type in the Author's collection.

222. **Vestalis nigrescens** Fraser.

Vestalis nigrescens Fraser, J. Bombay Nat. Hist. Soc. vol. xxxiii, pp. 584, 585 (1929).

Male.—Abdomen 46–50 mm. Hind-wing 35–37 mm.

Head: labium brownish-yellow, pruinosed white laterally and at base; labrum and epistome blackish-brown; cheeks, bases of mandibles, and basal segments of antennæ black; rest of head dark metallic green with blue reflections; eyes brownish-black above, paler below. *Prothorax* metallic green, borders of posterior lobe black. *Thorax* dark metallic green, dorsum with peacock-blue reflections, mid-dorsal carina finely black, humeral suture rather broadly outlined in black, antero-lateral suture finely black, postero-lateral suture cinereous bordered with black; beneath white, with a posterior black spot which may be obscured by pruinescence. *Legs* entirely black. *Wings* hyaline; apices of all, for nearly one-third of the distance from apex to node, deep black; nodal index similar to that of *V. apicalis*: 4 to 6 transverse nervures in the discoidal cell; 2 rows of cells between *Cuii* and *IA* at origins; *Riii* arising at or slightly proximal or distal to sub-node. *Abdomen* dark metallic green on segments 1 to 5, rest mat black, intersegmental nodes blackish-brown, base of segment 1 pruinosed white; 10 with a robust apical keel and spine on dorsum, and a lateral spine on each side. *Anal appendages* black; superiors and inferiors showing no marked differences from those of *V. apicalis*, lateral spines finer, apex of superior appendages more truncate, inferior appendages more sinuous and their apex with a fine point on the inner side.

Female.—Abdomen 43 mm. Hind-wing 38 mm.

Differs from male by the labium whitish, labrum yellow with a medio-basal black triangular spot; cheeks, except against the eyes, bases of mandibles and basal segments of antennæ yellow. *Thorax* very similar to that of male, but antero-lateral suture narrowly and postero-lateral suture broadly yellow. Beneath, yellow with a black triangular spot as in male. *Legs* entirely black. *Wings* hyaline, apices of all wings enfumed, this tinting vignetted off proximally. Details of venation similar to male. *Abdomen* with the metallic colouring confined to segments 1 and 2 and base of 3; 10 with a robust keel on dorsum, and lateral apical spines. *Anal appendages* short, conical, pointed, black; ovipositor dark yellow.

Distribution.—This beautiful species is confined to CEYLON, Nalande, 11. ix. 24, Kandy, 1. ix. 24, at 2,000 ft. (*F. Wall*), Belihul Oya in May–June, 1932 (*Fraser*). Laidlaw reports *V. apicalis* from Ceylon, but has probably confused this species with it. He gives the following localities and dates for his *V. apicalis* :—" Kandy, July, November ; Ratnapura, October ; Haragama, July ; Peradeniya, December 29."

The species is easily distinguished from others of the genus by its dark colouring, which gives it a close superficial resemblance to *Echo margarita*. The black labrum, antennæ, and legs, and the non-metallic abdomen, serve to distinguish it from *V. apicalis*.

Type in the British Museum.

223. **Vestalis amœna** Selys.

Vestalis amœna Selys, Syn. Cal. p. 25 (1853) ; id., Mon. Cal. p. 82 (1854) ; id., Bull. Acad. Belg. (2) vol. xxxv, p. 475 (1873) ; Kirby, Cat. Odon. p. 103 (1890) ; Karsch, Ent. Nachr. vol. xvii, no. 16, p. 243 (1891) ; Krüger, Stett. Ent. Zeit. p. 75 (1898) ; Laidlaw, Proc. Zool. Soc. Lond. p. 87 (1902) ; Williamson, Proc. U.S. Nat. Mus. vol. xxviii, p. 183 (1905) ; Laidlaw, Proc. Zool. Soc. Lond. pp. 30–31 (1915) ; id., ibid. p. 326 (1920).

Calopteryx amœna Walker, List Neur. Ins. Brit. Mus. iv, p. 611 (1853).

Vestalis amœna amœna Fraser, J. Bombay Nat. Hist. Soc. vol. xxxiii, p. 585 (1929).

Male.—Abdomen 38–52 mm. Hind-wing 31–38 mm.

Head : labium yellow, clouded with black in mature specimens ; labrum black, with a small yellow spot on each side at base, or in immature adults yellow, heavily bordered with black and with a median basal black mark ; bases of mandibles and base of second joint of antennæ yellow ; rest of head a beautiful dark metallic green ; eyes blackish-brown above, paler below. *Prothorax* metallic emerald-green, posterior border of posterior lobe finely yellow. *Thorax* metallic emerald-green, with humeral and first lateral sutures black ; posterior suture and posterior border of metepimeron yellow. Beneath, pale yellow spotted with black. *Legs* dark brown to black, posterior femora paler or brown towards proximal end. *Wings* hyaline, often with a pale yellowish-green tinge and iridescent viewed obliquely in a good light ; only a single row of cells between the proximal ends of *Cuii* and *IA*, and only 2 rows of cells between the latter and posterior margin of wing ; distal ends of main sectors curved down at a greater angle towards posterior margin of wing ; discoidal cell traversed by 2 to 3 nervures in fore-wing, by 3 to 4 in hind-wing ; 9 to 10 cubital nervures in all wings ; nodal index $\frac{55\text{–}25}{45\text{–}24}\Big|\frac{27\text{–}53}{25\text{–}42}$. *Abdomen* metallic green, passing to mat black after segment 4 or 5, segment 10 with a dorsal keel ending in a minute spine and with a small spine on each side of apical border as in *V. gracilis*. *Anal appendages* black. Superiors rather longer than segment 10, curving gradually towards one another and almost meeting at apices, subcylindrical at base but compressed at apex and bifid, so as to form two small branches, the upper, or outer, the longer, the lower, or inner, short and blunt, spined coarsely on the outer side. Inferiors about two-thirds the length of superiors, widely separated at base, subcylindrical, nearly straight, apex blunt, base tumid.

Female.—Abdomen 35–40 mm. Hind-wing 34–39 mm.

Similar to male in markings, colouring, and venation; nodal index $\frac{45\text{–}25 \mid 24\text{–}47}{35\text{–}18 \mid 20\text{–}41}$; 8 to 10 cubital nervures; discoidal cell traversed 2 to 3 times, usually twice in all wings. Borders and apices of wings in mature specimens enfumed with brown.

Distribution.—The *type* comes from Java, but the species appears to be widespread, as I have seen specimens from LOWER BURMA, Sumatra, Borneo, and Siam.

The markings vary considerably according to the age of specimens, especially the labrum and antennæ, which may be quite black when mature. The ground-colour is peacock-blue in teneral specimens, and changes gradually through brilliant emerald-green to coppery or golden-bronze. The colour of the labrum and venation will serve to distinguish it from other species.

Type in the Selys collection, Brussels Museum.

224. **Vestalis smaragdina** Selys.

Vestalis smaragdina Selys, Bull. Acad. Belg. (2) vol. xlvii, p. 362 (1879); Kirby, Cat. Odon. p. 103 (1890); Selys, Ann. Mus. Civ. Genova, (2) vol. x (xxx), p. 488 (1891); Williamson, Proc. U.S. Nat. Mus. vol. xxviii, p. 183 (1905); Ris (subsp. *velata*), Suppl. Ent. no. 1, p. 56, t. iv, fig. 2 (1912); Laidlaw, Rec. Ind. Mus. vol. xiii, pp. 29 & 30 (1917); Fraser, J. Bombay Nat. Hist. Soc. vol. xxxiii, p. 586 (1929).

Male.—Abdomen 42–45 mm. Hind-wing 32–35 mm.

Head: labium black; whole of rest of head, including labrum, bases of mandibles, cheeks, and basal segments of antennæ, brilliant metallic green or peacock-blue; eyes brown above, lower third yellowish-green. *Prothorax* metallic emerald-green; posterior lobe large, rounded. *Thorax* brilliant metallic emerald-green, with a blue reflection on dorsum; lower half of sides, including whole of metepimeron, under surface of thorax, also all coxæ bright yellow. *Legs* very long and slim, brown in teneral specimens, black when mature. *Wings* hyaline, in immature specimens tinted with yellow throughout, in mature specimens yellow at the base, otherwise with a pale greenish tinge; only a single row of cells between *Cuii* and *IA* as in *V. amœna*, and not more than 3 rows between the latter nervure and posterior border of wing ; cubital space with basal nervure (*ac*) usually isolated and with 2 or 3 nervures traversing it at its outer end ; discoidal cell traversed 2 or 3 times, usually only twice ; nodal index variable, $\frac{54\text{–}20 \mid 21\text{–}53}{49\text{–}17 \mid 17\text{–}48}$, $\frac{52\text{–}20 \mid 21\text{–}53}{45\text{–}18 \mid 18\text{–}46}$, $\frac{40\text{–}24 \mid 20\text{–}42}{38\text{–}18 \mid 17\text{–}37}$. (Discoidal cell traversed only once in all wings of one specimen.) *Abdomen* metallic emerald-green on dorsum and sides, black beneath ; segment 1 broadly citron-yellow on sides and base of dorsum, terminal segments more dully metallic, and often pruinosed white on dorsum. Segment 10 neither keeled nor spined. *Anal appendages* black. Superiors rather longer than segment 10, curving in gradually to meet at apices, but slightly angulated inwards at about the middle, outer border finely spined, base broad, and with a very robust dorsal spine inclining outward, then subcylindrical and finally broadening out at apex which is deeply bifid, outer branch much the longer, inner about half its length, and both rounded and blunt at apex. Inferiors about two-thirds the length of superiors, slim, cylindrical, ending in an acute inwardly directed spine.

Female.—Abdomen 34–38 mm. Hind-wing 30–34 mm.

Similar to male in colouring, but sides of abdominal segment 3 and lower parts of sides of 8 to 10 yellow. Segment 10 with a dorsal keel ending in an apical spine and a small spine on each side of the apical border as in *V. gracilis*. Vulvar scale robust, yellow. *Wings* more highly coloured and evenly saffronated except in very old specimens, the nervures rich ochre ; venation similar to that of male. *Anal appendages* short conical, pointed, brown or blackish-brown.

Distribution.—ASSAM, BURMA, and Tibet. The *type* comes from Shillong, Khasi Hills, Assam ; recent specimens from the same district.

Type in the Selys collection, Brussels Museum.

Genus **ECHO** Selys (1853). (Fig. 39.)

Echo Selys, Syn. Cal. p. 19 (1853) ; Walker, List Ins. Neur. Brit. Mus. iv, p. 604 (1853) ; Selys, Mon. Cal. p. 67 (1854) ; Kirby, Cat. Odon. p. 101 (1890) ; Laidlaw, Fascic. Malayenses (Zool.), part i, pp. 191–192 (1903) ; Williamson, Proc. U.S. Nat. Hist. Mus. vol. xxviii, p. 170 (1905) ; Laidlaw, Rec. Ind. Mus. vol. xiii, pp. 26–28 (1917) ; Fraser, J. Bombay Nat. Hist. Soc. vol. xxxiii, pp. 586, 587 (1929).

Ground-colour dark metallic green and black ; thorax short, robust ; legs long and slim ; abdomen slim, cylindrical, considerably longer than wings ; wings closely reticulated, hyaline or partly opaque black in both sexes ; apices of all wings rounded ; pterostigma opaque white or black, short and broad, and of the same shape in both sexes, wider below, with lower margin markedly rounded, situated much nearer apex of wings than usual; median (basal) space traversed by many nervures; *Rii* near its origin confluent with radius; *Riii* usually arising a little proximal to subnode; arc angulated, sectors of arc arising from it separately a little below its middle; discoidal cell about equal to length of median space, traversed by several nervures; anal area moderately simple; *IA* strongly forked a little after its origin; none of the sectors pectinate or branched, but numerous intercalated sectors between all main nervures; node situated nearer to base of wing than to pterostigma.

Larva somewhat similar to that of *Neurobasis.*

Genotype, *Echo margarita* Selys.

Distribution.—ASSAM and UPPER BURMA. Selys gives "China ?" as the locality of the genotype, but this is probably an error.

Laidlaw (Rec. Ind. Mus. vol. xiii, pp. 26 & 28) considers that his genus *Climacobasis* is synonymous with *Echo*, but this is due to an error in his interpretation of the Selysian description of *E. margarita*, and secondly to a wrong impression of the

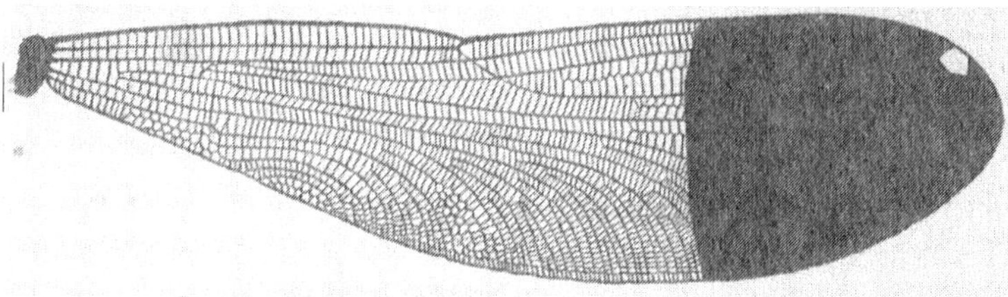

Fig. 39.—Hind-wing of *Echo margarita tripartita* Selys, female.

shape of the pterostigma in a male in the Indian Museum collection. He states that: "When Selys described the male of *E. margarita*, he did not call attention to the difference in the shape of the pterostigma in the two sexes"; but as the shape is identical in the two sexes, there was no call for Selys to make any comment on this point. As regards the specimen in the Indian Museum, I have made a re-examination of it and find that its pterostigma does not differ from that of the female. For reasons mentioned under the genus *Climacobasis*, I consider it a good genus. Dr. Laidlaw also suggests that Kirby's *Archineura* and Martin's *Echo maxima* belong to genus *Echo*. As regards the former, the basal neuration is so different that one cannot reconcile it with *Echo*; and in regard to the latter, I have made an examination of the type and find important differences in the venation; thus *Rii* is not confluent with the radius at its origin, but is similar to the condition found in the genus *Matrona*; the pterostigma is elongate in both sexes and situated in the usual place, viz. moderately far back from the apex ; the wings are longer, the apices less rounded ; the anal area of wings more complicated and the node situated nearer the pterostigma than base of wing as in *Climacobasis*. Thus the genus *Echo* contains but two species, *E. margarita* Selys and *E. uniformis* Krüg., of which only the former, with its race *tripartita* Selys, has been reported from within Indian limits.

These insects breed in montane streams, the adults being found along the banks settled on ferns or overhanging vegetation, and with a slow flitting flight similar to *Vestalis*.

225. **Echo margarita margarita** Selys.

Echo margarita Selys, Syn. Cal. p. 19 (1853); id., Mon. Cal. p. 67 (1854); id., Bull. Acad. Belg. (2) vol. xlvii, p. 356 (1879); Kirby, Cat. Odon. p. 101 (1890); Laidlaw, Rec. Ind. Mus. vol. xiii, pp. 26–28 (1917); Fraser, J. Bombay Nat. Hist. Soc. vol. xxxiii, pp. 587, 588 (1929).

Male.—Abdomen 44–46 mm. Hind-wing 36–37 mm.

Head : labium black ; labrum glossy black ; cheeks, bases of mandibles, anteclypeus, and antennæ mat black ; postclypeus glossy metallic dark bronzy-green ; eyes black above, pale brown to dove-grey below. Rest of head dull metallic dark bronzy-green. *Prothorax* shiny metallic dark green, pruinosed beneath. *Thorax* on dorsum and whole of sides dark metallic-green with a coppery reflex, beneath black and thinly pruinosed. *Legs* black ; femora pruinosed thinly on flexor surface. *Wings* hyaline, apical fifth dark blackish-brown ; a slight clouding of brown at node in all wings ; inner border of the dark area running almost straight back from costa to posterior margin of wing ; hyaline area with a beautiful bluish-purple iridescence ; pterostigma milky white in all wings, about twice as long as broad, distal end rounded, proximal end pointed, posterior border markedly rounded, the space distal to the pterostigma reticulated with a double row of cells ; discoidal cell traversed 7 to 9 times ; basal space traversed 6 to 9 times ; 16 to 18 cubital nervures ; nodal index $\frac{69\text{–}38}{65\text{–}28}\left|\frac{38\text{–}73}{29\text{–}64}, \frac{56\text{–}30}{56\text{–}30}\right|\frac{37\text{–}58}{30\text{–}58}$. *Abdomen* uniformly dark blackish-brown, non-metallic, thinly pruinosed beneath. *Anal appendages* black. Superiors tumid at base, then subcylindrical, finally compressed at apex and curving gradually in towards one another ; outer border coarsely spined ; apices blunt. Inferiors very broad at base, very narrow and slightly sinuous thereafter, apices slightly clubbed and with a minute inwardly directed spine.

Female.—Abdomen 37–41 mm. Hind-wing 35–37 mm.

Closely similar to male both as regards colouring of body and wings. Mature specimens have the hyaline area of wings variably enfumed, often of a warm brown, and then lose their bluish iridescence. Pterostigma usually slightly shorter and broader than in male, so that it is subrotund or pyriform in some specimens. Venation similar to that of male. Vulvar scale very robust, extending to end of abdomen, black ; *anal appendages* short, acutely conical, black.

Distribution.—Cherrapunji, ASSAM. I have seen but one specimen of this beautiful insect, a very dilapidated male in the Indian Museum collection, taken by Dr. Kemp in October. The *type* is a female in the Selysian collection labelled "China ?," and there is also a male in the same collection from Cherrapunji ; probably both came from the latter locality.

226. **Echo margarita tripartita** Selys.

Echo margarita race *tripartita* Selys, Bull. Acad. Belg. (2) vol. xlvii, p. 356 (1879); Kirby, Cat. Odon. p. 101 (1890); Laidlaw, Rec. Ind. Mus. vol. xiii, pp. 26–28 (1917); Fraser, J. Bombay Nat. Hist. Soc. vol. xxxiii, p. 588 (1929).

Differs from *E. margarita margarita* in the greater extent of black on the wing, which covers roughly the outer third of all wings but is subject to slight variations. In some specimens it extends more than half-way from pterostigma to node, in others slightly less than this distance.

Distribution.—The *type*, from the Khasi Hills, is a male in the Selysian collection ; it is probably from Shillong, where Mr. Bainbrigge Fletcher has found this form moderately common from June to October. He writes : "Weak flight, sits on bushes overhanging banks and on rocks in midstream. Also along pebbly bottomed swift streams. Never over muddy water. Never seen pairing or ovipositing. Occurs from June to August and struggles into September, last specimen seen 12. ix. 19."

Genus **CLIMACOBASIS** Laidlaw (1902).

Climacobasis Laidlaw, Proc. Zool. Soc. Lond. (1) p. 85 (1902); Williamson, Proc. U.S. Nat. Mus. vol. xxviii, p. 170 (1905); Laidlaw, Rec. Ind. Mus. vol. xiii, p. 26 (1917); Fraser, J. Bombay Nat. Hist. Soc. vol. xxxiii, pp. 588, 589 (1929).

Resembles *Echo*, but wings entirely hyaline, with apices less rounded, pterostigma in male elongate and of normal shape, situated more proximally, much reduced and more or less vestigial in female ; node situated much nearer pterostigma than base of wing ; *Rii* fused with radius in fore-wing but not in hind-wing ; sectors of arc arising from a common point and occasionally fused at origin ; abdomen much more slender and comparatively longer.

Genotype, *Echo modesta* Laid.

Distribution.—Borneo, Siam, ASSAM, and probably widely but sparsely distributed over the intervening region and throughout S.E. Asia.

As mentioned under genus *Echo*, Dr. Laidlaw came to the conclusion that his genus was synonymous with *Echo*, but the position of the node, the similarity of the pterostigma in the sexes of *Echo*, the separated sectors of the arc, and the shorter more robust abdomen of the latter genus seem definitely to establish the independence of *Climacobasis*.

227. **Climacobasis modesta** (Laid.). (Fig. 40.)

Echo modesta Laidlaw, Proc. Zool. Soc. Lond. (1), p. 84, pl. v, fig. 6 (1902); id., Rec. Ind. Mus. vol. xiii, p. 28 (1917).
Climacobasis lugens Laidlaw, Proc. Zool. Soc. Lond. (1) p. 85, pl. vi, fig. 5 (1902).
Echo (Climacobasis) modesta Laidlaw, Fascic. Malayenses (Zool.), pt. i, p. 191 (1903).
Climacobasis modesta Williamson, Proc. U.S. Nat. Mus. vol. xxviii, p. 186, fig. 17 (1905); Fraser, J. Bombay Nat. Hist. Soc. vol. xxxiii, p. 589 (1929).

Male.—Abdomen 52–54 mm. Hind-wing 37–39 mm.

Head: labium black; cheeks and bases of mandibles glossy black; anteclypeus dark brown; postclypeus and rest of head dark metallic green, frons and ocellar space in mature specimens pruinosed snowy white; eyes dark brown above, paler beneath. *Prothorax* and *thorax* dark metallic green with a coppery, bronzy, or golden reflection, beneath black (usually with white pruinescence when mature). *Legs* long,

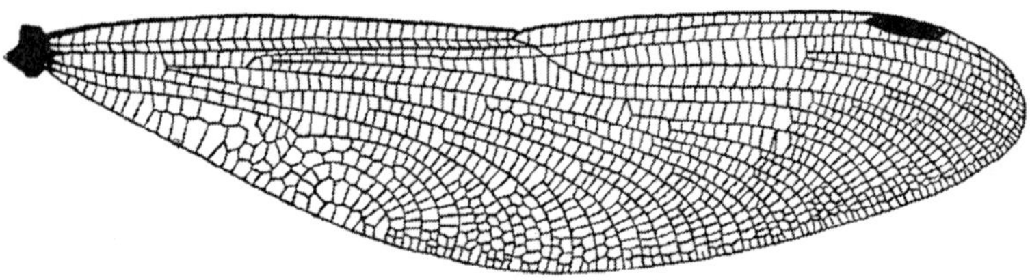

Fig. 40.—Wing of *Climacobasis modesta* (Laid.), male.

slim, black. *Wings* entirely hyaline, iridescent with purplish-blue; when mature with apices enfumed, especially that of fore-wing, and sometimes a cloudy fascia extending over distal two-thirds of the space between node and base of wing, especially in fore-wing; pterostigma black, elongate, narrow, oblique at both ends but more pointed proximally; basal (median) space traversed by 8 to 9 nervures in all wings; discoidal cell traversed by 7 to 10 nervures, usually by 7; 16 to 18 cubital nervures; nodal index $\frac{42\text{–}39 \mid 40\text{–}43}{33\text{–}31 \mid 34\text{–}40}$, $\frac{39\text{–}36 \mid 35\text{–}40}{33\text{–}31 \mid 33\text{–}32}$.

Abdomen dark metallic green on segments 1, 2, and base of 3, black for the remaining segments. *Anal appendages* black. Superiors slightly longer than segment 10, broad at base, curving gradually in almost to meet at apices, subcylindrical after base and slightly compressed at apex, which is truncate and obtuse; outer border coarsely spined. Inferiors almost two-thirds the length of superiors, broad at base, then cylindrical, narrow, apex blunt but furnished with a minute inner spine.

Female.—Abdomen 46 mm. Hind-wing 41 mm.

Closely similar to male, but wings with pterostigma much reduced, about one-third the length of that of male, its posterior border nearly twice the length of the costal and somewhat rounded; two rows of cells in the space between costa and radius distal to pterostigma; about 50 postnodal nervures, the increase in number over that found in male being due to the shorter pterostigma. Vulvar scale robust, extending to end of abdomen. *Anal appendages* about as long as segment 10, black, pointed.

Distribution.—King Island, Mergui, LOWER BURMA (*Elton Bott*).

Type in the Hope Museum, Oxford.

Genus **MNAIS** Selys (1853). (Fig. 41.)

Mnais Selys, Syn. Cal. p. 20 (1853); Walker, List Neur. Ins. Brit. Mus. iv, p. 605 (1853); Selys, Mon. Cal. p. 63 (1854); Kirby, Cat. Odon. p. 101 (1890); Williamson, Proc. U.S. Nat. Mus. vol. xxviii, p. 184 (1905); Ris, Suppl. Ent. no. v, pp. 8 & 9 (1916); Laidlaw, Rec. Ind. Mus. vol. xiii, p. 29 (1917); Munz, Mem. Amer. Ent. Soc. no. 3, p. 44 (1919); Fraser, J. Bombay Nat. Hist. Soc. vol. xxxiii, pp. 589, 590 (1929).

Ground-colour dark metallic green or black, often pruinosed on head, dorsum of thorax, and terminal segments of abdomen. Thorax short, robust; legs long and slim; abdomen cylindrical, longer than wings; wings very closely reticulated, hyaline or partly opaque black, or tinted with greenish-yellow or bright golden yellow; apices rounded; pterostigma usually red in male, white or cinereous in female, often vestigial, especially in female, in which sex it may be entirely absent; arc angulated; sectors of arc arising from a common point a little below the middle of arc; median (basal) space entire; *Rii*, near its origin, confluent with the radius; *Riii* arising at or a little distal to subnode; discoidal cell convex towards the costa, traversed by many nervures; anal area simple; *IA* forked a little after its origin, the hinder branch of fork weakly or not branched; main nervures rarely forked; many intercalated sectors; node situated nearer to base of wing than to pterostigma.

Larva unknown.

Genotype, *Mnais pruinosa* Selys.

Distribution.—NORTHERN INDIA, BURMA, Siam, Tonkin, Annam, South China, and Japan.

Some obscurity still exists as to the number of species or races belonging to this genus, Selys being of opinion that there were but two species, with a number of races or possibly merely age phases of the same insect. It seems fairly clear, however, that such is not the case, and that there are two well-defined

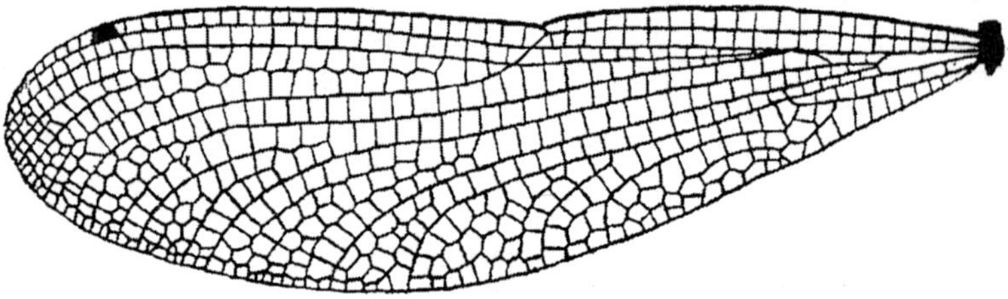

Fig. 41.—Hind-wing of *Mnais earnshawi* Will., male.

groups, one with black neuration and uncoloured membrane of wing, and the other with reddish neuration and bright golden yellow wing-membrane. To the former group we may add *Mnais maclachlani* Fras., in which the wing is partly opaque black, with an outer bordering to the dark area milky opalescent white. Williamson split up the species into two groups—Palæarctic and Oriental—but the characters he gave are not always constant, even in the same species.

Key to Indian Species of Mnais.

1.	Venation reddish; wings tinted with golden yellow	2.
	Venation black; wings untinted, or at the most with a faint greenish tint	*andersoni* McLach., [p. 140.
2.	Pterostigma quadrate	*earnshawi* Will., p. 142.
	Pterostigma elongate, more than twice as long as broad	*icteroptera* Fras., p. 143.

228. **Mnais andersoni** McLachlan.

Mnais andersoni McLachlan, Bull. Acad. Belg. (2) vol. xxxv. p. 472 (1873); Kirby, Cat. Odon. p. 101 (1890); Selys, Ann. Mus. Civ. Genova, (2) vol. x (xxx), p. 485 (1891); Williamson, Proc. U.S. Nat. Mus. vol. xxviii, p. 184 (1905); Ris, Suppl. Ent. no. v, pp. 8, 9, & 11 (1916); Laidlaw, Rec. Ind. Mus. vol. xiii, p. 29 (1917); Fraser, J. Bombay Nat. Hist. Soc. vol. xxxiii, pp. 590, 591 (1929).

Male.—Abdomen 35–42 mm. Hind-wing 26–33 mm.

Head: labium black; labrum and clypeus metallic emerald-green; bases of mandibles, adjacent parts of cheeks, and base of second segment of antennæ bright yellow; rest of head metallic green, with coppery or golden reflections; eyes brown above, paler below. *Prothorax* and *thorax* brilliant metallic emerald-green, latter with coppery or golden reflections on dorsum; antero-lateral suture finely black, postero-lateral suture narrowly yellow, as also the posterior half of metepimeron. Beneath, yellow, with a large black spot, pruinosed in mature specimens. *Legs* black, coxæ and trochanters pruinosed white. *Wings* hyaline, tinted with pale greenish-yellow or quite colourless; reticulation black; pterostigma small, about as long as broad, outer border straight, inner border slightly oblique, posterior border slightly rounded, variably black or dark brownish; only a single row of cells between costa and radius after the pterostigma; 11 to 13 cubital nervures; discoidal cell traversed 4 to 7 times; nodal index $\frac{30\text{–}23 \mid 22\text{–}34}{31\text{–}20 \mid 25\text{–}30}$; *IA* forked shortly after its origin, its branches streaming distally; anal area simple; *Riii* arising variably at or well distal to subnode. *Abdomen* black, segments 1 to 3 or 4 metallic bronzy-green or blue in immature examples, remainder black, but segments 8 to 10 in fully mature specimens pruinosed chalky white on dorsum; black beneath. *Anal appendages* black. Superiors tumid at base, subcylindrical at middle, compressed and slightly expanded at apex, which is obtuse, coarsely spined along outer border. Inferiors about two-thirds the length of superiors (which are slightly longer than segment 10), thick at base, then cylindrical and tapering to apex, which is curved slightly inward and ends in a minute point.

Female.—Abdomen 32–37 mm. Hind-wing 30–34 mm.

Closely similar to male; wings limpid; pterostigma smaller, red, brown, or cinereous, varying in shape and size, even in the wings of a single specimen. Second segment of antennæ wholly bright yellow; abdomen entirely metallic green from base to apex, black beneath. *Anal appendages* black, conical, pointed at apex, about as long as segment 10. Vulvar scale robust, extending to end of abdomen.

Distribution.—BURMA, South China, Formosa, Siam, and Tonkin; probably widely distributed over southern Asia. I have a female taken at an altitude of 5,500 ft. in April. Localities given by Selys are: Leito, Burma (April–May), Cobapo (September–November), Meteleo and Puepoli (in the same months), and Iado (April). Williamson gives Toungoo, Burma.

Nothing is known of its habits, but it is certainly a riverine breeder and probably submontane in habitat.

Type in the McLachlan collection.

229. **Mnais earnshawi** Williamson. (Fig. 41.)

Mnais earnshawi Williamson, Proc. U.S. Nat. Mus. vol. xxviii, pp. 185, 186 (1905); Ris, Suppl. Ent. no. v, pp. 8 and 10 (1916); Laidlaw, Rec. Ind. Mus. vol. xiii, p. 29 (1917); Fraser, Journ. Siam Soc. Nat. Hist. vol. iii, no. 4, pp. 460, 461 (1919); Fraser, J. Bombay Nat. Hist. Soc. vol. xxxiii, pp. 501, 502 (1929).

Male.—Abdomen 40–44 mm. Hind-wing 33–37 mm.

Head: labium black; labrum and clypeus glossy metallic emerald-green; bases of mandibles and adjacent portion of cheeks bright yellow, rest of cheeks glossy black; antennæ black, second segment bright yellow from base to its middle or nearly to apex; rest of head dull dark green metallic; eyes dark brown above, paler below. *Prothorax* and *thorax* dark metallic green, dorsum of latter chalky white between the humeral sutures and including the antealar sinus; postero-lateral suture and rather more than one-third the hinder surface of metepimeron bright yellow. Thorax black, beneath pruinosed white. *Legs* black, cilia numerous and very fine. *Wings* hyaline, tinted from base to apex with bright golden or amber yellow; reticulation reddish, costal half of wings more deeply tinted, costa somewhat darker from node to base. Pterostigma very small, rather less than 1 mm. in length, variable in size and shape, usually with the costal border about two-thirds the length of the rounded posterior border, outer end straight, proximal end very oblique; deep blood-red in colour, this colour sometimes spreading beyond the bordering nervures of the pterostigma distally and proximally. Discoidal cell rather convex costalwards, traversed 8 to 9 times in fore-wing, 6 in hind-wing; 12 to 15 cubital nervures; usually only a single row of cells in the space after pterostigma, but occasionally two; *Riii* arising widely distal to subnode in fore-wing, a shorter distance in hind-wing; *IA* forked shortly after its origin, but its branches running distally; anal area simple; some of the main nervures occasionally branched. *Abdomen* with segments 1 and 2 dark glossy metallic green, remaining segments bronzy-black metallic, especially 9 and 10; 8 to 10 pruinosed in mature specimens. *Anal appendages* black. Superiors curving gradually inwards to meet at apices, thick at base, then subcylindrical, with apices compressed, slightly broadened, and obtuse, outer border coarsely spined. Inferiors slightly shorter than superiors, broad at base, then cylindrical and with a minute spine at inner side of apex.

Female.—Abdomen 34–41 mm. Hind-wing 30–36 mm.

Very similar to male, wings much paler yellow or entirely colourless; reticulation reddish-brown, pterostigma white or cinereous, or occasionally quite uncolured or entirely absent, usually much smaller than in male and abnormal in shape, sometimes triangular, framed by two nervures which meet either at costa or at radius, sometimes represented by a mere bordering of opaque white on one or both sides of a postnodal nervure. *Thorax* and *abdomen* without pruinosed areas. *Anal appendages* small, pointed, conical, black. Vulvar scale robust, extending to end of abdomen, blackish-brown in colour.

Distribution.—BURMA, Siam, Tonkin, Annam, Formosa, and South China. The above description is made from specimens from Siam and Annam, which agree entirely with Williamson's description of Burmese specimens. The specimens from South China and Formosa described by Ris are decidedly larger, as is also the pterostigma ; they have also a much greater extent of yellow on the sides of thorax, so that I consider them a distinct race.

Type in the Williamson collection, Michigan University Museum.

230. **Mnais icteroptera** Fraser.

Mnais icteroptera Fraser, J. Bombay Nat. Hist. Soc. vol. xxxiii, pp. 592, 593 (1931).

Male.—Abdomen 47 mm. Hind-wing 36 mm.

Head : labium black ; labrum, clypeus, and occiput brilliant metallic green, labrum with bronze or golden reflections ; upper part of head emerald-green, but space between antennæ, clypeus, and middle ocellus pruinosed chalky white ; antennæ black, second segment and base of third bright yellow ; bases of mandibles and cheeks adjoining mandibles bright yellow, cheeks against eyes glossy black. Behind occiput on either side a very robust pointed spine, and behind each eye a second conical eminence not amounting to a spine. *Prothorax* metallic green, with golden reflections and patchy pruinescence, especially on the posterior lobe ; densely pruinosed beneath. *Thorax* brilliant metallic green, with golden or coppery reflections ; the dorsum, including antealar sinus as far out as humeral sutures, chalky white with pruinescence. Lateral sutures finely black, the postero-lateral suture with 2 to 3 small yellow spots, rather obscured by pruinescence when mature, ventro-posterior border very narrowly yellow : under surface black, obscured by pruinescence. *Legs* black, coxæ and trochanters pruinosed. *Wings* uniformly rich golden yellow throughout, with satiny reflections ; reticulation red ; pterostigma 2 mm. in length, nearly three times as long as broad, outer border straight, inner border oblique pointed, posterior border rounded, colour deep blood-red ; nodal index $\frac{34\text{–}29 \mid 31\text{–}33}{30\text{–}26 \mid 25\text{–}31}$; discoidal cell traversed 8 to 9 times ; 10 to 14 cubital nervures ; anal area simple ; *Riii* arising slightly distal to subnode ; only a single row of cells in the space distal to pterostigma. *Abdomen* black, segments 1 to 4 metallic green, with a bronze or coppery reflection on 3 and 4; segments 8, 9, and 10 obscurely metallic green, not pruinosed; under surface black, pruinosed on segments 1 and 8. *Anal appendages* black. Superiors rather longer than segment 10, forcipate, curving in gradually to meet at apices, broad at base, then subcylindrical, apices obtuse, compressed and a little expanded. Inferiors about three-fourths the length of superiors, broad at base, then cylindrical and tapering to apex, which is curled slightly inward and ends in a fine point.

Female unknown.

Distribution.—Kalaw, Chin Hills, BURMA.

This very beautiful insect appears to be closely related to *M. earnshawi* by the pruinescence on dorsum of thorax and the colour of the wings; it is distinguished from the latter by its larger size, the much longer pterostigma, the pruinosed spot on vertex, and the relatively longer abdomen. It appears to link up the Oriental with the Palæarctic forms by the large pterostigma and the single row of cells following it, the former character being common to *M. costalis* and *M. pruinosa*.

Type in the Author's collection.

Genus **MATRONA** Selys (1853). (Fig. 42.)

Matrona Selys, Syn. Cal. p. 17 (1853); id., Mon. Cal. p. 52 (1854); Kirby, Cat. Odon. p. 100 (1890); Williamson, Proc. U.S. Nat. Mus. vol. xxviii. p. 170 (1905); Ris, Suppl. Ent. no. v, pp. 5–8 (1916); Munz, Mem. Amer. Ent. Soc. no. 3, p. 44 (1919); Fraser, J. Bombay Nat. Hist. Soc. vol. xxxiii, p. 593 (1929).

Head wide, eyes widely separated and globular; epistome flat, frons low; thorax robust; legs of great length, slim, with numerous fine spines; abdomen of great length, slim,

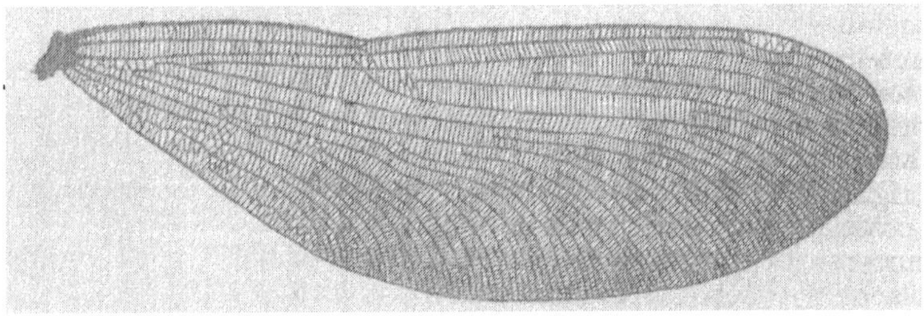

Fig. 42.—Hind-wing of *Matrona basilaris nigripectus* Selys, male.

cylindrical, depressed at end; body-colouring brilliant metallic green; wings long, very broad, apices rounded, reticulation very close, especially in the anal field, opaque in both sexes; pterostigma absent in male, but a false reticulated creamy white one in all wings of female ; median (basal) space reticulated with 2 rows of cells ; discoidal cell traversed by many nervures, equal in length to median space ; numerous cubital nervures ; anal area complex, *IA* bifurcated shortly after its origin and sending a branch basally ; great numbers of intercalated nervures ; *Riii* arising slightly or well proximal to subnode ; *Rii* not confluent with radius after its origin ; node situated nearer to base of wing than to apex.

Larva very similar to that of *N. chinensis*.

Genotype, *Matrona basilaris* Selys.

Distribution.—ASSAM, BURMA, Siam, Tonkin, Hainan, Annam, and North and South China.

Breeds in montane streams ; habits of the imago closely similar to those of *Calopteryx*, to which the genus is related. Great variability is exhibited in specimens from different localities, so that it would appear that there are but one or two species with a number of subspecies or races.

231. **Matrona basilaris basilaris** Selys.

Matrona basilaris Selys, Syn. Cal. p. 17 (1853) ; id., Mon. Cal. p. 53 (1854) ; Selys, Compt. Rend. Soc. Ent. Belg. vol. xxxii. p. lii (1888) ; Kirby, Cat. Odon. p. 100 (1890) ; McLachlan, Ann. Mag. Nat. Hist. (6) vol. xvii, p. 370 (1896) ; Kirby, ibid. (7) vol. v, p. 536 (1900) ; Martin, Mission Pavie, p. 15 (1904) ; Bartenef, Ann. Mus. Zool. Acad. St. Petersb. vol. xvii, p. 304 (1913) ; Ris, Suppl. Ent. no. v, p. 6 (1916) ; Laidlaw, Rec. Ind. Mus. vol. xiii, pp. 28–29 (1917) ; Fraser, J. Bombay Nat. Hist. Soc. vol. xxxiii, pp. 593, 594 (1929).

Calopteryx basilaris Walker, List Neur. Ins. Brit. Mus. iv, p. 601 (1853).

Male.—Abdomen 42–55 mm. Hind-wing 36–42 mm.

Head: labium black, lateral lobes paler or brownish ; labrum and anteclypeus steely blue ; postclypeus brilliant metallic bluish-green, rest of head metallic dark green ; antennæ black ; eyes dark brown above, paler below. *Prothorax* and *thorax* metallic emerald-green with bluish reflections on dorsum, humeral and antero-lateral sutures black, postero-lateral suture and posterior ventral border of metepimeron bright yellow, as also the under surface of thorax, which is spotted with black. (In a male from Tonkin the under surface of thorax and postero-lateral suture are entirely black, but there is a longish yellow streak bordering the lower part of metepimeron.) *Wings* rounded at apices, very broad, varying from a dark brown to blackish steely blue according to age, outer fifth of fore-wings comparatively hyaline, but the nervures in this part all framed in opaque brown ; reticulation and nervures (including costa) black, but the transverse nervures for a variable distance from base, usually up to a short distance distal to node, bluish-white (this part of the wing viewed horizonatally looks quite milky or bluish-white); nodal index $\frac{150\text{–}63 \mid 60\text{–}148}{138\text{–}58 \mid 59\text{–}145}$; discoidal cell traversed 18 to 25 times; 26 to 28 cubital nervures; *Riii* arising well proximal to subnode; basal (median) space traversed by 2 rows of cells, often in a network or with 10 to 12 traversing nervures; anal area complex, filled with a very close network of nervures; under side of wings similar, but differing in the shorter extent of bluish-white transverse nervures, which usually do not extend as far as the node. *Abdomen* brilliant glossy emerald metallic green on dorsum, black below. *Anal appendages* black. Superiors thick at base, then subcylindrical, finally abruptly expanded and angulated inward at a little beyond their middle, apices blunt, outer border with 3 to 5 coarse spines. Inferiors about two-thirds the length of superiors, tumid at base, then cylindrical as far as apex, which is hooked slightly in. *Legs* of great length, finely spined, black, distal ends of hind femora brown.

Female.—Abdomen 50–54 mm. Hind-wing 44–46 mm.

Differs from the male as follows:—Labrum, bases of mandibles, outer sides of second segment of antennæ, and labrum bright yellow, latter bordered with black and with a basal line and medio-basal point of the same colour; the yellow on sides of thorax of greater extent, coxæ and trochanters spotted with the same colour; under side of thorax almost entirely yellow; abdomen with the sides of segments 8 to 10 yellowish-brown and the whole dorsum rather dull brown, segment 1 slightly metallic; segment 10 with a well-developed dorsal keel ending in a robust apical spine, and with a small tubercle surmounted with minute teeth on either side as in *Vestalis*. *Anal appendages* short, pointed, brown. *Wings* uniform dull dark brown, apices of fore-wings much as in male; in all wings a moderately large creamy white pterostigma traversed by a variable number of nervures (in a specimen from the Chin Hills this varies from 2·75 to 3 mm. in length, but in others it may be as large as 3·5 mm.). The milky white or bluish-white transverse reticulation is also very variable, being hardly evident in some specimens, but as well marked as in the male in others. Vulvar scale robust, yellowish or pale brown, extending nearly to end of abdomen.

Distribution.—The *type*, a male in the British Museum, is from Sylhet, ASSAM, but the species extends as far as Formosa and Shanghai. It has also been reported from Tibet, BURMA (Chin Hills), Hainan, and Tonkin.

232. **Matrona basilaris nigripectus** Selys.

Matrona basilaris race *nigripectus* Selys, Bull. Acad. Belg. (2) vol. xlvii, p. 355 (1879); Kirby, Cat. Odon. p. 100 (1890); Williamson, Proc. U.S. Nat. Mus. vol. xxviii, p. 187 (1905); Ris, Suppl. Ent. no. v, p. 7 (1916); Laidlaw, Rec. Ind. Mus. vol. xiii, pp. 28, 29 (1917); Fraser, ibid. vol. xvi, pp. 463–464, pl. xxxv, fig. 1, and pl. xxxvii, fig. 1 (1919) (larva); id., J. Bombay Nat. Hist. Soc. vol. xxxii, pp. 594, 595 (1929).

Male.—Abdomen 52–54 mm. Hind-wing 38–40 mm.

Female.—Abdomen 43–53 mm. Hind-wing 39·45 mm.

Closely similar to *M. basilaris*, but wings all uniformly dark as far as the apices; sometimes the whole wing is dark metallic steely, bronze, or bluish, the pleating of the wings showing this off to fine advantage; the bluish-white nervures on basal half of wings not nearly as pronounced as in *basilaris*; pterostigma in female considerably smaller and more variable, on fore-wing 1·75 to 2 mm., on hind-wing 0·75 to 1·25 mm.; labium with lateral lobes dark and middle lobe entirely black; labrum of female with a small yellow spot on either side, otherwise black; under side of thorax entirely black, as is also the postero-lateral suture and that part of the metepimeron which is yellow in *basilaris*; abdomen of male a much more brilliant metallic emerald-green.

Distribution.—Khasi Hills, ASSAM; also UPPER BURMA, for which latter district Selys gives Puepoli, June, and Leito, September.

I am indebted to Mr. Bainbrigge Fletcher for the following notes:—"Common in Shillong throughout August, sexes in about equal numbers. A female was seen ovipositing in lower Trout Lake at the Fruit Gardens. Flapping along, settling on reeds and grasses. Sat on stem well above water and palpated it with ovipositor. As a rule found along streams, so this was rather remarkable, as the lake is so open (but has three streams flowing through it). Another female seen ovipositing below water-level, abdomen about two-thirds below water. Bases of wings of males bluish-grey, very noticeable in flight. . . . Occurs usually over tiny swift streams, usually clear pebbly streams, never over stagnant water except as mentioned above. Flies for only a short distance as a rule." The latest date on which the insect was seen by Mr. Fletcher was 18th October.

Type, a male in the Selys collection from Shillong, Assam.

Subfamily CALIPHÆINÆ. (Fig. 43.)

Caliphinæ Fraser, J. Bombay Nat. Hist. Soc. vol. xxxiii, p. 595 (1929).

Head broad: eyes hemispherical, tumid behind; occiput strongly ridged at its posterior border; labium notched deeply and narrowly for about one-third its length, lateral lobes shorter than middle lobe; face sloping to vertex; epistome not projecting. Thorax robust, elongate, mesothoracic triangle absent; abdomen cylindrical, comparatively short, but longer than wings in both sexes; legs long, hind femora extending to beyond middle of segment 2, slim, spines long and hair-like. Wings long and narrow, apices rounded, reticulation moderately close, tetragonal in type; petiolation marked, extending nearly to level of arc; discoidal cell elongate, convex above, broadening gradually distally, outer end

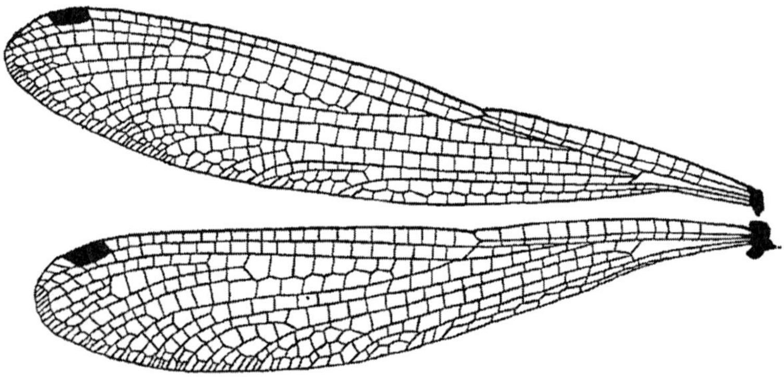

Fig. 43.—Wings of *Caliphæa confusa* Selys, male.

oblique, about half the length of median space, which is entire; antenodal nervures numerous, costal and subcostal series coinciding, no primary antenodals; arc situated between the fourth and fifth, or sixth and seventh antenodals, oblique, not angulated; sectors of arc arising from its middle, separated at origin; *Rii* shortly after its origin confluent for a long distance with radius, especially in hind-wing; *IRiii* arising from this confluence; *Cuii* and *Riv+v* forked, latter pectinate; no intercalated sectors between *Cuii* and *IA* and only a single row of cells between the latter and posterior border of wing, so that the anal field scarcely exists and is very simple; intercalated sectors between *MA* and *Cuii* and between the remaining principal sectors at apical half of wing; node situated well proximal to middle of wing; pterostigma present in all wings of both sexes. Anal appendages closely resembling those of the AGRIINÆ, but the inferiors deeply bifurcate at apex; segment 10 not keeled; ovipositor robust.

Distribution.—Tibet, South-western China, BENGAL, NEPAL, and ASSAM. Only a single species belonging to this subfamily has so far been described.

Genus **CALIPHÆA** Selys (1859).

Caliphæa Selys, Bull. Acad. Belg. (2) vol. vii, p. 439 (1859); Kirby, Cat. Odon. p. 108 (1890); Williamson, Proc. U.S. Nat. Mus. vol. xxviii, p. 167 (1905); Laidlaw, Rec. Ind. Mus. vol. xiii, p. 30 (1917); Munz, Mem. Amer. Ent. Soc. no. 3, p. 46 (1919); Fraser, J. Bombay Nat. Hist. Soc. vol. xxxiii, pp. 595, 596 (1929).

Notholestes McLachlan, Ent. Month. Mag. vol. xxiv, p. 31 (1887); Kirby, Cat. Odon. p. 111 (1890); McLachlan, Ann. Mag. Nat. Hist. (6) vol. xiii, p. 435 (1894).

Wings hyaline, tinted when mature; pterostigma about twice as long as broad, oblique at both ends; *Riii* arising far distal to subnode; cubital nervures numerous but far less than in the AGRIINÆ, usually 4 to 6 in number; discoidal cell traversed once in all wings; 1 or 2 intercalated sectors between *Cuii* and *MA*, but none between *Cuii* and *IA*.

The position of the genus is obscure and must remain so until the larva has been discoverd. On the whole I am disposed to place it near the AGRIINÆ on account of its metallic colouring, the shape of the labium and anal appendages, the confluence of *Rii* with the radius, and the short pterostigma. On the other hand, the extremely simple anal field, the extraordinary long petiolation of the wings, and the comparatively short abdomen are all foreign to the AGRIINÆ, and seem to indicate a relationship with the genus *Dicterias*.

233. **Caliphæa confusa** Selys. (Fig. 44.)

Caliphæa confusa Selys, Bull. Acad. Belg. (2) vol. vii, p. 440 (1859); Kirby, Cat. Odon. p. 108 (1890); Laidlaw, Rec. Ind. Mus. vol. xiii, pp. 30–31 (1917); Fraser, J. Bombay Nat. Hist. Soc. vol. xxxii, pp. 596, 597, text-fig. 2 (1929).

Notholestes elwesii McLachlan, Ent. Month. Mag. vol. xxiv, p. 32 (1887); Kirby, Cat. Odon. p. 111 (1890); McLachlan, Ent. Month. Mag. (6) vol. xvii, p. 371 (1896); Laidlaw, Rec. Ind. Mus. vol. xiii, p. 30 (1917).

Caliphœa consimilis McLachlan, Ann. Mag. Nat. Hist. (6) vol. xiii, p. 434 (1894).

Male.—Abdomen 36–40 mm. Hind-wing 30–32 mm.

Head: labium black; labrum brilliant metallic coppery; clypeus brilliant metallic coppery with a golden or green reflection; bases of mandibles, adjacent portions of cheeks, and second segment of antennæ bright yellow; rest of head dark blackish-brown with an obscure coppery glow (teneral specimens have the upper surface of head metallic green or peacock-blue). Eyes brown above, paler olivaceous below. *Prothorax* and *thorax* metallic coppery green with crimson or gold reflections in some lights (metallic emerald-green in teneral specimens), sides along ventral border, and whole of metepimeron save a small triangular area at its middle, bright yellow; under surface of thorax yellow. Mid-dorsal carina and humeral and antero-lateral sutures narrowly black. *Legs* black, coxæ and trochanters bright yellow. *Wings* uniformly pale chlorine-yellow, hyaline in teneral specimens; pterostigma reddish-brown (pale brown in teneral specimens),

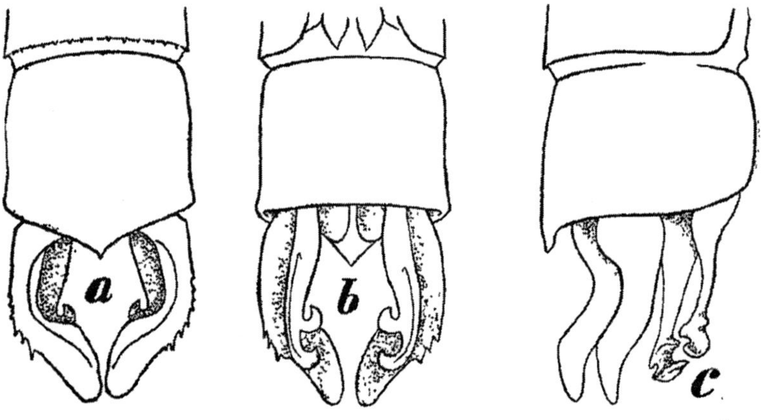

Fig. 44.—Anal appendages of *Caliphæa confusa* Selys, male. *a*, dorsal view; *b*, viewed from beneath; *c*, right lateral view.

covering 1½ to 2 cells; venation as for the genus; the second basal antenodal often incomplete; nodal index $\frac{29\text{–}16 \mid 15\text{–}28}{27\text{–}14 \mid 12\text{–}26}$, $\frac{28\text{–}17 \mid 17\text{–}28}{25\text{–}15 \mid 14\text{–}25}$. *Abdomen* dull coppery metallic when mature, brilliant metallic emerald-green in teneral specimens; segments 8 to 10 in mature specimens pruinosed white on dorsum, black beneath. *Anal appendages* black. Superiors curving in nearly to meet at apices, outer borders coarsely spined; tumid at base as seen from the side, then subcylindrical, and with a well-marked keel on dorsum extending nearly from base to apex, which is markedly broadened and rounded. Inferiors about three-fourths the length of superiors, flattened, tapering slightly to apex, which is bent in at a right angle as a robust pointed spine; on the inner border a second and equally robust spine situated as far from apex as its own length, so that the appendage appears to be deeply bifurcate.

Female.—Abdomen 34–37 mm. Hind-wing 31–32 mm.

Similar to male, apart from the usual sexual differences; nodal index usually slightly higher. Anal appendages short, conical, pointed, brown. Vulvar scale very robust, extending to extreme end of abdomen, brown.

Distribution.—As noted for the subfamily. Mr. Bainbrigge Fletcher found this species common in Shillong, ASSAM, 6,000 ft., during May and June, and I have found it in the DARJEELING DISTRICT at 3,000–5,000 ft. in May. It breeds in small brooks meandering through marshes on steep, heavily wooded hill-sides, and the adults are found perched on ferns or grasses overhanging the stream, or, if teneral, hiding in scrub-jungle near by.

McLachlan's types of *C. consimilis* differ only in having two traversing nervures to the discoidal cells and their teneral colouring.

Type in the Selys collection, Brussels Museum; types of *C. consimilis* and *N. elwesi* in the McLachlan collection.

Suborder ANISOZYGOPTERA.

Dr. R. J. Tillyard has described a single larva * which he and Dr. F. F. Laidlaw believe to belong to the suborder ANISOZYGOPTERA. This larva was discovered by Dr. S. Kemp in a stream a short distance below Ghoom, Darjeeling; but, although both Mr. Charles Inglis and I have searched the same spot, no others have been found, and the imago still remains to be discovered.

The suborder ANISOZYGOPTERA possesses characters linking up the other two suborders. The fore- and hind-wings are petiolated and closely similar in shape and size, as in the ZYGOPTERA; the discoidal cell is four-sided and closely resembles that of the LESTIDÆ. On the other hand, the imago rests with its wings expanded, as in the ANISOPTERA; the eyes, although separated, are not nearly so far apart as in the ZYGOPTERA, and the frons is ridged and markedly raised as in most ANISOPTERA.

Epiophlebia superstes (Selys), the only representative of the suborder known in the adult stage, resembles a small *Cordulegaster* in its general facies and colouring; it is known only from Japan.

* *Epiophlebia laidlawi* Tillyard, Rec. Ind. Mus. vol. xxii, pp. 93–107, pl. xiii (1921).

Suborder ANISOPTERA.

Head in archaic genera broad, with eyes separated, in others more or less globular, with eyes confluent to a variable extent on vertex; vesicle more or less developed, occasionally highly specialized, and with the ocelli arranged in a triangle around it; labium with the lateral lobes usually larger than the middle lobe, the latter with or without a median fissure. Wings

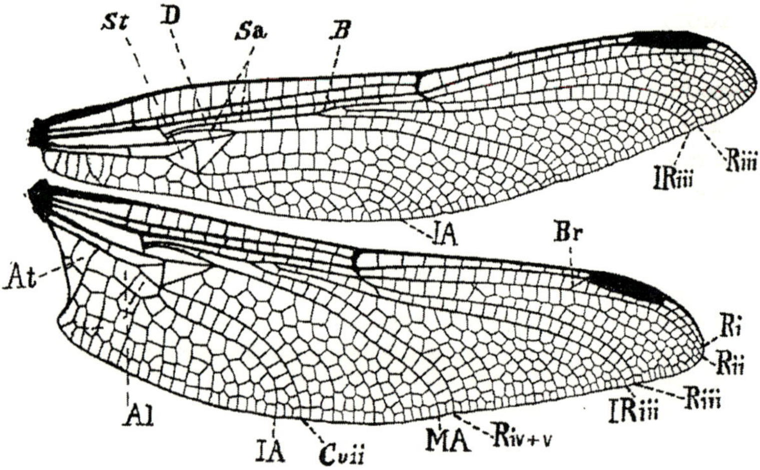

Fig. 45.—Wings of *Megalogomphus superbus* Fraser, male.

Al., anal loop; *At*, anal triangle; *B*, forking of *Rs*; *Br*, brace of pterostigma; *D*, discoidal cell; *Sa*, sectors of the arc; *St*, subtrigone; *Ri*, *Rii*, *Riii*, *IRiii*, *Riv+v*, *MA*, *Cuii*, and *IA*, the main sectors or nervures.

usually dissimilar (only approximately of the same shape in very archaic forms); hind-wing generally broadly dilated at the base and differing markedly in details of basal venation from the fore-wing; discoidal cell split into two triangular cells, a superior (hereafter called the *hypertrigone*) and an inferior (hereafter called the *discoidal cell*); pterostigma always present, narrow and of variable length, nearly always braced;

antenodal nervures usually numerous, the upper (costal) and the lower (subcostal) series coinciding or not (when the latter condition is present, then the proximal nervure, and another, about the fifth to seventh from the base, are much more robustly built than the rest, and represent the two primitive antenodal nervures of the Cœnagriidæ); wings in position of rest held horizontally outwards or even deflected strongly downwards, never closed over dorsum as in the Zygoptera. Tenth abdominal segment with a pair of superior anal appendages present, but only a single inferior appendage which may be simply triangular, or bifurcated into two branches, the latter in close apposition or more or less strongly divaricate. Female ovipositor in some archaic genera formed as in the Zygoptera, in other genera small and inconspicuous.

Larvæ breathing by rectal gills, caudal gills entirely absent; breeding in either still or running waters.

The Anisoptera comprise six families, of which only four are represented within the limits of India, Burma, and Ceylon. These are as follows :—

Key to Indian Families of Anisoptera.

1.	Eyes separated or meeting only at a point....	2.
	Eyes more or less broadly confluent on vertex.	3.
2.	Eyes only very slightly separated or meeting at a point; discoidal cells of fore- and hind-wings equal in size and shape, or, if dissimilar, then the median space traversed by one or more veins......................	**Cordulegasteridæ.**
	Eyes widely separated; discoidal cells unequal, that of hind-wing more elongate than that of fore-wing; median space never traversed.	**Gomphidæ,** p. 154.
3.	Discoidal cells approximately of the same size and shape in fore- and hind-wings and situated equally distant from the arc; costal and subcostal antenodal nervures not coinciding, the two robust primary antenodals present; middle lobe of labium large and fissured	**Æschnidæ.**
	Discoidal cells differing in size and shape in fore- and hind-wings, that of fore-wing situated far distal of the arc; middle lobe of labium very small, not fissured, broadly overlapped by the lateral lobes; costal and subcostal antenodal nervures coinciding, the robust primary antenodals absent	**Libellulidæ.**

Family GOMPHIDÆ. (Figs. 45, 45 *a* (A, B, C, D).)

Dragonflies of large or medium size, in colour yellow or black, marked with yellow or green, never metallic. Eyes always well separated; vesicle low and not usually well defined. Wings moderately broad, never petiolate, hyaline, never coloured; hind-wing in male strongly angulated at

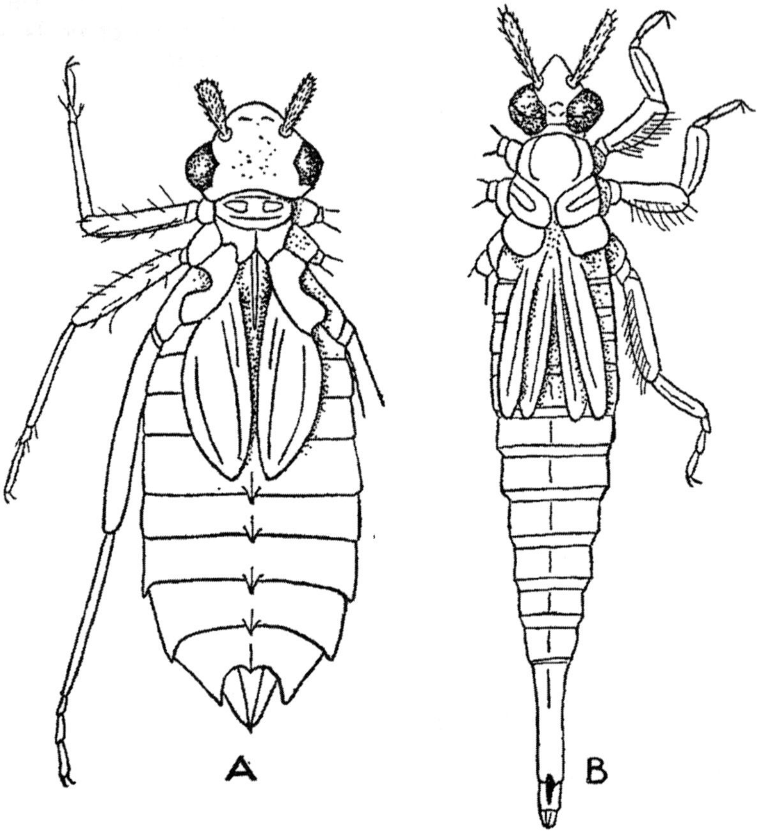

Fig. 45 *a*.—Larvæ of (A) *Microgomphus*; (B) *Macrogomphus*.

tornus, base more or less deeply excavate, in female rounded (rarely rounded in both sexes); reticulation close or moderately so; antenodal nervures numerous, those of the costal and subcostal series not coinciding, the two primary robust antenodals always present; discoidal cells traversed or entire, that of fore-wing subequilateral, that of hind-wing elongate in length of wing; median space never traversed; anal loop absent or rudimentary; membrane more or less rudimentary. Anal appendages often highly specialized, the inferior strongly bifid, with its branches closely apposed or more or less strongly divaricate; male genitalia very variable in different genera and species; female ovipositor represented by inconspicuous scales at apical end of segment 8.

Larvæ variable in form, cylindrical or extremely flattened, elongate or short and compact; crepuscular in habits, living in mud, débris, or buried in sand at the bottom of pools in streams or lakes, more commonly in the former.

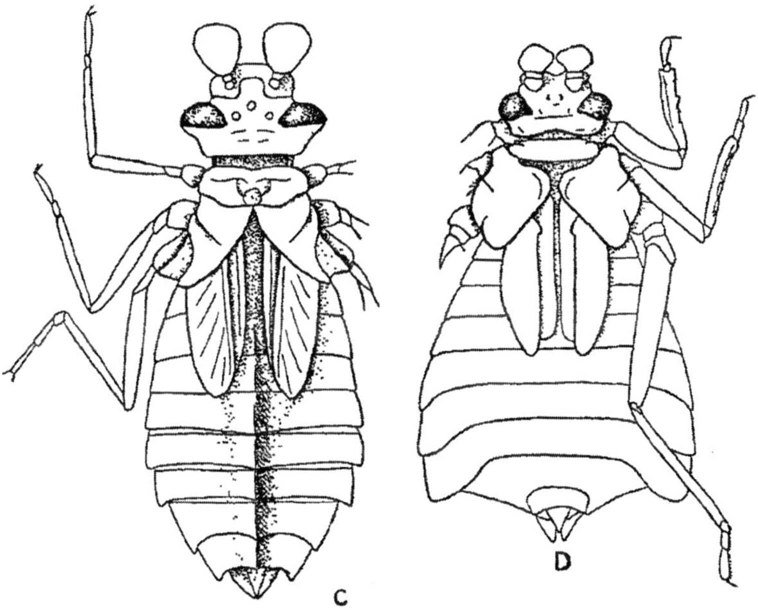

Fig. 45 *a*.—Larvæ of (C) *Lamelligomphus*; (D) *Sieboldius*.

Distribution.—Cosmopolitan. This is a very large and well defined family, represented within our limits by 108 species belonging to 25 genera.

Species of the family are remarkable for the archaic separation of the eyes, for the slight differences between the sexes (in contrast with the ZYGOPTERA), and, lastly, by their homogeneous colouring of black and yellow. Species from dry zones and desert areas are mainly yellowish, whilst those from regions of heavy rainfall, such as the Western Ghats of India or the hilly tracts of Assam and Burmah, are glossy black, with rather restricted bright greenish-yellow markings. In their natural habitat these colours are markedly cryptic, and serve to protect the insects.

Most GOMPHIDÆ, on emerging from the streams which give them birth, depart immediately into heavy jungle, and often to a great distance from the streams, this being particularly the case with the females. Thus most collections contain few examples, and females are particularly rare. Moreover, they have a very brief seasonal range and restricted habitats, except in the case of the commoner species. Although they wander far on emergence, their instincts appear to enable them to return to their parent streams.

Owing to the extremely small range of venational differences, upon which systematic writers have almost entirely depended, the classification of this family offers exceptional difficulties. This absence of venational characters has led to the adoption of other characters for specific differentiation, such as the length of the hind femora and their armature, and the form of the genitalia and anal appendages of the male. The determination of females is often difficult or impossible, and not uncommonly we are in doubt as to which genus a female may belong. It is for this reason that I have given the definition of each genus very fully.

The late Mr. E. B. Williamson was the first to attempt a classification based entirely on the venation of the wings, and he was followed by Dr. F. Laidlaw; it is the latter's "Series" which form my subfamilies, of which we have representatives of all three within Indian limits.

The larvæ of this family fall into three categories: (1) those with cylindrical or fusiform bodies, with robust fore-legs adapted for burrowing in the sandy bottoms of rivers in which they live; their colour is sandy and their antennæ are club-shaped: e. g., *Onychogomphus*, *Merogomphus*, *Acrogomphus* (fig. 45 *a*, A, B); (2) those with extremely flattened and broadened bodies, with comparatively feeble fore-legs; their colour is dark brown or black and their antennæ are triangular and flattened; these are found living among the rotting vegetation and detritus at the bottoms of deep pools: e. g., *Lamelligomphus*, *Sieboldius* (fig. 45 *a*, C, D); (3) those whose body is limpet- or tent-shaped, with flattened or concave under surface adapted for clinging to rocks or other flat surfaces in streams or deep pools: e. g., *Ictinus*, *Gomphidia*.

Examples of the first two types are found equally in the Gomphinae and Epigomphinæ, whilst the third type is found exclusively in the Ictiniinæ. Whilst the characters of this last group serve well to define the subfamily Ictiniinæ, those of the first two groups are shared so promiscuously that they are useless for other than generic definitions.

Very few of the Indian larvæ of the Gomphidæ have so far been described, and until we know more of their characters they must play a secondary part to venational characters as a means of classifying the families, genera, and species.

Key to Subfamilies of Gomphidæ.

1.	Discoidal cell, hypertrigone, and subtrigone of fore-wing traversed or reticulated	ICTININÆ, p. 370.
	Discoidal cell, hypertrigone, and subtrigone of fore-wing always entire	2.
2.	At least four transverse nervures between sectors of arc in fore-wing from arc to bifurcation of *Rs*	EPIGOMPHINÆ, [p. 315.
	Usually only two (rarely three or four) transverse nervures between sectors of arc in fore-wing from arc to bifurcation of *Rs*	GOMPHINÆ, p. 157.

Subfamily GOMPHINÆ.

Gomphus series Laidlaw, Rec. Ind. Mus. vol. xxiv, pp. 371, 386, 387 (1922).

This subfamily contains the bulk of the species of the family, and it is characterized by the reduction of the number of transverse nervures between the sectors of the arc from the level of the arc to the point of bifurcation of *Rs*, a character quite unique in the suborder ANISOPTERA. These nervures in the case of the hind-wing number only 1, but there may be as many as 4 in the fore-wing, although 2 is the usual number. The bifurcation of *Rs* is usually symmetrical. Generic characters are found in the shape of the anal appendages of the male, the genitalia, the presence or absence of a rudimentary anal loop, the presence or absence of an incomplete basal antenodal nervure, and lastly, the length and armature of the hind femora.

Distribution. Cosmopolitan. Sixteen genera are represented within Indian limits.

Key to Indian Genera of Gomphinæ.

1.	Discoidal cell of hind-wing traversed by a nervure running from costal to distal side; *IA* and *Cuii* in hind-wing widely divergent at the border of wing	DAVIDIUS Selys, p. 160.
	Discoidal cell of hind-wing traversed by a nervure running from costal to distal side; *IA* and *Cuii* in hind-wing running parallel as far as the wing-border	DAVIDIOIDES Fraser, [p. 225.
	Discoidal cell in hind-wing never traversed; *IA* and *Cuii* rarely, or but slightly, divergent at margin of hind-wing	2.

2.	Hind-wing not excavated at base, with tornus rounded in the male	ANORMOGOMPHUS Selys, [p. 172.
	Hind-wing more or less excavated at base, with tornus more or less angulate	3.
3.	Anal loop absent; first postanal cell not extending proximal to base of subtrigone in hind-wing; anal triangle nearly always 3-celled	4.
	Anal loop present, although often rudimentary; first postanal cell extending more or less proximal to base of subtrigone in hind-wing; anal triangle nearly always 4-celled	11.
4.	Lobe of genitalia enormously enlarged and very conspicuous; an incomplete basal antenodal nervure present in all wings; pterostigma very long, equal to nearly half the length of distance from node to proximal end of pterostigma	CYCLOGOMPHUS Selys, [p. 178.
	Lobe of genitalia not greatly enlarged and not over conspicuous; an incomplete basal antenodal nervure rarely present*; pterostigma shorter, equal to less than one-third or only one-fourth the length of distance from node to proximal end of pterostigma	5.
5.	Superior anal appendages and branches of inferior appendage of equal length and equally divaricate............	6.
	Superior anal appendages variable, usually much longer than branches of inferior appendage, or, if not, then furnished with a ventral process or spine beneath	8.
6.	Discoidal field with two rows of cells to beyond level of node; only two rows of cells between *IA* and border of fore-wing	BURMAGOMPHUS Will., [p. 211.
	Discoidal field with three rows of cells at level of node; three or more rows of cells between *IA* and margin of fore-wing	7.
7.	Anal triangle well formed; segments 8 and 9 not markedly dilated laterally.	GOMPHUS Leach, p. 197.
	Anal triangle poorly formed; segments 8 and 9 greatly dilated laterally	PLATYGOMPHUS Selys, [p. 207.

* An incomplete basal antenodal nervure is present in many species of *Anisogomphus* and *Merogomphus*, and irregularly so in one or more wings of *Gomphus personatus* Selys.

8.	Superior anal appendages as long as branches of inferior appendage and furnished beneath with a black robust spine or process	ANISOGOMPHUS Selys, [p. 187.
	Superior anal appendages longer than branches of inferior appendage, simple and without spines or processes beneath..................	9.
9.	Superior anal appendages very closely apposed, the apical ends curling strongly downwards; inferior appendage much shorter, its branches closely apposed and curled up hookwise; segments 8 and 9 with lateral foliate dilatations	MESOGOMPHUS Förster, [p. 228.
	Superior anal appendages widely separated, forcipate, the apical ends curling inwards so as to nearly meet or actually overlap and enclose between themselves an oval space (more rarely the superior appendages simple like those of *Gomphus*)	10.
10.	Inferior anal appendage with widely divaricate branches; hind femora with two rows of about five pairs of very long, widely-spaced spines; segment 9 elongate	MEROGOMPHUS Martin, [p. 309.
	Inferior anal appendage with parallel, short, closely apposed branches; hind femora with two rows of very short, closely set, gradually lengthening spines; segment 9 not elongate	STYLOGOMPHUS Fraser, [p. 288.
11.	Branches of inferior anal appendage ending in two fine branches	MEGALOGOMPHUS Campion, p. 291.
	Branches of inferior anal appendage ending in an acute or obtuse point .	12.
12.	Ground-colour a fine grass-green, with very reduced black markings; anal loop well developed, formed of three cells; anal appendages short, simple, of about equal length	OPHIOGOMPHUS Selys, [p. 305.
	Ground-colour black or bright yellow, with extensive markings; anal loop present, but poorly developed, formed of two cells; anal appendages variable	13.
13.	Anal triangle 3-celled; only two rows of cells between *IA* and margin of fore-wing; lobe of penis swollen and markedly vesiculate	NEPOGOMPHUS, gen. nov., [p. 282.
	Anal triangle 4-celled; three or more rows of cells between *IA* and margin of fore-wing; lobe of penis not vesiculate	14.

14.	Ground-colour preponderatingly black; superior anal appendages enormous curled hooks, which enclose a broad cordate or oval space by meeting the similarly curled branches of inferior appendage, which slightly overlap the superiors	[Fraser, p. 269. LAMELLIGOMPHUS
	Ground-colour preponderatingly yellow; superior anal appendages much straighter and curled only at apices; branches of inferior appendage slightly shorter than superiors, between which they may project ..	[p. 239. ONYCHOGOMPHUS Selys,

Genus **DAVIDIUS** Selys. (Fig. 46.)

Davidius Selys, Bull. Acad. Belg. (2) vol. xlvi, p. 667 (1878); Kirby, Cat. Odon. p. 75 (1890); Williamson, Proc. U.S. Nat. Mus. vol. xxxiii, pp. 273, 286, 287 (1907); Ris, Suppl. Ent. no. v, pp. 45–47 (1916); Laidlaw, Rec. Ind. Mus. vol. xxiv, p. 388 (1922); Fraser, J. Bombay Nat. Hist. Soc. vol. xxix, pp. 61, 329 (1923); id., ibid. vol. xxxi, pp. 164, 165 (1926); Laidlaw, Trans. Ent. Soc. Lond. vol. lxxviii, p. 190 (1930); Needham, Zool. Sinica, ser. A, vol. xi, fasc. 1, pp. 19, 32 (1930); id., Rec. Ind. Mus. vol. xxiv, pp. 217, 218, 224 (1932).

Size rather small, hind-wings with the discoidal cells traversed by a nervure, and nervures *Cuii* and *IA* widely divergent at the margin; colour black, marked with some shade of yellow or dull ochreous.

Head rather large; frons angulated or moderately rounded; occiput straight or emarginate, usually simple. Wings with reticulation close; membrane obsolete; tornus strongly angulated; base of hind-wing very oblique, strongly excavate and sinuous; anal triangle 3-celled; arc situated between second and third antenodal nervures; 2 to 3 transverse nervures between sectors of arc and bifurcation of *Rs* in fore-wing, only 1 in hind-wing; only a single row of postanal cells in fore-wing, four rows in hind-wing; first postanal cell in hind wing not extending beyond the base of subtrigone proximally; anal loop absent; no incomplete basal antenodal nervures present; nodal index high; primary antenodals the first and the fourth, fifth, or sixth; discoidal cell of fore-wing usually entire, obliquely and slightly elongate, distal side slightly longer than costal, latter one-third as long again as basal side; discoidal cell of hind-wing nearly always traversed by a nervure running from costal to distal side, latter slightly longer than former and at least twice as long as basal side; distal side of both discoidal cells sinuous or angulate, cells themselves quite frequently joined to lower sector of arc by a distinct but short stalk; pterostigma short and markedly swollen, that of fore-wing always shorter than that of hind-wing, braced or not, in the fore-wing, this organ about one fourth the length of distance from node to proximal end of pterostigma; *IA* in fore-wing zigzagged but not pectinate, only a double row of cells between it and margin of wing; *Cuii* and *IA* in hind-wing very widely divergent at margin, 4 to 5 cells between their ends at this point; *MA* and *Riv+v* in the same wing also divergent, 3 to 4 cells between their ends at margin; only 1 cubital nervure in all wings; subtrigones and hypertrigones entire in all wings. Legs rather long, hind femora extending to apical end of abdominal segment 1 or to the middle of segment 2, armed with two rows of short, moderately closely-set spines and a single much longer one at the extreme distal end, these spines more robust, more widely-set, and less numerous in the female: tibial spines

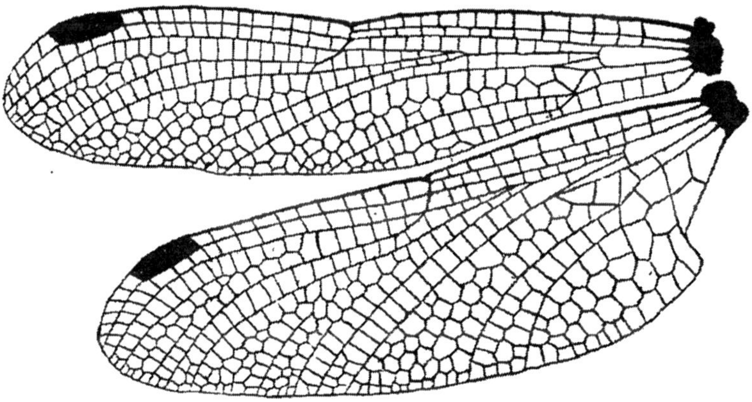

Fig. 46.—Wings of *Davidius zallorensis delineatus* Fraser, male.

very short, closely resembling those on the femora; abdomen short and rather stout, or slim and moderately long, tumid at base, cylindrical from segment 3 to 7, then gradually dilated again as far as the end; segments 7 and 8 occasionally with a curious tubercle or tubercles on the ventral surface. Anal appendages: superiors shortly conical, divergent, often bearing a curled process or spine on the ventral surface; inferior deeply and broadly bifid, the branches short and obtuse or longer, finer, and acutely pointed at apex. Genitalia: lamina depressed, emarginate; anterior hamules spatulate and strongly curled at apices; posterior hamules more robust, compressed, and bearing a short recurved spine at apex; lobe inconspicuous, flask-shaped.

Genotype, *Davidius davidi* Selys, by designation of Kirby.

Distribution.—Fourteen species are known, of which eight occur within Indian limits, these being confined to the montane areas of ASSAM, BENGAL, and SIKKIM: the remainder occur in Japan, Indo-China, and China. I have been able to examine the types of all save that of *D. zallorensis zallorensis* Selys.

The genus is especially interesting on account of the curious blending of archaic with modern characters; thus we find in the venation an angulated discoidal cell, and in the hind-wing a very much elongated one which requires a cross-nervure to strengthen it. The level of the arc is far out, but there is a marked reduction of the cross-nervures between the sectors of arc as far as the bifurcation of *Rs*. The habits of the species and general facies closely copy those of *Burmagomphus*, but with no knowledge of the larvæ of *Davidius* it is impossible to say how far this apparent relationship is real.

Key to Indian Species of Davidius.

1.	Superior anal appendages branched	2.
	Superior anal appendages simple, unbranched	6.
2.	Abdominal segments 3 to 8 with very broad complete basal annules	*kumaonensis* Fras., [p. 171.
	Abdominal segments 3 to 8 with paired yellow basal spots	3.
3.	Both humeral and antehumeral yellow markings present on dorsum of thorax.	4.
	Humeral and antehumeral markings absent or only a humeral cuneiform spot present	5.
4.	Superior branch of superior anal appendages less than half the length of segment 10	*zallorensis zallorensis* [Selys, p. 163.
	Superior branch of superior anal appendages equal in length to segment 10 ...	*zallorensis delineatus* [Fras., p. 164.
5.	An upper cuneiform humeral yellow spot present	*davidi davidi* Selys, [p. 168.
	No upper cuneiform spot present	*davidi assamensis* [Laid., p. 168.
6.	A narrow antehumeral yellow stripe lying close to and parallel with the mid-dorsal stripe	7.
	Antehumeral stripe absent and replaced by an upper oval yellow spot, which by confluence with the mid-dorsal stripe forms a T-like marking	*malloryi* Fras., p. 169.
7.	Labrum and face entirely yellow........	*aberrans aberrans* Selys, [p. 165.
	Labrum and face black	*aberrans senchalensis* [Fras., p. 167.

234. **Davidius zallorensis zallorensis** Selys.

Davidius zallorensis Hagen, in Selys, Bull. Acad. Belg. (2) vol. xlvi, p. 667 (1878); Williamson, Proc. U.S. Nat. Mus. vol. xxxiii, pp. 286, 287 (1907); Ris, Suppl. Ent. no. v. p. 47 (1916); Laidlaw, Rec. Ind. Mus. vol. xxiv, pp. 371, 388, 389 (1922); Fraser, J. Bombay Nat. Hist. Soc. vol. xxix, pp. 61, 330 (1923); Needham, Rec. Ind. Mus. vol. xxxiv, p. 224 (1932).

Davidius ? zallorensis Kirby, Cat. Odon. p. 75 (1890).

Davidius zallorensis zallorensis Fraser, J. Bombay Nat. Hist. Soc. vol. xxxi, pp. 165, 166, pl. i, fig. 3 (1926); Laidlaw, Trans. Ent. Soc. Lond. vol. lxxviii, p. 190 (1930).

Male.—Abdomen 31 mm. Hind-wing 27 mm.

Head: labium pale yellow; labrum greenish-yellow, its extreme base black; face and frons greenish-yellow, base of latter black above; vertex and occiput black, a pale median yellow spot on former, hinder part of latter yellow. *Prothorax* black, its base and posterior lobe yellow. *Thorax* black, marked with greenish-yellow as follows:—A complete mesothoracic collar confluent with a carinal stripe of the same colour, a narrow sinuous antehumeral stripe dilated abruptly above, a very fine humeral line and an upper point, laterally entirely yellow save for a narrow black line on the hinder suture. *Legs* black, rather long, robust; femora brownish, armed with a row of short spines. *Wings* hyaline, in teneral specimens a little enfumed at base as far as outer end of trigones and along costa as far as node. Costa black; pterostigma yellowish in teneral specimens, dark brown when mature, 2·5 to 3 mm. long, covering four cells, braced poorly; discoidal cell of fore-wings not traversed; nodal index 8/12—11/10; 3 cells in anal triangle. *Abdomen* black, marked with yellow as follows:—Segments 1 and 2 with a pale yellow mid-dorsal stripe and the sides broadly yellow, 3 to 6 with a small baso-lateral and a similar apico-lateral spot, borders of 8 and 9 yellowish. *Anal appendages* black, superiors bifid nearly to base, upper branches divaricate, shorter than half the length of segment 10, apex rounded, ending in a short point above, inner branch straight, nearly at a right angle to the outer, inclined abruptly downward, its end curved back slightly towards the base and resting on the inferior appendage; the latter paler, slightly longer, triangular, excavate above, with a transverse tooth on each side, apex blunt and slightly bifid. *Genitalia*: hamules robust, anteriors long curved spines, posteriors more robust and ending in a tooth which curves inwardly towards its fellow.

Female unknown.

Distribution.—Selys gives "Le col de Zallore," Himalayas, but I have been unable to trace any such place either on ordnance maps or by local inquiries.

Type in the Hagen collection.

235. **Davidius zallorensis delineatus** Fraser. (Fig. 47.)

Davidius zallorensis delineatus Fraser, J. Bombay Nat. Hist. Soc. vol. xxxi, p. 166, pl. i, fig. 4, text-fig. 3, iii, viii (1926); Laidlaw, Trans. Ent. Soc. Lond. vol. lxxviii, p. 191 (1930).

Davidius delineatus Needham, Rec. Ind. Mus. vol. xxxiv, pp. 224, 225 (1932).

Male.—Abdomen 26 mm. Hind-wing 25 mm.

Head : labium and labrum glossy black ; anteclypeus dirty yellow ; mandibles bright citron-yellow ; face and lower part of frons glossy black ; frons broadly yellow above, this colour slightly overlapping the fore-border, its extreme base above narrowly black ; rest of head mat black ; occiput with sinuous border, fringed with long black hairs. *Prothorax* black, with a broad anterior collar and a lateral posterior triangular spot citron-yellow. Above and just in front of the posterior lobe a tiny geminate spot. *Thorax* black, marked with yellow as follows :—A complete mesothoracic collar,

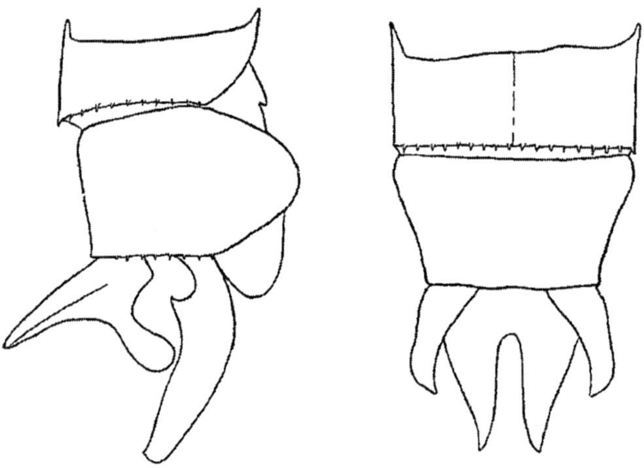

Fig. 47.—Anal appendages of *Davidius zallorensis delineatus* Fraser, male. Right lateral and dorsal views.

long, narrow antehumeral spots not confluent with the collar below, nor meeting the alar sinus above, oblique and converging above ; a small but prominent upper humeral spot and the sides broadly yellow. Posterior suture narrowly mapped out in black ; anterior suture with a vestigial similar line, broadly interrupted in its upper part. *Legs* black, unmarked ; hinder femora with a row of moderately closely-set robust spines, the distal ones the longer ; tibial spines short. *Wings* hyaline, bases bright saffron halfway from base to node. Pterostigma palest brown, framed in darker brown, short and dilated, covering 3 cells, poorly braced as a rule ;

discoidal cell of fore-wing entire, that of hind-wing traversed once, or very rarely entire: only 1 cubital nervure in all wings; 4 to 5 rows of postanal cells in hind-wing; usually only 2 nervures between the sectors of arc in the fore-wing, and only 1 in the hind-wing: nodal index $\frac{12\text{–}14}{10\text{–}9} \Big| \frac{13\text{–}12}{8\text{–}10}$. *Abdomen* black, marked with yellow as follows:—Segment 1 with sides broadly yellow; 2 with the dorsal carina very finely so, and two large subtriangular lateral spots nearly confluent with each other, the basal one of which involves the oreillet: 3 with a large triangular baso-lateral spot narrowing apically, where it is limited by the jugal suture, apical to the jugum with a narrow elongate subapical spot; 4 to 8 with a small baso-lateral and a similar apico-lateral spot on each side, which become progressively smaller from 4 to 8; remaining segments unmarked. *Anal appendages* black (fig. 47). *Genitalia* similar to those of *D. zallorensis zallorensis.* Lamina very depressed, bulb of penis funnel-shaped, notched at its lip, very prominent.

Female.—Abdomen 27 mm. Hind-wing 27 mm.

Almost exactly similar to the male. The yellow markings of abdomen more extensive; yellow areas on sides of segment 2 confluent, on 3 only narrowly divided by the jugal suture, on 4 to 6 an additional spot at basal side of jugal suture and a similar narrow linear spot placed subapically as on segment 3 of the male. Vulvar scale small, triangular, about half the length of segment 9, slightly bifid at apex.

Distribution.—Gangtok, DARJEELING DISTRICT; May (*C. M. Inglis*).

Distinguished easily from *D. zallorensis zallorensis* by the thoracic markings and by the glossy black face. Possibly the face of the latter, in fully mature specimens, is darker than as described by Selys. Venation in both species identical. *Type* in the British Museum.

236. **Davidius aberrans aberrans** (Selys). (Fig. 48.)

Hagenius ? aberrans Selys, Bull. Acad. Belg. (2) vol. xxxvi, p. 506 (1873); Kirby, Cat. Odon. p. 75 (1890).

Davidius aberrans Selys, Bull. Acad. Belg. (2) vol. xlvi, p. 669 (1878); id., Ann. Soc. Ent. Belg. vol. xxxviii, p. 178 (1894); Williamson, Proc. U.S. Nat. Mus. vol. xxxiii, pp. 286, 287 (1907); Laidlaw, Rec. Ind. Mus. vol. xxiv, pp. 371, 388, text-fig. 9 (1922); Fraser, J. Bombay Nat. Hist. Soc. vol. xxix, pp. 61, 329 (1923); Needham, Rec. Ind. Mus. vol. xxxiv, pp. 224, 225 (1932).

Davidius aberrans aberrans Fraser, J. Bombay Nat. Hist. Soc. vol. xxxi, pp. 169, 170, text-fig. 3, ii, pl. i, fig. 1 (1926); Laidlaw, Trans. Ent. Soc. Lond. vol. lxxviii, p. 191 (1930).

Male.—Abdomen 27 mm. Hind-wing 25 mm.

Head: labium, labrum, and face entirely yellow; frons yellow, its base above narrowly black; vertex black; occiput yellow, sinuous, fringed with hairs. *Prothorax* black, a large spot on each side and the posterior lobe bright yellow. *Thorax* black, marked with bright greenish-yellow as follows:—A short uninterrupted mesothoracic collar, which is confluent with two contiguous dorsal stripes which lie close to and parallel with the mid-dorsal carina, a sinuous, moderately narrow antehumeral stripe. Laterally broadly yellow, postero-lateral suture finely lined with black, lastly the remnants of a similar stripe on the anterior suture extending only as far

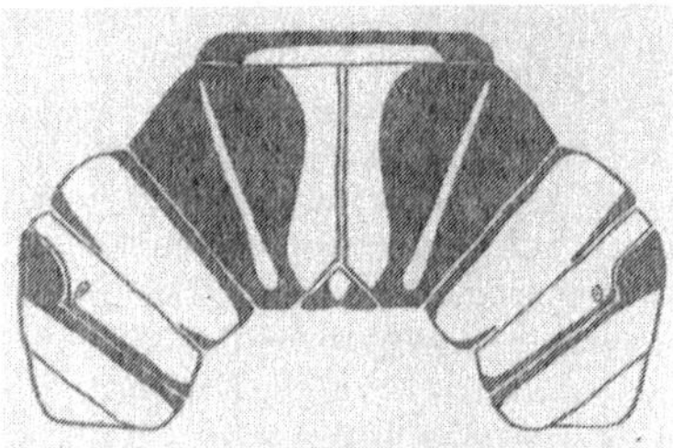

Fig. 48.—Thoracic markings of *Davidius aberrans aberrans* (Selys), male.

as spiracle. The dorsal carina has some dark shading on each side. *Legs* black, coxæ yellow, femora armed as in genus with a row of closely-set, evenly-spaced spines, the last 3 or 4 slightly longer and less closely set. *Wings* hyaline, pale saffron, diffusely so and of a deeper tint at base; pterostigma reddish-brown, swollen, short, and broad, covering 3 cells; 5 rows of postanal cells in hind-wing, 3 cells in anal triangle; discoidal cell in fore-wing entire, in hind-wings traversed once; nodal index $\frac{10\text{–}10}{7\text{–}8}\Big|\frac{10\text{–}10}{8\text{–}7}$. *Abdomen* black, marked with yellow as follows:—A mid-dorsal stripe on segment 1, continued over segment 2 to the basal half of 3; sides of 1 and 2 broadly yellow, including the oreillets; 3 with a subtriangular baso-lateral spot and a narrow linear subapical lateral spot; 4 with similar but much smaller spots; 5 and 6 with the basal spots only; 7 with a baso-lateral "T"-shaped spot; 8 with a linear lateral spot broken at the jugal suture and extending downwards at either end; 9 with a minute subbasal lateral spot and its ventral border narrowly; 10 with a large lateral spot not quite reaching base. *Anal appendages* yellow; superiors divaricate, simple, equal in length to segment 10, unbranched, the apices pointed and turning a little inward; inferior appendage triangular, slightly bifid at the blunt apex. *Genitalia*: lamina depressed, lobe tumid, corrugated, but glossy black; posterior hamules very robust, ending in a short, stout, slightly upturned recurved spine which turns in towards its fellow.

Female.—Abdomen 29 mm. Hind-wing 27 mm.

Very similar to the male, but nodal index slightly higher; discoidal cell of fore-wing traversed or entire; pterostigma slightly longer, covering 4 to 5 cells and slightly longer in the hind-wing than in the fore-wing. Vertex with a small median spot of yellow, occiput low, emarginate, and with two prominences behind. Prothorax with a small geminate spot just in front of the posterior lobe (this probably present also in most males).

Distribution.—*Type* a female in the Mclachlan collection, collected by Capt. Lang in NORTH INDIA. Allotype male in the Pusa collection, collected by Mr. T. Bainbrigge Fletcher at Muktesar, Kumaon, ASSAM, 7,000 ft., May 9, 1923.

The species differs from all others except *D. malloryi* in the simple nature of the anal appendages, and from *D. malloryi* in its markings. The two species are closely related, but, I think, must be considered as distinct rather than subspecies.

237. **Davidius aberrans senchalensis** Fraser.

Davidius aberrans senchalensis Fraser, J. Bombay Nat. Hist. Soc. vol. xxxi, p. 170 (1926); Laidlaw, Trans. Ent. Soc. Lond. vol. lxxviii, p. 191 (1930); Needham, Rec. Ind. Mus. vol. xxxiv, pp. 224, 225 (1932).

Male unknown.

Female.—Abdomen 28 mm. Hind-wing 27 mm.

This subspecies closely resembles typical *D. aberrans*, but differs in the following particulars:—Nodal index higher, $\frac{14\text{–}12 \mid 12\text{–}12}{12\text{–}10 \mid 10\text{–}13}$. Face black, except for a fine obscure submarginal line of yellow on labrum and an obscure yellow line on postclypeus. Occiput indented on either side and not emarginate. Humeral stripe entirely absent. Discoidal cell of fore-wings is entire. Bases of all wings markedly saffronated. Pterostigma in all wings covering only 4 cells. Dorsal abdominal stripe is arrested at extreme base of segment 3; segments 3 to 7 each with a basal and a subapical spot; basal spot on 3 and 4 elongate and dumb-bell-shaped, on 5 to 7 not dilated at apical end, on 8 appearing only as a small baso-lateral spot, and a smaller triangular dorsal spot at base; 9 and 10 unmarked.

Distribution.—A single female from Senchal, DARJEELING DISTRICT, 8,000 ft., May 19, 1924.

The characteristic parallel antehumeral stripes, almost confluent with the yellow mid-dorsal carina, show a close relationship to *D. aberrans aberrans*.

Type in the Author's collection.

Davidius davidi davidi Selys.

Davidius davidii Selys, Bull. Acad. Belg. (2) vol. xlvi, p. 671 (1878); Kirby, Cat. Odon. p. 76 (1890); Selys, Ann. Soc. Ent. Belg. vol. xxxviii, p. 179 (1894); Williamson, Proc. U.S. Nat. Mus. vol. xxxiii, p. 286 (1907); Needham, Zool. Sinica, ser. A, vol. xi, fasc. 1, pp. 34, 35 (1930).

Davidius davidi davidi Fraser, J. Bombay Nat. Hist. Soc. vol. xxix, pp. 61, 329 (1923); id., ibid. vol. xxxi, pp. 166, 167, pl. i, fig. 6 (1926); Laidlaw, Trans. Ent. Soc. Lond. vol. lxx–viii, p. 191 (1930).

Davidius davidi Needham, Rec. Ind. Mus. vol. xxxiv, p. 224 (1932).

The typical form of this species is not known to occur in India.

Male unknown.

Female.—Abdomen 34 mm. Hind-wing 32 mm.

Head black, except the frons, which has a broad dark yellow transverse stripe on its crest; occiput low, fringed with hair behind. *Prothorax* black, its base and posterior lobe dark yellow. *Thorax* black, marked with yellow as follows:—A short unbroken mesothoracic collar; lower part of mid-dorsal carina, which is confluent with the yellow collar; an upper isolated cuneiform humeral spot, and laterally two very broad yellow stripes separated by a narrow black stripe on the postero-lateral suture. Interalar space also yellow. *Legs* black, femora armed as in *D. zallorensis*, rather long (8 mm.). *Wings* hyaline, slightly saffronated, costa black. Nodal index 13/15—15/15; discoidal cell of fore-wing entire, that of hind-wing traversed once; pterostigma blackish-brown, 3 mm. long, stout, covering 3½ to 4 cells. *Abdomen* black, dorsum and sides of segment 1 and a dorsal stripe and the sides of segment 2 yellow; 3 to 5 with baso-lateral and apico-lateral yellow spots, 6 to 7 with basal spots only. *Anal appendages* small, as long as segment 10, which is very short. Vulvar scale slightly notched.

Distribution.—Tibet. Two females, one of which is the *type*, in the Paris Museum, collected by the Abbé David.

238. **Davidius davidi assamensis** Laidlaw. (Fig. 49.)

Davidius davidi assamensis Laidlaw, Rec. Ind. Mus. vol. xii, pp. 135, 136, text-fig. 2 (1916); id., ibid. vol. xxiv, pp. 371, 388, 389, text-fig. 10 (1922); Fraser, J. Bombay Nat. Hist. Soc. vol. xxix, pp. 61, 329 (1923); id., ibid. vol. xxxi, p. 167, pl. i, fig. 2 (1926); Laidlaw, Trans. Ent. Soc. Lond. vol. lxxviii, p. 191 (1930); Needham, Rec. Ind. Mus. vol. xxxiv, pp. 224, 225 (1932).

Male and *female.*—Abdomen 34 mm. Hind-wing 32 mm.

Differs from the last by the total absence of the antehumeral cuneiform spot on thorax. In the male the last five abdominal segments are unmarked; segment 7 has on its ventral aspect, a little apical to the middle of segment, a small tubercle-like process on the ventral tergite, coated with tiny backwardly directed spines : the sides on the eighth segment have a series of larger spines, and the sternite has also a small obtuse projection close after the base. The markings of the body are a rich ochreous- rather than greenish- or citron-yellow as in other species. (This may be due to postmortem changes in

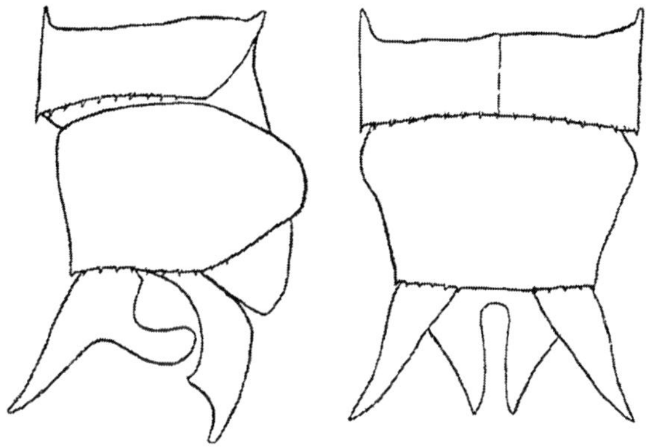

Fig. 49.—Anal appendages of *Davidius davidi assamensis* Laidlaw, male. Right lateral and dorsal views.

the type.) *Wings* hyaline, slightly enfumed at the bases. Venation corresponding to that of *D. zallorensis* : discoidal cells traversed only on hind-wings. *Anal appendages* (fig. 49) very similar to those of *D. zallorensis*.

Distribution.—Gopaldhara, DARJEELING DISTRICT ; one male (*type*) and two females in the Indian Museum.

239. **Davidius malloryi** Fraser. (Figs. 50 & 51.)

Davidius malloryi Fraser, J. Bombay Nat. Hist. Soc. vol. xxxi, pp. 167, 168, text-fig. 3, iv, pl. i, fig. 5 (1926); Laidlaw, Trans. Ent. Soc. Lond. vol. lxxviii, p. 191 (1930); Needham, Rec. Ind. Mus. vol. xxxiv, p. 224 (1932).

Male.—Abdomen 29 mm. Hind-wing 21–23 mm.

Head : labium dirty brown or dirty yellow ; labrum and face greenish-yellow, unmarked : frons greenish-yellow, its base above broadly black ; vertex black ; occiput greenish-yellow, notched at the middle, fringed with long black hairs. *Prothorax* black, its posterior lobe and a large subdorsal spot, which is confluent with it, yellow. *Thorax* black, marked with greenish-yellow as follows :—A very narrow mesothoracic collar, confluent with a median dorsal stripe, which narrows above, and is confluent here with two subdorsal, longitudinal oval spots, the three markings together shaped like a " T." The extreme upper, keeled part of the dorsal carina black, this alone separating the adjacent dorsal oval spots. A tiny upper antehumeral point. Laterally greenish-yellow, postero-lateral suture very finely black, also antero-lateral suture below level of spiracle ; from this a line runs back to meet the posterior suture, thus forming a black inverted "Y." Underside and dorsum very hairy. *Legs* black, femora armed as for genus, very hairy, the distal spine considerably longer. *Wings* hyaline, palely enfumed, not saffronated ; pterostigma bright ochreous between heavy black nervures, braced, but the brace at an angle to the oblique end of pterostigma, covering 3–4 cells, the hind pterostigma slightly the longer, not nearly as swollen as in other species. Nodal index $\frac{8\text{–}11}{8\text{–}8}\Big|\frac{9\text{–}8}{8\text{–}8}$, anal triangle with 3 cells ; 2 cross-nervures between the sectors of arc in fore-wing, only 1 in hind-wing ; only 1 cubital nervure in all wings ; discoidal cell in fore-wing always entire, in hind-wing more often

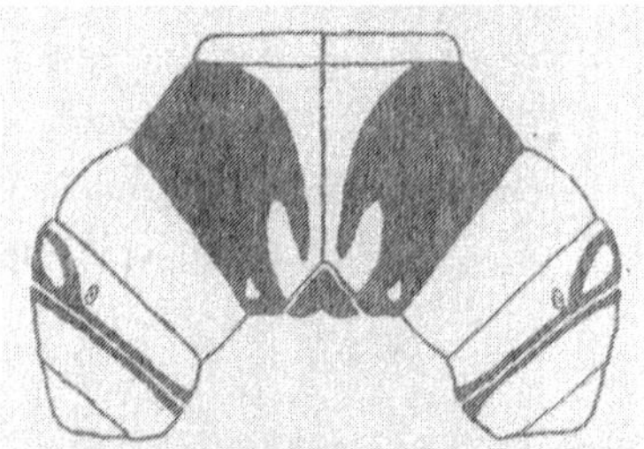

Fig. 50.—Thoracic markings of *Davidius malloryi* Fraser, male.

entire than traversed. *Abdomen* black, marked with yellow as follows :—Segment 1 with a narrow mid-dorsal stripe, continued over 2 nearly to apical border of 3 ; sides of 1 to 3 broadly, including oreillets on 2, after which the yellow is almost interrupted by an invasion of black from above. On segment 3 the lateral yellow broadly interrupted by the black jugal suture, 4 to 7 with small triangular baso-lateral spots, these segments also with the mid-dorsal carina finely yellow, but not extending quite to apex on 5 and 7, latter segment with ventral border narrowly yellow ; 8 to 10 with this same border more broadly yellow ; 10 with a dorso-apical spot ; the intersegmental joints between the last four segments bright yellow. *Anal appendages* (fig. 51) black, the inferior yellow within. *Genitalia* : lamina depressed ; anterior hamules fine, long, black spines, markedly curled, so that the points almost meet the stem again to enclose a space ; posteriors very robust, tumid, paler, sloping downwards and backwards, contracting rapidly near the apex into a short, stout, robust, recurved spine ; lobe inflated, funnel-shaped, glossy black.

Female.—Abdomen 29 mm. Hind-wing 24 mm.

Almost entirely similar to the male ; abdominal markings rather more extensive ; nodal index a little higher ; discoidal cell of hind-wings traversed or entire. Vulvar scale triangular, slightly bifid, apex blunt.

Distribution.—ASSAM. Several specimens, adult males and a single rather teneral female, collected by Mr. T. Bainbrigge Fletcher at Laitlyngkot, Khasi Hills, Assam, 21. iv. 24.

The species is closely allied to *aberrans* by its appendages, but is easily distinguished from it and from all other species by the remarkably specialized dorsal thoracic markings, unique in the family GOMPHIDÆ, and recalling the bizarre patterns of *Cœliccia*.

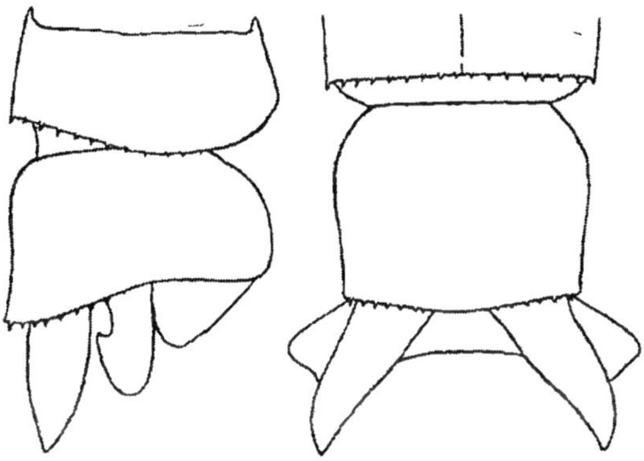

Fig. 51.—Anal appendages of *Davidius malloryi* Fraser, male. Right lateral and dorsal views.

The species is found settled flat on rocks and stones in the beds of rivers, and lies so close that Mr. Fletcher states that he had to " shovel them off with the rim of his net." These habits closely correspond to those of *Burmagomphus*, the genus which most closely approaches the genus *Davidius*, at least in considering the Asiatic genera. The narrower pterostigma and the simple form of the anal appendages might justify a removal of this species and *D. aberrans* to a separate genus. In their simple appendages they resemble *Burmagomphus* more closely than do other species of the genus.

Type in the British Museum.

240. **Davidius kumaonensis** Fraser.

Davidius kumaonensis Fraser, J. Bombay Nat. Hist. Soc. vol. xxxi, pp. 170, 171, pl. i, fig. 6 (1926); Laidlaw, Trans. Ent. Soc. Lond. vol. lxxviii, p. 191 (1930); Needham, Rec. Ind. Mus. vol. xxxiv, pp. 224, 225 (1932).

Male unknown.

Female.—Abdomen 27 mm. Hind-wing 22 mm.

Head: labium dirty yellow, labrum black, face and frons glossy black, upper surface and fore-border of latter greenish-yellow, vertex and occiput black, the latter sinuous, notched at its middle. *Prothorax* black, its posterior lobe, a geminate spot just in front of it, and a largish lateral spot yellow. *Thorax* black, marked with bright yellow as follows:—A very short but complete mesothoracic collar, oblique narrow antehumeral stripes not confluent with the collar, a small upper humeral spot; laterally almost entirely greenish-yellow; posterior suture finely mapped out in black. *Legs* black, unmarked, hind femora with armature common to the genus. *Wings* hyaline, slightly saffronated at the bases. Discoidal cell of fore-wings entire, of hind-wings usually traversed; nodal index $\frac{13\text{–}13}{11\text{–}10}\Big|\frac{15\text{–}13}{10\text{–}11}$; pterostigma pale yellow, stout and short, covering 3 cells, braced indifferently. *Abdomen* black, ringed with yellow as follows:—Basal half of segments 1 and 2, latter with a prolongation along dorsal carina and another along ventral border; segments 3 to 8 with rather more than the basal third yellow, but the ring narrowing considerably on 7 and 8; 9 and 10 unmarked; sutures between last four segments bright yellow. *Anal appendages* and the conical process between them yellow, shortly conical. Vulvar scale small, in poor condition owing to the teneral condition of the specimens.

Distribution.—KUMAON, 7,000 ft., in May (*T. Bainbrigge Fletcher*). Two females, one of which is the *type*, in the British Museum.

This interesting species is distinguished from all others of the genus by the broad rings on the abdomen. The markings on the thorax closely resemble those of *D. zallorensis delineatus*, but without the evidence of the male it cannot be said whether they are nearly related.

Genus **ANORMOGOMPHUS** Selys. (Fig. 52.)

Anormogomphus Selys, Bull. Acad. Belg. vol. xxi, pt. 2, p. 60 (1854); id., Mon. Gomph. p. 102 (1857); Kirby, Cat. Odon. p. 70 (1890); Williamson, Proc. U.S. Nat. Mus. vol. xxxiii, p. 298 (1907); Laidlaw, Rec. Ind. Mus. vol. xxiv, p. 396 (1922); Fraser, J. Bombay Nat. Hist. Soc. vol. xxix, pp. 61, 330 (1923); id., ibid. vol. xxxi, pp. 744, 745 (1926); Laidlaw, Trans. Ent. Soc. Lond. vol. lxxviii, p. 190 (1930); Needham, Rec. Ind. Mus. vol. xxxiv, pp. 217, 225 (1932).

Size rather small; colouring pale, uniform, and almost without markings.

Head moderately large, frons angulated, rather prominent, occiput flat and simple. Wings: reticulation open, base of hind-wing rounded in both sexes, the tornus absent; membrane almost absent: anal triangle single-celled; arc situated between the first and second antenodal nervures; only 1 to 2 nervures between sectors of arc, from arc to bifurcation of *Rs*, in fore-wing, 1 in hind-wing; 1 row of postanal cells in fore-wings, 2 to 3 in hind-wing; first postanal cell in hind-wing extending proximally as far as the centre of base of subtrigone only; anal loop absent; basal incomplete postcostal nervures absent; nodal index low; primary antenodal nervures the first and fourth; discoidal cells entire, that of fore-wing with costal and proximal sides equal, distal slightly longer, that of hind-wing with costal side longer than proximal and distal longer than costal; pterostigma large, expanded at the middle,

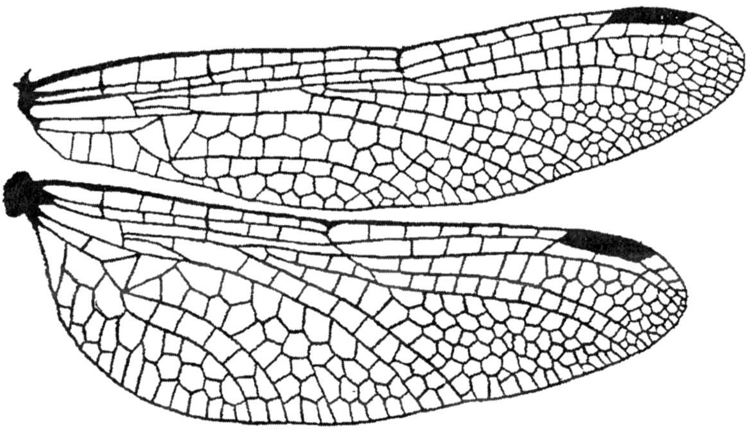

Fig. 52.—Wings of *Anormogomphus heteropterus* Selys, male.

equal to rather more than one-fourth the distance from node to distal end of pterostigma, strongly braced; *IA* in fore-wing not pectinate; *Cuii* and *IA* in hind-wing nearly parallel to termen; only 1 cubital nervure in all wings; hypertrigones and subtrigones entire in all wings. Legs moderately long, hind femora with a group of short, closely-packed, robust spines and a long single spine at distal end. Abdomen slightly tumid at base, terminal segments in male slightly dilated, in female of even width and cylindrical throughout. Anal appendages very short; superiors conical, widely divaricate, and furnished with a robust basal ventral spine; inferior deeply cleft into two widely divaricate branches. Genitalia: lamina depressed, minutely emarginate; anterior hamules small, fine; posterior hamules much longer, projecting almost perpendicularly, tapering to an acute hook-like apex; lobe funnel-shaped, moderately large; vulvar scales two, very small, triangular, widely divaricate plates.

Genotype, *Anormogomphus heteropterus* Selys.

Distribution.—Generally desert and arid areas of INDIA. Iraq, and Persia, but one species is known from BENGAL.

Larvæ undescribed.

Key to Indian Species of Anormogomphus.

Head and thorax with pale brown markings..	*heteropterus* Selys, [p. 174.
Head and thorax without markings	*kiritschenkoi* Bart., [p. 176.

241. **Anormogomphus heteropterus** Selys. (Fig. 53.)

Anormogomphus heteropterus Selys, Bull. Acad. Belg. vol. xxi, pt. 2, p. 61 (1854); id., Mon. Gomph. p. 103 (1857); id., Bull. Acad. Belg. (2) vol. xxxvi, p. 500 (1873); Kirby, Cat. Odon. p. 70 (1890); Williamson, Proc. U.S. Nat. Mus. vol. xxxiii, p. 298, text-fig. 26 (1907); Laidlaw, Rec. Ind. Mus. vol. xxiv, pp. 371, 396 (1922); Fraser, J. Bombay Nat. Hist. Soc. vol. xxix, pp. 61, 330 (1923); id., ibid. vol. xxxi, pp. 745, 746, text-figs. 2, xii, & 5 *b* (1926); Needham, Rec. Ind. Mus. vol. xxxiv, p. 225 (1932).

Male.—Abdomen 25 mm. Hind-wing 22 mm.

Head pale whitish-yellow, eyes and ocelli brown and contrasting strikingly with the pale ground-colour; an obscure brown line at base of frons above and another in front of occiput; latter simple, depressed, its border angulated obtusely. *Prothorax* sandy-yellow, with obscure brownish transverse lines. *Thorax* darker yellow with a grass-green tinge laterally; dorsum with obscure vestiges of brownish lines; a median bordering each side of mid-dorsal carina and convex inwards, an antehumeral stripe, more distinct, running obliquely from above downwards and outwards, lastly a humeral stripe only visible at its middle part. *Wings* pale, reticulation open, black, costa bright yellow; pterostigma pale yellow between black nervures, short, and stout, covering 1 and 2 cells, braced; nodal index $\frac{5\text{–}9}{6\text{–}7}\Big|\frac{10\text{–}5}{7\text{–}6}$; two rows of cells in discoidal field to beyond level of node. *Legs* sandy yellow with black spines. *Abdomen* pale sandy yellow, marked with dark brown as follows:—A spot on each side of segment 1, a baso-lateral and a broad apico-lateral on 2, jugal sutures and articulations of 3 and 6, these segments also having a small blackish-brown spot on each side of apical border; 7 with a basal ring, 8 and 9 with a dorsal stripe on carina pale yellow; sides of 8, 9, and 10 darker yellow; apical border of segment 10 rounded and not turned down between anal appendages as in *A. kiritschenkoi*. *Anal appendages* (fig. 53), yellow. *Genitalia*: lamina depressed, shallowly notched; anterior hamules small, conical, tapering; posterior hamules more robust, projecting perpendicularly to long axis of abdomen, of even thickness to apex, where they are abruptly conical, and end in a short, fine, inwardly curved black spine; lobe large, tumid, deeply bifid at lip, scoop-shaped.

Female.—Abdomen 27 mm. Hind-wing 25 mm.

Very similar to but larger than male. Closely resembles in many respects the female of *A. kiritschenkoi*, but reticulation of wings is much more abundant, as in the male. *Head* (missing in the only known specimen of this sex). *Prothorax* similar to that of male. *Thorax* yellow laterally, grass-green on dorsum with some obscure dark brown stripes: upper part of mid-dorsal carina; a well-defined antehumeral stripe

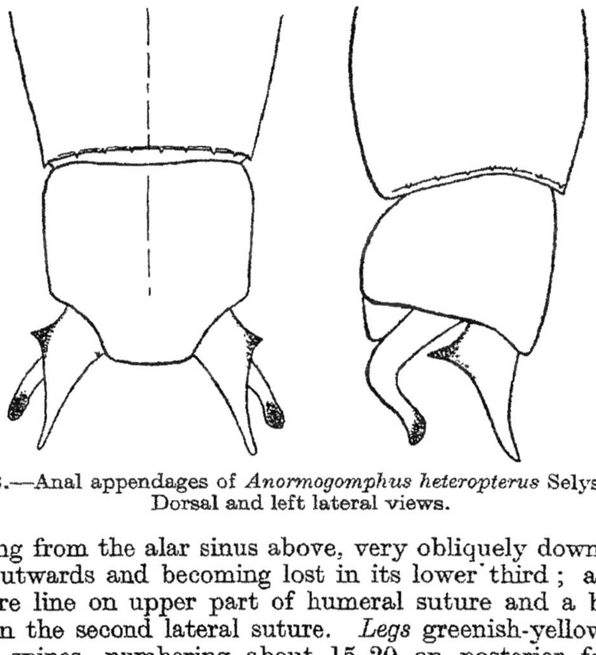

Fig. 53.—Anal appendages of *Anormogomphus heteropterus* Selys, male. Dorsal and left lateral views.

running from the alar sinus above, very obliquely downwards and outwards and becoming lost in its lower third; a short obscure line on upper part of humeral suture and a broken line on the second lateral suture. *Legs* greenish-yellow with black spines, numbering about 15–20 on posterior femora. *Wings* hyaline, similar to those of male; *Cuii* pectinate in fore-wing. *Abdomen* yellow with a greenish tinge, marked obscurely as follows: segments 2 and 7 with a subapical, subdorsal, oblique linear spot on each side convergent apically; 8 and 10 ferruginous, the subdorsum and subapical borders clouded with black; 10 with the base similarly clouded on dorsum. *Anal appendages* ferruginous, very small, conical, separated by a similarly coloured conical structure; vulvar scale very minute; two tiny triangular processes on each side of hinder border of segment 8.

Distribution.—*Type*, a male in the Selys collection, from India, without further data as to locality, collected by Stevens. There is a male from Lahore, PUNJAB, in the Indian Museum, and a third male in my own collection from BIHAR, the latter very teneral. The only female known is a damaged but fully coloured specimen in my collection from Baghwonie, BIHAR.

A. heteropterus appears to be much rarer than *A. kiritschenkoi*, but may be overlooked on account of its weak flight and colourless body and wings. The above description has been taken from the Bihar specimen.

242. **Anormogomphus kiritschenkoi** Bartenef. (Fig. 54.)

Anormogomphus kiritschenkoi Bartenef, Revue Russe d'Ent. vol. xiii, p. 179 (1913); Morton, Ent. Month. Mag. (3) vol. v, p. 149 (1919); id., ibid. (3) vol. vi, p. 87 (1920); id., Ann. Mag. Nat. Hist. (9) vol. v, pp. 296, 297 (1920); Fraser, J. Bombay Nat. Hist. Soc. vol. xxxi, pp. 746, 747, text-figs. 2, xi, & 5 *a* (1926); Needham, Rec. Ind. Mus. vol. xxxiv, p. 225 (1932).

Male.—Abdomen 29–31 mm. Hind-wing 24–25 mm.

Very similar to the last species, but rather more colourless, and the reticulation of wings more simple and open. *Head*: lips, face, and vertex palest greenish-white or creamy; ocelli and eyes dark brown and strongly contrasted against the pale ground-colour; occiput pale, its hinder border obtusely angled. *Prothorax* palest yellow, rather darker on dorsum; posterior lobe and a transverse ridge in front of it brighter greenish-yellow. *Thorax* pale green, yellowish in some specimens, but very variable and almost grass-green in others, unmarked; dorsum darker olivaceous. *Legs* yellow, all femora with a longitudinal brown line on the outer sides, dark and well defined distally, paler and diffuse proximally, and much darker and better defined on the anterior than posterior femora; spines shorter, but more numerous and more closely set than in *A. heteropterus*. *Wings* hyaline, costa and most of the nervures pale straw-coloured; pterostigma pale with a pinkish tinge, between black nervures, covering three cells, braced; nodal index of two specimens $\frac{7\text{–}8}{6\text{–}6}\Big|\frac{9\text{–}6}{6\text{–}5}$, $\frac{6\text{–}10}{7\text{–}7}\Big|\frac{10\text{–}6}{7\text{–}7}$; discoidal field in fore-wing with two rows of cells nearly to level of node. *Abdomen* sandy-yellow, with subapical subdorsal spots on segments 2 to 6 exactly similar to those of *A. heteropterus*; segments 7 to 10 unmarked. *Anal appendages* (fig. 54) yellow, superiors with a pale green tint, inferior ferruginous. *Genitalia* very similar to those of *A. heteropterus*;

lamina very depressed, its surface with a narrow deep fissure dividing it into two convex areas ; anterior hamules conical and short in profile, sloping backwards and downwards, parallel, flattened as seen from below ; posterior hamules much more robust, conical, ending in a tiny forwardly directed spine, cream-coloured, projecting very slightly backwards, and almost perpendicularly so, to long axis of abdomen ; lobe dark olivaceous, scrotal-shaped, narrowly but deeply notched at apex.

Female.—Abdomen 29–31 mm. Hind-wing 26–27 mm.

Similar in all respects to male ; two rows of cells in discoidal field of fore-wing to level of node ; pterostigma slightly longer ; vulvar scale as for genus.

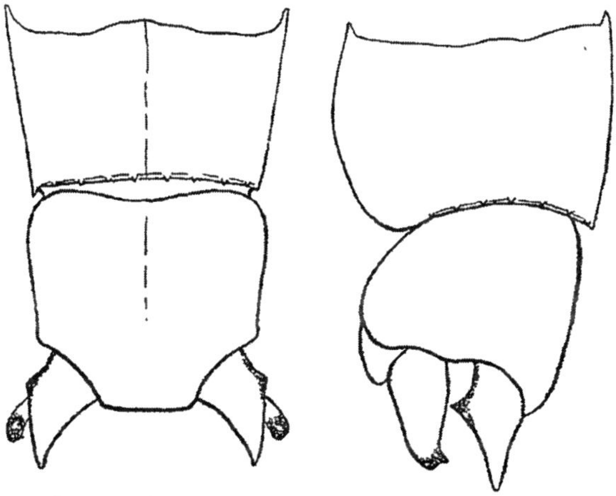

Fig. 54.—Anal appendages of *Anormogomphus kiritschenkoi* Bartenef, male. Dorsal and left lateral views.

Distribution.—Iraq, Persia, along the shores of the Persian Gulf, the Mekran coast, and SIND. I took this species in large numbers at Zobeir, Mesopotamia, in April 1915. Baghdad is the most northern locality from which I have received specimens ; it occurs later there, towards the end of August. It is probably not uncommon along the banks of the lower Indus.

Its flight is weak and not sustained. It may be found near its breeding places settled gregariously on low bushes or in grass. At Zobeir the inhabitants dig holes in the sand for their water supply, fresh pits being dug as others become fouled. It is in these pits that *A. kiritschenkoi* breeds.

Type in the Leningrad Museum ; other specimens in the British Museum, Morton and Author's collections.

Genus CYCLOGOMPHUS Selys. (Fig. 55.)

Cyclogomphus Selys, Bull. Acad. Belg. vol. xxi, pt. 2, p. 61 (1854); id., Mon. Gomph. p. 105 (1857); Kirby, Cat. Odon. p. 69 (1890); Williamson, Proc. U.S. Nat. Mus. vol. xxxiii, pp. 274, 296–298 (1907); Laidlaw, Rec. Ind. Mus. vol. xxiv, p. 390 (1922); Fraser, J. Bombay Nat. Hist. Soc. vol. xxix, pp. 64, 332 (1923); id., ibid. vol. xxxi, pp. 158, 159 (1926); Laidlaw, Trans. Ent. Soc. Lond. vol. lxxviii, p. 186 (1930); Needham, Rec. Ind. Mus. vol. xxxiv, p. 218 (1932).

Size medium; flight slow and unsustained; colour pale yellow, with linear black markings.

Head triangular, moderately large for the size of the insect, frons angulate, occiput simple, flat or slightly concave. Wings: reticulation very open; tornus angulate; base of hind-wing rather deeply excavate; membrane obsolete; costa brightly coloured; anal triangle 3-celled; arc situated between first and second antenodal nervures; only 2 transverse nervures between sectors of arc from latter to bifurcation of

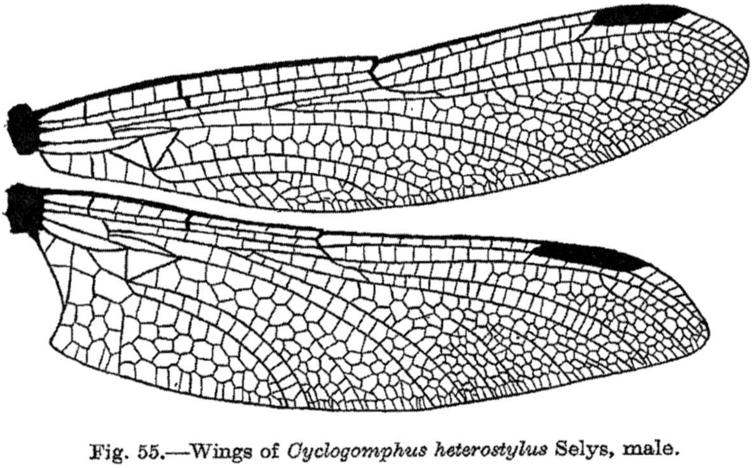

Fig. 55.—Wings of *Cyclogomphus heterostylus* Selys, male.

Rs in fore-wing, only 1 in hind-wing; fore-wing with a single row of postanal cells, 3 to 4 rows in hind-wing; anal loop absent, first postanal cell not extending proximal to inner end of base of subtrigone; incomplete basal antenodal nervure present in all wings; nodal index moderately high; first and fifth antenodal nervures the primaries; discoidal cell of fore-wing entire, with costal side slightly longer than basal and distal slightly longer than costal, entire; that of hind-wing also entire, elongate in length of wing, costal and distal sides about equal and nearly twice the length of basal; pterostigma long and stout, distinctly less than twice the distance between node and proximal end of pterostigma, braced; *IA* in fore-wing short and strongly arched, not pectinate; *IA* and *Cuii* in hind-wing parallel nearly to wing-border; only 1 cubital nervure in all wings; subtrigones and hypertrigones all entire. Legs robust, hind femora very long, extending to middle of segment 2, and furnished with a group of short numerous spines on the proximal half and two rows of longer, more robust spines on the distal half; tibial spines short, numerous, very closely set. Abdomen relatively short and stout, tumid at base, cylindrical from segment 3 to base of 7, the latter and 8 and 9 markedly dilated, 10 very short and narrow. Anal appendages: superiors as long as segment 10, parallel, broad at base, then constricted and again expanded at apex, which tapers to a fine point (the superior appendages of *C. gynostylus* are aberrant); inferior broad at base, deeply bifid into two long, narrow, widely divaricate branches, with an inner spine near the middle. Genitalia: lamina small, rounded, deeply emarginate; anterior hamules very small; posterior hamules very long and projecting prominently from the genital sac, their apices pointed and turned inwards; lobe enormously swollen, a massive, scrotal-shaped organ.

Genotype, *Cyclogomphus ypsilon* Selys.

Distribution.—India, Burma, Ceylon, Malaysia, and Siam.

The larvæ (as yet undescribed) breed in sluggish submontane or Deccan streams, or in marshy spots at the origin of such streams. The imago rests in grass or on sedges, and has an extremely weak flight.

Key to Indian Species of Cyclogomphus.

1.	Two black Y-shaped markings on each side of thorax	2.
	One Y-shaped marking on each side of thorax	3.
2.	Superior anal appendages black, closely apposed and curved and arched strongly downwards	*gynostylus* Fras., [p. 185.
	Superior anal appendages pale yellow, short, straight, slightly separated, and directed straight backwards	*heterostylus* Selys, [p. 182.
3.	Black stripe on antero-lateral suture of thorax complete; yellow mid-dorsal stripes on abdominal segments 4 to 7	*wilkinsi* Fras., p. 183.
	Black stripe on antero-lateral suture of thorax vestigial, directed obliquely backwards but not fusing with stripe on postero-lateral suture; yellow oval spots on abdominal segments 4 to 7	*ypsilon*, Selys, p. 180.

243. **Cyclogomphus ypsilon** Selys.

Cyclogomphus ypsilon Selys, Bull. Acad. Belg. vol. xxi, pt. 2, p. 62 (1854); id., Mon. Gomph. pp. 107, 406 (1857); Williamson, Proc. U.S. Nat. Mus. vol. xxxiii, p. 297 (1907); Fraser, J. Bombay Nat. Hist. Soc. vol. xxix, pp. 64, 332 (1923); id., Rec. Ind. Mus. vol. xxvi, pp. 427, 475 (1924); id., J. Bombay Nat. Hist. Soc. vol. xxx, text-fig. 3, viii, pl. facing p. 854, fig. 1 (1925); id., ibid. vol. xxxi, pp. 159–160, text-fig. 1 (1926); Laidlaw, Trans. Ent. Soc. Lond. vol. lxxviii, p. 186 (1930); Fraser, Rec. Ind. Mus. vol. xxxiii, p. 447 (1931).

(?) *Cyclogomphus vesiculosus* Selys, Bull. Acad. Belg. (2) vol. xxxvi, p. 300 (1873); Kirby, Cat. Odon. p. 69 (1890); Williamson, Proc. U.S. Nat. Mus. vol. xxxiii, p. 297 (1907); Laidlaw, Rec. Ind. Mus. vol. xxiv, pp. 371, 390 (1922); Fraser, J. Bombay Nat. Hist. Soc. vol. xxix, pp. 64, 332 (1923); id., ibid. vol. xxxi, p. 161 (1926); Laidlaw, Trans. Ent. Soc. Lond. vol. lxxviii, p. 186 (1930).

Cyclogomphus hypsilon Laidlaw, Rec. Ind. Mus. vol. xxiv, pp. 371, 390, 391, text-fig. ii (1922).

Male.—Abdomen 30 mm. Hind-wing 27 mm.

Head : labium yellow, its middle lobe bordered with black ; labrum, ante- and postclypeus greenish-yellow, former with a very fine bordering and base black ; frons above and in front greenish-yellow, with a fine transverse black band across its lower front border, and a narrow basal border above, rather more extensively so in middle ; vertex black ; occiput pale yellow, simple. Eyes bottle-green, yellow below and behind, glossy black above and behind. *Prothorax* black, with a geminate spot on mid-dorsum of posterior lobe and a large yellow spot on each side. *Thorax* black and greenish-yellow ; two broad black mid-dorsal stripes narrowly separated by the finely yellow median ridge, which is itself confluent below with a broad yellow complete mesothoracic collar, the yellow thus forming an inverted " Y " ; the two black stripes widely confluent above through the medium of the black alar sinus, and also outwardly with a broad black humeral stripe which crosses the humeral suture rather obliquely and encloses a small yellow spot above its upper part, thus being split into two arms and forming an upright " Y " ; laterally a vestigial fine black stripe on the upper part of the first lateral suture, and a complete black stripe on the second lateral suture. *Legs* bright yellow, marked with black ; tibiæ and tarsi black, but the hind tibiæ with a small yellow spot on the distal end of the flexor surface ; hind femora with an inner and outer stripe, broad and confluent distally, rapidly tapering and ending well before the base ; middle and anterior femora entirely black on the outer side. *Wings* hyaline, or in old specimens evenly enfumed ; pterostigma pale brown, or in some male specimens brown at the centre, pale at either end, and well-braced, covering 3 to 4 cells ; nodal index very variable :

$\frac{8-11}{7-9} \Big| \frac{11-8}{9-8}$, $\frac{9-14}{10-11} \Big| \frac{12-9}{11-9}$; four rows of postanal cells in hind-wings. *Abdomen* black, marked with greenish-yellow as follows : segment 1 almost entirely yellow, a fine basal black border broadening very slightly subdorsally ; 2, including the large oreillets, yellow, with broad longitudinal subdorsal black stripes which enclose a mid-dorsal bilobed spot, the black broadening at level of transverse suture and almost confluent with that of the other side ; 3 similar to 2, the black, however, confluent at apical border of segment and almost so at level of transverse suture, thus cutting the dorsal yellow into a smallish basal spot and a much larger apical spot ; 4 to 6 with the subdorsal black stripe not extending nearly to base of segments, there being a broadish yellow complete ring here, at apex of each segment the black broadly confluent over the dorsum, and almost equally so at the transverse sutures ; 7 similar, but the black stripe of even width at transverse suture and not nearly confluent, at apical border only slightly confluent ; 8 similar, but the black subdorsal stripe broader and extending almost to base of segment ; 9 similar, but the black stripes well separated throughout their entire length ; 10 almost entirely black, with a small arrowhead-shaped mark on the dorsum and the ventro-lateral borders narrowly yellow. *Anal appendages* : superiors resembling those of LIBELLULINÆ, almost equal in length to segment 10 ; seen from the side cylindrical in basal half, broadened in apical half, extreme apex curved up, stem of appendage curved down, broadened part expanded below into a robust blunt tooth ; seen from above thickened at extreme base, then constricted and again considerably dilated and finally tapered to an acute point ; these appendages moderately closely apposed, enclosing a small foramen between their stems, yellow. Inferior appendage nearly one-third longer, deeply bifid, the branches widely divaricate, more slim than the superiors, bluntly acute at the apices, tapering, presenting a large blunt tooth at the middle third which is directed down and somewhat inwards, yellow changing to black in the outer half. *Genitalia* prominent ; lamina depressed ; hamules very large, projecting almost perpendicularly from genital sac, apices directed slightly forward, pale yellow ; lobe enormously swollen, pyriform, globular, overlapping basal third of third abdominal segment, greenish-yellow, surmounted by a black penis.

Female similar to male, differing as follows : usually larger, abdomen 32 mm., hind-wing 29 mm. ; nodal index as variable as in male ; basal marking on abdominal segment 1 broader ; segments 2 and 3 similar ; 4 to 7 with the black broadly confluent over dorsum at transverse suture and apical border, thus enclosing long oval yellow spots ; segment 10 broadly yellow on dorsum. Vulvar scale small and rudimentary, two tiny triangular processes in close apposition at base of segment 8, followed immediately afterwards by two raised folds on the ventral aspect of segment 9 which are widely divaricate.

Distribution.—CENTRAL INDIA and the DECCAN. *Type*, a male in the British Museum ; a paratype in the Selys collection is labelled " Cuna," obviously an error for " Guna," in the Central Provinces, India.

I found this insect moderately plentiful in the marshland at the head of the Katraj Lake, near Poona, Deccan. It rests in long grass, and has to be put up by beating. Its flight is short and weak, and it falls an easy prey to the collector's net.

The type of *C. vesiculosus*, a male from an unrecorded locality in India, with the last five abdominal segments missing, was formerly in the Moore collection, but appears to have been lost. It was probably merely a small specimen of *C. ypsilon.*

244. **Cyclogomphus heterostylus** Selys. (Fig. 56.)

Cyclogomphus heterostyla Selys, Bull. Acad. Belg. vol. xxi, pt. 2, p. 62 (1854).

Cyclogomphus heterostylus Selys, Mon. Gomph. p. 106 (1857); id., Bull. Acad. Belg. (2) vol. xxxv, p. 757 (1873); Kirby, Cat. Odon. p. 69 (1890); Williamson, Proc. U.S. Nat. Mus. vol. xxxiii, pp. 297, 298, text-fig. 23 (1907); Laidlaw, Rec. Ind. Mus. vol. xxiv, pp. 371, 390, 391 (1922); Fraser, J. Bombay Nat. Hist. Soc. vol. xxix, pp. 64, 332 (1923); id., Rec. Ind. Mus. vol. xxvi, pp. 427, 474 (1923); id., J. Bombay Nat. Hist. Soc. vol. xxxi, pp. 160, 161 (1926); Laidlaw, Trans. Ent. Soc. Lond. vol. lxxviii, p. 186 (1930); Fraser, Rec. Ind. Mus. vol. xxxiii, p. 447 (1931).

Male.—Abdomen 32 mm. Hind-wing 28 mm.

This species is very similar to the last. The differences indicated in Selys's description are found on examination of a large number of specimens to be inconstant. Thus no dependence can be placed on the nodal index, which varies widely in this as in *C. ypsilon* and *C. wilkinsi.* The bicolorous character of the pterostigma is also found to be shared by fully mature specimens of *C. ypsilon.* Size also is very variable. Other minor differences given by Dr. Hagen are also unimportant. The only reliable characters which I have been able to find are the following :—Black band on lower part of frons distinctly thicker ; labrum entirely yellow ; stripe on first lateral suture of thorax oblique, and confluent at an angle with stripe on second lateral suture, so as to form a second black " Y " on sides. Finally, the anal appendages (fig. 56) present some differences : the superiors are more closely apposed and their apices are turned outward as well as downward; the inferior is relatively much longer, the branches being stouter and tapering more gradually, the tooth at the middle third is smaller, more acute, and directed straight up.

Female unknown.

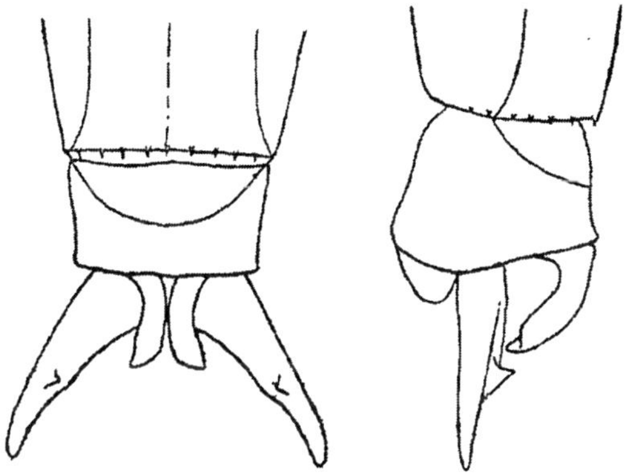

Fig. 56.—Anal appendages of *Cyclogomphus heterostylus* Selys, male. Dorsal and left lateral views.

Distribution.—Katraj Lake, Poona, DECCAN, and MADRAS, in marshes along the course of the Coomb River.

Habits similar to those of *C. ypsilon*, and, like it, found on the wing during September and October.

Type in Saunders's collection (British Museum), from the north of India, locality not stated.

245. **Cyclogomphus wilkinsi** Fraser. (Fig. 57.)

Cyclogomphus wilkinsi Fraser, J. Bombay Nat. Hist. Soc. vol. xxx, text-fig. 3, v, vi, & vii, pl. facing p. 854, fig. 2 (1925); id., ibid. vol. xxxi, pp. 161, 162 (1926); Laidlaw, Trans. Ent. Soc. Lond. vol. lxxviii, p. 186 (1930); Fraser, Rec. Ind. Mus. vol. xxxiii, p. 447 (1931).

Male.—Abdomen 34 mm. Hind-wing 28 mm.

Head: labium bright citron-yellow, mid-lobe somewhat greenish; labrum bright citron-yellow, base very narrowly black; face bright yellow, with a narrow well-defined black band separating frons from postclypeus; frons unmarked save for a very fine black basal line; vertex black, occiput bright yellow, flat, a little concave. *Prothorax* black, marked with an anterior yellow collar, a mid-dorsal geminate spot, and a large lateral spot on each side. *Thorax* bright greenish-yellow, marked vividly with black as follows: two broad dorsal stripes which taper to a point below and outwards, but converging and confluent above, the complete yellow mesothoracic collar sending a fine carinal prolongation upwards which separates them except near the alar sinus; a humeral black "Y," the arms of which are of even thickness and equal length; two lateral narrow black stripes on the lateral sutures, parallel, not extending below the level of the spiracle, finely connected above by a bordering line. Unmarked beneath. *Legs* bright citron-yellow, marked with black; inner sides of femora almost entirely black, outer sides with a fine longitudinal black stripe, almost obsolete in anterior femora. Hind femora armed with pairs of moderately widely spaced, fine, short, black spines; tibial spines short, tibiæ with a black stripe on both inner and outer sides. *Wings* hyaline, costa wholly bright citron-yellow except at pterostigma, latter brown

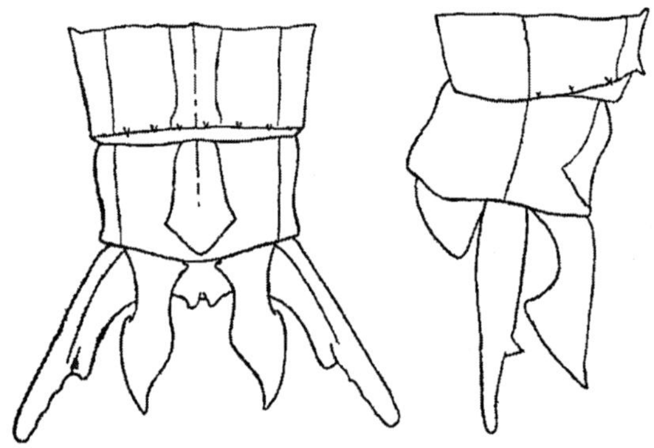

Fig. 57.—Anal appendages of *Cyclogomphus wilkinsi* Fraser, male. Dorsal and left lateral views.

between black nervures, braced, covering four cells; a basal incomplete antenodal nervure in all wings; nodal index $\frac{10\text{–}11}{10\text{–}10} \left| \frac{14\text{–}9}{9\text{–}9} \right.$; one row of cells between *Rii* and *IRii*. *Abdomen* short and tumid, segments 3 to 6 narrower and cylindrical; black, marked with yellow as follows: segment 1 with sides and dorsum broadly, enclosing a subdorsal stripe of black; 2 similar, the subdorsal black stripes enclosing a trilobed carinal dorsal yellow stripe; 3 very similar, but the dorsal yellow stripe more narrow and nearly divided by the fine transverse black suture; 4 to 6 each with a basal yellow ring, which is prolonged in a narrow fusiform dorsal stripe to the apical border of segment and laterally for a very short distance along the ventral border, being separated by a considerable interval from an oval lateral spot; 7 similar, but the basal ring much narrower and the ventro-lateral oval spot expanding into a broad lateral fascia, especially towards the apex of segment; 8 and 9 with a narrow even dorsal stripe running from apex to base, and the sides along the ventral border broadly yellow; segments 8 and 9 and the greater part of 7 laterally expanded as in *C. heterostylus*; segment 10 with narrow dorsal ventro-lateral yellow stripes; intersegmental joints from 1 to 7 finely black, the remaining finely yellow. *Anal appendages* (fig. 57) very similar to those of *C. ypsilon*, but branches of inferior longer, and black from the lateral spine as far as apex. *Genitalia*: lobe rather more tumid; lamina slightly larger; hamules bright yellow, longer, less recurved.

Female.—Abdomen 35 mm. Hind-wing 31 mm.

Similar to male, but the black more extensive. The subdorsal black stripes extend basally as far as the extreme base of segments, thus cutting the basal rings into dorsal and lateral spots. Laterally the ventral oval spots are much more extensive, and may be actually confluent with the basal yellow or merely separated by a fine prolongation from the transverse suture.

Distribution.—One male and two females taken by Mr. Wilkins and myself along the banks of a small stream at Hunse, MYSORE, 12–13. x. 1924, settled in long grass, after the usual habit of species of the genus.

C. wilkinsi is closely allied to both *C. heterostylus* and *C. ypsilon*, but differs in its larger size, in the continuous dorsal carinal stripes on segments 3 to 7, and more especially in the complete antero-lateral black stripe on sides of thorax (vestigial in the other species), and in the form of the humeral black "Y." Lastly the extreme apices of the superior appendages are turned distinctly inwards.

Type and allotype at present in my own collection, but will be deposited eventually in the British Museum.

246. **Cyclogomphus gynostylus** Fraser. (Fig. 58.)

Cyclogomphus heterostylus Laidlaw, Spolia Zeylan. vol. xii, p. 341 (1924).

Cyclogomphus gynostylus Fraser, J. Bombay Nat. Hist. Soc. vol. xxx, text-fig. 3, iv, pl. facing p. 854, fig. 3 (1925); id., ibid. vol. xxxi, pp. 162, 163 (1926); Laidlaw, Trans. Ent. Soc. Lond. vol. lxxviii, p. 186 (1930); Fraser, Ceylon J. Sci. vol. xviii, p. 26 (1933).

Male.—Abdomen 26 mm. Hind-wing 23 mm.

Head: labium dirty yellow; labrum bright citron-yellow, its base finely black; ante- and postclypeus yellow; frons above yellow traversed by a black band across its crest, from which a short black tongue runs downward; vertex black; occiput dark ochreous, its border straight, fringed with black hairs. *Prothorax* black, with a geminate spot on dorsum of posterior lobe and a large lateral spot. *Thorax* black, marked with greenish-yellow as follows: a complete mesothoracic collar; a small median spot on dorsal carina; a broad very oblique antehumeral stripe which runs from near alar sinus downwards, steadily broadening; a small upper humeral spot which bisects the black, dividing the antehumeral stripe from the lateral yellow, and converts it into a black "Y." Laterally yellow, with a broad black stripe on the postero-lateral suture which sends a short oblique branch forwards at its upper part and so forms a second black "Y" on the sides. *Legs* short, hind femora not extending beyond the apical border of segment 1, black; inner sides of the two anterior femora greenish-yellow studded with minute black spines; posterior femora largely yellow, clouded with black on the outer sides, these armed with two rows of 7–8 black, robust, widely-spaced spines.

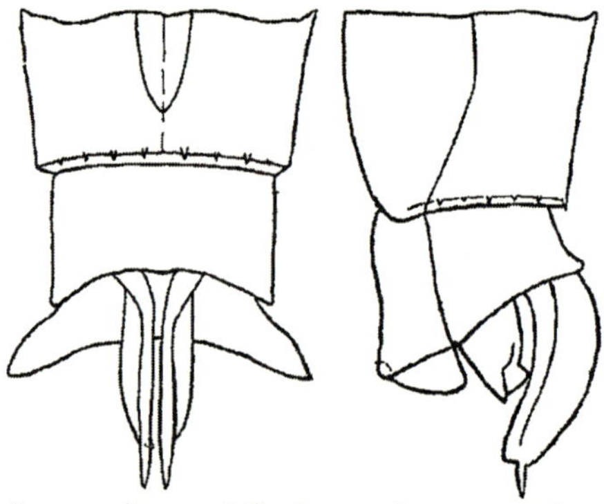

Fig. 58.—Anal appendages of *Cyclogomphus gynostylus* Fraser, male. Dorsal and left lateral views.

Wings hyaline. A single row of postanal cells in fore-wing, 4 in hind-wing; sectors of arc parallel from origin; trigone of hind-wing elongate as in genotype; 2 rows of cells in discoidal field of fore-wing almost to level of node; first postanal cell in hind-wing entire and not nearly extending inwards as far as proximal angle of subtrigone; anal triangle of 3 cells; base of wing slightly excavate; pterostigma yellow, swollen, short, braced, covering 2 to 3 cells; a basal incomplete antenodal nervure present in all wings; nodal index $\frac{8\text{–}11}{8\text{–}9}\Big|\frac{13\text{–}8}{9\text{–}9}$. *Abdomen* short, tumid, black, marked with yellow as follows: sides of segments 1 and 2, including the large swollen lobe and a lanceolate mid-dorsal stripe on segment 2; sides of 3 (but narrowly broken here at the transverse suture); narrow basal complete rings on 4 to 6; a considerably broader ring on 7, occupying rather more than its basal fifth, and expanding ventro-laterally as far as its apical border; sides of segments 8 to 10, on the latter also a confluent basal ring; segments 7 to 9 rather dilated. *Anal appendages* (fig. 58) black. Oreillets large, denticulate behind, yellow. *Genitalia* very prominent; lamina small, projecting; hamules narrow, long, projecting down, with the apex recurving forward; lobe of penis of enormous size, globular.

Female.—Abdomen 28 mm. Hind-wing 29 mm.

Does not differ in any marked respect from the female of *C. ypsilon*. Anal appendages yellow, shortly conical; vulvar scale triangular, very short and deeply emarginate at apex.

Distribution.—CEYLON only. Three specimens are known, a teneral male (the *type*) from Kandy Lake, May, and an adult male from Ambatenne, September, in my collection, and a female from Kandy in the Colombo Museum.

The curious anal appendages of the male will serve to distinguish this species from all others of the genus.

Type in the Author's collection.

Genus **ANISOGOMPHUS** Selys. (Fig. 59.)

Anisogomphus Selys, Mon. Gomph. p. 102 (1857); id., Bull. Acad. Belg. (2) vol. xlvi, p. 451 (1878); Kirby, Cat. Odon. p. 69 (1890); Williamson, Proc. U.S. Nat. Mus. vol. xxxiii, pp. 274, 298 (1907); Laidlaw, Rec. Ind. Mus. vol. xxiv, pp. 391, 392 (1922); Fraser, J. Bombay Nat. Hist. Soc. vol. xxix, pp. 63, 331 (1923); id., ibid. vol. xxxi, pp. 420, 421 (1926); Laidlaw, Trans. Ent. Soc. Lond. vol. lxxviii, p. 185 (1930); Needham, Zool. Sinica, vol. xi, ser. A, fasc. 1, pp. 20, 67 (1930); id., Rec. Ind. Mus. vol. xxxiv, p. 218 (1932).

Temnogomphus Laidlaw, Rec. Ind. Mus. vol. xxiv, p. 394 (1922); Fraser, J. Bombay Nat. Hist. Soc. vol. xxix, pp. 64, 332 (1923); id., ibid. vol. xxxi, pp. 424, 425 (1926); Laidlaw, Trans. Ent. Soc. Lond. vol. lxxviii, p. 185 (1930); Needham, Rec. Ind. Mus. vol. xxxiv, p. 218 (1932).

Size medium, colour black, marked with bright greenish-yellow. Head with frons slightly rounded, flattened above, occiput flat, posterior border straight or slightly convex. Wings: reticulation close; tornus rounded or with the slightest angulation; base of hind-wing very oblique; anal triangle 3-celled; arc situated opposite second antenodal nervure or between second and third; 3 to 4 transverse nervures between sectors of arc from arc to bifurcation of *Rs* in fore-wing, only 1 or rarely 2 in hind-wing; 2 rows of postanal cells in fore-wing, 4 rows in hind-wing; anal loop absent, first postanal cell of hind-wing not extending proximal to middle of base of subtrigone; basal incomplete antenodal nervure present or absent in both fore- and hind-wings (nearly always present in *A. bivittatus* and *A. orites*, but absent in one or all wings of 50 per cent. of *A. occipitalis*); nodal index high; primary antenodals the first and the sixth or seventh; discoidal cells entire (very rarely traversed in the hind-wing of some specimens of *A. orites*), that of fore-wing with the costal and basal sides equal in length, the distal slightly longer, or all sides equal, that of hind-wing with the distal side nearly twice the length of basal and slightly longer than costal, the connection between this cell and the lower sector of arc in hind-wing nearly always stalked as in *Merogomphus* ; pterostigma short and swollen, equal to about one-third the distance from node to proximal end of pterostigma, braced ; *IA* in fore-wing markedly pectinate ; *Cuii* and *IA* in hind-wing, but slightly divergent at wing-margin ; 1 to 2 cubital nervures in fore-wings, 1 in hind-wings (*A. occipitalis* has two cubital nervures in the fore-wings almost invariably, other species but one) ; all subtrigones and hypertrigones entire. Legs long and slender ; hind

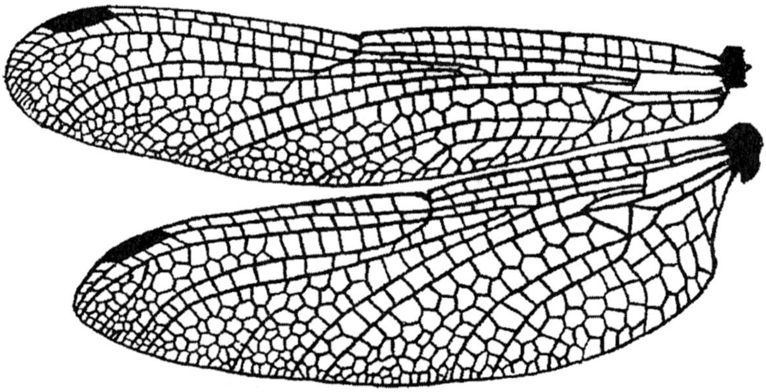

Fig. 59.—Wings of *Anisogomphus occipitalis* (Selys), male.

femora extending to middle or slightly beyond middle of segment 2, armed with a group of short, numerous, closely-set spines at the proximal half or third and with two rows of four or five very long, very widely set spines at the distal half or two-thirds, very similar to the arrangement met with in *Merogomphus* (*A. occipitalis* has these spines shorter and less numerous than in *A. orites* and *A. bivittatus* ; none of the species have any short interposed spines between the pairs of long ones) ; tibial spines long, slim, and widely spaced. Abdomen tumid at base, then slim and cylindrical as far as the base of segment 7, from which point the abdomen dilates again, especially at segments 8 and 9, as far as the end. Anal appendages : superiors white, with a ventral black spine-like protuberance, these appendages directed straight backwards or but slightly divaricate ; the inferior deeply and very broadly bifid, the two branches splayed out to such an extent as to be nearly in a straight line. Genitalia : lamina depressed, deeply and broadly arched ; anterior hamules short, slim, acutely pointed processes ; posterior hamules differing greatly in the various species ; lobe funnel- or purse-shaped, with emarginate lip.

Genotype, *Gomphus occipitalis* Selys.

Distribution.—Confined to N.E. India, ASSAM, BENGAL, and SIKKIM. Four species found within these limits, all closely similar in appearance.

The imago is found resting on bushes or foliage in the neighbourhood of montane streams, in which the larvæ (at present unknown) probably breed.

The genus is closely related to *Merogomphus*, and, except for the characteristic anal appendages, can hardly be separated therefrom. Thus the venation of the wings and the armature of the femora agree, including the stalked character of the discoidal cell of the hind-wing and the long legs and excessively long widely spaced spines on the hind femora, these latter being most evident in the female. Laidlaw separated *A. bivittatus* from the genus, erecting the genus *Temnogomphus* to accommodate it, because he was under the impression that *A. occipitalis*, and perhaps *A. orites*, did not possess an incomplete basal antenodal nervure ; but in long series of both these species I find that exactly half the specimens of *A. occipitalis* have the incomplete antenodal present and usually duplicated in the fore-wings, whilst all of *A. orites* have it present. For this reason I have decided to suppress the genus *Temnogomphus*.

Key to Indian Species of Anisogomphus.

1.	An incomplete basal antenodal nervure always present; rarely more than a single cubital nervure present in the fore-wings	2.
	An incomplete basal antenodal nervure very rarely present; nearly always two cubital nervures in the fore-wings	3.
2.	A sinuous humeral stripe present, not confluent above with the antehumeral stripe; postclypeus at its lower part and the occiput yellow	*bivittatus* (Selys), [p. 192.
	Antehumeral and humeral stripes connected above by an oblique transverse spot; postclypeus and occiput black................	*orites* Laid., p. 194.
3.	Only an upper humeral spot present; antehumeral stripes confluent with the middles of each half of the mesothoracic collar, so as to form inverted T-shaped figures	*occipitalis* (Selys), [p. 190.
	An upper humeral spot and a stripe present; antehumeral stripes confluent with the outer ends of the mesothoracic collar, so as to form L-shaped figures	*caudalis* Fras., [p. 196.

247. **Anisogomphus occipitalis** (Selys). (Figs. 59 & 60.)

Gomphus occipitalis Selys, Bull. Acad. Belg. vol. xxi, pt. 2, p. 45 (1854); id., Mon. Gomph. p. 166 (1857).
Anisogomphus occipitalis Kirby, Cat. Odon. p. 69 (1890); Williamson, Proc. U.S. Nat. Mus. vol. xxxiii, p. 298, text-figs. 24, 25 (1907); Laidlaw, Rec. Ind. Mus. vol. xxiv, pp. 371, 392, text-figs. 12, 12 *a* (1922); Fraser, J. Bombay Nat. Hist. Soc. vol. xxix, pp. 64, 332 (1923); id., ibid. vol. xxxi, p. 421, 423, pl. ii, fig. 2, text-figs. 3, v, 4, 5 *c* & *d* (1926); Laidlaw, Trans. Ent. Soc. Lond. vol. lxxviii, p. 185 (1930); Needham, Rec. Ind. Mus. vol. xxxiv, p. 226 (1932).

Male.—Abdomen 52 mm. Hind-wing 30 mm.

Head : labium yellow, middle lobe blackish-brown ; labrum glossy black, with two small, basal, greenish-yellow spots ; bases of mandibles citron-yellow, rest of face glossy black ; frons broadly greenish-yellow ; vertex and occiput black ; eyes bottle-green. *Prothorax* black, with a large lateral spot and a smaller median twin spot on middle lobe, citron-yellow. *Thorax* black, marked with citron-yellow or greenish-yellow as follows :—A slightly interrupted mesothoracic collar ; slightly oblique antehumeral stripes confluent below with the middle

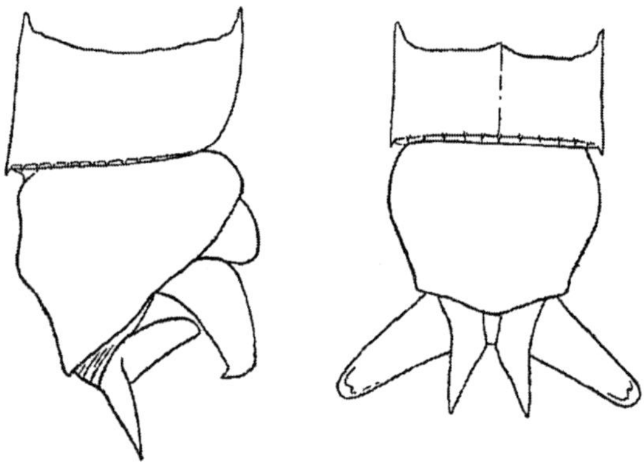

Fig. 60.—Anal appendages of *Anisogomphus occipitalis* (Selys), male. Right lateral and dorsal views.

of each half of the mesothoracic collar ; an upper triangular spot, and the vestiges of a humeral stripe well below this. Laterally, two narrow black stripes on the sutures, the first of which curls forwards above and sends a short prolongation backwards below at level of thoracic spiracle ; the stripe on the second suture sends a short prolongation forwards and downwards below level of the spiracle. *Wings* usually hyaline or more or less enfumed brown according to age of specimens ; pterostigma dark reddish-brown, covering 3 to 4 cells, lying between thick black nervures ; 2 cubital nervures in fore-wing, only 1 in hind-wing; membrane almost obsolete, greyish; nodal index of two specimens $\frac{10\text{–}15}{12\text{–}12}\Big|\frac{16\text{–}12}{12\text{–}10}$, $\frac{12\text{–}18}{12\text{–}11}\Big|\frac{16\text{–}13}{13\text{–}11}$. *Legs* black, inner sides of anterior femora yellow. *Abdomen* black, marked with yellow as follows: segment 1 with a large triangular dorsal spot tapering basally, and its sides very broadly; 2 with a trilobed mid-dorsal stripe extending whole length of segment, middle lobe of stripe globular, the sides, the oreillets, and a very large apical spot which is prolonged along ventral border towards base of segment; 3 with a baso-lateral triangular spot and its dorsal carina finely, but more broadly at the base; 4 to 6 finely yellow along the mid-dorsal carina; 7 similar, but the yellow stripe broader; 8 with a mere basal vestige of this stripe; 9 and 10 unmarked; 10 globular, compressed dorsally at apex. *Anal appendages* (fig. 60) pale greenish-white at base, reddish thereafter. *Genitalia*: lamina depressed, broadly arched; anterior hamules short, broad at base, tapering to a fine point which is curled back; posterior hamules very robust, thick, sinuously curved, S-shaped, apex shortly pointed and directed slightly forwards, the surface coated with minute spines and hairs, the latter especially thick near the apex; lobe shaped like the spout of a jug, long and tapering, coated with long hairs.

Female.—Abdomen 33 mm. Hind-wing 32 mm.

Very similar to the male, differs as follows:—Thoracic stripe more evident (sometimes confluent with the upper spot, especially in teneral specimens of both sexes); abdominal markings more extensive, on segment 2 the dorsal stripe broader and the lateral yellow stripes uninterrupted, on segment 3 is a broad lateral stripe extending the whole length, and tapering towards the apex; stripes on 4 to 7 similar but not reaching the apical border and broadly interrupted at level of transverse suture, stripe on 8 vestigial. The femoral spines are more numerous, more robust, and more widely-spaced (6 to 7 in number); nodal index $\frac{14\text{–}19}{13\text{–}14}\Big|\frac{19\text{–}14}{13\text{–}13}$; the first and eighth are the primary antenodals; occiput more deeply notched than in the male; vulvar scale of great length, nearly as long as segment 9, very narrow, tapering to a blunt point, split nearly to its base into two closely apposed scales, which closely resemble those of *Onychogomphus M-flavum*.

Distribution.—NORTHERN BENGAL and ASSAM, from June to August.

The shape of the anal appendages of the male and of the vulvar scale of the female and the presence of two cubital nervures in the fore-wing will distinguish this species from others of the same genus.

Type in the British Museum. Other specimens in the British Museum, Pusa, and Indian Museums, and in the Laidlaw and Fraser collections.

248. **Anisogomphus bivittatus** (Selys). (Fig. 61, **a**.)

Gomphus bivittatus Selys, Bull. Acad. Belg. vol. xxi, pt. 2, p. 46 (1854); id., Mon. Gomph. pp. 168–170 (1857).
Anisogomphus bivittatus Kirby, Cat. Odon. p. 69 (1890); Williamson, Proc. U.S. Nat. Mus. vol. xxxiii, p. 298 (1907).
Temnogomphus bivittatus Laidlaw, Rec. Ind. Mus. vol. xxiv, pp. 371, 394, 395, text-fig. 15 (1922); Fraser, J. Bombay Nat. Hist. Soc. vol. xxix, pp. 64, 332 (1923); id., ibid. vol. xxxi, pp. 425, 426, text-figs. 3, viii, & 5 *a*, pl. ii, fig. 4 (1926); Needham, Rec. Ind. Mus. vol. xxxiv, p. 225 (1932).

Male.—Abdomen 36 mm. Hind-wing 33 mm.

Head : labium greenish-yellow, border of middle lobe black ; labrum and bases of mandibles greenish-yellow ; anteclypeus glossy black ; postclypeus greenish-yellow, bordered with black above ; frons greenish-yellow, its lower part in front black, this black confluent with the black of upper border of

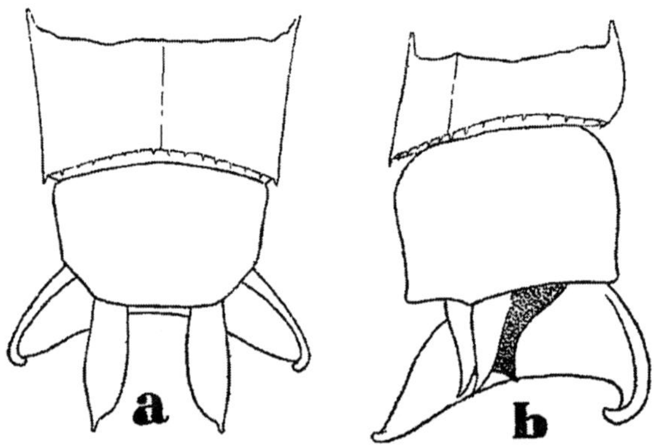

Fig. 61.—Anal appendages of (**a**) *Anisogomphus bivittatus* (Selys), male, dorsal view ; (**b**) *Anisogomphus caudalis* Fraser, male, right dorso-lateral view.

postclypeus ; the face is thus crossed by a succession of alternating greenish-yellow and black bands ; vertex black ; occiput markedly convex, yellow fringed with black hairs. *Prothorax* black, marked with a moderately large subdorsal spot on each side of the posterior lobe, and a large geminate spot at its middle above. *Thorax* black, marked with bright greenish-yellow as follows :—An interrupted mesothoracic collar confluent with very oblique antehumeral stripes on each side, which form with it inverted figures 7; long sinuous humeral stripes slightly expanded at upper end and continuous with the yellow at the base of middle pair of legs below. Laterally bright greenish-yellow marked with two fine sinuous black lines on the sutures. The yellow space between the black humeral stripe and that on the first suture sends a hook-like extension above, this curious marking being invariably present. The wavy black stripes on the yellow ground-colour of the sides give a curiously tigrine effect to this insect. *Legs* black; hind femora with a linear yellow stripe running the whole length of the outer side, middle femora with a distal stripe at the inner sides, femora entirely yellow on the inner sides. *Wings* hyaline, costa greenish-yellow in subadults; pterostigma light brown between black nervures, covering 3 to 4 cells; nodal index of two specimens $\frac{9\text{–}16}{8\text{–}10} \bigg| \frac{14\text{–}10}{9\text{–}8}, \frac{11\text{–}16}{9\text{–}11} \bigg| \frac{15\text{–}11}{11\text{–}9}$. *Abdomen* black, marked with citron-yellow as follows:—A mid-dorsal stripe on segments 1 to 7, broad and triangular on 1, broad and trilobed on 2, then narrow and fine to end of 7, and interrupted only narrowly at the apical border of each segment; sides of segments 1 and 2 broadly, including the oreillets; 3 with a broad lateral stripe tapering apically, and interrupted slightly at the jugal suture; 4 to 7 with a small triangular baso-lateral spot, and 7 with an additional mid-ventro-lateral spot; 8 with a very large baso-lateral spot which runs narrowly along ventral border nearly to apical border and ascends some way up the dorsum; 9 and 10 unmarked. (In one male there is a much larger spot on the side of segment 8, extending as far as the apex of the segment, and a similar spot on the side of 9; segment 10 has a much smaller lateral spot. In all other respects it resembles the type.) *Anal appendages*: structure as shown in fig. 61, a; superiors black at base, otherwise dark reddish-brown; inferior black or very dark brown. *Genitalia*: the posterior hamules of unique shape, black, or bright yellow in some, cylindrical at base, and broadened at end so as to be roughly T-shaped, the hinder limb long and rounded at end, the anterior end bearing a sharp robust spine tipped with black.

Female.—Abdomen 38 mm. Hind-wing 33 mm.

Colour and markings as in the male, except that there is a continuation of the lateral stripe on segments 4, 5, and 6, and that 8 and 9 are marked as described above for an aberrant male. Segment 10 is broadly yellow laterally and the *anal appendages* are also reddish-yellow, very short and conical.

Vulvar scale highly specialized; an elliptical ridge at ventro-apical border of segment 8, followed by a flatter ridge, from the border of which springs the vulvar scale proper; this is narrow, about half the length of segment 9, very slightly bifid at its apex, which is squared and nearly as broad as its base; the surface is raised on either side into two long ridges tapering to a point which projects slightly beyond the apex of scale; these two ridges appear to be the usual bilateral scales united across the mid line. Femora armed with longer, more widely spaced, and more numerous spines (about 8 or 9 long ones and some shorter near the base). Nodal index $\frac{13\text{–}17}{12\text{–}10}\Big|\frac{15\text{–}10}{11\text{–}11}$; pterostigma longer, covering 4 to 5 cells; 5 rows of postanal cells in hind-wing.

Distribution.—EASTERN HIMALAYAS: Darjeeling district, Pashok (May) and Mungpoo (August). WESTERN HIMALAYAS: Kumaon (May).

A. bivittatus may be at once recognized by the alternating three black bands and three bright yellow ones on the face, by the dilated terminal segments of abdomen, and lastly by the incomplete antenodal nervure in all wings.

Type in the Selys collection, Brussels Museum.

249. **Anisogomphus orites** Laidlaw. (Fig. 62.)

Anisogomphus orites Laidlaw, Rec. Ind. Mus. vol. xxiv, pp. 371, 393, text-figs. 13, 14 (1922); Fraser, J. Bombay Nat. Hist. Soc. vol. xxix, pp. 64, 332 (1923); id., ibid. vol. xxxi, pp. 422, 423, pl. ii, fig. 7 (1926); Needham, Rec. Ind. Mus. vol. xxxiv, p. 225 (1932).

Male.—Abdomen 31 mm. Hind-wing 30 mm.

Head: labium with middle lobe black, lateral lobe citron-yellow; labrum black, marked with two small transversely oval basal greenish-yellow spots; frons and bases of mandibles broadly citron-yellow; rest of head black. *Prothorax* black, marked with citron-yellow as follows:—An anterior collar; a large spot on each side; a median spot on posterior lobe; and a small geminate spot just in front of it. *Thorax* black, marked with bright citron-yellow as follows:—A broadly interrupted mesothoracic collar; narrow oblique antehumeral stripes confluent below with outer part of collar; a small triangular upper humeral spot, its long axis across the dorsum of thorax and its apex confluent or almost confluent with upper part of antehumeral stripe; a vestigial fine humeral line situated well below humeral spot. Laterally, two narrow black stripes on lateral sutures, the hinder of which is continued back for some distance along the ventral border. *Legs* black, fore femora yellow internally. *Wings* hyaline, sometimes palely enfumed when mature; pterostigma dark brown between thick black nervures, covering 5 to 6 cells; nodal index $\frac{14\text{–}20 \mid 17\text{–}13}{13\text{–}13 \mid 12\text{–}12}$; discoidal cells normally entire (in one specimen those of both hind-wings traversed by a nervure which runs from the basal to the outer side); *only one cubital nervure in all wings*. *Abdomen* black, marked with citron-yellow as follows:—Segment 1 with a mid-dorsal spot and the sides broadly; 2 with a trilobed mid-dorsal stripe extending the whole length of segment; 3 to 7 with the mid-dorsal carina finely yellow, more conspicuous on 7; 2 laterally, including the oreillets, and a baso-lateral spot on 3: 8 to 10 unmarked. *Anal appendages*: superiors greenish-white; inferior black. Structure as shown in fig. 62.

Female.—Abdomen 34 mm. Hind-wing 32 mm.

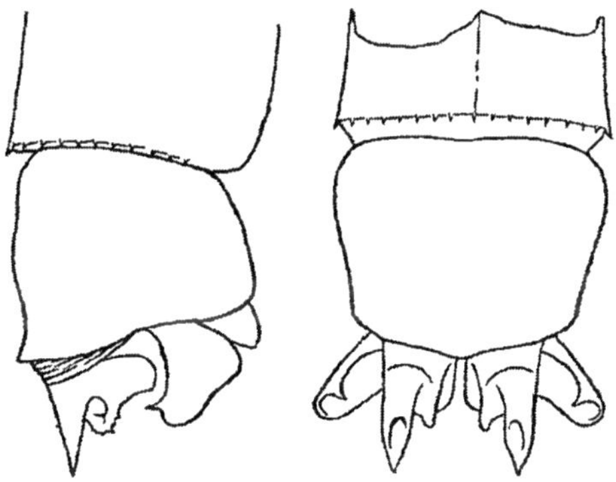

Fig. 62.—Anal appendages of *Anisogomphus orites* Laidlaw, male. Dorsal and right lateral views.

Very similar to male. The antehumeral stripe nearly always confluent with the humeral spot, sometimes broadly so. Segment 2 of abdomen more broadly yellow laterally; 3 to 5 with lateral stripes as in *A. occipitalis*, 6 and 7 with only a basal spot left of this stripe, 7 with a rather large basal dorsal spot formed by an expansion of the mid-dorsal carinal stripe. *Anal appendages* small, greenish-yellow; vulvar scale very different from that of *A. occipitalis*, very short and broad, especially at base, its apex about half the breadth of base and very shallowly concave, not split. Hind femora with a row of 6 long, robust, widely-spaced spines, this femur extending to middle of segment 2 of abdomen.

Distribution.—NORTHERN BENGAL, ASSAM, and SIKKIM.

Distinguished from *A. occipitalis* by having only a single cubital nervure in all wings in both sexes. The female is at once distinguished by its short broad vulvar scale, contrasting strongly with the long, narrow, cleft scale of *A. occipitalis*. The distinctive dorsal thoracic markings will separate it from the other two species.

Frequents and breeds in mountain streams, its habits apparently similar to those of *Burmagomphus*. A female taken by Mr Fletcher was captured flying along a hedge some distance from any stream.

Type in the Indian Museum, from Shillong, Assam, 6,000 ft. (*Bainbrigge Fletcher*). Several males and females, including one with traversed discoidal cells, in my own collection.

250. **Anisogomphus caudalis** Fraser. (Fig. 61, b.)

Anisogomphus caudalis Fraser, J. Bombay Nat. Hist. Soc. vol. xxxi, pp. 423, 424, text-figs. 3, xii, & 5 *b*, pl. ii. fig. 3 (1926); Laidlaw, Trans. Ent. Soc. Lond. vol. lxxviii, p. 185 (1930); Needham, Rec. Ind. Mus. vol. xxxiv, pp. 225, 226 (1932).

Male.—Abdomen 36 mm. Hind-wing 30 mm.

Head: labium with lateral lobes greenish-white, middle lobe black; labrum black, with a narrow linear basal streak greenish-white, nearly divided at its middle; bases of mandibles pale greenish-white; ante- and postclypeus glossy black with a small greenish-white spot at sides of latter against the eyes; frons black in front, broadly greenish-yellow above; vertex and occiput black, latter nearly straight. *Prothorax* black, marked with a large transversely oval yellow spot on each side. *Thorax* black, marked with greenish-yellow as follows:—A broadly interrupted mesothoracic collar; narrow nearly parallel antehumeral stripes forming inverted 7s by confluence with the outer ends of the mesothoracic collar below; an upper triangular spot and a lower vestigial stripe representing a broken humeral stripe. Laterally, two narrow black stripes on sutures, the hinder of which fuses with an elongate triangular spot beneath thorax; ground-colour at the sides greenish-yellow. *Legs* black, unmarked; hind femora with a group of tiny spines on ventral surface of the proximal three-fifths and an inner and outer row of 5 robust spines on the distal two-fifths. *Wings* hyaline; pterostigma black, short; 2 to 3 cubital nervures in fore-wing, only 1 in hind-wing; 4 to 5 rows of postanal cells in hind-wing, the first entire; nodal index $\frac{12\text{–}17}{13\text{–}13}\Big|\frac{16\text{–}12}{12\text{–}13}$. *Abdomen* black, marked with citron-yellow as follows:—Segment 1 with a dorsal triangular spot tapering basally, and its sides very broadly; 2 with a trilobed mid-dorsal stripe extending the whole length of segment, the oreillets and the lower part of its apical border and the apical two-thirds of its ventral border broadly; 3 with its mid-dorsal carina finely and a large baso-lateral triangular spot; 4 to 6 similar, but the lateral spots very small; 7 with the basal half of dorsum narrowly, the end of this spot square; 8 with a small triangular spot on its dorsum at base; 7 and 8 each with a small round apico-lateral spot, rather larger on segment 7; 9 and 10 unmarked. *Anal appendages*: superiors greenish-yellow above on dorsum, black below and at sides. Structure as shown in fig. 61, **b**. *Genitalia*: lobe hood-shaped, rather depressed; anterior hamules exactly similar to those of *A. occipitalis*; posterior also shaped similarly, but much smaller and not spined on surface; lobe similar to that of *A. occipitalis*.

Distribution.—ASSAM.

This species closely resembles *A. orites*, but differs in the following points:—Lateral spots present on postclypeus; humeral spots further removed from upper ends of ante-humeral stripes; lateral spots on segments 4 to 6, the basal half of 7 yellow on dorsum, apico-lateral spots on segments 7 and 8; lastly the formation of the superior anal appendages. It agrees in the extra cubital nervures in the fore-wings and in the thoracic markings. The nodal index is almost the same, especially that of the hind-wings. The abdominal markings appear to be the best guide for separating the two species.

Type in the British Museum, from Shillong, 6,000 feet, June 18, 1924 (*Bainbrigge Fletcher*).

Genus **GOMPHUS** Leach. (Fig. 63.)

Æshna Fabricius, Syst. Ent. p. 424 (1775); Latreille, Gen. Crust. Ins. vol. iii, p. 286 (1802); Kirby, Cat. Odon. p. 64 (1890).

Gomphus Leach, Edinb. Encycl. vol. ix, p. 137 (1815); Rambur, Ins. Névrop. p. 154 (1842); Selys, Bull. Acad. Belg. vol. xxi, pt. 2, p. 44 (1854); id., Mon. Gomph. p. 115 (1857); Williamson, Proc. U.S. Nat. Mus. vol. xxxiii, pp. 275, 304, 305 (1907); Kennedy, op. cit. vol. lii, pp. 549, 550 (1917); Laidlaw, Rec. Ind. Mus. vol. xxiv, p. 396 (1922); Fraser, J. Bombay Nat. Hist. Soc. vol. xxix, pp. 62, 330 (1923); id., ibid. vol. xxx, pp. 657, 658 (1925); Laidlaw, Trans. Ent. Soc. Lond. vol. lxxviii, p. 187 (1930); Needham, Zool. Sinica, ser. A, vol. xi, fasc. 1, pp. 19, 46–48 (1930); id., Rec. Ind. Mus. vol. xxxiv, pp. 217, 218 (1932).

Diastatomma (pars) Burmeister, Handb. Ent. vol. ii. p. 831 (1839); Charpentier, Lib. Eur. p. 15 (1840).

Size large or moderate, colour yellow marked with black, or (in the case of Indian species) black marked with yellow. Head of robust build, frons well angulated, occiput simple, straight or shallowly concave. Wings: reticulation close; tornus angulated; base of hind-wing moderately excavated; membrane obsolete; anal triangle 3-celled (in Indian species);

arc situated between the first and second antenodal nervures; 1 to 2, rarely 3 nervures between the sectors of arc from arc to bifurcation of *Rs* in fore-wings, only 1 or rarely 2 in hind-wings; 2 rows or a single row with a few double cells in the postanal area of fore-wings, 4 to 6 rows in hind-wings; anal loop absent, the first postanal cell not extending proximally beyond the middle of base of subtrigone; basal incomplete antenodal nervures usually absent (present in many specimens of *G. personatus*), nodal index high; primary antenodals the first and fourth or fifth (the fifth in Indian species); discoidal cells entire, in fore-wing with basal and costal sides equal or slightly subequal, the distal side considerably longer, in hind-wing elongate in the length of wing, its junction with lower sector of arc sometimes with a short but definite stalk, distal side more than half as long again as basal side and

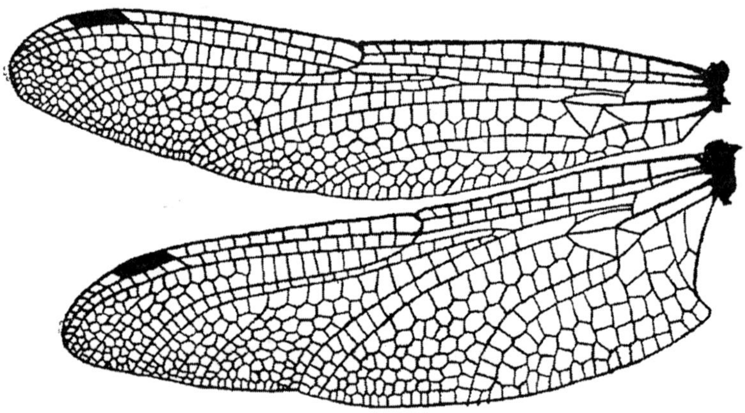

Fig. 63.—Wings of *Gomphus personatus* Selys, male.

slightly longer than costal; pterostigma small, braced or not, swollen at middle, equal to about one-third the distance from node to proximal end of pterostigma; *IA* in fore-wing pectinate; *IA* and *Cuii* in hind-wing more or less divergent at the wing-border; only 1 cubital nervure in all wings; subtrigones and hypertrigones entire in all wings. *Legs* robust, hind femora extending to apical border of segment 1 and furnished with two rows of closely-set short spines, which are slightly increased in length and size distally; tibial spines serrate, very short, robust, closely set, with two or three longer ones at the distal end. Abdomen (in Indian species) tumid at basal segments, then slim and cylindrical as far as the base of segment 7, from which point it is again expanded as far as the end, segments 8 and 9 having the lateral borders curled slightly upwards and more expanded than the others, segment 10 rather broad and square; anal appendages simple, the superiors shortly conical and widely divaricate, the inferior broadly and shallowly bifid, its two branches equally divaricate as the superiors; genitalia variable; lamina usually depressed, arched or slightly emarginate; anterior hamules short and slim; posterior hamules very robust, broad hooked processes projecting markedly from the genital sac, with sinuous anterior border and a robust recurved hook at apex; lobe rather long and narrow flask-shaped.

Genotype, *Libellula vulgatissima* Linn.

Distribution.—The genus is largely north-temperate in range, the greater number of species occurring in Europe, Northern Asia, and North America. Only three Indian species are known, from the WESTERN GHATS, BURMA, BENGAL, and ASSAM.

The genus contains a large number of rather heterogeneous species, which have not yet been satisfactorily classified, but which will have to be arranged in separate genera eventually. Needham, for example, has grouped no less than twenty-five species under this genus from China alone, including in it also those of *Platygomphus* and *Burmagomphus*. The above definition of the genus has been made from Indian species as being most helpful to Indian students, and comparisons have been made with the genotype.

The Indian species breed in clean gravelly or rocky-bottomed submontane streams, the females depositing their ova in shallow swift runnels and the males awaiting the females perched on prominent rocks in the course of streams in dense primary jungles. This is rather foreign to the habits of the genotype, which prefers brooks running through open grassy meadow-lands.

Key to Indian Species of Gomphus.

1.	Segments 7 and 8 of abdomen unmarked; humeral stripe entirely absent. (Western Ghats.)	*nilgiricus* Laidlaw, [p. 202.
	Segments 7 and 8 of abdomen broadly marked with yellow; humeral stripe or spot present. (Assam, Bengal, and Burma.)	2.
2.	Occiput yellow above; abdominal segment 9 with a broad apical yellow triangle on dorsum; two narrow black stripes on each side of thorax	*personatus* Selys, [p. 200.
	Occiput black above; abdominal segment 9 with a large yellow spot on each side, the apical border unmarked; a broad black stripe bearing two or three yellow spots on each side of thorax	*o'doneli* Fraser, [p. 205

251. **Gomphus personatus** Selys.

Gomphus personatus Selys, Bull Acad. Belg. (2) vol. xxxvi, p. 497 (1873); id., ibid. (2) vol. xlvi, p. 454 (1878); Martin, Mission Pavie Indo-Chine, vol. iii, p. 214 (1904); Williamson, Proc. U.S. Nat. Mus. vol. xxxiii, pp. 305, 308 (1907); Laidlaw, Rec. Ind. Mus. vol. xxiv, pp. 396, 398 (1922); Fraser, J. Bombay Nat. Hist. Soc. vol. xxix, pp. 62, 330 (1923); id., ibid. vol. xxx, pp. 658–660, pl. i, figs. 1 & 2 (1925); Laidlaw, Trans. Ent. Soc. Lond. vol. lxxviii, p. 187 (1930); Needham, Rec. Ind. Mus. vol. xxxiv, p. 226 (1932).

Æshna personata Kirby, Cat. Odon. p. 68 (1890).

Gomphus xanthenatus Williamson, Proc. U.S. Nat. Mus. vol. xxxiii, pp. 305–308, text-figs. 32, 33 (1907); Laidlaw, Rec. Ind. Mus. vol. xxiv, pp. 371, 398 (1922); Fraser, J. Bombay Nat. Hist. Soc. vol. xxix, pp. 61, 330 (1923).

Male.—Abdomen 42–45 mm. Hind-wing 37–39 mm.

Head: labium with middle lobe blackish, lateral lobes yellow; labrum glossy black, marked at base with two oblong bright yellow spots; bases of mandibles yellow, ante- and postclypeus black, latter with a large lateral yellow spot against the eye and a variable small medial spot of the same colour; frons bright greenish-yellow except the lower part in front and its base above, which is marked with a fine crenulate black basal line; rest of head black, but occiput variably bright to dull yellow at its centre, black laterally, straight, fringed with very long black hairs. *Prothorax* black, marked with bright citron-yellow; a broad anterior collar, a small medial spot, and a much larger lateral one on the posterior lobe. *Thorax* black, marked variably with citron-yellow as follows:—A mesothoracic collar slightly broken at its middle by the black mid-dorsal carina, and confluent at its outer end with straight antehumeral stripes which run from the antealar sinus parallel with the dorsal carina, thus forming an inverted 7 on each side; a variable humeral marking, represented in a slight majority of specimens by a small upper spot, but in others by a sinuous stripe which expands into an upper spot and may be broken at its middle. In Shillong specimens both forms are met with; in a pair taken by Col. F. Wall at Maymyo, Upper Burma, the stripe is well formed, in Williamson's variety *xanthenatus* from Burma, the stripe is absent in three specimens, incomplete and broken in a fourth. Laterally more or less broadly yellow, marked with two narrow black lines on sutures; in Williamson's *xanthenatus* these two lines are more or less confluent and almost obliterate the included yellow; in the pair from Maymyo the lines are well separated throughout their entire length by the yellow; in Shillong specimens the line on the anterior suture is invariably broadly broken, being represented by a small upper section and a longer lower which begins at the spiracle. The line on the posterior suture is much finer and the lateral yellow correspondingly much more extensive. *Legs* black, anterior femora with a yellow stripe on the inner side, posterior femora with a row of short, robust, moderately widely spaced, gradually lengthening spines. *Wings* hyaline, very palely and evenly enfumed ; pterostigma dark brown between black nervures ; costa reddish-brown ; 2 rows of cells between *Rii* and *IRii* at level of outer end of pterostigma, but 4 rows at apex of wing ; a basal incomplete antenodal nervure frequently present on one or more wings. Anal triangle of 3 to 5 cells : 4 to 5 rows of postanal cells in the hind-wing ; 3 rows of cells at level of node in the discoidal field ; nodal index of two specimens $\frac{14\text{–}18}{13\text{–}11}\Big|\frac{17\text{–}15}{11\text{–}13}$, $\frac{12\text{–}14}{14\text{–}10}\Big|\frac{15\text{–}12}{11\text{–}13}$; membrane obsolete. *Abdomen* tumid at base and from segment 7 to 10, thin and cylindrical in middle, black, marked with bright citron-yellow as follows :—A broad continuous stripe on sides of segments 1 and 2, extending shortly on to base of 3 ; a large apico-dorsal spot on 1 : a trilobed stripe on dorsum of 2 extending the whole length of segment ; a fine dorsal carinal stripe on 3 tapering apically and with a slightly crenate edge : 4 to 6 with small baso-lateral spots and larger baso-dorsal triangular spots which taper along the carinal ridge for a variable distance ; on 7 these spots are confluent to form a basal ring which extends for a short distance apically along the carina ; markings on 8 extremely variable, usually a narrow apical ring which expands on the sides into a triangular point directed basally, a small triangular baso-dorsal spot, and a largish ventro-lateral elongate spot ; on this segment the apical ring may be completely absent or with two irregularly shaped lateral spots at the apex, the ventral spot may be very minute or entirely absent, and the carinal spot may extend as far as the apical border and show a medial expansion; segment 9 with a broad apical ring of variable width, extending nearer base on dorsum than on sides, in some specimens extending as far as base of segment, in others quite fine with its basal border serrated ; usually, however, it covers about half the segment ; 10 entirely unmarked, or more rarely with a fine broken apical ring. *Anal appendages* as long as segment 10, of equal length and equally divaricate, black ; superiors tapering to a point, curling evenly down nearly to the apex, which is directed abruptly straight backwards ; inferior deeply cleft into two branches which lie parallel to the superiors, hollowed out above, curling evenly upwards. *Genitalia*: lamina cowl-shaped, projecting slightly; hamules very robust and projecting almost perpendicularly from the genital sac, their apices turning rather abruptly forward, black; lobe mat black, very large and tumid, projecting as a massive funnel-shaped organ.

Female.—Abdomen 42–45 mm. Hind-wing 38–42 mm.

Almost exactly similar to male. Occiput more extensively yellow in centre, fringed with shorter hairs, slightly sinuous and usually with one or two small spines on either side. Abdominal markings more extensive; carinal stripe almost complete from segments 1 to 6 and often confluent with the baso-lateral spots; lateral stripe extending almost to apical end of segment 3; segment 8 with a fine apical ring and less rarely a small triangular baso-dorsal point; apical border of segment 10, anal appendages, and a conical prominence between them yellow. In some specimens the anal appendages are black and marked with yellow on the dorsum only. Vulvar scale not quite half the length of segment 9, bifid for about half its length, narrow and subtriangular.

Distribution.—ASSAM: Khasi Hills, Shillong (*Bainbrigge Fletcher*). BURMA: Maymyo (*Col. F. Wall*); Toungoo and Karen districts (*Williamson*).

The above description has been made from a large number of specimens collected at Shillong during May and June, and from a pair collected at Maymyo in June. Mr. Williamson's specimens grade almost imperceptibly through the Maymyo specimens into the Shillong ones, the only real point of difference being the lateral sutural black lines, which are complete in Burmese material but invariably broken in Assam specimens.

Type, a specimen with complete humeral stripe, in Selys's collection from the Khasi Hills. Type of *G. xanthenatus* in the Williamson collection.

252. **Gomphus nilgiricus** Laidlaw. (Fig. 64.)

Gomphus nilgiricus Laidlaw, Rec. Ind. Mus. vol. xxiv, pp. 371, 397, 398, text-fig. 16 (1922); Fraser, J. Bombay Nat. Hist. Soc. vol. xxix, pp. 62, 330 (1923); id., Rec. Ind. Mus. vol. xxvi, pp. 427, 476, 477 (1924); id., J. Bombay Nat. Hist. Soc. vol. xxx, pp. 660, 661, pl. i, fig. 3 (1925); Laidlaw, Trans. Ent. Soc. Lond. vol. lxxviii, p. 188 (1930); Fraser, Rec. Ind. Mus. vol. xxxiii, pp. 444, 447 (1931); Needham, Rec. Ind. Mus. vol. xxxiv, p. 226 (1932).

Male.—Abdomen 40 mm. Hind-wing 35 mm.

Head: labium pale brownish-yellow; labrum, bases of mandibles, and anteclypeus dark brown; postclypeus and front of frons black, upper part of frons bright greenish-yellow, rest of head black. Occiput black, slightly raised in the middle. *Prothorax* black, marked with citron-yellow as follows:—An anterior collar, a transversely oval spot on the dorsum of posterior lobe, and a small paired spot just in front of and confluent with latter. *Thorax* mat black marked with greenish-yellow as follows:—Straight antehumeral stripes running parallel with the mid-dorsal carina, gradually broadening below, where they are confluent with a widely interrupted mesothoracic collar. . laterally broadly greenish-yellow, marked with two moderately broad black stripes on sutures, often confluent with one another so as to split up the intervening yellow into two or three spots. *Legs* black, robust, posterior femora extending well on to second abdominal segment, minutely and densely spined. *Wings* hyaline; pterostigma braced, covering 4 to 5 cells, dark brown between black nervures; nodal index $\frac{13\text{–}13}{15\text{–}9}\Big|\frac{15\text{–}12}{10\text{–}13}$; membrane obsolete;

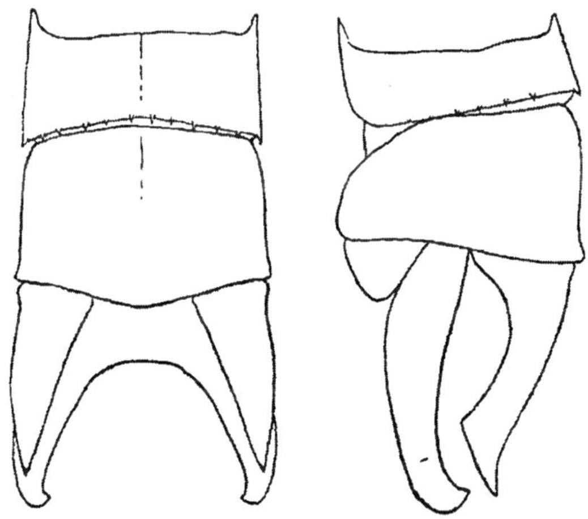

Fig. 64.—Anal appendages of *Gomphus nilgiricus* Laidlaw, male. Dorsal and left lateral views.

1 to 2 rows of cells between *Rii* and *IRii* at outer end of pterostigma; 5 to 6 rows of postanal cells in hind-wing, the first entire, extending basally for a little more than half the length of base of subtrigone; a basal incomplete antenodal nervure variably present (more often present in fore-wing than hind-wing); 2 to 3 rows of cells in discoidal field at level of node, usually 3; only 1 cubital nervure in all wings. *Abdomen* tumid at base, moderately and squarely dilated from segment 7 to 10, slender and cylindrical between these parts;

black marked with bright primrose-yellow as follows :—The sides of segments 1 and 2 at and below level of oreillets, latter margined with black ; a dorsal spot on segment 1 widening apically ; a trilobed band on mid-dorsum of 2, extending the whole length of segment ; a mid-dorsal line on 3 tapering rapidly from its base and not quite reaching apical border of segment ; a triangular basal spot on the sides of this segment ; paired basal dorsal spots on 4 to 7 and confluent with one another at the base ; 8 and 10 unmarked but 9 with its apical half yellow, the border of this wide annule markedly crenate or serrate and resembling a similar marking in *G. personatus*. *Anal appendages* (fig. 64) black. *Genitalia* : lamina depressed, broad and shallowly arched ; hamules very robust, broad at base, tapering to a point, projecting markedly from the genital sac with a slight posterior slope, black ; lobe large and prominent, funnel-shaped, black.

Female.—Abdomen 43 mm. Hind-wing 38–40 mm.

Closely similar to the male save for sexual differences. Abdomen thick and robust, cylindrical throughout. Wings rather broader ; pterostigma pale brown, covering 4 to 6 cells ; nodal index considerably higher. Abdominal markings broader, the dorsal basal spots entirely confluent and continued along the mid-dorsal carina as fine lines. The basal lateral spot on segment 3 much larger, confluent with the lateral yellow area on the two previous segments and continued nearly to apex of segment. Segments 8 and 10 are unmarked as in the male, but 9 bears a similar broad apical ring. Unlike *G. personatus* the markings of this insect are remarkably free from variation. Vulvar scale of great length, projecting from the abdomen at an angle of about 45 degrees, consisting of two closely apposed scales which taper from a moderately broad base to a fine point. Anal appendages small, conical, black, the conical protuberance between them yellow.

Distribution.—A rare insect found sparsely throughout the WESTERN GHATS. A few specimens have been taken on streams flowing down the Mettapalayam and Gudalur Ghats, NILGIRIS. I have taken five males at the head of the Sampaji River and on a small stream above the Hatti River, COORG. Major Frere found it in larger numbers on the Bear and Cinchona streams, Kodaikanal, PALNI HILLS. Lately I took a few specimens in the Mudis Hills, Coimbatore district, at 4,000 feet.

It rests on rocks in mid-stream ; females are rarely seen, and then only when they are ovipositing in some quiet shaded pool.

Type male and allotype female in the British Museum.

253. **Gomphus o'doneli** Fraser.

Gomphus o'doneli Fraser, Rec. Ind. Mus. vol. xxiv, pp. 420, 421, pl. xi, fig. 6 (1922); id., J. Bombay Nat. Hist. Soc. vol. xxix, pp. 61, 330 (1923); id., ibid. vol. xxx, pp. 662, 663, pl. i, fig. 4 (1925); Laidlaw, Trans. Ent. Soc. Lond. vol. lxxviii, p. 188 (1930); Needham, Rec. Ind. Mus. vol. xxxiv, p. 226 (1932).

Male.—Abdomen 42 mm. Hind-wing 30 mm.

Head: labium pale yellow, middle lobe bordered with black; labrum glossy black, with two small basal yellow spots; rest of face and head black except for a narrow transverse yellow band on crest of frons. Occiput emarginate, fringed thickly with black hairs. *Prothorax* black, posterior lobe, a small spot on either side of it, a duplicated spot just in front of it, and an anterior collar yellow. *Thorax* black, marked with yellow as follows:—Oblique antehumeral stripes which are confluent with a slightly interrupted mesothoracic collar; a small upper humeral spot. Laterally yellow, marked by a broad medial black band which is itself marked by three small yellow spots. *Legs* short, robust, entirely black; hind femora armed with small spines and with a single larger spine at the distal end. *Wings* hyaline; pterostigma dark brown, braced, rather small; membrane almost obsolete, dark brown; only 1 row of cells between *Rii* and *IRii* at level of distal end of pterostigma; *Cuii* and *IA* nearly parallel in the hind-wing; nodal index $\frac{9\text{–}16}{11\text{–}10}\Big|\frac{14\text{–}9}{11\text{–}10}$; 3 rows of cells in discoidal field at level of node in fore-wing; 3 to 4 rows of postanal cells in hind-wing, the first postanal cell very small, entire, extending inwards towards base of wing for only half the length of the base of subtrigone; no incomplete basal antenodal nervure in either sex. *Abdomen* tumid at base; apex of segment 7 and segments 8 and 9 broadly dilated although not winged; segments 3 to 7 narrow and cylindrical; colour black, marked with yellow as follows:—A triangular dorsal spot at apex of segment 1 and a broad lateral spot on each side; 2 with a trilobed dorsal stripe, the oreillet and a large lateral spot beyond it; 3 with the dorsum narrowly at the base and a large lateral basal spot; 4 to 6 with dorsal basal spots rather widely confluent across the carina; 7 with a broad basal annule prolonged apically along the dorsal carina; 8 with a small spot on the mid-dorsal carina at the base and another larger angulated spot on the middle of each side; 9 with the whole of the ventro-lateral border broadly yellow; 10 unmarked. *Anal appendages* black, widely and equally divaricate; superiors slightly longer than branches of inferior and about equal in length to segment 10, very robust, very thick at base, at first directed obliquely out, the apex then sharply bent backwards so that the outer border presents a prominent angle, the apex itself rapidly tapering to a sharp point; inferior deeply and broadly cleft into two strongly divaricate branches, short, broad, and with the apex ending in a minute upturned spine. *Genitalia*: lamina depressed, broad; hamules very robust, projecting almost perpendicularly from the genital sac, ending in a robust, slightly forwardly directed spine; lobe of great size, bulbous, funnel-shaped, black.

Female.—Abdomen 42 mm. Hind-wing 32 mm.

Resembles the male in its robust stature and general colouring. Abdomen a little stouter than that of the male at middle, but not of such even width as usual in female Gomphines. Labrum, in addition to the two small basal spots, with another smaller but brighter spot on the outer side of each; anteclypeus very dark brown; postclypeus with two small median yellow spots on the lower border and a very obscure spot at the border of eyes. Occiput fringed with short black hairs, simple, its extreme edge brown, but behind (as seen by tilting the head forward) bright yellow. *Prothorax* and *thorax* similar to those of male, but mesothoracic collar barely interrupted and upper humeral spot nearly obsolete. *Legs* black, hind femora with a row of gradually lengthening, closely set, very robust spines, the distal one of great length. *Wings* similar to those of male. *Abdomen* with the basal spots so confluent as to form rather broad basal rings on segments 3 to 7; mid-dorsal stripe on 2 tapering apically, more so than in the male; lateral spot on 8 not angulate but quadrate and extending to base of segment; an additional small baso-dorsal spot on 9, lower part of the sides of 9 and 10 yellow. *Anal appendages* very small, pointed, yellow. Vulvar scale rudimentary, merely a small broad short projection of the apical border of segment 8, but slightly overlapping 9; the expanded borders of the segments here function for the vulvar scale.

Distribution.—Two males and a female only of this rare species are known, collected by Mr. H. V. O'Donel at Hasimara Tea Estate, Duars, BENGAL. In general facies they recall strongly species of the same genus from North America, and cannot be readily confused with either *G. personatus* or *G. nilgiricus*.

Type male and allotype female in my collection, paratype male in the British Museum.

Genus **PLATYGOMPHUS** Selys. (Fig. 65.)

Platygomphus Selys, Bull. Acad. Belg. vol. xxi, pt. 2, p. 44 (1854); id., Mon. Gomph. p. 112 (1857); Kirby, Cat. Odon. p. 70 (1890); Williamson, Proc. U.S. Nat. Mus. vol. xxxiii, pp. 275, 303 (1907); Laidlaw, Rec. Ind. Mus. vol. xxiv, p. 398 (1922); Fraser, J. Bombay Nat. Hist. Soc. vol. xxix, pp. 61, 330 (1923); id., ibid. vol. xxxi, pp. 415, 416 (1926); Laidlaw, Trans. Ent. Soc. Lond. vol. lxxviii, p. 188 (1930); Needham, Rec. Ind. Mus. vol. xxxiv, p. 218 (1932).

Gomphus (pars) Needham, Zool. Sinica, ser. A, vol. xi, fasc. 1, pp. 46, 47 (1930).

Size rather above the average; colour bright sandy yellow, marked with black or brown: hind-wings of both sexes rounded at the base and barely excavate in male. Head comparatively small: frons rounded; face rather oblique: occiput simple, flat, and with straight posterior border. Wings: reticulation close; base of hind-wing in male barely excavated; tornus slightly rounded; membrane obsolete: anal triangle poorly formed, 3- to 4-celled; arc situated between the first

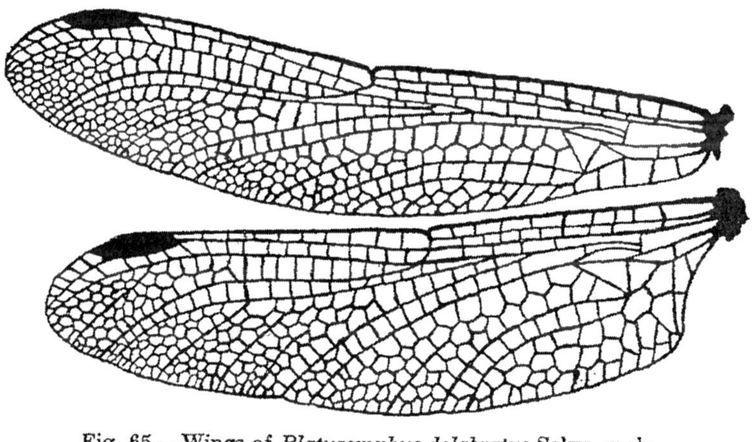

Fig. 65.—Wings of *Platygomphus dolabratus* Selys, male.

and second or opposite the second antenodal nervure; 2 transverse nervures between the sectors of arc from arc to bifurcation of *Rs* in fore-wings, only 1 in hind-wings; a single row of postanal cells in fore-wings, 3 rows in hind-wings; anal loop absent, first postanal cell of hind-wing extending only as far as the middle of base of subtrigone; no basal incomplete antenodal nervures present; nodal index rather high; primary antenodals the first and the fifth; discoidal cells entire, that of fore-wing subequilateral, but the distal side slightly longer than the two others; that of hind-wing elongate in length of wing, the distal side twice as long as the basal and slightly longer than the costal; pterostigma short and swollen, about one-third the distance from node to proximal end of pterostigma, braced; *IA* in fore-wing pectinate; *Cuii* and *IA* in hind-wing divaricate at border of wing; only 1 cubital nervure in all wings; subtrigones and hypertrigones entire in all wings. Legs short, hind femora extending but slightly beyond end of thorax, furnished with a group of short, robust, and numerous spines on the proximal half of limb and two rows of much longer spines on the distal half, these latter in female being much longer than in male; hind tibial spines slim, numerous, and very closely set. Abdomen tumid on basal segments, then slim and cylindrical as far as segment 6, 7 from its middle and 8 and 9 greatly dilated, 8 being more than four times the width of 6 and bearing lateral expansions which are minutely spined along the borders; anal appendages closely similar to those of genus *Gomphus*, superiors shortly conical and markedly divaricate, inferior broadly and shallowly cleft into two widely divaricate branches. Genitalia: lamina depressed; anterior hamules short, slim; posterior hamules much more robust, projecting and ending in a small crochette-like hook; lobe scrotal-shaped, short, deeply emarginate.

Genotype, *Platygomphus dolabratus* Selys.

Distribution.—BENGAL, BIHAR, BURMA, and China (?). Two species occur in India.

In general facies and colouring these species resemble closely those of *Onychogomphus* (group *M-flavum*), but the appendages are as in *Gomphus*, s. str., as well as the venation, so that they are probably closely allied to this genus.

The larvæ are unknown, but the teneral imago has been taken along the banks of rivers in Bihar and has habits very similar to those of *Cyclogomphus*, being slow in flight and settling on low herbage.

Key to Indian Species of Platygomphus.

Legs marked with yellow; abdominal segments 5, 6, and 7 with mid-dorsal yellow spots confluent with basal yellow rings	*dolabratus* Selys, p. 208.
Legs entirely black; abdominal segments 5, 6, and 7 without mid-dorsal yellow spots	*feæ* Selys, p. 210.

254. **Platygomphus dolabratus** Selys. (Fig. 65)

Platygomphus dolabratus Selys, Bull. Acad. Belg. (2) vol. xxi, p. 44 (1854); id., Mon. Gomph. p. 113 (1857); Kirby, Cat. Odon. p. 70 (1890); Williamson, Proc. U.S. Nat. Mus. vol. xxxiii, pp. 303, 304, text-figs. 30, 31 (1907); Laidlaw, Rec. Ind. Mus. vol. xxiv, pp. 371, 398, 399 (1922); Fraser, J. Bombay Nat. Hist. Soc.

vol. xxix, pp. 62, 330 (1923); id., ibid. vol. xxxi, pp. 416, 417, text-figs. 2, 3, xi, pl. ii, fig. 5 (1926); Laidlaw, Trans. Ent. Soc. vol. lxxviii, p. 188 (1930); Needham, Rec. Ind. Mus. vol. xxxiv, p. 225 (1932).

Male.—Abdomen 41 mm. Hind-wing 30 mm.

Head: labium, labrum, and face bright yellow; labrum with a small brown median stripe projecting down from the base; postclypeus with a broad transverse black band, the upper border of which is slightly convex and encroaches on the anterior surface of frons, the lower produced into two quadrate projections, one on each side near the middle line; frons entirely yellow above and for greater part of front; vertex black, with a large round yellow spot behind ocelli; occiput bright yellow, its border very slightly convex at the middle. Eyes bottle-green. *Prothorax* yellow at sides, black on dorsum except for a pair of twin spots at middle of posterior lobe and the whole of the prominent anterior collar. *Thorax* yellow, marked with black as follows:—Dorsum, where the black encloses a complete bright yellow mesothoracic collar and two broad triangular antehumeral duller yellow spots; behind these a moderately broad and even, humeral, bright yellow stripe which turns in a little at its upper part. On the sides a short fine vestigial black line on the first lateral suture and a complete very fine black line on the second suture. *Wings* hyaline, a little tinted with yellow occasionally; pterostigma yellow between black nervures, braced strongly, covering 3 cells; nodal index $\frac{9\text{–}13 \mid 14\text{–}9}{9\text{–}10 \mid 10\text{–}10}$. *Legs* yellow, marked with black; a narrow distal streak on the outer side of the hind femora and the whole of the outer sides of the other two pairs; all tibiæ black; hind femora with a row of short robust spines. *Abdomen* yellow, marked with black as follows:—Segment 1 with a small subdorsal spot and a fine black line running obliquely backwards and outwards from it; 2 with a broad subdorsal stripe extending the whole length of segment and enclosing an acorn-shaped spot, the stalk of which is directed towards base of segment; 3 to 7 black on dorsum; in mature specimens this black encloses a lanceolate mid-dorsal spot, broader at base and pointed towards apical border; basally the spot is confluent with a basal ring in immature specimens (in all the specimens from Pusa the black is present as an apical ring with a subdorsal branch running towards the base of segment on either side, enclosing a prolongation of the ground-colour which runs from a basal ring); 8 and 9 dark brown above, changing to yellow on the sides; 8 with a fine mid-dorsal line at base; 10 paler brownish-yellow. *Anal appendages* yellow or yellowish-brown, bordered with black, as long as segment 10 ; superiors flat, divaricate, inner sides straight, outer right-angled with prominent outer point, apex seen from above pointed, slightly bifid seen in profile ; inferior deeply forked, its branches rather more divaricate than the superiors, their apices turned up a little. *Genitalia* : lamina slightly raised, rounded, broadly arched ; posterior hamules very long, projecting almost perpendicularly from abdomen, subcylindrical, the apex strongly curled and ending in a hooked spine ; lobe very bulky, subrounded, deeply cleft, its border somewhat sinuous as seen in profile.

Female.—Abdomen 40 mm. Hind-wing 33 mm.

Coloured and marked similarly to male. Terminal segments of abdomen less dilated, more broadly yellow along the sides, segment 10 yellow. Vulvar scale very short, emarginate. Occiput low, its free border a little tumid, a little rounded at the middle.

Distribution.—BIHAR and BENGAL.

Type in the Selys collection, paratype females in the British Museum, and specimens of both sexes in the Pusa and my own collections.

255. **Platygomphus feæ** Selys.

Platygomphus feæ Selys, Ann. Mus. Civ. Stor. Nat. Genova, (2) vol. x, pp. 479, 480 (1891); Williamson, Proc. U.S. Nat. Mus. vol. xxxiii, pp. 303, 304 (1907); Laidlaw, Rec. Ind. Mus. vol. xxiv, pp. 371, 399 (1922); Fraser, J. Bombay Nat. Hist. Soc. vol. xxix, pp. 62, 330 (1923); id., ibid. vol. xxxi, pp. 417–419, pl. ii, fig. 6 (1926); Laidlaw, Trans. Ent. Soc. Lond. vol. lxxviii, p. 188 (1930); Needham, Rec. Ind. Mus. vol. xxxiv, p. 225 (1932).

Male.—Abdomen 35–37 mm. Hind-wing 30–31 mm.

Head : labium, labrum, face, and frons dull yellow, unmarked except for some brownish suffusion of face and a basal black line on frons which extends forwards as far as the crest at the middle ; vertex black ; occiput yellow, fringed with greyish hairs, black behind, straight. *Prothorax* yellow, black at middle of dorsum. *Thorax* yellow, marked with black as follows :—Dorsum similar to that of *P. dolabratus*, the black enclosing a broad antehumeral oval or subtriangular spot on each side and a moderately narrow humeral yellow stripe. Laterally a single narrow black stripe on the second lateral suture, which divides above to form a " Y," the anterior arm of which runs towards the base of fore-wing. *Legs* short, black, inner surfaces of femora pale yellow. *Wings* hyaline ; pterostigma brown, between thick black nervures, covering 3 cells ; nodal index $\frac{6\text{–}13}{10\text{–}9}\Big|\frac{12\text{–}8}{10\text{–}9}$; costal border brownish. *Abdomen* black, marked with yellow as follows :—Segment 1 with a mid-dorsal stripe and the sides broadly; 2 with a lanceolate stripe on the mid-dorsum extending from base to apex, and the sides very broadly; 3 to 6 with a narrow basal ring prolonged along the sides below; on 3 and 4 a small median dorsal oval spot; 7 with its basal half yellow; 8 to 10 coloured as in *P. dolabratus*. *Anal appendages* black, equal in length to the tenth abdominal segment. Superiors markedly flattened, the outer border strongly angulate but the point situated much nearer apex than in *P. dolabratus*. Apex shortly acuminate so that, seen, in profile, it appears to be bifid. Inferior strongly forked, with the apices of its branches blunt and turned slightly up.

Female unknown.

Distribution.—Bhamo, BURMA, July, August.

Distinguished from *P. dolabratus* by the face being unmarked, the vertex without a yellow spot, the abdomen more broadly marked with black, the anal appendages and legs black, the forking of the hinder stripe on the sides of thorax, segments 5 and 6 of the abdomen without a dorsal spot, and lastly by the shape of the superior anal appendages, which appear more bifid in profile than those of *P. dolabratus*.

Type in the Selys collection.

Genus **BURMAGOMPHUS** Williamson. (Fig. 66.)

Burmagomphus Williamson, Proc. U.S. Nat. Mus. vol. xxxiii, pp. 275, 298–301 (1907); Laidlaw, Rec. Ind. Mus. vol. xxiv, p. 399 (1922); Fraser, J. Bombay Nat. Hist. Soc. vol. xxix, pp. 62, 330 (1923); id., ibid. vol. xxxi, pp. 408, 409 (1926); Laidlaw, Trans. Ent. Soc. Lond. vol. lxxviii, p. 189 (1930); Needham, Rec. Ind. Mus. vol. xxxiv, p. 218 (1932).

Gomphus (pars) Needham, Zool. Sinica, ser. A, vol. xi, fasc. 1, pp. 46, 47 (1930).

Size medium, colour black, marked with bright citron-yellow. Head of moderate size and proportions; frons angulate, rounded above; occiput simple, shallowly concave. Wings: reticulation very close; base of hind-wing very oblique, tornus angulate and prominent; anal triangle 3-celled; arc situated between the first and second antenodal nervures; only 2 transverse nervures between the sectors of arc from arc to bifurcation of *Rs* in fore-wings, only 1 in hind-wings; anal loop absent; the first postanal cell in hind-wing extending proximally as far as middle of base of sub-trigone; basal incomplete antenodal nervures always absent; nodal index high; primary antenodal nervures usually the first and fifth; discoidal cell in fore-wing entire, costal and basal sides equal, the distal slightly longer; that of hind-wing with the costal side nearly half as long again as the basal but distinctly shorter than the distal, entire; pterostigma relatively short, dilated, braced, equal to rather more than one-third the distance from node to proximal end of pterostigma; *IA* in fore-wing pectinate; *Cuii* and *IA* in hind-wing only slightly divaricate at wing-border; only a single cubital nervure in all wings; subtrigones and hypertrigones all entire. Legs moderately long, extending as far as the base of segment 2; hind femora furnished with a group of closely-set, small even spines, which at the distal third of the limb become arranged into two single rows of slightly more robust spines; hind tibial spines of same limb short, robust, and very closely set. Abdomen tumid at basal segments, then narrow and cylindrical as far as segment 7, from which point it again expands markedly, especially at segment 8; terminal segments

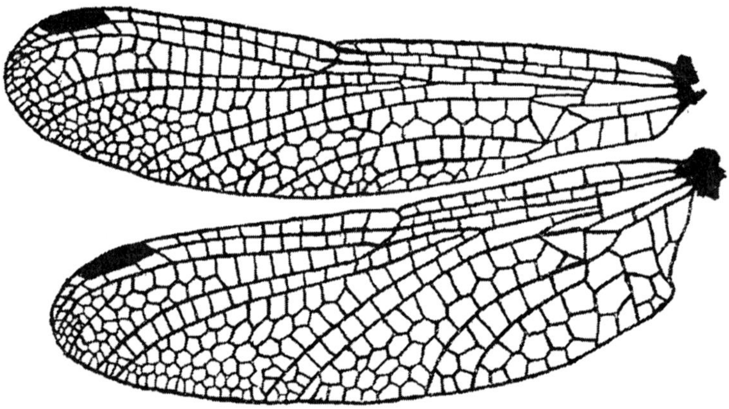

Fig. 66.—Wings of *Burmagomphus pyramidalis* Laidlaw, male.

gradually decreasing in length and not elongate as in genus *Merogomphus*. Anal appendages: superiors simple, short, and markedly divaricate as in genus *Gomphus*; inferior bifurcated into two equally long and divaricate branches. Genitalia: lamina markedly depressed, arched; anterior hamules short, slim; posterior hamules variable, broad flattened plates usually furnished with one or more marginal spines.

Genotype, *Burmagomphus williamsoni* Fraser.

Distribution.—INDIA, BURMA, CEYLON, Malaysia, Indo-China, and Java. Eight species are known from within our limits, all of which are closely similar in size and appearance.

The larvæ burrow in sand in montane streams; they are sandy in colour, fusiform or cylindrical in shape; the neck is quadrate, without setæ, and furnished with a robust, movable hook.

The adult insects are arboreal by nature and are curiously fond of settling on bright green foliage, in which position their rather cryptic shagreened markings make them very inconspicuous.

The position of the genus is doubtful, but Laidlaw places it near *Gomphus*.

Key to Indian Species of Burmagomphus.

1.	Only upper and lower halves of antehumeral and humeral thoracic stripes respectively present, the two halves joined obliquely to form a sinuous stripe on each side of dorsum of thorax	2.
	Antehumeral and humeral stripes separated, the latter often incomplete or represented by an upper spot only	4.
	Antehumeral and humeral thoracic stripes confluent at their lower ends to form V-shaped markings on each side of dorsum of thorax.	*V-flavum* Fras., [p. 223.
2.	A yellow spot on vertex; mesothoracic collar not interrupted; segment 8 broadly yellow at sides, finely so at base	*pyramidalis* Laid., [p. 213.
	Vertex entirely black; mesothoracic collar finely interrupted; segment 8 entirely black.	3.
3.	Labium yellow; a continuous mid-dorsal yellow stripe on segments 3 to 5............	*williamsoni* Fras., [p. 216.
	Middle lobe of labium black; segment 5 without a yellow mid-dorsal stripe	*hasimaricus* Fras., [p. 216.
4.	Humeral thoracic stripe incomplete	5.
	Humeral stripe complete; three lateral black stripes on each side of thorax	*sivalikensis* Laid., [p. 217.
5.	Humeral stripe made up of an upper spot and a short lower stripe	6.
	Humeral stripe a short lower one only, without any upper spot	*cauvericus* Fras., [p. 222.
6.	Antero-lateral thoracic stripe complete	*laidlawi* Fras., [p. 220.
	Antero-lateral stripe incomplete, its upper half directed obliquely backwards so as nearly to join the postero-lateral stripe	*sinuatus* Fras., [p. 219.

256. **Burmagomphus pyramidalis** Laidlaw. (Fig. 67.)

Burmagomphus pyramidalis Laidlaw, Rec. Ind. Mus. vol. xxiv, pp. 371, 399–401, text-fig. 17 (1922); Fraser, J. Bombay Nat. Hist. Soc. vol. xxix, pp. 62, 331 (1923); id., Rec. Ind. Mus. vol. xxvi, pp. 427, 476 (1924); id., J. Bombay Nat. Hist. Soc. vol. xxxi, pp. 409–410, text-figs. 1, 3, 9, pl. i, fig. 3 (1926); Laidlaw, Trans. Ent. Soc. Lond. vol. lxxviii, p. 189 (1930); Fraser, Rec. Ind. Mus. vol. xxxiii, p. 447 (1931); Needham, Rec. Ind. Mus. vol. xxxiv, p. 226 (1932).

Male.—Abdomen 30 mm. Hind-wing 23–24 mm.

Head: labium yellow, margined with black; labrum greenish-yellow, with its anterior border narrowly and base more broadly black, and a short medial black prolongation running from base; bases of mandibles black, genæ, anteclypeus, and postclypeus black, latter marked with a greenish-yellow median spot and a lateral triangular larger spot; frons greenish-yellow, its base above black; vertex and occiput black, latter slightly concave. *Prothorax* black, marked with a greenish-yellow spot on each side. *Thorax* black, marked with greenish-yellow as follows:—A complete mesothoracic collar, a small spot in the alar sinus above, a sinuous dorsal stripe running from near the alar sinus above downwards and outwards to the second pair of legs. This, stripe is formed by a union of an antehumeral with a humeral stripe, the upper antehumeral portion is very narrow and broadens abruptly at its middle, where it takes a bayonet turn outwards to fuse with the humeral portion. A small upper humeral spot representing the upper part of an obsolete humeral stripe and the sides broadly yellow. Sides marked

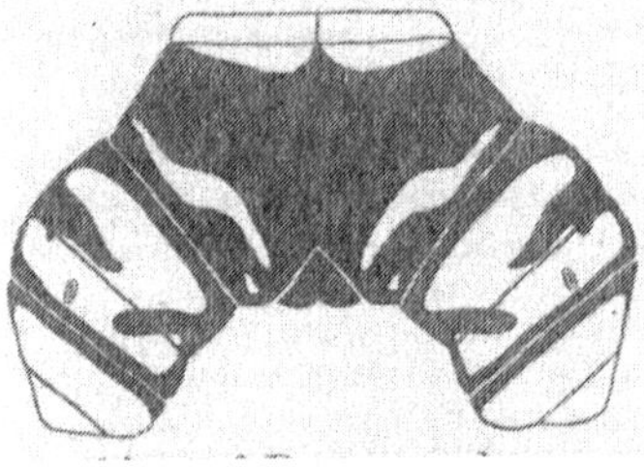

Fig. 67.—Thoracic markings of *Burmagomphus pyramidalis* Laidlaw, male.

with a narrow, complete black stripe on the postero-lateral suture and an incomplete one on the lower half of the anterior suture. *Legs* black, coxæ yellow, as also the inner surface of the anterior pair of femora; armed as for genus. *Wings* hyaline, often slightly tinted with saffron at bases; nodal index of two specimens $\frac{10\text{–}12}{10\text{–}9}\Big|\frac{13\text{–}9}{9\text{–}9}, \frac{10\text{–}14}{11\text{–}9}\Big|\frac{12\text{–}11}{9\text{–}10}$; pterostigma yellow between black nervures; 4 rows of postanal cells in hind-wing. *Abdomen* black, marked with yellow as follows:—Segment 1 with a triangular mark on dorsum at apex and a broad baso-lateral spot; 2 with a lanceolate dorsal stripe not quite reaching apical border, and a very broad spot on sides; 3 to 8 with narrow basal rings, rather broader on 7 and very narrow on 8, where, however, it is produced laterally as an elongated spot on the basal half of the segment; 3 also with a large baso-lateral triangular spot confluent with the dorsal ring; 8 often with an additional apico-lateral spot; 9 with nearly its apical half yellow, but laterally this mark much narrower and confluent with a ventral yellow border at its apex; at the mid-dorsum a very robust spine, also yellow (this spine was mentioned by Martin in his short description of *B. vermiculatus,* but escaped the notice of Laidlaw and of Williamson; it appears to be a group character of the species allied to *B. vermiculatus*); 10 entirely black. *Anal appendages* black; superiors as long as segment 10, divaricate, tapering to a point, to the outer side of which is a stunted spine which gives a bifid appearance to the apices when viewed a little to one side; inferior deeply cleft into two widely divaricate branches, which are rather longer than the superiors and taper to a fine point, the apex of which curls rather abruptly upward. *Genitalia*: lamina depressed, broadly arched; anterior hamules short, stiletto-shaped, with an inconspicuous backwardly turned point; posterior hamules very robust, broad, projecting markedly, pyriform, with a robust forwardly directed spine at apex; lobe tumid, rounded, shallowly grooved in front.

Female.—Abdomen 33 mm. Hind-wing 27 mm.

Similar to male, differing as follows:—Nodal index usually slightly higher; the basal ring on segment 2 more regular and the yellow on its sides more extensive; on the sides of 3 an elongate yellow spot, not confluent with the basal triangular spot; 9 with only its apical fourth yellow; 10 with a pair of small dorso-apical spots. *Anal appendages* small, conical, black, the protuberance between them yellow. Vulvar scale glossy black, its apical third broadly bifid, the whole scale broadly triangular.

Distribution.—Confined to WESTERN INDIA from south-west of the Deccan to Malabar, Coorg, Nilgiris, Kanara, and Coimbatore district.

Specimens from Poona, Deccan, have the ground-colour bright citron-yellow, whilst from the moister zones of Malabar and Coorg specimens have the ground-colour more grass-green. The measurements given are of those from Poona (type-locality), those from the Western Ghats are from 3 to 4 mm. larger.

They emerge from May to July in the Western Ghats and from August to September at Poona. At the latter place I took most specimens settled fairly high up on evergreens near the river-side, whilst in the Western Ghats they are rarely seen except settled on rocks in mid-stream, and then only during bursts of sunshine. It is a shy creature, but not uncommon.

Type in the Indian Museum, allotype female and other specimens in the British Museum and the Author's collection.

257. **Burmagomphus williamsoni** Fraser.

Burmagomphus vermiculatus Williamson (nec Martin), Proc. U.S. Nat. Mus. vol. xxxiii, pp. 298–303, text-fig. 29 (1907); Laidlaw, Rec. Ind. Mus. vol. xxiv, p. 399 (1922); Fraser, J. Bombay Nat. Hist. Soc. vol. xxix, pp. 62, 331 (1923).

Burmagomphus williamsoni Fraser, J. Bombay Nat. Hist. Soc. vol. xxxi, pp. 410, 411, pl. i, fig. 2 (1926); Laidlaw, Trans. Ent. Soc. Lond. vol. lxxviii, p. 189 (1930); Needham, Rec. Ind. Mus. vol. xxxiv, pp. 226, 227 (1932).

Male.—Abdomen 28 mm. Hind-wing 23 mm.

Closely resembles *B. pyramidalis* Laid. and *B. vermiculatus* Mart. Differs from the former as follows :—Nodal index $\frac{8\text{–}10}{10\text{–}8}\Big|\frac{10\text{–}9}{8\text{–}8}$; labium not bordered with black; labrum black, with two large transversely elongate greenish spots at base; mesothoracic collar slightly interrupted; segment 2 with a long trilobed dorsal stripe; narrow mid-dorsal lines on segments 3 to 5 confluent basally with the basal rings; segment 6 with a fine triangular dorsal spot at base; *genitalia* specialized, posterior hamules broader, the spine more forward and smaller, and, in addition, two smaller spines on the outer border, lobe more prominent and more deeply notched.

Female unknown.

Distribution.—BURMA. Described by Williamson from three males collected by Mr. R. A. Earnshaw; the specimens were recorded as *B. vermiculatus* Martin, but this I consider to be a distinct species. The type of *B. vermiculatus* was from Tonkin.

Type in Williamson's collection, Michigan University Museum.

258. **Burmagomphus hasimaricus** Fraser.

Burmagomphus hasimaricus Fraser, J. Bombay Nat. Hist. Soc. vol. xxxi, p. 411 (1926); Laidlaw, Trans. Ent. Soc. Lond. vol. lxxviii, p. 189 (1930); Needham, Rec. Ind. Mus. vol. xxxiv, p. 226 (1932).

Male.—Abdomen 32 mm. Hind-wing 27 mm.

Closely similar to *B. pyramidalis* Laid., but differs markedly by its genitalia. I note also the following differences :—Middle lobe of labium entirely black; labrum black, with two elongate basal spots, as in *B. williamsoni*, epistome with a mere vestige of the central spot and the large lateral spots against the eyes entirely absent; occiput and vertex entirely black, the large yellow spot entirely absent, borders of occiput flat, not curled up as in *B. pyramidalis*; mesothoracic collar very finely divided; the mid-lateral black stripe longer and nearly confluent with the short oblique stripe which runs towards the hinder stripe above;

pterostigma darker, pale brown; a trilobed dorsal stripe on segment 2 and a fine mid-dorsal line on 3 and 4, 8 entirely black; superior *anal appendages* longer and finer, and without the outer spine, the outer side bevelled for its apical two-thirds; inferior appendage with the apices of branches scarcely upturned. Posterior hamules very bulky, base a little constricted, apical portion produced squarely back with distinct angles, its anterior spine more robust and directed more forward.

Female.—Differs less in some respects and more in others than does the male from *B. pyramidalis.* The labrum is similar to that of *B. pyramidalis* and the large triangular lateral spots of epistome are present. The median spot on this organ is present as two fine transversely linear spots. The occiput is entirely black. The thorax is much less broadly black than in the male and approximates to that of *B. pyramidalis,* the lateral stripes being very narrow and the median short and not nearly reaching to the upper segmental stripe. *Abdomen*: the dorsal surface of segment 2 entirely without the mid-dorsal stripe and with only a basal ring and the sides broadly yellow; segments 3 to 6 have the basal rings produced slightly apicalwards, and in addition to the lateral elongate spot on segment 2 there is a similar but smaller one on 3; segment 8 is unmarked as in the male. The protuberance between the anal appendages is black. Lastly, the vulvar scale is longer and is cleft very deeply, ending in two long fine spine-like processes, very different from what is seen in *B. pyramidalis.*

Distribution.—One male, the *type,* and two females from Mr. H. V. O'Donel, from Hasimara, Duars, BENGAL, at present in the Author's collection.

259. **Burmagomphus sivalikensis** Laidlaw.

Burmagomphus sivalikensis Laidlaw, Rec. Ind. Mus. vol. xxiv, pp. 401, 402 (1922); Fraser, J. Bombay Nat. Hist. Soc. vol. xxix, pp. 62, 331 (1923); id., ibid. vol. xxxi, p. 414, pl. i, fig. 4 (1926); Laidlaw, Trans. Ent. Soc. Lond. vol. lxxviii, p. 189 (1930); Needham, Rec. Ind. Mus. vol. xxxiv, pp. 226, 227 (1932).

Male.—Abdomen 34 mm. Hind-wing 25 mm.

Head: labium pale greenish-white, the middle lobe narrowly bordered with black; labrum black, marked with two very large, broad, greenish-white spots; bases of mandibles coloured similarly; anteclypeus black, with a median linear spot and a large spot on each side palest green; frons black in front, greenish above, its base narrowly black, this colour prolonged forwards at the middle as a very short triangle; vertex black; occiput greenish-white, slightly notched

at its middle and convex on either side, fringed with pale hairs. *Prothorax* black, with an anterior collar, a large lateral spot, and a small median dorsal spot citron-yellow. *Thorax* black, marked with greenish-yellow as follows:— A mesothoracic collar finely divided by the dorsal black carina; very oblique long antehumeral spots, well separated from the collar below and the alar sinus above; a long irregular humeral stripe, constricted just below its upper part and continued below on to the base of the middle pair of legs; laterally two narrow black lines on sutures, rather narrower than the black stripe separating the humeral stripe from the yellow stripe following it; the first lateral stripe sinuous above and the second constricted above and angulated backwards below; a fine black line on the upper and back part of metepimeron. *Legs* black, inner surface of the anterior pair of femora yellow, armature of hind femora as for genus. *Wings* hyaline; pterostigma pale yellowy brown between thick black nervures, over 3 to 4 cells; nodal index $\frac{7\text{–}12}{6\text{–}9} \Big| \frac{13\text{–}8}{9\text{–}7}$.
Abdomen black, marked with pale citron-yellow as follows:— Segment 1 with a triangular spot on dorsum, its base against the apical border of segment; 2 with a lanceolate mid-dorsal stripe extending from base to apex; sides of these two segments broadly; 2 with an extension of black behind the oreillet and a narrow black apical ring; 3 to 7 with a narrow basal ring, rather broader on 3, and, on all, constricted at middle by an invasion of black; 3 also with an elongate lateral spot each side; 8 unmarked; 9 with a very large triangular spot on its apical three-fourths, more extensive on the dorsum than on the sides; 10 unmarked. *Anal appendages* black, as long as segment 10; superiors nearly parallel, pointed at apex and with a stout stunted spine near the apex on outer side as in *B. pyramidalis*, which gives a bevelled appearance to this part of appendage; inferior deeply cleft into two very widely divaricate branches with upturned apices. *Genitalia*: lamina slightly raised, hood-like, deeply cleft; anterior hamules short stiletto-shaped organs; posterior broad and robust, longer and narrower than in other species, and tapering outwardly to a blunt point, from the apex of which springs a long fine spine; lobe projecting, shaped like the spout of a milk-jug.

Female unknown.

Distribution.—Dehra Dun, N.W. PROVINCES, and Hasimara, Duars, BENGAL.

The above description differs in several respects from that of Laidlaw, which was made from an old, broken, and faded specimen. The present description has been made from a fresh specimen sent me by Mr. H. V. O'Donel, who took it at Hasimara.

The character of the genitalia, as well as its markings, will easily distinguish this species from others of the genus.

Type in the Indian Museum, from Dehra Dun.

260. **Burmagomphus sinuatus** Fraser. (Fig. 68.)

Burmagomphus sinuatus Fraser, Ceylon J. Sci. vol. xviii, pp. 33–35 (1933).

Male unknown.

Female.—Abdomen 34 mm. Hind-wing 27 mm.

Head: labium pale yellow, middle lobe clouded with black; labrum glossy black, with two larger bright citron-yellow spots narrowly separated at the middle line; bases of mandibles citron-yellow; anteclypeus black; postclypeus black, with a medial triangular and a larger rounded spot on each side bright citron-yellow; frons black, with a broad transverse citron-yellow stripe on its crest, very narrowly

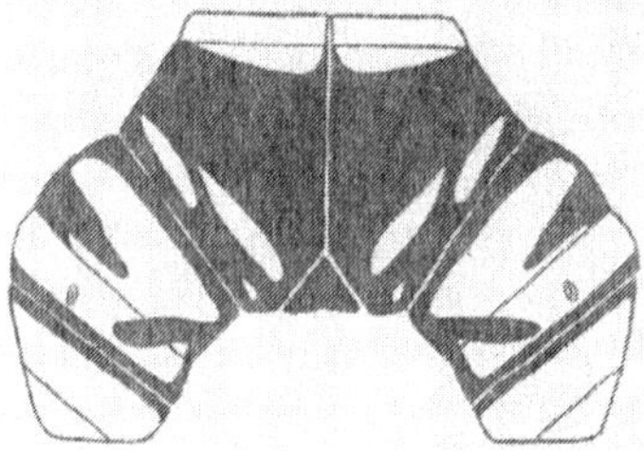

Fig. 68.—Thoracic markings of *Burmagomphus sinuatus* Fraser, female.

interrupted at the middle; vertex and occiput black, the latter with a large rounded greenish-yellow spot at its anterior part; eyes bottle-green during life. *Prothorax* black, marked with citron-yellow as follows:—An anterior collar; on the middle lobe a narrow anterior linear spot, a tiny geminate spot on the mid-dorsum, and a much broader one on each side. *Thorax* black, marked with citron-yellow as follows:—A narrow mesothoracic collar nearly interrupted at the middle; oblique antehumeral stripes, strongly divergent below and not confluent with the mesothoracic collar; an upper humeral spot followed by a short stripe below, which latter overlaps the upper part of the antehumeral stripe *en echelon* (thus forming a link between the *pyramidalis* and *laidlawi* groups, in which the two stripes are confluent at the middle and widely separated respectively); the sides broadly yellow, marked with two black stripes, an anterior broadly interrupted one, the upper portion of which is obliquely displaced so as to nearly join a posterior stripe, which lies on the postero-lateral suture and is complete. *Legs* black, inner surface of anterior pair of femora and external surface of coxæ and trochanters yellow. *Wings* hyaline; pterostigma golden-yellow framed in black nervures, braced, covering 4 to 5 cells; nodal index $\frac{9\text{–}12}{9\text{–}9} \Big| \frac{14\text{–}10}{9\text{–}11}$. *Abdomen* black, marked with bright citron-yellow as follows:—Segment 1 broadly so on sides and dorsum, a longitudinal black stripe separating these two areas; 2 with a bilobate mid-dorsal stripe, tapering to a point apically, its sides and under surface broadly yellow, including the rudimentary oreillets; 3 with a very small basal mid-dorsal point, a large triangular baso-lateral spot, and a narrow lateral stripe slightly separated from the former; 4 to 7 with only the baso-lateral spots, which are confluent at a point over the dorsum; 8 with a linear spot on each side at the base and a similar spot on each side, the two nearly in continuation; 9 with a small subdorsal apical spot; 10 unmarked. *Anal appendages* shortly conical, black. Vulvar scales very short, triangular, emarginate at apex.

Distribution.—Urugalla, CEYLON, during May.

As mentioned above, *B. sinuatus* forms a connecting-link between the two groups of the genus, but seems to be more nearly related to *B. laidlawi* and *B. cauvericus*. In the former the black stripe on the antero-lateral suture of thorax is complete, whilst in the latter the upper portion of the same stripe is confluent with the stripe on the postero-lateral suture.

Type, a unique female, in the Author's collection. It was found resting on an evergreen bush near the side of the river, and was the only one seen.

261. **Burmagomphus laidlawi** Fraser. (Fig. 69.)

Gomphus sp. Fraser, Rec. Ind. Mus. vol. xxiv, pp. 419 (1922); id., J. Bombay Nat. Hist. Soc. vol. xxix, pp. 62, 330 (1923).

Burmagomphus laidlawi Fraser, Rec. Ind. Mus. vol. xxvi, pp. 427, 475, 476 (1924); id., J. Bombay Nat. Hist. Soc. vol. xxxi, pp. 412–413, pl. i, fig. 5 (1926); Laidlaw, Trans. Ent. Soc. Lond. vol. lxxviii, p. 189 (1930); Fraser, Rec. Ind. Mus. vol. xxxiii, p. 447 (1931); Needham, Rec. Ind. Mus. vol. xxxiv. pp. 226, 227 (1932).

Male.—Abdomen 33 mm. Hind-wing 27 mm.

Head: labium with middle lobe black, lateral lobes greenish-white; labrum black, marked with two transversely oval yellow spots; bases of mandibles yellow; ante- and postclypeus black, with or without a small linear spot at lower border of latter; frons bright greenish-yellow, rest of head black; occiput slightly raised at the middle, fringed with long black hairs; eyes bottle-green. *Prothorax* black, with a large lateral spot and an anterior collar of yellow. *Thorax* marked with greenish-yellow or greenish-white as follows :—A complete mesothoracic collar ; oblique antehumeral stripes extending from alar sinus but not meeting the mesothoracic collar ; a minute upper humeral point and traces of a lower humeral stripe ; and on alar sinus. Sides greenish-yellow, marked with two narrow black stripes on the sutures. (These two stripes are occasionally confluent above and below, thus enclosing an elongate yellow spot, but usually only the upper parts are confluent. In one specimen, from South Kanara, the stripes are so confluent as to almost obliterate the included yellow, and the labial spots are also absent.) *Wings* hyaline, slightly tinted with yellow at base ; 4 rows of postanal cells in hind-wing ; pterostigma dark brown, covering 4 to 5 cells ;

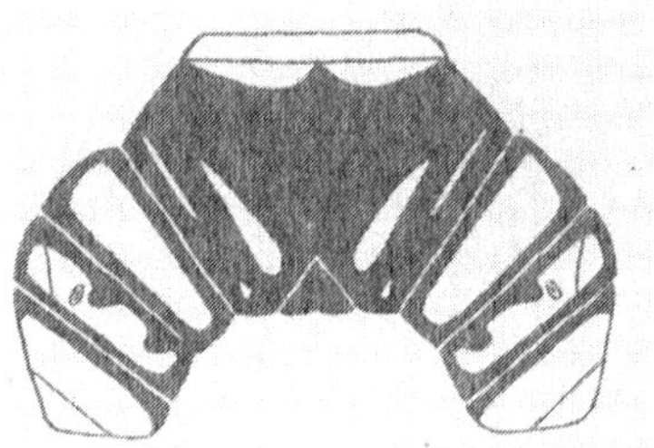

Fig. 69.—Thoracic markings of *Burmagomphus laidlawi* Fraser, male.

nodal index of two specimens $\frac{13\text{–}16}{11\text{–}10}\Big|\frac{15\text{–}12}{10\text{–}12}, \frac{10\text{–}14}{10\text{–}9}\Big|\frac{12\text{–}10}{10\text{–}11}$. *Legs* black ; inner side of anterior femora yellow ; hind femora armed as for genus. *Abdomen* black, marked with yellow as follows :—Segment 1 with a dorsal spot extending from base to apex, and its sides broadly ; 2 with a bilobed mid-dorsal stripe extending from base to apex, and the sides broadly, the subdorsal black descending as a vertical stripe behind the oreillets ; 3 with the mid-dorsal carina finely yellow, but usually only at the base, and a large baso-lateral spot ; 4 to 6 with basal dorsal triangular spots and baso-lateral lunules confluent with the former ; 7 with a broad basal ring occupying about one-fourth the length of segment ; 8 unmarked or with a minute basal dorsal triangular spot, and, in some specimens, with a still smaller apical spot ; 9 with its apical half yellow, this colour sometimes extending irregularly along the dorsal crest ; 10 unmarked. *Anal appendages* black, both the superiors and especially the branches of the inferior markedly divaricate ; apices of superiors long and thin and twisting inwards and a little downwards ; branches of inferior curling upwards ; slightly shorter than segment 10. *Genitalia* : lamina very narrow.

very depressed; anterior hamules short stylets with an inconspicuous hook at apex; posterior hamules very robust, projecting, squared, and with a robust spine on the anterior corner projecting forward, a smaller one on the hinder angle projecting vertically. In some specimens there are two or three such posterior spines, but usually only one; lobe very tumid and prominent, rounded, deeply cleft.

Female.—Abdomen 32 mm. Hind-wing 29 mm.

Very similar to the male; markings better defined; the humeral spot larger and continued as a broken line below, or similar to that of the male; often a small spot on each side of the postclypeus. A very prominent horn on each side of head just behind the lateral ocelli; occiput simple, nearly straight. *Abdomen*: baso-lateral spot on sides of segment 3 continued after a slight interval by a broad lateral stripe, which extends nearly to apex of segment; 8 with a very fine basal yellow ring; the yellow on 9 more restricted; 10 with its apical border lined with yellow. Superior *anal appendages* small, conical, pointed, black with a yellow spot above. Vulvar scale very broad, of even width, slightly notched at its middle, glossy black. (The Kallar female has segment 10 unmarked and the horns on the vertex poorly developed.) Wings, when mature, enfumed pale brown, in immature examples rather deeply saffronated at the base.

Distribution.—Western Ghats only. The original pair was taken *in copula*, flying along the bed of the Kallar River, Nilgiris, 2,000 ft., August 13, 1922. A female was taken at Gudalur, Nilgiri-Wynaad, 3,500 ft., and several males in Coorg and South Kanara during September 1923. It is also not uncommon in the Bolovumpatti Forest, Coimbatore district.

Its habits are similar to *B. pyramidalis*, with which it is occasionally found in company.

Type in the British Museum.

262. **Burmagomphus cauvericus** Fraser.

Burmagomphus cauvericus Fraser, J. Bombay Nat. Hist. Soc. vol. xxxi, p. 413, pl. i, fig. 6 (1926); Laidlaw, Trans. Ent. Soc. vol. lxxviii, p. 189 (1930); Fraser, Rec. Ind. Mus. vol. xxxiii, p. 447 (1931); Needham, Rec. Ind. Mus. vol. xxxiv, pp. 226, 227 (1932).

Male.—Abdomen 35 mm. Hind-wing 28 mm.

Very similar to *B. laidlawi*, differing as follows:—Middle lobe of labium greenish narrowly bordered with black; a small triangular spot of greenish-yellow on each side of postclypeus; upper humeral spot usually entirely absent, or, if present, then a mere point. Lateral markings on thorax similar to those of *B. pyramidalis*; the stripe on the first suture is incomplete above, but extends higher than it does in *B. pyramidalis*, and is almost confluent with a short oblique stripe which runs back from below the forewing to join the upper part of the stripe on the second suture, enclosing a small spot above. (In some specimens this short stripe is so broadly confluent that the stripe on the second suture appears to be expanded above, and marked with a tiny point of citron-yellow.) Hinder border of metepimeron narrowly bordered above with black. Segment 2 with a dorsal lanceolate spot extending to about the middle of segment 8 unmarked; 10 with a fine vestigial line along its mid-dorsal ridge. *Genitalia*: posterior hamules less broad than in *B. laidlawi*; the anterior spine smaller and with two smaller spines situated close together at the hinder angle; lobe smaller, less deeply notched.

Female.—Abdomen 35 mm. Hind-wing 31 mm.

Very similar to the male, differing as follows:—Black markings on sides of thorax more restricted, the enclosed spot above front of second lateral suture very large and the black stripe on the suture at this level reduced to a fine line; basal ring on segment 3 produced laterally as a broad triangular spot, followed by a large elongate lateral isolated spot; segment 4 with a similar lateral but smaller spot. Occiput notched in the middle and rounded on either side of the notch. Vertex with a pair of horns as in *B. laidlawi*, but these shorter, with a broader base, and situated a little to the outer side of the hinder ocelli, instead of directly behind them. Vulvar scale more broadly notched than in *B. laidlawi*, so that its border exhibits two broad triangular processes. *Anal appendages* entirely black.

Distribution.—COORG only, along the banks and tributaries of the River Cauvery.

This handsome species may be immediately recognized by the combination of the anterior thoracic markings of *B. laidlawi* with the lateral thoracic markings of *B. pyramidalis*.

Type in the British Museum.

263. **Burmagomphus V-flavum** Fraser. (Fig. 70.)

Burmagomphus V-flavum Fraser, J. Bombay Nat. Hist. Soc. vol. xxxi, pp. 414, 415, pl. i, fig. 7 (1926); Laidlaw, Trans. Ent. Soc. Lond. vol. lxxviii, p. 189 (1930); Needham, Rec. Ind. Mus. vol. xxxiv, pp. 226, 227 (1932).

Male unknown.

Female.—Abdomen 31 mm. Hind-wing 30 mm.

Head: labium pale greenish-white; labrum black, marked with two large basal transversely oval citron-yellow spots; bases of mandibles yellow; anteclypeus black; postclypeus glossy black, narrowly bordered below with citron-yellow, and with a large spot of the same colour on each side against the eyes; frons black, with a broad citron-yellow stripe above, which is cut by a median black line into two large oval spots; vertex black; occiput yellow encircled with black, the hinder border narrowly black, posterior surface yellow, slightly concave. *Prothorax* black, broadly marked with citron-yellow on the whole posterior lobe, a small geminate spot in front of it, a very large lateral spot nearly confluent with the yellow on the posterior lobe, and a broad anterior collar. *Thorax* black, marked with citron-yellow as follows:—A complete mesothoracic collar, slightly notched behind at its centre; antehumeral and humeral oblique stripes, which converge and become broadly confluent below so as to form a large yellow V-shaped spot on each side of dorsum of thorax; a small spot in the front angle of the alar sinus; laterally broadly yellow, with an incomplete black stripe on the first suture extending up as far as the

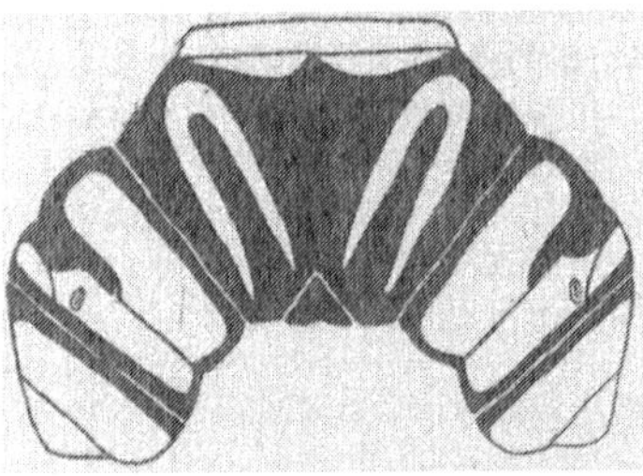

Fig. 70.—Thoracic markings of *Burmagomphus V-flavum* Fraser, male.

spiracle and a narrow complete black line on the second suture. *Legs* black, inner sides of fore femora greenish-yellow; hind femora with a row of 9 to 10 short robust spines gradually lengthening distally and numerous smaller spines at the proximal third; the femora overlapping the base of segment 2 when apposed. *Wings*: hyaline; pterostigma pale yellow between black nervures, rather large, more than one-fourth the distance between node and proximal end of pterostigma, covering 4 to 5 cells; nodal index $\frac{11\text{–}14}{11\text{–}11} \Big| \frac{13\text{–}11}{12\text{–}11}$; *Cuii* and *IA* rather divergent in hind-wing from a little more than halfway to wing-border; no basal incomplete antenodal nervures. *Abdomen* black, marked with citron-yellow as follows:—Segment 1 with a small subdorsal apical black spot, otherwise entirely yellow; 2 with a broad mid-dorsal trilobed stripe and the sides very broadly; 3 with a basal ring confluent with a fine dorsal stripe which extends to apex of segment, and laterally with a very broad stripe nearly interrupted at the jugal suture and not extending to apex of segment; 4 to 7 somewhat similar, but the sides broadly black, only segments 4 to 5 having a small oval lateral yellow spot; 8 with a small yellow triangle on its dorsum at the base; 9 with its apical half bright citron-yellow; 10 with its apex narrowly bordered with three triangular yellow spots. *Anal appendages* short, conical, black. Vulvar scale very short and broad, less than one-fourth the length of segment 9, very shallowly notched at apex.

Distribution.—Maymyo, N. Shan States, UPPER BURMA, June 15, 1924, two females collected by Col. F. Wall, I.M.S.

This species is not quite typical of the genus *Burmagomphus*, as the pterostigma is too large and the legs too short, and the armature of the hind femora differs a little. The large swollen pterostigma recalls that of *Cyclogomphus*, but there is no incomplete basal antenodal nervure present in any of the wings of the two specimens. The formation of yellow V's on the dorsum of thorax, by the fusion of the humeral and antehumeral stripes below, is sufficiently characteristic to distinguish *B. V-flavum* from any other Indian species of the genus.

Type in the Author's collection.

Genus **DAVIDIOIDES** Fraser. (Fig. 71.)

Davidioides Fraser, Rec. Ind. Mus. vol. xxvi, p. 472 (1924); id., J. Bombay Nat. Hist. Soc. vol. xxxi, p. 419 (1926); Laidlaw, Trans. Ent. Soc. Lond. vol. lxxviii, p. 188 (1930); Needham, Rec. Ind. Mus. vol. xxxiv, pp. 217, 218 (1932).

Closely similar to *Burmagomphus*; size moderate; colour mat or glossy black, marked with bright citron-yellow.

Head rather small, triangular; frons well angulated; occiput simple, concave. Wings broad and long, nearly as long as abdomen, tornus markedly angulated, base of hind-wing deeply excavated; membrane obsolete; anal triangle 3-celled; arc situated between the second and third antenodal nervures; 3 transverse nervures between the sectors of arc from arc to bifurcation of *Rs* in fore-wing, only 1 in hind-wing; only 1 row of postanal cells in fore-wings, but occasional double cells, 4 rows in hind-wing; first postanal cell in hind-wing extending nearly to the proximal end of base of subtrigone, this cell very broad; anal loop absent; basal incomplete antenodal nervures absent; nodal index moderately high; primary antenodals the first and sixth; discoidal cell in fore-wings nearly equilateral, but the basal side slightly shorter than the two others, the cell distinctly separated from the lower sector of arc by a very short stalk, entire; discoidal cell of hind-wing traversed by a nervure running from the middle of costal side to the distal, which two sides are equal and nearly double the length of the basal, the cell separated from the lower sector of arc by a much longer stalk than in the fore-wings ; pterostigma short and swollen at its middle, less than one-third as long as the distance between the node and proximal end of pterostigma, braced ; *IA* in fore-wing pectinated ; *IA* and *Cuii* running parallel to hinder border of wing ; only 1 cubital nervure in all wings ; subtrigones and hypertrigones all entire. Legs moderately long, hind femora extending to base of segment 2, but not overlapping that segment, furnished with a group of numerous, closely set, short spines on the proximal two-thirds of the flexor surface, which later form into two rows of longer more robust spines at the distal third of the limb ; tibial spines moderately long and closely set. Abdomen tumid at base, then very narrow and cylindrical from segment 3

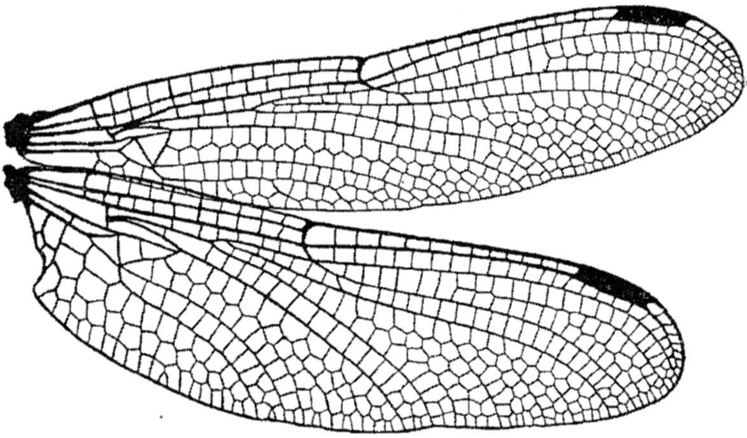

Fig. 71.—Wings of *Davidioides martini* Fraser, male.

to apical end of segment 7, where it again becomes abruptly broadened, segment 8 especially broad and almost winged, as in *Platygomphus*. Anal appendages short, equal in length to segment 10, conical, pointed at apex, slightly divaricate, the branches of the inferior equal to the superior appendages and equally divaricate. Genitalia : lamina slightly projecting, arched ; anterior hamules long slender processes ; posterior hamules very broad robust sinuous processes, projecting markedly and ending in a recurved point ; lobe moderately large and prominent, scoop-shaped and strongly lipped.

Genotype, *Davidioides martini* Fraser.

Distribution.—MALABAR only. A monotypic genus.

The species is remarkable on account of the traversed discoidal cells in the hind-wing, resembling in this the genus

Davidius, but it differs in most other respects from the latter. The venation is similar to that of *Burmagomphus*, with the exception of the traversed discoidal cells and the stalking of the latter, especially of that of the hind-wing, in which the venation resembles that of *Merogomphus*. The shape of the abdomen also recalls that of the latter genus. It is unfortunate that the female of *D. martini* is unknown, so that we are at present ignorant of the character of the armature of the femora in this sex, which might go far to determine the true position of the genus.

264. **Davidioides martini** Fraser. (Fig. 71 and Pl. II, fig. 4.)

Davidioides martini Fraser, Rec. Ind. Mus. vol. xxvi, pp. 427, 472, 473, text-fig. 2 (1924); id., J. Bombay Nat. Hist. Soc. vol. xxxi, pp. 419, 420, text-fig. 3, vi, pl. ii, fig. 1 (1926); Laidlaw, Trans. Ent. Soc. Lond. vol. lxxviii, p. 188 (1930); Fraser, Rec. Ind. Mus. vol. xxxiii, p. 447 (1931); Needham, Rec. Ind. Mus. vol. xxxiv, p. 226 (1932).

Male.—Abdomen 38 mm. Hind-wing 33 mm.

Head: labium dirty yellow; labrum black, with a large yellow basal spot on each side, the pair widely separated; bases of mandibles and anteclypeus yellow; postclypeus black; frons black on the lower part of front, yellow on the upper part and above, its base above broadly black especially at the middle; vertex and occiput black, the latter simple, slightly concave. *Prothorax* black, marked with yellow as follows:—A minute spot on the centre of the posterior lobe; a large lateral spot on each side of the middle lobe; a twin spot on the mid-dorsum and an anterior collar. *Thorax* black on dorsum, yellow on the sides, a slightly interrupted yellow mesothoracic collar, slightly oblique antehumeral stripes separated widely from the collar below and not quite reaching the alar sinus above, a minute upper humeral spot. Laterally two narrow parallel black stripes on the sutures enclosing an equally narrow yellow line; a narrow black posterior border to the metepimeron, confluent above and below with the posterior black stripe. Yellow beneath. *Legs* black, anterior and middle pairs of femora yellow on the inner side, armed with a row of very closely set, short, even spines. *Wings* hyaline; pterostigma blackish-brown, braced, covering 3 to 4 cells; nodal index $\frac{12\text{–}14}{12\text{–}10}\Big|\frac{16\text{–}12}{10\text{–}13}$.

Abdomen black, marked with yellow as follows:—Segment I with a large spot on the dorsum and the sides broadly; 2 with a mid-dorsal fusiform spot not extending to base or apex, the ventral and lower part of apical borders narrowly; 3 to 7 with basal rings, narrow on 3 to 6, occupying the basal half of 7, where also there is a small apico-lateral spot on each side; 8 to 10 unmarked. *Anal appendages*: superiors as long as segment 10, bright yellow, narrowly black at base, widely divaricate, conical, tapering to a point, with 5 or 6 minute black spines on the ventral surface; inferior deeply cleft, its branches widely divaricate and equally so as the superiors but shorter, curled slightly up at apex, black. *Genitalia* described under generic characters.

Distribution.—WESTERN GHATS: Kunnoth, N. Malabar, May 18, 1923.

Type in the Author's collection.

Genus **MESOGOMPHUS** Förster. (Fig. 72.)

Mesogomphus Förster, Wiesbaden Jahr. Ver. Natk. vol. lix, p. 323, (1906); Ris, Ann. S. Afr. Mus. vol. xviii, p. 343 (1920); Fraser, J. Bombay Nat. Hist. Soc. vol. xxix, p. 991 (1924); Laidlaw, Trans. Ent. Soc. Lond. vol. lxxviii, p. 192 (1930); Needham, Rec. Ind. Mus. vol. xxxiv, pp. 217, 218, 221 (1932).

Size medium, colour for the most part sandy yellow marked with shades of brown and black, more rarely black marked with bright yellow; characterized by the long curled anal appendages, the short inferior appendage having its branches very closely apposed as in *Onychogomphus* and *Lamelligomphus*.

Head moderately large, triangular, frons angulated, occiput simple, usually slightly and shallowly concave. Wings: reticulation close; tornus angulated and base of hind-wing moderately excavated; membrane very narrow, almost obsolete; anal triangle 4-celled; arc situated midway between the first and second antenodal nervures; only 2 transverse nervures between the sectors of arc from arc to bifurcation of *Rs* in the fore-wings, 1 in the hind-wing; 2 rows of postanal cells in fore-wings, 5 in the hind-wing; anal loop absent, the first postanal cell extending proximally to about the middle of base of subtrigone; nodal index high; primary antenodal nervures the first and fifth; discoidal cells entire, that of fore-wing almost equilateral, but the distal side very slightly longer than the two others, that of the hind-wing elongate in the length of wing, the distal side longer than the costal and the costal at least one-third longer than the basal; pterostigma rather long, expanded at its middle, well braced, slightly longer than one-third the distance from node to proximal end of pterostigma; *IA* in fore-wings pectinate, with a maximum of four cells between it and the margin of wing; *Cuii* and *IA* parallel to margin of hind-wing; only a single cubital nervure in all wings; subtrigones and hypertrigones entire in all wings; discoidal field with 2 rows of cells to the level of node or to slightly distal of that point. Legs short, hind femora extending only to posterior margin of thorax and furnished with a dense group of short spines on proximal half of flexor surface which become more or less arranged into two rows at distal end of femur ; tibial spines moderately short and closely set. Abdomen of male tumid at base, narrow and cylindrical from segment 3 to base of segment 7, from which point the abdomen is again dilated, especially at segments 8 and 9, which bear broad lateral wing-like projections very similar to those found in species of the genus *Ictinus*. Abdomen of female more or less cylindrical throughout, segments 8 and 9 without lateral dilatations. Anal appendages: superiors nearly double the length of

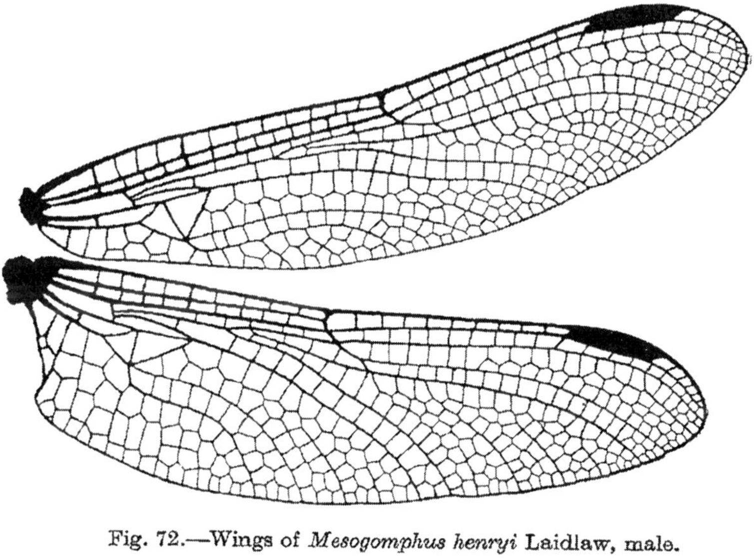

Fig. 72.—Wings of *Mesogomphus henryi* Laidlaw, male.

segment 10, tapering to apex, the terminal half curled regularly downwards and very closely apposed ; inferior less than half the length of superiors, strongly curled in a downward and then an upward direction, bifid nearly to base, the two branches closely apposed like the superiors. Genitalia : lamina very broadly and deeply arched, depressed ; anterior hamules rather short, narrow, pointed processes ; posterior hamules much broader, longer, and more robust processes, ending in an incurved spine ; lobe spout-like, projecting, deeply emarginate.

Genotype, *Gomphus cognatus* Ramb.

Distribution.—Tropical Africa and Madagascar, INDIA, BURMA, and CEYLON, Malaysia, Java, and the Celebes ; four or five species are found within Indian limits.

Species of the genus breed in open rocky streams, the males haunting shallow runnels where the water flows swiftly over a clean sandy or pebbly bottom, or they may be found settling on sandy beaches beside streams, indulging in frequent rapid flights up and down stream; *M. henryi* usually settles on rocks in mid-stream, and I have seen several occupying adjacent rocks.

The larvæ resemble those of *Burmagomphus* and *Onychogomphus*, and have simllar habits.

The genus is closely related to *Onychogomphus*, from which it is distinguished by the dilated and winged lateral borders of segments 8 and 9, as well by the absence of an anal loop in the hind-wing. The larvæ burrow in sand and are stream-dwellers.

Key to Indian Species of Mesogomphus.

1.	Anal appendages yellow	*lineatus* (Selys), p. 230.
	Anal appendages black	2.
2.	Mesothoracic collar confluent with antehumeral stripes	*lindgreni* Fras., p. 235.
	Mesothoracic collar separated from antehumeral stripes	3.
3.	Occiput yellow above; hind femora black: smaller species (abdomen with appendages not more than 27 mm.)	*risi* Fras., p. 237.
	Occiput black or dark brown; hind femora broadly marked with yellow: larger species (abdomen with appendages 30 to 39 mm.)	4.
4.	Face, including labrum, entirely yellow..	*frontalis* (Selys), p. 238.
	Labrum with a black basal triangular spot; postclypeus black, rest of face yellow	*henryi* (Laid.), p. 234.

265. **Mesogomphus lineatus** (Selys). (Fig. 73.)

Gomphus lineatus Selys, Rev. Odon. p. 386 (1850).
Onychogomphus lineatus Selys, Bull. Acad. Belg. vol. xxi, pt. 2, p. 36 (1854); id., Mon. Gomph. p. 48 (1857); Williamson, Proc. U.S. Nat. Mus. vol. xxxiii, pp. 309, 310 (1907); Morton, Trans. Ent. Soc. Lond. part ii, Sept. p. 305 (1907); Fraser, Rec. Ind. Mus. vol. xvi, p. 463 (1919) (larva); Laidlaw, ibid. vol. xxiv, pp. 371, 404, 405 (1922); Fraser, ibid. vol. xxiv, p. 426 (1922); id., J. Bombay Nat. Hist. Soc. vol. xxix, pp. 65, 333 (1923).
Lindenia lineata Kirby, Cat. Odon. p. 59 (1890).
Mesogomphus lineatus Fraser, Rec. Ind. Mus. vol. xxvi, pp. 427, 477 (1924); id., J. Bombay Nat. Hist. Soc. vol. xxix, pp. 991–994, text-fig. xii, pl. i, fig. 1 (1924); Laidlaw, Spolia Zeylanica (*M. henryi*, nec *lineatus*), vol. xii, pp. 340, 341, text-fig. 1 (1924); id., Trans. Ent. Soc. Lond. vol. lxxviii, p. 192 (1930); Fraser, Rec. Ind. Mus. vol. xxxiii, p. 448 (1931); Needham, Rec. Ind. Mus. vol. xxxiv, p. 221 (1932).

Male.—Abdomen (with appendages) 32–37 mm. Hind-wing 24–27 mm.

Head: labium pale yellow; labrum, face, and frons sandy yellow, the latter with a more or less ill-defined brownish-black line across the crest; vesicle and occiput yellow, a narrow transverse dark brown streak separating them; in some specimens a narrow diffuse basal brown line on upper surface of frons, and base of labrum may be clouded with the same colour. Occiput raised into a slight point at its centre and usually bearing a line of from two to four minute spines on either side of this eminence. *Prothorax* blackish-brown, posterior lobe narrowly and a large spot on either side yellow. *Thorax* sandy yellow, marked with dull or dark brown or even blackish-brown (according to age of specimens) as follows:—A dorsal line bordering the mesothoracic collar closely and turning abruptly up on either side of the mid-dorsal carina, which is black in its upper part; an oblique antehumeral line beginning from lower end of dorsal line and running upwards and inwards to join this line in its upper part, thus enclosing a thin stripe of the ground-colour; a line on the humeral suture and two lateral lines close together, all three parallel, of the two latter one crosses the spiracle and the other maps out the postero-lateral suture. *Legs* yellow, marked with black; a stripe on inner side of all femora, an outer stripe on distal half of middle femora, and a rounded distal spot or stripe on outer side of hind femora; tarsi black. Two rows of robust, evenly-spaced, short spines on hind femora which converge to form a group of spines proximally; middle femora with more numerous and more closely set spines. *Wings* hyaline; costa yellow as far as pterostigma, which is pale reddish-brown heavily bordered with black especially on the costal border, covering 4–6 cells, usually well braced; nodal index rather variable:— $\frac{10\text{–}15 \mid 16\text{–}9}{10\text{–}11 \mid 10\text{–}9}$, $\frac{8\text{–}15 \mid 15\text{–}7}{8\text{–}9 \mid 9\text{–}7}$, $\frac{8\text{–}13 \mid 14\text{–}7}{7\text{–}9 \mid 9\text{–}8}$; 5–6 rows of postanal cells; 2 nervures between the sectors of the arc in fore-wing, only 1 in hind-wing; 2 rows of discoidal cells as far as node; outer border of discoidal cell in hind-wing well angulated. *Abdomen* black marked with yellow, or yellow marked with black, the two colours occupying a variable space according to the age of specimens; in old specimens the yellow reduced to basal rings, in young specimens, the black present as apical rings. Segment 1 with the sides broadly yellow and a large dorsal apical spot of the same colour; 2 with a subdorsal black line on each side enclosing a dorsal bilobed yellow spot, the black curving down on each side narrowly behind the oreillets, and an apical black ring of variable width; 3 to 7 usually with broad black apical rings occupying a variable length of the segments, but generally less on 7 than on the others ; a lateral black line runs from the apical ring on each side and extends almost to the base ; the dorsal carina and jugal sutures finely black ; in some specimens the lateral stripe shortens as traced from 3 to 7, in others the jugal suture is so broadly black that it cuts off two dorsal yellow spots, one in front and the other behind ; 8 and 9 with wide dilatations at their sides, black on the dorsum except for a fine basal ring, and an irregular spot of the same colour on the mid-dorsum, more evident on segment 8 than 9 ; 10 sandy yellow, with the basal half or two-thirds of the dorsum black. The terminal three segments show considerable variation ; thus the foliate lateral processes may be broadly edged with black, or almost entirely blackish-brown, and the dorsal surfaces may be entirely deep black save for a fine basal ring on segment 9 and an apical ring on 10, the latter sending a slight prolongation along the dorsal carina. There are many

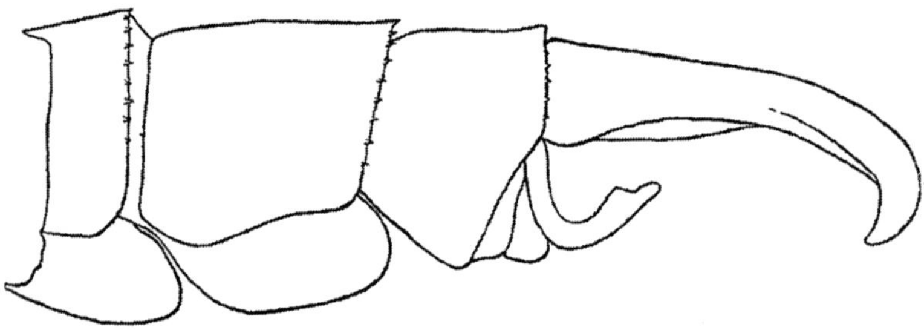

Fig. 73.—Terminal abdominal segments and anal appendages of *Mesogomphus lineatus* Selys, male, viewed from the left side.

variations of this pattern, but the differences appear to be purely individual rather than racial. *Anal appendages* (fig. 73) dull sandy yellow. *Genitalia* very similar to those of *M. cognatus* ; lamina pale yellow, very shallow, broadly notched ; inner hamules long and narrow, slightly convergent, the apices ending in a short, sharp, and slightly outwardly-turned hook, black ; outer hamules much longer and stouter, broadish at base but rapidly tapering to a point, turning directly in and nearly meeting at their apices, each bearing on the inner side, somewhat below the apex, a short robust spine, black, yellow at the base ; lobe deeply excavate, hood-shaped, its apex strongly indented, directed almost straight back, black.

Female.—Abdomen 31–36 mm. Hind-wing 24–27 mm. Closely resembles the male except for shape of abdomen ; colour as a rule much paler than in male and less marked with brown or black. The lateral prolongations of black extend on to the second segment, segments 8 and 9 are usually deep black and unmarked, but in some specimens the abdominal markings are much restricted and 8, 9, and 10 are entirely yellow; segments 3 to 6 ventral to the lateral black stripes are pure shining silvery white, a colour rarely met with in other members of the genus. Occiput bearing four spines on either side of the middle line, very evenly disposed in all specimens I have examined. (Laidlaw states that they are entirely absent in one of my Poona female specimens.) Wings palely enfumed or saffronated at the base and even more faintly along the costal margin; costal and many of the nervures, especially the nodal, light yellow; nodal index of two specimens $\frac{6\text{–}13}{6\text{–}9}\Big|\frac{12\text{–}5}{9\text{–}6}$, $\frac{7\text{–}14}{8\text{–}10}\Big|\frac{16\text{–}8}{10\text{–}8}$.

Larva.—Length 25 mm., abdomen 17 mm., hind femur 5 mm. Cylindrical. Head moderately large; antennæ of four segments, clubbed, the last segment very minute. Mask very broad, almost square, the base a little constricted, its outer surface coated with short hairs; middle lobe straight, not projecting, fringed with fine bristles; lateral lobe furnished with a long movable hook, somewhat squat, saw-shaped, its inner border with a row of blunt molar-like teeth. Abdomen tapering gradually to the end, tenth segment short, not hollowed out above and not furnished with spines. All segments except the ninth and tenth furnished with a blunt mid-dorsal spine. Legs very short, femora fringed with long coarse hairs. The larva thus differs strikingly from that of *Lamelligomphus nilgiriensis* in its narrow and cylindrical shape and in the clubbed antennæ.

Distribution.—Found throughout India. I have specimens from Burma, Bengal, Dehra Dun, Poona, Bangalore, Khandala (Deccan), Madras, Palghat (Malabar), Trichinopoly, Coorg, and the Nilgiris.

M. lineatus breeds both in still and running waters. In Poona and Coorg the females deposit their eggs in shallows either at the edge of a sandy beach or over ripples flowing over a gravelly bottom; the males await them there, settled on the sandy foreshore, where by reason of their colour they are well nigh invisible, occasionally rising and patrolling backwards and forwards over the shallow ripples. In Coorg, at Hoskoti, I found numerous specimens emerging in a dense reedy swamp, and can only conjecture that ova had been deposited in a tiny brook with a sandy bottom which flowed into the marsh; the site was quite unusual, and may have been forced on the females by the extraordinarily swollen state of the rivers, or larvæ may possibly have been swept from the rivers and left stranded in such marshy retreats.

M. lineatus is a very wary and shy insect and somewhat difficult to catch. It occurs almost throughout the year but is most common from September to November.

M. lineatus varies greatly in colour according to the climatic conditions under which it occurs, pale forms occurring in the arid zones and dark forms in regions of heavy rainfall.

Type in the Selys collection, Brussels Museum.

266. **Mesogomphus henryi** Laidlaw.

Mesogomphus lineatus Laidlaw (*henryi* Laid., nec *lineatus* Selys), Spolia Zeylanica, vol. xii, pp. 340, 341, text-fig. 1 (1924).

Mesogomphus henryi Laidlaw, Proc. Zool. Soc. Lond. pp. 131, 132 (1928); Needham, Rec. Ind. Mus. vol. xxxiv, p. 221 (1932); Fraser, Ceylon J. Sci. vol. xviii, p. 33 (1933).

Male.—Abdomen 32–36 mm. Hind-wing 24–27 mm.

Head: labium bright yellow, middle lobe dark brown; labrum citron-yellow, narrowly bordered with black and with a rounded black area at middle extending from base halfway across the lip; ante- and postclypeus and lower third of anterior surface of frons bright citron-yellow, with an elongate black spot on the centre of postclypeus; frons black, with two large transversely oval citron-yellow spots above; bases of mandibles yellow; eyes bottle-green during life; vertex and occiput black, the latter concave and fringed with short black hairs. *Prothorax* black, with a tiny yellow point on each side. *Thorax* black, marked with citron-yellow as follows:—A rather broadly interrupted mesothoracic collar; short pointed antehumeral stripes; a tiny upper humeral spot; two broad oblique stripes on each side, one on mesepimeron, the other on metepimeron, with one or three small spots on the black area between them, a large oval spot just above posterior trochanters. Underside and extreme posterior border of thorax bright yellow. *Legs* black, the anterior two pairs of femora striped on outer sides with citron-yellow, coxæ and trochanters also yellow. *Wings* hyaline; pterostigma black, with a dark ochreous centre, braced, covering 4 cells; nodal index of two specimens $\frac{8\text{–}14}{9\text{–}9} \Big| \frac{14\text{–}7}{9\text{–}9}$, $\frac{6\text{–}13}{7\text{–}9} \Big| \frac{12\text{–}7}{9\text{–}7}$; 4 cells in anal triangle. *Abdomen* black, marked with citron-yellow as follows:—Segment 1 with an obscure mid-dorsal apical spot and a large triangular latero-apical spot on each side; 2 with a quadrate latero-apical spot on each side which extends basally to include the oreillets, and with a narrow trilobate mid-dorsal stripe, the median lobe of this clubbed; 3 to 6 with paired subdorsal basal spots, on 3 extending to apical side of jugal suture; 7 with an elongate subdorsal spot covering the basal two-thirds of the segment, interrupted by the black jugal suture and well separated from its fellow spot on the opposite side; 8 and 9 unmarked, bearing well developed but narrow black lateral foliations; 10 with an obscure lateral spot and a bright geminate apical dorsal spot. *Anal appendages* blackish or dark reddish-brown, almost similar in shape to those of *M. lineatus*, but the inferior more extended and nearly half the length of superiors instead of only one-third or less.

Female.—Abdomen 30–34 mm. Hind-wing 26–28 mm.

More robust than the male and differing in a few particulars: labium entirely yellow; black area on labrum more restricted; a well-defined humeral stripe present, consisting of a large upper spot and a more or less well-defined stripe below slightly separated from it; the yellow spots on the medial black band on sides of thorax fused to form an almost complete stripe, or two large slightly separated elongated spots; legs more extensively yellow, the femora entirely yellow save for the distal extremities of the two posterior pairs on the outer side; wings sometimes palely enfumed; pterostigma almost black, longer, covering 5 to 6 cells; markings of abdomen similar, but the ventral borders of segments 1 and 2 yellow and the subdorsal apical stripe on segment 2 prolonged to the base. *Anal appendages* yellow or dark brown, tapered, cylindrical; vulvar scale similar to that of *M. lineatus*.

Distribution.—CEYLON only. It is the commonest Gomphine found in the island, and is met with in numbers on most streams in submontane areas. At Madugoda I saw specimens every few yards perched on prominent rocks in mid-stream; it was also common around Kandy.

The species is very closely related to *M. lineatus*, but is easily distinguished by the great preponderance of the black ground-colour over the yellow markings, the opposite condition obtaining in *M. lineatus*. It is the only species of the genus *Mesogomphus* found in Ceylon.

Type in the British Museum; specimens in the Colombo Museum and my own collection.

267. **Mesogomphus lindgreni** (Fraser).

Onychogomphus lindgreni Fraser, J. Bombay Nat. Hist. Soc. vol. xxix, pp. 65, 332 (1923).
Mesogomphus lindgreni Fraser, J. Bombay Nat. Hist. Soc. vol. xxix, pp. 995, 996 (1924); Laidlaw, Trans. Ent. Soc. Lond. vol. lxxviii, p. 193 (1930); Needham, Rec. Ind. Mus. vol. xxxiv, p. 221 (1932).

Male.—Abdomen with appendages 39 mm. Hind-wing 29 mm.

Head: labium whitish-green; labrum greenish-yellow, broadly black at base; mandibles similarly coloured; anteclypeus and postclypeus greenish-yellow, the latter clouded with black on either side of the middle line; frons greenish-yellow, with a black streak along its crest and the base above broadly black; vertex black; occiput very dark brown, *its border bearing no spines*, entirely smooth save for a few hairs at each end. *Prothorax* black. *Thorax* greenish-yellow, the black markings similar to those of *M. lineatus* but much more extensive; antehumeral and humeral stripes coalescent, obliterating the ground-colour between them save for a small upper spot; similarly the two lateral stripes coalescent, forming one broad black stripe and leaving a mere vestige of the ground-colour above. *Wings* evenly enfumed, *costa and pterostigma black*; reticulation black; 3 nervures between the sectors of arc in fore-wing, 1 or 2 in hind-wing; discoidal field of even width till well beyond the node; nodal index $\frac{8\text{–}13}{8\text{–}9}\Big|\frac{13\text{–}8}{9\text{–}8}$. *Legs* black save for a yellow spot behind the proximal end of each femur, coxæ and trochanters yellow. *Abdomen* black, marked with yellow as follows:—Segment 1 entirely black save for a small postero-lateral spot; 2 with a broad line on the mid-dorsal carina and the sides including the oreillets, the black sending a prolongation down behind the latter structures as in *M. lineatus*; 3 to 7 with the basal half yellow, the lateral border of this colour running obliquely from the mid-dorsum to reach the venter of each segment just proximal to the jugal suture; 8 with only a largish baso-lateral spot and a fine apical yellow ring, the large, lateral foliate dilatations with an apical yellow spot; 9 with the lateral dilatations entirely yellow and an angulated spot on the ventro-lateral and baso-lateral border; 10 with a greenish-yellow diamond-shaped spot on the mid-dorsum, otherwise deep black. *Anal appendages* black; superiors as long as segments 9 and 10 together, exactly similar to those of *M. lineatus* as seen in profile, but seen from above there is a distinct linear gap between them for rather more than the basal half, the apical portions closely apposed, the apposed surfaces quite flat, thus resembling a pair of closed forceps; inferior smaller than in *M. lineatus*, more curled, and passing up between the superiors and enclosing a tiny round foramen. *Genitalia*: lamina shallow, more depressed than in *M. lineatus*, its border not curling outwards, black; inner hamules more conical, broad at base and tapering to a terminal hook; outer hamules stout hooks, more robust than in *M. lineatus, and without the subapical spine*; lobe directed outwards at right angles to axis of body, more prominent, its outlet more closed in and less indented at the apex than in *M. lineatus*, black.

Female unknown.

Distribution.—Turzum, DARJEELING DIST. (*O. Lindgren*).

Very easily distinguished from *M. lineatus*, to which it is very closely related, by the great extent of black markings, by the black costa and pterostigma, by the absence of spines on the occiput and by the lack of specialization in the neuration between the sectors of arc, by the genitalia (lamina black, stout hamules, and projecting lobe), and lastly by the anal appendages.

Type, a unique male, in the Author's collection.

268. **Mesogomphus risi** Fraser.

Mesogomphus risi Fraser, J. Bombay Nat. Hist. Soc. vol. xxix, pp. 996, 997, pl. ii, fig. 3 (1924); Laidlaw, Trans. Ent. Soc. Lond. vol. lxxviii, p. 193 (1930); Needham, Rec. Ind. Mus. vol. xxxiv, p. 221 (1932).

Male.—Abdomen (with appendages) 27 mm. Hind-wing 26 mm.

Head: labium pale yellow, middle lobe blackish; labrum pale yellow, its base and lateral borders deep black; anteclypeus and postclypeus pale yellow, the latter with a broad black border at its central third; mandibles pale yellow; frons deep black in front, bright citron-yellow above, its base deep black and sending a fine medial prolongation forwards in the sulcus to join the black on front of frons; vertex and occiput black, the latter with a broad yellow stripe traversing it, fringed with whitish hairs and a few minute black spines. *Prothorax* blackish-brown, yellow on the sides. *Thorax* black, with bright citron-yellow markings as follows:—A mesothoracic collar broken in the middle line; oblique pyriform dorsal spots, pointed below and not joining the mesothoracic collar; a minute humeral spot on each side above and a vestige of a stripe below separated from it, but in the same straight line; laterally a broad medial black stripe dividing up the sides into three areas of even width, two yellow and one black. *Legs* black, anterior femora yellow on the inner side, hind femora with a row of very small, very closely-set, numerous black spines, mid-femora with a row of much more robust, less closely-set, less numerous spines. *Wings* (badly damaged and shrivelled in the type) hyaline; costa finely yellow, reticulation and pterostigma black, the latter braced; 2 nervures between the sectors of arc in fore-wing, only one in hind-wing; discoidal field of even width to beyond level of node; nodal index $\frac{7\text{–}12}{7\text{–}9}\Big|\frac{13\text{–}6}{9\text{–}6}$. *Abdomen* shaped as in *M. lineatus* but much shorter, jet black marked vividly with bright yellow rings; segment 1 with the apico-lateral border yellow; 2 with a mid-dorsal stripe expanding apically but not quite meeting the apex of segment, its ventro-lateral borders rather broadly yellow; 3 to 7 with the basal halves yellow, this colour running obliquely from the mid-dorsal carina to meet the ventral border just proximal to the jugal suture; 8 and 9 with only the foliate dilatations bright yellow; 10 black, with a small mid-dorsal subapical spot. *Anal appendages* black, superiors yellow near base on outer side; very similar in shape to those of *M. lineatus*. Superiors as long as segments 9 and 10 together, rather straighter, the apex curled more abruptly and the final point directed back towards the abdomen; seen from above they are separated in rather more than the basal half, the ends being closely apposed. Inferior rather more than one-third as long, rather less curled, and the apex longer than in *M. lineatus*. *Genitalia*: lamina very shallow, very depressed, its border barely concave, yellow; inner hamules comparatively short, broad at base, and rapidly tapering to a point, black; outer hamules very long, almost twice the length of inner, very robust, their ends converging and curled in a very robust spine just below the apex on its inner side (much more easily seen than in *M. lineatus*), black; lobe very similar to *M. lindgreni*, black, directed somewhat outwards, deeply indented at apex, outlet squarish but not nearly so deep as in *M. lindgreni*.

Female unknown.

Distribution.—Kalaw, S. Shan States, BURMA (*G. Dingavan*).

The species is very distinct from others by its small size and by its brightly contrasted markings, black pterostigma, etc.

Type, male, unique, in the Author's collection.

269. **Mesogomphus frontalis** (Selys).

Onychogomphus frontalis Selys, Bull. Acad. Belg. (2) vol. xlvi, p. 428 (1878); Williamson, Proc. U.S. Nat. Mus. vol. xxxiii, p. 309 (1907); Laidlaw, Rec. Ind. Mus. vol. xxiv, pp. 371, 409–410 (1922); Fraser, J. Bombay Nat. Hist. Soc. vol. xxix, pp. 65, 333 (1923); Needham, Rec. Ind. Mus. vol. xxiv, p. 223 (1932).

Lindenia frontalis Kirby, Cat. Odon. p. 60 (1890).

Mesogomphus frontalis Laidlaw, Trans. Ent. Soc. Lond. vol. lxxviii, p. 192 (1930).

Male unknown.

Female.—Abdomen 29 mm. Hind-wing 25 mm.

Head: labium, labrum, and epistome pale yellow; frons black in front, yellow above, a black area at base running forwards and cutting the yellow area into two halves; vertex black; occiput yellow, its crest dark, sinuous, fringed with long hairs. *Thorax* black, marked with yellow as follows:—A mesothoracic collar interrupted in the middle line by the black dorsal carina; an antehumeral stripe, squared above where it fails to reach the alar sinus, pointed below and well separated from the mesothoracic collar ; a vestigial humeral stripe ; laterally two broad yellow bands, one at the level of fore-wing, the other covering the middle of the metepimeron, *Legs* very robust, short, yellow, outer sides of femora striped with black, hind femora with short spines. *Wings* hyaline ; reticulation black, pterostigma brown or blackish brown, 4 mm. in length, covering 5 cells ; discoidal cell of fore-wing followed by a row of 3 cells, then rows of 2 ; fore-wing with 13–14 antenodals, 7–8 postnodals. *Abdomen* black, marked with yellow as follows :—Segment 1 with a dorsal and a lateral spot ; 2 with a mid-dorsal trilobed stripe and two parallel superimposed stripes on each side ; 3 to 7 with broad basal rings occupying rather less than the basal half of each segment (probably considerably less than half except on segment 7), all these rings nearly cut in half by an invasion of the black of dorsal carina ; 8 with a lateral basal spot ; 9 and 10 wholly black, very short. *Anal appendages* slender, yellow tipped with black, separated by a conical yellow tubercle. Vulvar scale about half the length of segment 9, lanceolate, cleft at apex.

Distribution.—Moolai, BURMA. Kirby, in his Catalogue, gives the Nicobars as a locality, but this is an obvious error, as the type is the only specimen known.

I have modified Selys's description, to allow for the development of full adult colouring, as the original description would be very misleading.

Until the male has been discovered the correct generic position of this insect is uncertain. It resembles the female of *Nepogomphus modestus* rather closely, but its larger size, longer pterostigma, etc., will serve to distinguish it.

Type, a teneral specimen, in the Selys collection.

Genus **ONYCHOGOMPHUS** Selys. (Fig. 74.)

Lindenia (pars) De Haan, Bijd. Nat. Wetensch. vol. i, (2) p. 47 (1826); Hoeven, ibid. vol. iii, p. 338 (1828); Kirby, Cat. Odon. p. 57 (1890).

Diastatomma (pars) Burmeister, Handb. Ent. vol. ii, p. 831 (1839); Charpentier, Lib. Eur. p. 15 (1840).

Onychogomphus Selys, Bull. Acad. Belg. vol. xxi, pt. 2, p. 30 (1854); id., Mon. Gomph. p. 15 (1857); Williamson, Proc. U.S. Nat. Mus. vol. xxxiii, pp. 275, 308–312 (1907); Ris, Ann. S. Afr. Mus. vol. xviii, p. 343 (1920); Laidlaw, Rec. Ind. Mus. vol. xxiv, pp. 402, 403 (1922); Fraser, J. Bombay Nat. Hist. Soc. vol. xxix, pp. 64, 332, 997–999 (1923); Needham, Zool. Sinica, ser. A, vol. xi, fasc. 1, pp. 19, 38 (1930); Laidlaw, Trans. Ent. Soc. Lond. vol. lxxviii, p. 194 (1930); Needham, Rec. Ind. Mus. vol. xxxiv, pp. 217, 218, 222 (1932).

Size medium ; colour yellow marked with black, or black marked with bright yellow ; anal appendages long, strongly curled and forcipate, the branches of the inferior closely apposed.

Head of moderate size, frons angulated, occiput usually simple and slightly concave posteriorly. Wings: reticulation close; tornus angulated; base of hind-wing oblique and deeply excavate; membrane obsolete; anal triangle 4-celled; arc situated between the first and second antenodal nervures, or more rarely opposite the second; only 2 transverse nervures between sectors of arc from arc to bifurcation of *Rs* in fore-wing, 1 in hind-wing; a single row of postanal cells in fore-wings, with occasional double cells, 4 rows in hind-wings; a rudimentary anal loop present, formed by a splitting of the first postanal cell into two cells, the loop extending slightly proximal to base of subtrigone; no incomplete basal antenodal nervures present; nodal index high; primary antenodal

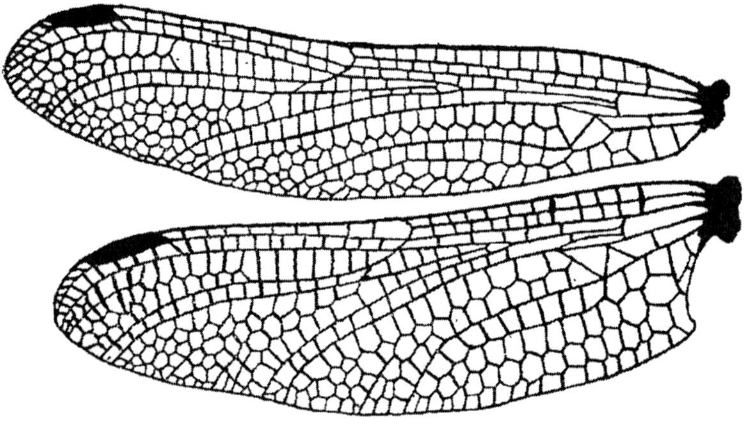

Fig. 74.—Wings of *Onychogomphus saundersi* Selys, male.

nervures the first and the fifth; discoidal cells entire, on fore-wings with costal and basal sides equal, distal side but slightly longer and slightly angulated, on hind-wings but moderately elongate in the length of the wing, with the distal side slightly longer than the costal and both one-third again as long as the basal; pterostigma rather short and swollen at its middle, braced, equal to slightly less than one-third the distance from node to the proximal end of pterostigma; *IA* in fore-wing pectinated, 3 rows of cells between it and border of wing; *Cuii* and *IA* in hind-wing closely parallel to the border of wing or rarely slightly divergent; only 1 cubital nervure in all wings; subtrigones and hypertrigones entire in all wings; discoidal field with 2 rows of cells to level of node. Legs short, hind femora extending to apical border of abdominal segment 1 and furnished with a group of numerous very short, closely-set spines which, near the distal end of limb, become arranged into two rows; hind tibial spines moderately long and fine. Abdomen tumid at basal segments, thin and cylindrical from segment 3 to 7, from which point it becomes somewhat dilated again, segments 8 and 9 often having pseudo-lateral dilatations. Anal appendages: superiors about double the length of segment 10, forcipate, the apices curled strongly downwards; inferior bifid almost to base, the two branches very closely apposed and the apical ends curled strongly up to meet the superiors. Genitalia: lamina depressed, narrowly arched; anterior hamules rather short and slim; posterior hamules very robust, compressed, projecting markedly from the genital sac and ending in a robust spine; lobe flask-shaped, rather small, beaked.

Genotype, *Libellula forcipata* Linn.

Distribution.—Europe, N. Africa, INDIA, BURMA, NICOBARS, Malaya, Sumatra, Java, Indo-China, and China. The genus is a large one, and no fewer than fourteen species are known from within Indian limits. Several other species, formerly included in the genus, have since been removed to the new genera *Lamelligomphus* and *Nepogomphus*.

The larvæ breed in streams at all altitudes below 6,000 feet; they are similar to those of *Burmagomphus* and *Mesogomphus*, and have similar habits.

Key to Indian Species of Onychogomphus.

1.	Face and frons not marked with black; thorax grass-green, marked with warm reddish-brown	*pulcherrimus* Fras., [p. 267.
	Face and frons marked with black; thorax greenish-yellow or yellow marked with black	2.
2.	Yellow humeral stripe absent	3.
	Yellow humeral stripe present, but vestigial or interrupted	5.
	The humeral stripe present and complete.	8.
3.	Only a single broad black stripe on sides of thorax	*maclachlani* Selys, [p. 266.
	Two narrow black stripes on each side of thorax, separated or joined at intervals	4.
4.	Occiput black; segments 3 to 5 with basal triangular dorsal spots, but without mid-dorsal oval spots	*echinoccipitalis* Fras., [p. 257.
	Occiput yellow; segments 3 to 5 with basal rings and mid-dorsal oval spots.	*circularis* Selys, p. 261.

5.	A single broad black stripe on each side of thorax	*annularis* Selys, p. 265.
	Two narrow black stripes on each side of thorax, of which the first may be incomplete or interrupted	6.
6.	Antero-lateral black stripe on thorax interrupted and made up of three elongate spots	*earnshawi* Fras., p. 263.
	Antero-lateral black stripe on thorax nearly obsolete, represented by a mere upper vestige......................	*aureus* Laid., p. 254.
	Antero-lateral black stripe not interrupted	7.
7.	Inferior anal appendage without an upper basal spine	*saundersi* Selys, p. 242.
	Inferior anal appendage with a distinct but obtuse upper basal spine ...	*duaricus* Fras., p. 245.
8.	Antehumeral stripes not confluent with mesothoracic collar	9.
	Antehumeral stripes confluent with mesothoracic collar	10.
9.	Antehumeral stripes long and very narrow; abdominal segments 3 to 6 with mid-dorsal oval yellow spots	*cerastes* (Selys), p. 260.
	Antehumeral stripes short and oval in shape; abdominal segments 3 to 6 without mid-dorsal oval yellow spots .	*grammicus* (Ramb.), [p. 256.
10.	Antehumeral and humeral stripes separated above; two narrow black stripes on each side of thorax	11.
	Antehumeral and humeral stripes confluent above; only a single black stripe on each side of thorax, the antero-lateral one obsolete or present as a mere upper vestige......................	*dingavani* Fras., p. 258.
11.	Many of the transverse nervures proximal to node and on anterior part of wings bright yellow	*bistrigatus* Selys, p. 247.
	Transverse nervures black throughout ..	12.
12.	Superior anal appendages with a black line on dorsum of apical third; humeral stripes separated from the yellow on lower part of thorax	*striatus* Fras., p. 249.
	Superior anal appendages entirely yellow; humeral stripes confluent with the yellow on lower part of thorax	*M-flavum* Selys, p. 250.

270. **Onychogomphus saundersi** Selys. (Pl. III, fig. 3; text-figs. 74 & 75, c.)

Onychogomphus saundersii Selys, Mon. Gomph. p. 22 (1851); id., Bull. Acad. Belg. vol. xxi, pt. 2, p. 31 (1854); id., Ann. Mus. Civ. Genova, (2) vol. x (xxx), p. 474 (1891); Krüger, Stett. Ent. Zeit. lxix, p. 318 (1898); Martin, Mission Pavie, Neuropt. vol. iii, p. 212 (1904); Williamson, Proc. U.S. Nat. Mus. vol. xxxiii, pp. 311, 312, text-fig. 36 (1907); Fraser, J. Bombay Nat. Hist. Soc. vol. xxix, pp. 65, 333 (1923); Campion, J. F.M.S. Mus. vol. viii, p. 163 (1925); Fraser, J. Siam. Soc., Nat. Hist. Suppl. vol. ix, pp. 139, 141 (1933).

Lindenia saundersii Kirby, Cat. Odon. p. 58 (1890).
Onychogomphus geometricus var. *nigrescens* Laidlaw, Proc. Zool. Soc. Lond. (1) p. 80 (1902).
Onychogomphus nigrescens Williamson, Proc. U.S. Nat. Mus. vol. xxxiii, p. 310 (1907).
Onychogomphus saundersi Laidlaw, Rec. Ind. Mus. vol. xxiv, pp. 371, 405, 406 (1922); Fraser, J. Bombay Nat. Hist. Soc. vol. xxx, pp. 106–108, text-fig. 1, pl. i, fig. 2 (1924); Ris, Zool. Meded. Leiden, vol. x, pp. 23, 30, 45–46 (1927); Laidlaw, Trans. Ent. Soc. Lond. vol. lxxviii, p. 194 (1930); Needham, Rec. Ind. Mus. vol. xxxiv, p. 222 (1932).

Male.—Abdomen 40 mm. Hind-wing 31 mm.

Head: labium brownish-yellow; labrum citron-yellow, bordered narrowly with black and traversed by a vertical line of the same colour at its middle, this splitting the yellow area into two large spots; anteclypeus yellow; postclypeus black, with a large triangular spot on each side; bases of mandibles yellow; frons black below anteriorly, its crest and upper surface yellow, with a vestige of black at the base of the sulcus; eyes bottle-green during life; occiput and vertex black, the former slightly convex and fringed with short stiff hairs. *Prothorax* black, with an anterior collar and a small lateral spot of yellow. *Thorax* black on dorsum, marked with yellow as follows:—A broad slightly interrupted mesothoracic collar, very broad and short antehumeral stripes which are broadly confluent with the collar, a narrow humeral stripe broadening into a triangular spot above, beneath which it is interrupted for a short distance, the alar sinus, and some spots on the tergum. Sides yellow, with a thick black stripe on each lateral suture, these stripes being thicker than the included stripe of the ground-colour. *Legs* black, rather short, inner surface of fore femora yellow, a spot of the same colour on the inner sides of the two posterior pairs. *Wings* hyaline; pterostigma black, about 3 mm. in length, braced; anal triangle 3- to 4-celled; anal field 4 cells deep; nodal index $\frac{8\text{–}13}{10\text{–}11}\Big|\frac{14\text{–}9}{10\text{–}9}$; costa finely yellow. *Abdomen* black, marked with yellow as follows:—Segment 1 with an apical dorsal spot and the sides broadly; 2 with a lanceolate dorsal stripe not extending quite to apical end, the whole of the sides except the base, which is narrowly black; 3 to 6 with rings covering the basal fourth of each segment and prolonged along the ventral border, whilst dorsally the black dorsal carina almost divides them into two; in addition to these, each segment bears an oval mid-dorsal spot which grows progressively smaller from segment 3 to 6; 7 with the basal half yellow; 8 and 9 with lateral basal spots, larger on the former segment; 10 unmarked. Segments 8 and 9 are greatly broadened, far more so than in the related *O. duaricus*. *Anal appendages*: superiors yellow, with black apices; inferior black; structure as shown in fig. 75, c. *Genitalia*: lamina rounded, hood-like, projecting but slightly; anterior hamules long, narrow hooks bent at their middle and again at extreme apex, where the bent portion forms an acute spine; posterior hamules broadly conical, the apex slightly recurved; lobe bulky, narrowly emarginate, projecting rather markedly.

Female.—Abdomen 37 mm. Hind-wing 32 mm.

Closely similar to the male in colour and markings, but differs in the following points:—Yellow area on frons descending to a greater extent on to face; no vertical median black line on labrum; a small rounded ochreous spot on vertex; back of occiput yellow at its middle; posterior

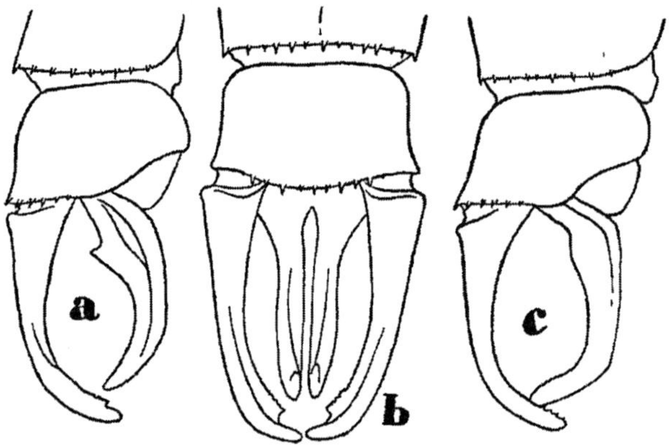

Fig. 75.—Anal appendages of (a) *Onychogomphus duaricus* Fraser, male, right lateral view; (b) the same, dorsal view; (c) *Onychogomphus saundersi* Selys, male, right lateral view. (Camera lucida studies from the types.)

lobe of prothorax finely bordered with yellow; segment 2 of abdomen with basal half yellow, 3–6 with mid-dorsal spots on larger, 10 with a yellow point on sides. *Anal appendages* yellow, shortly conical; vulvar scales short, deeply emarginate.

Distribution.—INDIA, locality of type unrecorded; Williamson has reported a pair from Bhamo, BURMA.

Selys considered this species to be closely related to *O. geometricus* from Java, and it is also a near relative of *O. duaricus*, differing principally by the absence of any trace of an upper subbasal spine on the inferior anal appendage of the male. An atypical specimen from Sumatra, described by Ris, has such a spine, and would therefore appear to be more nearly related to *O. duaricus*.

The *type* is a male in the Brussels Museum which I have been able to examine; the allotype female is in the Saunders collection in the British Museum and was referred to *O. M-flavum* by Selys.

271. **Onychogomphus duaricus** Fraser. (Fig. 75, a, b.)

Onychogomphus duaricus Fraser, J. Bombay Nat. Hist. Soc. vol. xxix, pp. 1001–1003, pl. ii, fig. 5 (1924); Ris, Zool. Meded. Leiden, vol. x, p. 46 (1927); Laidlaw, Trans. Ent. Soc. Lond. vol. lxxviii, p. 195 (1930); Needham, Rec. Ind. Mus. vol. xxxiv, p. 222 (1932).

Male.—Abdomen 39–41 mm. Hind-wing 31 mm.

Head: labium pale whitish-yellow; labrum greenish-yellow, base narrowly, anterior border broadly black and sending up a point which may nearly or entirely cut the yellow area in two by reaching the black base (in the majority of GOMPHINÆ a black tongue runs down from the base, the opposite condition found in this species being quite exceptional); anteclypeus and base of mandibles greenish-yellow, postclypeus black, with a small triangular greenish-yellow spot on each side against the eyes; frons greenish-yellow, its lower part black in front, this confluent with the black of postclypeus, its base very finely black and a number of fine black points scattered about its crest; vertex and occiput black, occipital border nearly straight, not raised, fringed with long black hairs. *Prothorax* black, with a small lateral yellow spot. *Thorax* black on dorsum, marked with yellow as follows:—A mesothoracic collar narrowly broken by the black dorsal carina, which is narrowly yellow at its middle; oblique antehumeral stripes connected with the collar below; humeral stripes variable, widened above, tapering to a fine point below (tapered part in some specimens cut off from upper part of stripe, in others broken up into a chain of fine points); laterally greenish-yellow, sutures boldly mapped out in black, anterior stripe connected to the black of dorsum by a line descending from latter. *Legs* black, hind femora with a broad yellow stripe on inner side, furnished with a row of closely-set, robust, short black spines, the mid-femora with a similar row. *Wings* distinctly tinted with greenish-yellow, especially towards the base; costa finely yellow to nearly as far as the pterostigma, which is deep brownish-black, well-braced, covering 5 cells; antenodal nervures and arc yellow; first postanal cell divided; 2 to 3 rows of cells between *Rii* and *IRii* at level of outer end of stigma; nodal index of two specimens $\dfrac{10\text{–}15 \mid 17\text{–}10}{11\text{–}12 \mid 12\text{–}11}$, $\dfrac{12\text{–}16 \mid 18\text{–}11}{12\text{–}13 \mid 12\text{–}11}$; only 2 cells between *Cuii* and

IA at wing-margin in hind-wing. *Abdomen* black, marked as follows :—Segments 1 and 2 with a broad continuous yellow dorsal stripe, tapering at apical end of 2 and bordered with a broad subdorsal black stripe, which is confluent with a narrow apical ring on 2 ; a short apical black mark occasionally runs along border of 1 and another behind the oreillets on 2 ; 3 to 6 with clear yellow basal rings covering rather less than one-fourth the length of segments ; on 3 to 5 mid-dorsal oval spots of yellow, almost obsolete on 5, absent entirely on 6 ; 7 with the basal half yellow, its crest finely black ; 8 and 9 with lateral yellow spots, bifid posteriorly ; 10 entirely black. Ventral border of segment 2 furnished with a row of small black spines. *Legs* black except for a stripe on the inner side of first femora. *Anal appendages* : superiors bright yellow, the apical third or more abruptly black ; inferior black, with the upper surface sometimes paler. Structure as shown in fig. 75, *a*, *b* ; *no apical tooth. Genitalia* differing rather from those of other species, most resembling those of *O. striatus*. Lamina black, keeled only in its posterior half, depressed, but shallowly notched ; inner hamules forming long tapering black spines, with apices only slightly everted, outer of the same length but much stouter, directed straight outwards, with apex curled and finally directed forwards, yellow at base, black at apex ; lobe rather prominent, borders tumid, similar in shape to *O. striatus*.

Female.—Abdomen 40 mm. Hind-wing 33 mm.

Very similar to the male, differing as follows :—Yellow on labrum not quite bisected by black ; black at base of frons extending a short way on to floor of sulcus ; occiput black and similar to male in shape ; yellow on mid-dorsal carina meeting the mesothoracic collar, which is uninterrupted ; humeral stripe complete below and connected with the elbow formed by the junction of the mesothoracic collar and antehumeral stripe. *Abdomen* with sides of segments 1 and 2 more broadly yellow, the black subdorsal bands narrower and with straight borders ; no dorsal spots on 4 ; basal yellow area on 7 more restricted ; spots on sides of 8 and 9 absent. *Anal appendages* pale yellow, almost white, conical, pointed, rather longer than segment 10, the protuberance between them pale yellow. *Legs* : all femora with yellow stripes on inner sides ; spines on hind femora much less numerous, longer, more widely-spaced, and more robust than in the male. *Wings* with the greenish tinting more marked, especially towards base ; 2 to 3 rows of cells between *Rii* and *IRii*, usually only 2 rows ; 2 rows of cells in the anal area of forewing. Vulvar scale only perceptible as two rudimentary rounded shiny tubercles.

Distribution.—Hasimara, Duars, BENGAL (*H. V. O'Donel*).

The following combination of characters will distinguish this species from all others :—The unusual distribution of black on the labrum, the black occiput, the wide space between *Rii* and *IRii*, the shape of the genitalia in both sexes, the striking coloration of the superior anal appendages, and the presence of only one tooth on the inferior.

Type male and allotype female in the British Museum.

272. **Onychogomphus bistrigatus** (Selys). (Fig. 76, **b**.)

Gomphus bistrigatus Selys, Bull. Acad. Belg. vol. xxi, pt. 2, p. 46 (1854).

Onychogomphus bistrigatus Selys, Mon. Gomph. pp. 24, 392 (1857); id., Bull. Acad. Belg. (2) vol. xxviii, p. 173 (1869); Williamson, Proc. U.S. Nat. Mus. vol. xxxiii, pp. 309, 311 (1907); Laidlaw, Rec. Ind. Mus. vol. xxiv, pp. 371, 410 (1922); Fraser, J. Bombay Nat. Hist. Soc. vol. xxix, pp. 66, 333, 999, 1000, pl. ii, fig. 4 (1923–1924); Needham, Rec. Ind. Mus. vol. xxxiv. p. 222 (1932).

Lindenia bistrigatus Kirby, Cat. Odon. p. 58 (1890).

Onychogomphus sp. Laidlaw, Rec. Ind. Mus. vol. xxiv, pp. 411, 412 (1922).

Male.—Abdomen 39 mm. Hind-wing 31 mm.

Head : labium pale whitish-yellow ; labrum yellow, very finely margined with black anteriorly and at base ; clypeus, base of mandibles, and frons yellow, a fine black line between frons and postclypeus, incomplete laterally and sending down two fine points on either side of middle line ; base of frons narrowly black with but the slightest prolongation in the middle line ; vertex black, a small yellow spot between posterior ocelli ; occiput slightly convex, yellow, with the postero-lateral angles black, fringed with yellow hairs ; behind occiput and eyes yellow, a narrow black line margining the latter above. *Prothorax* black, its anterior and posterior borders narrowly and two fine approximated points yellow. *Thorax* black on dorsum, marked with yellow as follows :—Median part of dorsal carina narrowly, a mesothoracic collar slightly interrupted in the middle line, oblique antehumeral stripes joined to the mesothoracic collar, complete humeral stripes slightly constricted just below their upper end ; laterally yellow, with two fine black lines mapping out the sutures. *Legs* yellow, fore femora broadly black on outer side, middle femora with a fine black line, hind pair with but a vestige of same on outer side ; tibiæ black on the flexor surface ; femora armed with a row of robust, short, closely-set black spines and a single larger one at the distal end. *Wings* hyaline, with a greenish-yellow tinge ; neuration black, *but a large number of cross-nervures proximal to node and on fore part of wings bright yellow* ; the pterostigma dark brown, well-braced, covering 3–4 cells ; costa to halfway over stigma pale yellow. Nodal index $\frac{10\text{–}13}{8\text{–}10} \left| \frac{14\text{–}9}{10\text{–}10} \right.$; 2 cross nervures between the sectors of arc in the fore-wing, only 1 in hind-wing; first postanal cell undivided in both hind-wings (but this is probably an aberration, as the network is very irregular in the type). *Abdomen* black and yellow, segment 1 with a quadrate subdorsal black spot not reaching the apex; 2 with broad subdorsal black stripes barely reaching the apical border and enclosing a dorsal lobed yellow spot, a tongue of black running down behind each oreillet; 3 yellow, with a broad black stripe on each side falling well short of the base, the pair coalescing beneath and with a subbasal and a subapical tongue running up over the dorsum, almost coalescing over the carina, and enclosing a median yellow spot; 4 to 6 with similar markings, but

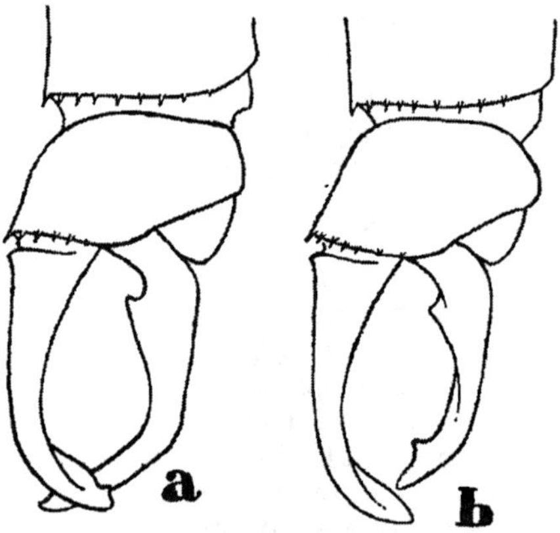

Fig. 76.—Anal appendages of (**a**) *Onychogomphus striatus* Fraser, male; (**b**) *Onychogomphus bistrigatus* (Selys), male, seen from the right side.

the black prolongations coalesce dorsally and the apical band extends right up to the apical border; 7 yellow, with a black dorsal stripe, very broad apically, where it sends a tongue-like process back along sides of segment, tapering rapidly on dorsum and extending to extreme base; 8 to 10 yellow, broadly black on dorsum, the yellow, however, constricting the black on segment 10 so as to almost cut it in two at the apical border; apical borders of last three segments finely black and bordered with black spines. *Anal appendages* (fig. 76, **b**) yellow. *Genitalia*: lamina yellow, depressed, its border angulated outwards so that it appears rather deeply cleft; inner hamules broad at base, deeply cleft into two branches, an outer short spine and an inner long one which is everted sinuously at its apex, seen from the front converging rapidly on each other, black; outer hamules very stout, short, yellow tipped with blackish-brown, robust, forwardly-directed spines; lobe funnel-shaped, with a broad opening, pale yellow narrowly margined with black.

Female.—Abdomen 39 mm. Hind-wing 31–34 mm.

Very similar to the male, with slight differences in the extent of the black markings, probably due to the varying age of the three specimens.

The type female does not appear to have a basal black line to the labrum, whilst on the other hand the specimen from Kumaon has no similar border anteriorly. The occiput in the type female is bordered with a row of six tiny spines, which are absent in the Kumaon specimen. *Prothorax* of type and of the male similar, but the two small dorsal spots absent in the Kumaon specimen. *Wings* hyaline, many pale yellow nervures in both females similarly situated as in the male; 13–16 antenodal nervures in fore-wings, 10 in hind-wing, 10 postnodals in fore-wings, 12 in hind-wings. *Abdomen*: segment 1 entirely yellow in the Kumaon specimen, marked with black in the type female and male; 2 variable, the subdorsal black stripes not meeting the apical border in the type, confluent with an articular black ring in the Kumaon female; 3 to 6 similar in the two sexes (4 to 6 missing in the Kumaon specimen), but the black extensions not quite meeting over the carina in the type; 7 very similarly marked to the others. *Anal appendages* as long as segment 10, conical, pointed, separated by a shorter conical protuberance, yellow. Vulvar scale short, about half the length of segment 9, deeply notched at apex into two subtriangular processes.

Distribution.—*Type* female in the Selys collection from "India," allotype from "North India" in the British Museum; there is also a male in my collection from Gopaldhara, ASSAM (*H. Stevens*, 21. ix. 20). Laidlaw's female is from Kumaon, WESTERN HIMALAYAS.

273. **Onychogomphus striatus** Fraser. (Fig. 76, a.)

Onychogomphus bistrigatus Fraser, Rec. Ind. Mus. vol. xxiv, p. 424 (1922).

Onychogomphus striatus Fraser, J. Bombay Nat. Hist. Soc. vol. xxix, p. 1000, text-fig. 2, vi (1924); id., Rec. Ind. Mus. vol. xxvi, pp. 428, 478 (1924); id., ibid. vol. xxxiii, p. 448 (1931); Needham, Rec. Ind. Mus. vol. xxxiv, p. 223 (1932).

Male.—Abdomen 37 mm. Hind-wing 28 mm.

Very similar to *O. bistrigatus*, differing as follows:—Size slightly smaller; labrum more broadly bordered with black anteriorly; postclypeus entirely black at its centre, with a very large yellow spot on each side ; occiput with its border quite straight and turned forwards, so that its posterior surface can be seen when the head is viewed from above, black with a narrow yellow border, fringed with long black hairs. *Prothorax* without the mid-dorsal spots. *Thorax* : humeral stripes tapering to a point below, black lines on lateral sutures rather thicker. *Wings* with black reticulation, *no pale yellow nervures present* ; nodal index $\frac{9\text{–}13}{9\text{–}9}\Big|\frac{12\text{–}10}{10\text{–}10}$; 2 rows of cells between *Rii* and *IRii* ; first postanal cell divided ; pterostigma pale brown, rather weakly braced. *Legs* as in *O. bistrigatus*, but tibiæ black except at proximal ends. *Abdomen* : segment 1 black, with a large lateral yellow spot ; 2 with a narrow complete basal black ring and broad subbasal black stripes broadly confluent with the apical black ring ; 3 to 6 with the yellow reduced to a complete but irregular basal ring and a median dorsal spot ; 7 with its basal half yellow ; 8 and 9 black, with a large yellow spot on each side ; 10 entirely black. *Anal appendages* yellow, the inferior rather darker coloured and the superiors with a fine black line along the upper surface of the apical third. Almost exactly similar to those of *bistrigatus*, but with the apices of superiors almost cylindrical, not flattened out lancet-wise as in that species (fig. 76, **a**). *Genitalia* : lamina blackish-brown, with a longitudinal dorsal keel, its border concave, neither everted nor angulated ; inner hamules more shallowly notched, the inner branch not much longer than the outer and much shorter than in *O. bistrigatus*, its apex tipped with yellow ; outer hamules dark brownish-yellow, similar in shape to those of *O. bistrigatus* ; lobe black, prominent but not markedly so, its border thickened, hollowed out medially, the extremities of the notch forming rounded bosses.

Distribution.—A single male from Kallar, Nilgiris, about 1,000 ft., May 1917.

The species differs from *O. bistrigatus* in the greater extent of black markings, in the shape and colouring of the superior anal appendages and genitalia, and in the reticulation being entirely black.

Type in the Author's collection.

274. **Onychogomphus M-flavum** Selys. (Fig. 77.)

Onychogomphus bistrigatus Selys, Mon. Gomph. p. 22 (adult female), p. 392 (teneral male) (1854); id., Bull. Acad. Belg. (2) vol. xxviii, p. 173 (adult male) (1869); Kirby, Cat. Odon. p. 58 (1890).

Onychogomphus M-flavum Selys, Ann. Soc. Ent. Belg. vol. xxxviii, p. 169 (1894); Williamson, Proc. U.S. Nat. Mus. vol. xxxiii,

p. 309 (1907); Laidlaw, Rec. Ind. Mus. vol. xxiv, pp. 371, 410, 411 (1922); Fraser, J. Bombay Nat. Hist. Soc. vol. xxix, pp. 66, 333, 1003–1005, pl. ii, fig. 6, & text-fig. 2, v (1923–1924); Needham, Rec. Ind. Mus. vol. xxxiv, p. 223 (1932).

Male.—Abdomen 38 mm. Hind-wing 30–33 mm.

FORM *a.*—*Head*: labium dirty yellow; labrum greenish-yellow, very narrowly bordered with black at base, more broadly along anterior border, a narrow black streak of even width running from its centre and not quite reaching the black anterior border; base of mandibles, anteclypeus, and a narrow border on lower edge of postclypeus yellow, latter black with a large greenish-yellow spot on each side; lower part of frons black, this confluent with the black of postclypeus, the upper part and above greenish-yellow, its base narrowly black; vertex and occiput black, but a broad greenish-yellow streak on posterior half of latter, its border slightly convex in the middle, and turned forward so that the hinder surface is

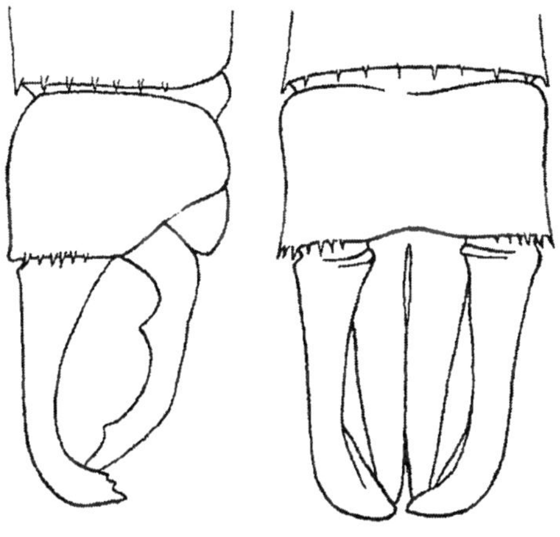

Fig. 77.—Anal appendages of *Onychogomphus M-flavum* Selys, male. Right lateral and dorsal views.

easily visible when the head is viewed from above; the back of eyes black and of occiput yellow. *Prothorax* yellow, with a large triangular spot of black on each side. *Thorax* black on dorsum, marked with bright greenish-yellow as follows:—A mesothoracic collar broken by the black mid-dorsal carina, which is itself bright yellow somewhat higher up; oblique antehumeral stripes confluent with the collar; complete humeral stripes constricted a little below the upper part; laterally greenish-yellow, the sutures outlined rather finely in black, which on the first suture shows two interruptions which isolate a small part of the line opposite the spiracle.

Legs black, fore femora yellow within, mid-femora similar but with a narrow interrupted streak on outer side, hind femora with broad streaks of yellow on both sides and furnished with a row of short, robust, closely-set black spines. *Wings* hyaline, faintly tinged with greenish at extreme base; pterostigma dark brown margined heavily with black nervures (much paler in teneral specimens, black when fully mature), covering 4–5 cells, well braced; nodal index $\frac{15\text{–}14}{15\text{–}12}\Big|\frac{15\text{–}13}{13\text{–}14}$, $\frac{12\text{–}15}{13\text{–}11}\Big|\frac{15\text{–}12}{11\text{–}13}$; 2 rows of cells in anal area of fore-wing, only 1 row between *Rii* and *IRii* except at extreme margin of wing; first postanal cell irregularly divided; costa finely yellow as far as pterostigma, but this colour lost in very adult specimens. *Abdomen* black, marked with bright yellow, becoming ochreous towards tip as follows:—Segments 1 and 2 with a dorsal lobed stripe extending from base of 1 to apical border of 2 and separated from the yellow sides by a moderately broad black stripe which sends a stripe down behind the oreillets; 3 with a complete basal ring, prolonged along both venter and dorsum and confluent on the latter with a large irregularly angulated spot; on 4 to 6 the basal rings less extensive laterally and partially bisected by an invasion of the black along the dorsal carina, the mid-dorsal spots smaller, oval, and isolated from the basal rings; 7 with the basal two-thirds yellow; 8 with the lower half of the sides; 9 with only a clouding of black on dorsum; 10 ochreous, its apical margin finely black. All segments from 3 to 10 with an apical ring of fine black spines. *Anal appendages* (fig. 77) golden yellow, the superiors a little darker. *Genitalia*: lamina very depressed, its border much everted, arched and folded back on itself, dark brown; inner hamules of great length and very attenuated, converging, the outer branch also elongate, black; outer hamules broad, of about the same length, yellow with a robust apical spine; lobe black, projecting prominently like an open spout.

FORM *b* (including *type* male).—Differing from form *a* above by the much greater extent of black. Anteclypeus, vertex, and occiput entirely black; no short black median stripe on labrum; prothorax almost entirely black; humeral stripe interrupted or entire (entire in the type); lateral lines of thorax complete; costa yellow or black (black in the type); pterostigma blackish-brown; yellow on sides of abdominal segment 2, nearly or quite cut into two by a prolongation of the subdorsal black stripe, medial spot on dorsum of segment 3 small and completely isolated, on 4 almost or quite obsolete; only narrow basal lunules on segments 5 and 6; 7 with only its basal half yellow and this divided by a narrow black dorsal carina; 8 and 9 broadly reddish-yellow on the sides, the small lateral wings black; 10 reddish-yellow, finely margined with black along apical border. *Legs* black, only the anterior femora marked with yellow within.

Female.—Abdomen 39 mm. Hind-wing 32–36 mm.

Very variable individually, differing also from both forms of the male described above. *Head*: labrum finely or broadly bordered with black, which may be more extensive than the yellow; anteclypeus dark brown or light greenish-yellow, postclypeus black, marked with a small lateral yellow spot, or almost entirely greenish-yellow, marked only with a small quadrate black spot at the middle of its posterior border; the black stripe at base of frons and that bordering it below in front narrow or broad; vertex black, with or without a small spot of yellow between the posterior ocelli; occiput brownish in the type, black with a small central spot of yellow or entirely bright yellow in others, its border almost straight and yellow behind as in male; humeral stripe complete or occasionally broken at its upper part, the first lateral line on thorax complete or widely interrupted in one specimen as in the male described above. *Legs* black, marked to a variable extent with yellow, the hind femora armed with a row of very robust, very widely spaced, long black spines. *Wings* slightly enfumed; pterostigma dark brown or blackish, covering $4\frac{1}{2}$–$5\frac{1}{2}$ cells; reticulation as in male, but in one female 2 rows of cells between *Rii* and *IRii* nearly as far as the pterostigma; costa black in one female, yellow in another. *Abdomen* very similar in all specimens examined to that of the type male, the black on segment 1 reduced to a mere subdorsal vestige or entirely wanting; segment 2 in two females examined with a large lateral isolated black spot; on 3 to 6 the basal yellow area very broadly confluent with the mid-dorsal spots and extending almost to apical border of segments as a broad dorsal stripe; remaining segments as in male, but the pale portions very dark reddish or ochreous. *Anal appendages* and the protuberance between them rather dark yellow, short, conical. Vulvar scale differing from that of any other species of the genus, shaped like an acuminate leaf, very long, extending to the middle of segment 10, cleft for rather more than its apical half into two closely contiguous halves, reddish-brown, black at apex, very thin when viewed in profile, its surface raised into two longitudinal folds.

Distribution.—DARJEELING DISTRICT. I possess two males and three females, and have seen a few others taken at

Gopaldhara by Mr. H. Stevens, and I have others from Kurseong taken by Mr. O. Lindgren.

At first sight it would appear that the two forms of the male differ so markedly that they cannot be conspecific, but all the variations met with in the males are reproduced in the females and the remarkable shape of the ovipositor of these latter leaves no doubt as to their specific identity.

The male of *O. M-flavum* is less easily distinguished. It differs from *O. duaricus* in the unicolorous superior anal appendages and in possessing two teeth on the inferior appendage; from *O. bistrigatus* in the all-black reticulation, the greater extent of yellow on segment 7, and the genital mat black lobe; from *O. striatus* in the complete humeral stripe, the much narrower lateral lines on sides of thorax, and the shape of the superior anal appendages and genital lobe.

Type in the McLachlan collection.

275. **Onychogomphus aureus** Laidlaw. (Pl. III, fig. 5; text-fig. 78, **b.**)

Onychogomphus aureus Laidlaw, Rec. Ind. Mus. vol. xxiv, pp. 405, 406 (1922); Fraser, J. Bombay Nat. Hist. Soc. vol. xxix, pp. 66, 333 (1923); id., ibid. vol. xxx, pp. 111–113, text-fig. 2, 4, pl. i, fig. 1 (1924); Needham, Rec. Ind. Mus. vol. xxxiv, p. 222 (1932).

Male.—Abdomen 38 mm. Hind-wing 31 mm.

Head: labium yellow; labrum black, with a pair of oval basal spots bright yellow; anteclypeus and bases of mandibles yellow; postclypeus black, with a small rounded spot on each side; frons greenish-yellow, bordered narrowly with black below in front and again at base above, where this colour projects as a small triangular point into the yellow; rest of head black; occiput with a raised point at its middle; eyes greenish during life, black behind, with a large yellow spot. *Prothorax* black, borders of anterior and posterior lobes, a large spot on each side middle lobe, and a small geminate dorsal spot bright yellow. *Thorax* black on dorsum, marked with yellow as follows:—A narrow mesothoracic collar slightly interrupted at its middle; antehumeral stripes broadly confluent with the collar below, but not quite extending to the alar sinus above, an incomplete humeral stripe made up of an upper rounded spot and a narrow line below. Laterally broadly yellow, the postero-lateral suture bearing a narrow black stripe, and vestiges of a similar stripe on the anterior suture consisting of short upper and lower portions. *Legs* blackish or dark reddish-brown, anterior two pairs of femora yellow on the inner side, hind femora entirely yellow. *Wings* hyaline; pterostigma dark ochreous framed in black ner-

vures, braced, covering 3 to $4\frac{1}{2}$ cells ; anal triangle 4-celled ; anal field in hind-wing 4 cells deep ; nodal index $\frac{10-17}{11-12} \Big| \frac{10-16}{13-11}$.
Abdomen : segments 1 and 2 bright yellow, with a broad continuous subdorsal black stripe on each side, enclosing a trilobate stripe of the ground-colour on the mid-dorsum of segment 2 ; oreillets finely margined with black ; 3 to 7 yellow, with broad apical black rings and an oval black spot on mid-dorsum ; 7 without this spot, but the black at apex prolonged slightly along the mid-dorsal carina ; 8 and 9 black ; 10 bright ochreous, with its apical border finely black and with a mid-dorsal black quadrate spot at its base. *Anal appendages* (fig. 78, **b**) bright yellow. *Genitalia* : lamina projecting markedly, broadly arched and cupped behind ; anterior

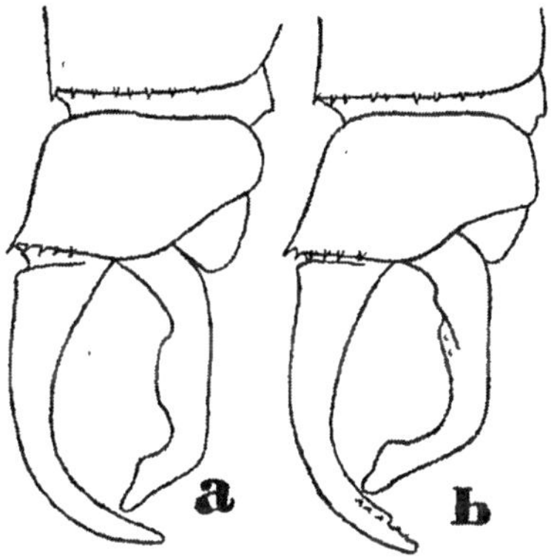

Fig. 78.—Anal appendages of (**a**) *Onychogomphus dingavani* Fraser, male; (**b**) *Onychogomphus aureus* Laidlaw, male. Right lateral views.

hamules very long and slender, remarkably angulated backwards, and ending in a spine ; posterior hamules broader and projecting straight out, crossing the anteriors and ending in a minute recurved black spine ; lobe narrow, spout-like, with a two-pronged process pointing from its apex.

Distribution.—Tura, Garo Hills, ASSAM. Three pairs taken in June and July between 1200 and 1500 ft. altitude, all in the Indian Museum, Calcutta, one being the *type*. I have a single male in my own collection also from these hills.

The very broad extent of the yellow ground-colour has no parallel amongst Indian species of the genus, and this character will serve to distinguish the species at a glance.

276. **Onychogomphus grammicus** (Rambur).

Gomphus grammicus Rambur, Ins. Névrop. p. 164 (1842).
Onychogomphus grammicus Selys, Bull. Acad. Belg. vol. xxi, pt. 2, p. 35 (1854); id., Mon. Gomph. p. 45 (1857); Williamson, Proc. U.S. Nat. Mus. vol. xxxiii, p. 309 (1907); Laidlaw, Rec. Ind. Mus. vol. xxiv, pp. 371, 403 (1922); id., Trans. Ent. Soc. Lond. vol. lxxviii, p. 195 (1930).
Lindenia grammicus Kirby, Cat. Odon. p. 59 (1890).
Mesogomphus grammicus Fraser, J. Bombay Nat. Hist. Soc. vol. xxix, pp. 994, 995 (1924); Needham, Rec. Ind. Mus. vol. xxxiv, p. 221 (1932).

Male.—Abdomen 33–39 mm. Hind-wing 29–30 mm.

Head: eyes bottle-green; labium yellowish; labrum, face, and frons sandy yellow, a fine black line below frons, another between the ante- and postclyepus, and an equally fine line at base of labrum; base of frons, vertex, and occiput black, vesicle and back of occiput yellowish; behind eyes bright yellow, bordered with glossy black above. *Prothorax* black; posterior lobe, two small points just in front of it, a narrow anterior collar, and the sides yellow. *Thorax* yellow, marked with black as follows:—Two dorsal bands converging above, widely divergent below, markedly convex towards one another (the mid-dorsal carina and collar below are yellow and confluent, the former tapering into the latter), a humeral stripe on each side connected with the dorsal bands above and below so as to enclose an oval spot of the ground-colour, a posthumeral stripe connected with the humeral at its upper part only. Both lateral sutures finely black. Tergum spotted with yellow. *Legs* yellow, femora marked with black or brownish-black on the extreme distal end of outer side of hind, the distal half of outer side of middle, and the whole length of outer side of front pair; tibiæ narrowly yellow on extensor surface. *Wings* hyaline, pale saffron, costa yellow; pterostigma yellow between black nervures, covering 5 cells, 4 mm. long, braced; 1 cubital nervure to all wings; nodal index $\frac{9\text{–}16}{9\text{–}10}\Big|\frac{14\text{–}9}{10\text{–}9}$; 3 to 4 rows of postanal cells in hind-wing; no vestige of an anal loop; 2 rows of postanal cells in fore-wing; membrane almost obsolete. *Abdomen* tumid at base, cylindrical and slender as far as segment 8, the latter and 9 dilated, 10 very small. Markings as follows:—Segment 1 yellow, with a basal black spot on each side separated by the dorsal carina; 2 with a trilobed yellow dorsal band tapering apically and lying between narrow black stripes, the sides broadly yellow; 3 to 6 yellow, with broad black apical rings and a median spot on the jugal suture tapering laterally, basally, and apically; on segments 4 to 6 the apical ring sends a prolongation forwards on either side which meets the jugal spot and encloses a yellow subdorsal spot; 7 to 10 ochreous or reddish-yellow; basal articulations finely black. *Anal appendages* yellow, superiors as long as segments 9 and 10 taken together, subcylindrical, tapering apically, where they curve downward, apices flattened, slightly bifid; inferior shorter, broad and flat at base, curling abruptly up in its basal half, where it bifurcates into two slender contiguous truncate branches.

Female.—Abdomen 37 mm. Hind-wing 30 mm.

Similar to male except for sexual differences in shape. The abdomen differs slightly in colouring as follows:—The mid-dorsal band on segment 2 is broader; the jugal spots are more restricted and not connected to apical rings on segments 4 to 6; segment 7 has a diffuse apical dorsal spot tapering basally; 8 to 10 are reddish-yellow and the sides of 8 and 9 are only slightly dilated. *Anal appendages* short, conical, pointed, yellow. Hind femora armed with a row of rather widely spaced, robust, gradually lengthening spines, mid-femora with similar but more closely-set and smaller spines, black, both pairs with a close group of spines on inner side.

Distribution.—CENTRAL and NORTH INDIA. Laidlaw records a male from Agra and I have examined a female taken at Pusa, 16. vii. 20, which is quite complete. (In Pusa Mus.) Rambur's *type* is an incomplete female labelled "India, Stevens."

Type and paratypes in the Selys collection; specimens in Indian Museum and Pusa collections.

277. **Onychogomphus echinoccipitalis** Fraser.

Onychogomphus echinoccipitalis Fraser, Mem. Dept. Agric. India, vol. vii, no. 7, pp. 74, 75 (1922); id., J. Bombay Nat. Hist. Soc. vol. xxix, pp. 65, 333 (1923); id., ibid. vol. xxx, p. 117 (1924); Laidlaw, Trans. Ent. Soc. Lond. vol. lxxviii, p. 196 (1930); Needham, Rec. Ind. Mus. vol. xxxiv, p. 223 (1932).

Male unknown.

Female.—Abdomen 34 mm. Hind-wing 35 mm.

Head: labium dark yellow; labrum similar, but its base narrowly black; bases of mandibles yellow, anteclypeus and postclypeus olivaceous-green; frons black, its crest bright yellow, its base black, from which springs a medial prolongation cutting the yellow in halves; rest of head black, including occiput, which is slightly indented at its middle and bordered with a row of about twelve small spines. *Prothorax* black, spotted with yellow on dorsum. *Thorax* black, marked with yellow as follows:—An interrupted mesothoracic collar, oblique antehumeral stripes separated from the mesothoracic collar (*humeral stripe entirely wanting*); tergum spotted with yellow. Laterally yellow, traversed by two rather diffuse black stripes, one medial, the other lining posterior border of metepimeron. *Legs* black, very short and robust, femora yellow on the outer side. *Wings* hyaline; pterostigma black, rather long (venation in left hind-wing of type aberrant; hypertrigone traversed twice, and there are vestigial nervures in the discoidal cell and subtrigone of same wing); 5 to 6 rows of postanal cells in hind-wing, 2 in fore-wing; nodal index $\frac{9\text{–}21}{10\text{–}11}\Big|\frac{18\text{–}10}{12\text{–}10}$. *Abdomen* very stout, compressed, of even width throughout, but segments 7 and 8 slightly dilated. Colour black marked with yellow as follows:—Segment 1 with a triangular apical spot on its dorsum and the sides broadly; 2 yellow on the sides, traversed obliquely by a black stripe, dorsally marked with a trilobed lanceolate stripe; 3 to 6 with large basal, subtriangular, dorsal spots covering about half of 3 and rather less of segments 4 to 6; 7 with its basal half occupied by a large quadrate spot; 8 and 9 with the base finely yellow, and small subdorsal basal lunules on either side; 10 wholly black. *Anal appendages* very small, conical, black.

Distribution.—A single female from the Shillong–Gauhati road, ASSAM, *ca.* 1,000 ft., 2. viii. 1919 (*T. Bainbrigge Fletcher*).

Type in the British Museum.

278. **Onychogomphus dingavani** Fraser. (Fig. 78, a.)

Onychogomphus dingavani Fraser, J. Bombay Nat. Hist. Soc. vol. xxx, pp. 1005–1006, pl. ii, fig. 7 (1924); Laidlaw, Trans. Ent. Soc. Lond. vol. lxxviii, p. 195 (1930); Needham, Rec. Ind. Mus. vol. xxxiv, p. 223 (1932).

Male.—Abdomen 35 mm. Hind-wing 28 mm. Pterostigma 3 mm.

Head: labium palest yellow; labrum greenish-yellow, its anterior border narrowly black and with an obscure central black line not quite reaching its base; anteclypeus yellow, postclypeus black with a large greenish-yellow spot on each side; frons greenish-yellow, margined below the front with black which is confluent with the black of postclypeus, its base broadly and evenly bordered with black; vertex black, with a small spot of yellow between the posterior ocelli; occiput bright greenish-yellow, broadly yellow behind, its border quite straight and fringed with long brownish hairs; eyes black behind. *Prothorax* black, with an anterior collar, the posterior lobe, a small duplicated spot on dorsum, and a linear one on sides yellow. *Thorax* black on dorsum, marked with greenish-yellow as follows:—A mesothoracic collar, broadly interrupted by the black dorsal carina, which is narrowly bright yellow above for about its middle third;

oblique antehumeral stripes confluent with the collar; complete humeral stripes curling in above, where they become confluent with the antehumeral stripes, but slightly disconnected from the yellow of thorax below. Sides greenish-yellow, with the remnants of an anterior lateral line on the upper part of the first lateral suture and a complete narrow black line mapping out the postero-lateral suture. *Legs* yellow marked with black, fore and mid pairs of femora entirely black on the outer side, hind pair with only a narrow stripe not extending to the base. Flexor surface of hind femora thickly covered with small but robust spines, the middle femora with a row of widely spaced, more robust, longer spines. *Wings* faintly and diffusely tinged with greenish-yellow: nodal index $\frac{11\text{–}14}{12\text{–}10}\Big|\frac{14\text{–}13}{10\text{–}12}$; only one row of cells between *Rii* and *IRii* to within two cells of wing-margin; 2 rows of cells in anal area of fore-wings; first postanal cell in hind-wings undivided; *Cuii* and *IA* with a single row of cells between, to within 4 cells of wing-margin; pterostigma well-braced, covering 4 to 5 cells, dark brown; costa yellow as far as pterostigma, as also are many of the cross-nervures in fore part of wings, the antenodals, the arc, and in the spaces running out from it. *Abdomen* black, marked with yellow as follows:—Segments 1 and 2 with broad subdorsal black stripes of even width enclosing a lobed dorsal stripe of yellow, sides of both segments broadly yellow, including oreillets, a black articular ring on 2; 3 with a narrow basal ring confluent with a short ventral stripe and a broad dorsal stripe with crenulate borders which extends to the apical border; 4 to 6 similar, but the basal rings limited below and the ventral stripes absent; 7 with the yellow extending apically along the dorsum for two-thirds the length of segment and black for the same length along the sides, the venter yellow; 8 and 9 with the sides broadly and the apical border of 9 narrowly yellow; 10 with its basal fourth black, the rest yellow; 9 and 10 with a fine apical border of black, fringed with minute spines. *Anal appendages* (fig. 78, a, p. 255) yellow; inferior deeply cleft, basal third cylindrical and then abruptly thickened to nearly as far as the apical tooth (the thickening begins at the site of the basal tooth in *O. M-flavum* and is really a continuation of this tooth). *Genitalia*: lamina low, shaped like that of *O. M-flavum*, deeply arched, brownish; inner hamules fine, divergent (this may be because the penis is erect between them), of great length, passing down between the outer hamules nearly as far as lobe; outer hamules robust, yellow at base, black at apex, perpendicular to the body axis, ending in a robust forwardly directed spine; lobe mat black, projecting prominently like the spout of a tea-pot.

Female unknown.

Distribution.—A single male (in slightly damaged condition) from Kalaw, S. Shan States, BURMA, collected by Mr. G. Dingavan.

The shape of the inferior anal appendages and the genitalia, and the confluence of the humeral and antehumeral stripes, will serve to distinguish this species from all others of the genus.

Type in the author's collection.

279. **Onychogomphus cerastes** (Selys).

Ophiogomphus cerastis Selys, Bull. Acad. Belg. vol. xxi, pt. 2, p. 41 (1854).
Onychogomphus cerastes Selys, Mon. Gomph. pp. 63, 398 (1854); id., Bull. Acad. Belg. (2) vol. xxviii, p. 173 (1869); Fraser, J. Bombay Nat. Hist. Soc. vol. xxix, pp. 65, 333 (1923).
Lindenia cerastis Kirby, Cat. Odon. p. 60 (1890).
Onychogomphus cerastis Williamson, Proc. U.S. Nat. Mus. vol. xxxiii, pp. 309, 311 (1907); Laidlaw, Rec. Ind. Mus. vol. xxiv, pp. 371, 411 (1922).
Indogomphus cerastes Fraser, J. Bombay Nat. Hist. Soc. vol. xxx, pp. 404, 405, pl. i, fig. 5 (1925); Needham, Rec. Ind. Mus. vol. xxxiv, p. 224 (1932).
Merogomphus cerastes Laidlaw, Trans. Ent. Soc. vol. lxxviii, p. 184 (1930).

Male.—Abdomen 40 mm. Hind-wing 32–34 mm.

Head: labium dull reddish-yellow; labrum yellow, bordered with black along anterior border and base, a tongue of black projecting from the base and cutting the yellow into two spots; anteclypeus yellow; postclypeus black, with a large yellow spot on each side against the eyes, also a small median yellow spot which is confluent with the yellow on anteclyepus; frons broadly yellow, its anterior surface and the base narrowly black; vertex black, with a small point of yellow just posterior to the ocelli; occiput yellow, bordered finely with black, fringed with long brown hairs, slightly notched at its centre. *Prothorax* black, with two small points at its middle, the posterior lobe and the sides broadly yellow. *Thorax* black on dorsum, yellow at sides. The following yellow marks on dorsum:—The mid-dorsal carina finely yellow; oblique antehumeral stripes, slightly separated from an uninterrupted mesothoracic collar; narrow humeral stripes and the tergum spotted with yellow. Lateral sutures finely mapped out in black, that on the last expanding into a black spot behind the hind legs and beneath the thorax. *Legs* black, femora yellow externally, their apical ends showing the beginnings of two brown lines; hind femora long, but not overlapping the third segment, furnished with 5 or 6 very long robust spines. Tibiæ black, flexor surfaces finely yellow. *Wings* hyaline, slightly tinted with yellow at the bases, costa finely yellow, as also are many of the transverse nervures, especially at base and near costa; nodal index $\frac{12\text{–}14}{11\text{–}12}\Big|\frac{16\text{–}12}{12\text{–}11}$; pterostigma reddish-brown bordered with black nervures, 3·5 mm. in length, over 5 cells; membrane pale, short, and narrow. (The condition and colour of the wings points to the specimen being somewhat teneral.) *Abdomen*: segment 1 yellow, its base above and a medio-lateral spot black; 2 yellow laterally, including the oreillets, black above, with a trilobed stripe of yellow on the dorsal carina tapering as far as the apical border; 3 with a subbasal ring nearly divided by the black dorsal carina, a large yellow spot on mid-dorsum and a lateral spot at the same level (4–10 missing in allotype).

Female.—Abdomen 43–45 mm. Hind-wing 35–37 mm.

Differs from the male as follows:—Yellow stripe on frons nearly divided into two by a tongue of black running from the base; occiput armed with two closely apposed medial spines which converge towards one another. *Abdomen* compressed, the last segments slightly dilated, proportionate length of segments 8 and 9 as in *Merogomphus*; first three segments marked as in the male; 4 to 6 similar to 3; 7 with its basal half yellow; 8 and 9 black, with a triangular yellow spot on the sides; 10 yellow, its base and apical border narrowly black. *Anal appendages* yellow, slightly longer than segment 10, slender, pointed.

Distribution.—NEPAL and NORTH INDIA.

Type a female in the Selys collection; allotype male and two females in the British Museum.

280. **Onychogomphus circularis** Selys. (Fig. 79.)

Onychogomphus circularis Selys, Ann. Soc. Ent. Belg. vol. xxxviii, p. 165 (1894); Martin, Mission Pavie, sep. p. 212 (1904); Williamson, Proc. U.S. Nat. Mus. vol. xxxiii, p. 312 (1907); Laidlaw, Rec. Ind. Mus. vol. xxiv, pp. 371, 412 (1922); Fraser, J. Bombay, Nat. Hist. Soc. vol. xxix, pp. 65, 333, 997 (1923–1924); id., ibid. vol. xxx, pp. 115, 116, pl. i, fig. 4 (1924); Needham, Rec. Ind. Mus. vol. xxxiv, p. 223 (1932).

Acrogomphus circularis Laidlaw, Trans. Ent. Soc. Lond. vol. lxxviii, p. 191 (1930).

Male.—Abdomen 41 mm. Hind-wing 32 mm.

Head: labrum yellow, bordered and traversed with black; anteclypeus yellow, encircled with black; postclypeus yellow, with a medial black spot; frons yellow, bordered with black below at its junction with the clypeus, its base above finely black; rest of head black save for a small yellow spot on occiput, the latter slightly raised at its middle and fringed with brown hairs. *Prothorax* yellow, black at its middle. *Thorax* black in front, marked with yellow as follows:—

The dorsal carina finely, a mesothoracic collar narrowly broken at its middle, an oblique isolated antehumeral stripe (which may be joined to the collar but is not confluent in the type), a small upper spot, the remnants of a humeral stripe. Laterally and beneath yellow, with two medial black bands. *Legs* short, femora yellow, armed with short spines, fore pair brown on the outer side, the hind pair black on the outer side; tibiæ and tarsi black. *Wings* hyaline, pale saffron, reticulation brown, costa finely yellow, pterostigma blackish-brown, 3·5 mm. in length, covering 5 cells, irregularly braced; incomplete

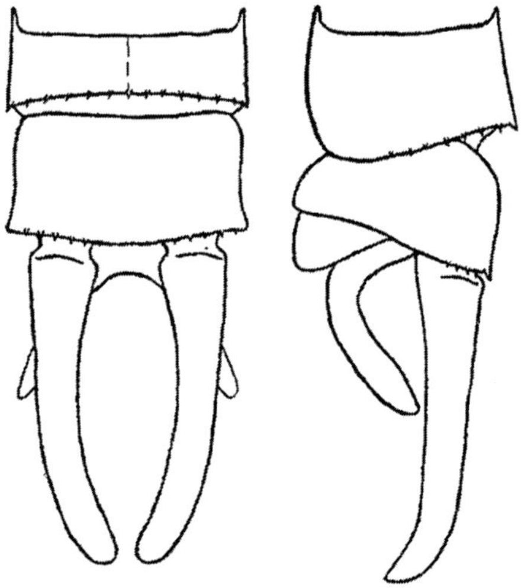

Fig. 79.—Anal appendages of *Onychogomphus circularis* Selys, male. Dorsal and left lateral views.

basal antenodal nervure absent; nodal index $\frac{9\text{–}16}{10\text{–}12}\Big|\frac{16\text{–}10}{12\text{–}10}$; anal triangle with 4 cells; tornus prominent. *Abdomen* moderately slender, black marked with yellow (probably very similar to that of the female described below, but faded in type); segments 8 and 9 slightly dilated, 10 very short. *Anal appendages* (fig. 79) yellow. (Inferior appendage absent in the type.)

Female.—Abdomen 41 mm. Hind-wing 38 mm.

Head and *thorax* similar to those of male; humeral spot better defined and antehumeral stripes confluent with the mesothoracic collar; first lateral black line nearly obsolete in its upper part. *Legs* and *wings* as in male; nodal index slightly higher. *Abdomen* compressed, a little tumid at its base, parallel-sided thereafter, except the borders of 8 and 9, which are perceptibly dilated. Colour black, marked with yellow as follows :—The whole of segment 1 ; segment 2 with a trilobed mid-dorsal stripe not extending as far as the apical border, its sides broadly ; 3 to 7 with complete basal rings, broadest on 3 and 7 and followed on 3 to 5 by a lanceolate mid-dorsal spot. Segment 8 with small dorsal and lateral spots ; 9 and 10 wholly black. Vulvar scale of medium size, made up of two contiguous tubercles. *Anal appendages* yellow, tipped with black, slender, tapering, separated by a conical yellow prominence which terminates the abdomen.

Distribution.—One pair in the McLachlan collection from UPPER BURMA, the male being the *type.*

A complete and well preserved male in my collection from Maymyo, Upper Burma, differs slightly from the type. In this insect the labrum is black, marked with two transversely oval yellow spots near the base, the yellow crest of the frons is confined to its upper surface, and the base is broadly black ; the occiput is entirely black. The prothorax has an anterior collar and a large spot on each side of the middle lobe citron-yellow ; the mid-dorsal carina of the thorax is unmarked with yellow ; the hind pair of femora are striped with yellow on the outer side ; the nodal index differs slightly, $\frac{11\text{–}17}{12\text{–}11} \Big| \frac{18\text{–}10}{13\text{–}12}$. The abdomen varies rather more widely from the type as follows :—Segment 1 has a mid-dorsal spot and its sides broadly yellow ; segment 2 has the sides broadly yellow and a mid-dorsal stripe which tapers very abruptly apically ; segments 3 to 5 have paired basal dorsal spots which are slightly confluent over the dorsum, and there are also oval mid-dorsal spots which decrease in size from segment 3 to 5, being very small on the latter segment ; segment 6 has the basal but not the mid-dorsal spot ; segment 7 has its basal half yellow and segment 8 has only a small baso-lateral spot. The anal superior appendages are similar to type except that only the extreme apices are yellow ; the inferior is about half the length of the superiors and is deeply bifid, its two branches being widely separated, directed straight back, but with the apices curved slightly outwards, a shape quite unique in the genus.

281. **Onychogomphus earnshawi** Fraser.

Onychogomphus sp. Williamson, Proc. U.S. Nat. Mus. vol. xxxiii, pp. 313–315, text-fig. 37 (1907).

Onychogomphus earnshawi Fraser, J. Bombay Nat. Hist. Soc. vol. xxx, pp. 113–114 (1924); Needham, Rec. Ind. Mus. vol. xxxiv, p. 222 (1932).

Acrogomphus earnshawi Laidlaw, Trans. Ent. Soc. Lond. vol. lxxviii, p. 192 (1930).

Male.—Abdomen 38 mm. Hind-wing 33 mm.

Head: labium pale yellow; labrum yellow, bordered in front and behind and traversed with black; anteclypeus yellow, bordered below on either side with black; postclypeus black, with a narrow margin below at its middle and a large spot on either side yellow; frons yellow, its lower part in front black, whilst above there is a triangular medial basal black spot almost dividing the yellow; occiput simple, nearly straight, yellow at its centre, black on either side against the eyes; back of head black. *Prothorax* black, margined with yellow. *Thorax* black in front, marked with yellow as follows:—The lower part of the mid-dorsal carina, an uninterrupted mesothoracic collar, oblique antehumeral stripes joined to the mesothoracic collar below and forming inverted 7's, a vestigial humeral stripe on either side, represented by an upper spot and a smaller lower one; laterally yellow, the sutures mapped out in black, that on the anterior suture interrupted in two places. *Wings* hyaline; costa yellow; pterostigma black; nodal index $\frac{10\text{–}13}{9\text{–}10}\Big|\frac{13\text{–}10}{10\text{–}9}$; 4 cells in anal triangle; 4 rows of postanal cells in hind-wing, the first cell divided; a single row, or occasional double-cells in anal area of fore-wing, *Cui* and *IA* widely divaricate at the wing-margin, 5 cells between them at their distal ends, only a single row of cells between *Rii* and *IRii* nearly to apex of wings. *Abdomen*: segment 1 with a dorsal interrupted crescent of brown, the ends of the crescent running backwards and downwards; 2 yellow, with subdorsal bands of brown which are confluent at the apical border and enclose a tapered stripe of yellow on the mid-dorsal carina; 3 and 4 yellow marked with black, a narrow basal ring, an interrupted median ring, and a lateral apical triangular spot which is confluent with its fellow dorsally; 5 to 7 similar, but the apical spots larger, confluent, and covering the apical half of segments; 7 has the median ring almost obsolete; 8 to 10 black, the former with a large lateral basal yellow spot and a trace of a dorsal basal spot; 9 with a mere trace of a similar latero-basal spot. *Anal appendages* brown at base, shading at once into pale yellow. Superiors twice as long as segment 10, slender, tapering, curved towards each other and downward, the apex with a shining black tooth, the lower external edge on the curve before the apex minutely denticulate. Inferior slightly more than half the length of superiors, broadly bifid for more than half its length, its branches simple, rounded, tapering, and continuously divaricate, only slightly recurved dorsally, and ending at the apex in a minute tooth.

Female unknown.

Distribution.—Toungoo, BURMA (*R. A. Earnshaw*).

As Mr. Williamson's description of the type is very detailed, and is moreover backed up by an excellent photograph of the wings, the species may be considered a good one, and so I have taken the liberty of naming it. The anal appendages, like those of *O. circularis*, are quite foreign to the genus, but apart from this the species is a true *Onychogomphus*, and is probably closely allied to *O. circularis*. It differs from the latter principally by its nodal index, which is much lower.

Type male (in fragments) in the Michigan University Museum collection; Mr. Williamson refrained from naming it on account of its damaged condition.

282. **Onychogomphus annularis** Selys.

Onychogomphus annularis Selys, Ann. Soc. Ent. Belg. vol. xxxviii, p. 166 (1894); Williamson, Proc. U.S. Nat. Mus. vol. xxxiii, pp. 308, 312, 313 (1907); Laidlaw, Rec. Ind. Mus. vol. xxiv, p. 411 (1922); Fraser, J. Bombay Nat. Hist. Soc. vol. xxix, pp. 66, 333 (1923); id., ibid. vol. xxx. pp. 110, 111, pl. i, fig. 3 (1924); Laidlaw, Trans. Ent. Soc. Lond. vol. lxxviii, p. 196 (1930); Needham, Rec. Ind. Mus. vol. xxxiv, p. 222 (1932).

Male.—Abdomen 37 mm. Hind-wing 32 mm.

Head glossy black, marked with citron-yellow as follows:—Labrum with two oval spots, the bases of mandibles, a small spot on either side against the eyes at level of post-clypeus, and a transverse band on upper surface and crest of frons, interrupted in the middle. Labium dirty yellow. Rest of head black, occiput simple, hinder border nearly straight. *Prothorax* black. *Thorax* glossy black, marked with greenish-yellow as follows:—A mesothoracic collar interrupted at its middle, oblique antehumeral stripes not extending upwards as far as the alar sinus, but confluent below with the mesothoracic collar; humeral stripes represented by an upper spot followed by a fine line below, at about the middle of the thorax. Laterally two rather broad bands. *Wings* hyaline slightly tinted with yellow, especially at bases; costa slightly yellow outwardly; pterostigma black, 3·5 mm. in length, braced; nodal index $\frac{10\text{–}14}{10\text{–}9}\Big|\frac{15\text{–}9}{10\text{–}10}$. *Abdomen* tumid at base, then slender and cylindrical as far as segment 7 (8–10 missing). Colour black, marked with yellow as follows:—Segments 1 and 2 with a longitudinal pyriform spot on dorsum, tapering nearly to apical border of segment 2; the sides of these segments broadly, including the oreillets; 3 to 6 with broad basal rings occupying from one-fourth to one-third the length of segments; 3 and 4 with an oblong oval mid-dorsal median spot on the carina. *Legs* black, short, hind femora 5 mm. long, furnished with a row of short spines.

Female unknown.

Distribution.—UPPER BURMA. Described from two males sent to Selys by McLachlan, and now in the McLachlan collection, one being the *type*.

Selys was of opinion that this species might be conspecific with *O. maclachlani* (described below), but in *O. annularis* the markings on the face are much more extensive ; there is also a vestigial humeral stripe on the thorax and mid-dorsal spots on segments 3 and 4 which are absent in *O. maclachlani*.

There are some strong points of similarity between this species and *O. saundersi* which do not appear to have been noticed by Selys or subsequent authors.

In *O. annularis* the yellow stripe on the frons is broken, in *O. saundersi* it is nearly so. The latter is described as having the sides of the thorax yellow with two broad black bands on the sutures, whilst *O. annularis* is described as having the sides black with two broad yellow bands. These two descriptions may be but two different ways of describing the same markings. The legs in *O. annularis* are entirely black, but marked with yellow in *O. saundersi*. Segments 5 and 6 in the latter have median dorsal spots in addition to those on 3 and 4 seen in *O. annularis*. With the exception of these small differences the two species are alike, and *O. annularis* may be merely a variety of *O. saundersi*.

283. **Onychogomphus maclachlani** Selys.

Onychogomphus maclachlani Selys, Ann. Soc. Ent. Belg. vol. xxxviii, p. 167 (1894); Williamson, Proc. U.S. Nat. Mus. vol. xxxiii, pp. 312, 313 (1907); Laidlaw, Rec. Ind. Mus. vol. xxiv, pp. 371, 411 (1922); Fraser, J. Bombay Nat. Hist. Soc. vol. xxix, pp. 65, 333 (1923); id., ibid. vol. xxx, pl 114, pl. i, fig. 5 (1924); Laidlaw, Trans. Ent. Soc. Lond. vol. lxxviii, p. 196 (1930); Needham, Rec. Ind. Mus. vol. xxxiv, p. 222 (1932).

Male.—Unknown.

Female.—Abdomen 43 mm. Hind-wing 38 mm.

Head glossy black, marked with bright citron-yellow on bases of mandibles and a transverse band across upper part of frons. Occiput simple, fringed with hairs. *Prothorax* black. *Thorax* black, marked with citron-yellow as follows :—A mesothoracic collar interrupted in the middle, oblique antehumeral stripes not extending as far upwards as the alar sinus, but confluent with the mesothoracic collar below. Laterally two equally broad yellow stripes. *Legs* short, brownish-black, hind femora 6 mm. in length, furnished with a row of short closely-set spines. *Wings* hyaline, costa yellow outwardly ; pterostigma pale yellow between black nervures, 4 mm. in length, covering 5 cells ; no basal antenodal of the second series present ; nodal index $\frac{10\text{–}13}{12\text{–}10}\Big|\frac{15\text{–}10}{10\text{–}12}$.

Abdomen slightly tumid at base, compressed, segments 8 and 9 slightly dilated laterally. Colour black, marked with yellow as follows :—Segment 2 with a tapered dorsal stripe not quite extending to the apical border, its sides, including the oreillets, broadly; 3 to 7 with lunules occupying about the basal fifth and confluent over the dorsum; 8 with only a small basal spot on each side; 9 and 10 wholly black, the latter very short. *Anal appendages* small, conical, brown. Vulvar scale rudimentary, not discernible.

Distribution.—UPPER BURMA.

As mentioned above, Selys remarks that this species may be the female of *annularis*, but the differences pointed out above preclude this possibility.

Type female, unique, in the McLachlan collection.

284. **Onychogomphus pulcherrimus** Fraser. (Pl. III, fig. 4; text-fig. 80.)

Onychogomphus pulcherrima Fraser, Rec. Ind. Mus. vol. xxix, pp. 78, 79 (1927); Laidlaw, Trans. Ent. Soc. Lond. vol. lxxviii, p. 196 (1930); Needham, Rec. Ind. Mus. vol. xxxiv, p. 222 (1932).

Male.—Abdomen 40 mm. Hind-wing 30 mm.

Head: labium yellowish, the anterior border narrowly reddish-brown; labrum pale green; face and frons vivid pea-green above as well as in front; vesicle and vertex dark blackish-brown; occiput grass-green, brownish at the sides. *Prothorax* dark reddish-brown, marked with yellow on posterior lobe, a large geminate spot in the middle line confluent with it and a narrow anterior collar. *Thorax* dark mahogany-brown, with two very broad green antehumeral stripes, confluent with a narrow mesothoracic collar, which latter is continuous with a narrow yellow mid-thoracic suture, merely separated from the antehumeral stripes by a clouding of warm reddish-brown. Sides greenish except for a moderately broad reddish-brown stripe on the second lateral suture, broadening anteriorly below. *Legs* blackish-brown, the extensor surfaces of fore-femora yellow, of the two posterior pairs reddish-brown. *Wings* hyaline, palely saffronated and distinctly tinted with yellow at the extreme base. Pterostigma dark reddish-brown, framed in blackish-brown, but with a fine yellow streak between the ground-colour and black frame, strongly braced, covering $3\frac{1}{2}$ to $4\frac{1}{2}$ cells; 4 cells in anal triangle; first post-anal cell entire, not extending basally beyond the centre of subtrigone; nodal index $\frac{10\text{–}14}{12\text{–}10}\left|\frac{14\text{–}9}{10\text{–}10}\right.$. *Abdomen* dark ochreous and blackish-brown; segments 1 to 2 marked with yellow; 1 with the sides broadly light greenish-yellow below;

2 with the same area, including the oreillets, yellow tinged with ochreous, the dorsum brown, this colour passing down on each side as a narrow stripe behind the oreillets; 3 with the base and dorsum broadly dark ochreous; 4 to 7 dark reddish-brown deepening to black apically: 8 to 10 gradually paling in colour from dark to bright reddish-brown; apical borders of all segments narrowly black. *Anal appendages* (fig. 80) yellow, reddish at apices. *Genitalia*: lamina projecting markedly, its border slightly emarginate and concave; anterior hamules long, fine, stylet-like hooks ending in a fine recurved spine, directed backwards and converging; posterior

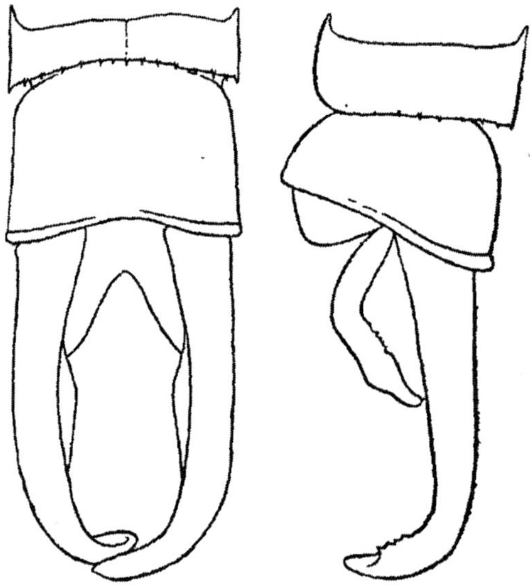

Fig. 80.—Anal appendages of *Onychogomphus pulcherrimus* Fraser, male. Dorsal and left lateral views.

hamules much stouter structures, broad, abruptly narrowed near apex, where they form a very robust slightly recurved spine; lobe deeply cleft into quadrate, slightly bifid branches, with a shallow groove on the outer surface, its base tumid.

Female unknown.

Distribution.—Three males taken by Col. F. Wall, I.M.S., at Maymyo, UPPER BURMA, 31st May, 1925.

The bright apple-green markings on a mahogany-red background are so unique in the genus as to distinguish it easily from all other species.

Type in the British Museum; paratypes in the Author's collection.

Genus **LAMELLIGOMPHUS** Fraser. (Figs. 45 *a*(C) & 81.)

Lamelligomphus Fraser, Rec. Ind. Mus. vol. xxiv, p. 426 (1922); Laidlaw, Trans. Ent. Soc. Lond. vol. lxxviii, p. 193 (1930).
Lamellogomphus Fraser, J. Bombay Nat. Hist. Soc. vol. xxix, p. 983 (1924); Needham, Rec. Ind. Mus. vol. xxxiv, pp. 217, 218, 223 (1932).

Size large, body glossy black, sparingly marked with bright greenish-yellow; anal appendages highly curled.

Head large, triangular; frons prominent and markedly angulated; occiput simple, slightly concave, short. Wings: reticulation very close; tornus markedly angulated; base of hind-wing deeply excavate; membrane obsolete; anal triangle 4-celled; arc situated between the first and second antenodal nervures or opposite the second; only 2 transverse nervures between the sectors of arc in fore-wing from arc to bifurcation of *Rs*, 1 in hind-wing; 2 rows of postanal cells in fore-wing, 4 in hind-wing; a rudimentary anal loop present, formed by a splitting of the first postanal cell which extends well proximal of the base of subtrigone; nodal index high; primary antenodal nervures the first and the fifth; discoidal cells entire, that of fore-wing with the distal side longer than the costal and basal, which are approximately equal, that of the hind-wing not elongated in the length of wing, the sides of the same dimensions as those of fore-wing or the costal and distal, but slightly longer than the basal; pterostigma moderately short, about one-third the distance from node to proximal end of pterostigma, braced; *IA* in fore-wing markedly pectinate, 3 rows of cells between it and margin of wing; *Cuii* and *IA* in hind-wing barely divergent at wing-border; only 1 cubital nervure in all wings; subtrigones and hypertrigones all entire; 2 rows of cells in discoidal field as far as slightly proximal of the node. Legs short, hind femora extending to the base of segment 1 and furnished with a group of closely-set numerous spines at the proximal half, which become arranged into two rows of very closely-set short spines at the distal half; hind tibial spines slim, moderately long and closely-set. Abdomen of male robust, variable in length as compared to the length of wings, tumid at base, moderately narrow and cylindrical as far as the base of segment 7, then again expanding as far as segment 10, which is somewhat squared. Anal appendages: superiors forcipate, remarkably curled, so that the tips are directed forwards (except in *L. acinaces*); the inferior bifid almost to base, the two branches very closely apposed and curled almost as much as the superiors, which they overlap at the apices, so that the two sets of appendages come to enclose a large cordate space. Genitalia: lamina projecting hood-like, angularly arched; anterior hamules slim sinuous processes ending in a minute spine; posterior hamules broad, long, compressed processes, ending in a short, robust, incurled spine; lobe small, very inconspicuous, flask-shaped.

Genotype, *Onychogomphus biforceps* Selys.

Distribution.—The WESTERN GHATS, at altitudes of about 3,000–4,000 ft., BENGAL, SIKKIM, BURMA, Indo-China, Malaysia, and Java. Most of the species mentioned from China by Needham as belonging to the genus *Onychogomphus* also evidently belong to this genus, for they bear the highly characteristic appendages and the very marked black colouring.

Six species are found within our limits, all confined to dense evergreen jungles and keeping in close proximity to the

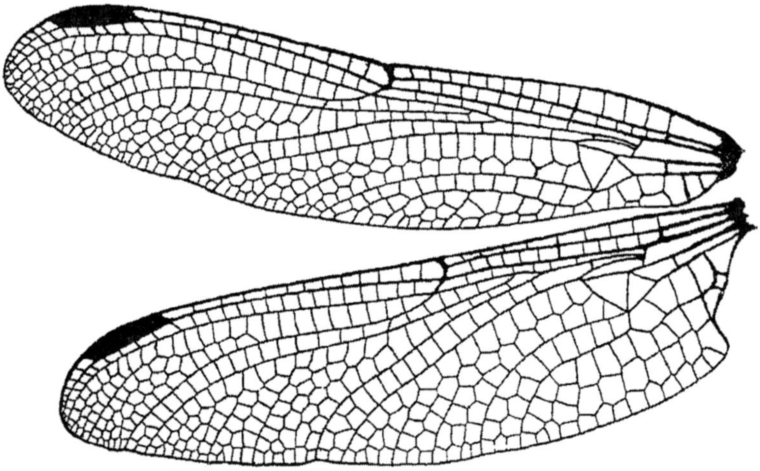

Fig. 81.—Wings of *Lamelligomphus nilgiriensis* (Fraser), male.

montane streams in which they breed. The larvæ (fig. 45 *a*, C) are to be found in deep pools lurking amongst leafy débris, and are curiously flattened, dark brown in colour, and closely resembling the decaying leaves amongst which they live. The genus derives its name from the peculiar flattened short lamellate antennæ of the larvæ, the latter differing entirely from those of *Onychogomphus*.

Key to Indian Species of Lamelligomphus.

1. { Antehumeral stripes confluent with the mesothoracic collar; humeral stripes absent .... 2.
   { Antehumeral stripes well separated from the mesothoracic collar .... 3.

2.	A single interrupted lateral black stripe on thorax; abdominal segments 9 and 10 yellow laterally	*cacharicus* Fras., [p. 275.
	A single broad uninterrupted lateral black stripe on thorax; segments 9 and 10 without markings	*nilgiriensis* (Fras.), [p. 276.
3.	Humeral stripe present and complete; abdominal segments 3 to 6 with yellow oval mid-dorsal spots..................	*biforceps* (Selys), [p. 271.
	Humeral stripe absent or represented only by a tiny upper spot	4.
4.	The two lateral black stripes on thorax confluent at one or two points only; abdominal segments 3 and 4 with mid-dorsal oval yellow spots	*malabarensis* Fras., [p. 279.
	The two lateral black stripes fused to form a single broad stripe on the sides of thorax; segments 3 and 4 without mid-dorsal oval spots	5.
5.	Superior and inferior anal appendages but slightly curved and directed nearly straight backwards; legs black; antehumeral stripes short and pointed below; superior humeral spot absent	*acinaces* (Laid.), [p. 280.
	Superior and inferior anal appendages markedly curled, so as to enclose together a broad cordate space; legs black marked with yellow; antehumeral stripes long and not pointed below; superior humeral spot present	*risi* (Fras.), p. 273.

285. **Lamelligomphus biforceps** (Selys). (Figs. 82, **a**, & 83, **c**.)

Onychogomphus biforceps Selys, Bull. Acad. Belg. (2) vol. xlvi, p. 420 (1878); Martin, Mission Pavie Indo-Chine, vol. iii, p. 212 (1904); Williamson, Proc. U.S. Nat. Mus. vol. xxxiii, p. 308 (1907); Laidlaw, Rec. Ind. Mus. vol. xxiv, p. 407, text-fig. 19 (1922).

Lindenia biforceps Kirby, Cat. Odon. p. 60 (1890).

Lamellogomphus biforceps biforceps Fraser, J. Bombay Nat. Hist. Soc. vol. xxix, pp. 64, 332, 983, 984 (1923–1924).

Lamelligomphus biforceps Laidlaw, Trans. Ent. Soc. Lond. p. 193 (1930).

Lamellogomphus biforceps Needham, Rec. Ind. Mus. vol. xxxiv, p. 224 (1932).

Male.—Abdomen 41 mm. Hind-wing 32 mm.

Head black; labium dark brown; labrum marked with a transverse yellow oval spot on each side; base of mandibles, anteclypeus, and a broad band across the upper surface of frons bright yellow; base of frons black, this colour extending as a short tongue into the floor of sulcus; occiput black, with a small median yellow spot, slightly rounded, ciliated along its free margin. *Prothorax* black, with a lateral yellow spot. *Thorax* black, marked with greenish-yellow as follows:— A complete mesothoracic collar confluent with a stripe on lower part of mid-dorsal carina; a dorsal oblique stripe not joined to the mesothoracic collar; and a humeral stripe. Laterally yellow, marked with two broad black stripes on the sutures, the black converging and becoming confluent at the middle of the stripes. Underside black. *Legs* black, the four posterior femora with an outer yellow band. *Abdomen* tumid at base, thin and cylindrical as far as segment 8, which is abruptly dilated, 9 and 10 narrowing again slightly; black, marked with bright yellow as follows:—Segment 1 on sides and a triangular spot on dorsum; 2 with two large spots (including the oreillets) on sides and a bilobate stripe on dorsum; 3 to 6 with narrow basal rings nearly divided by the black of the dorsal carina, occupying about one-third of the segments, and with an oval spot at the middle; basal half of 7 and a small rounded spot near the base of both sides of 8.

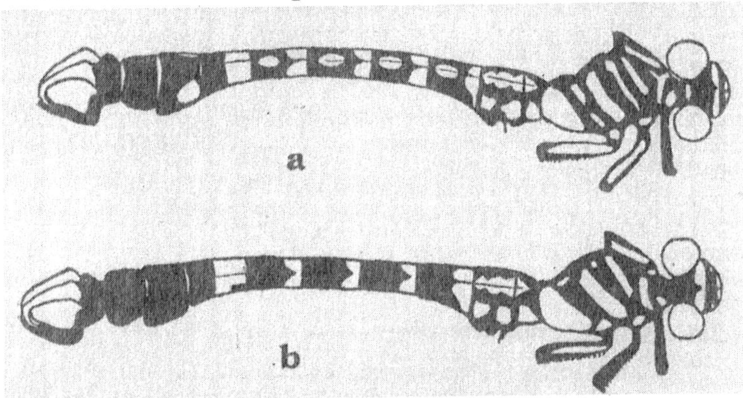

Fig. 82.—Body markings of (a) *Lamelligomphus biforceps* (Selys), male; (b) *Lamelligomphus risi* (Fraser), male.

Anal appendages (fig. 83, c): superiors yellow, changing to black at the tips, thick at base, where they are rather widely separated, converging, tapering, and finally meeting at the apices; branches of inferior black. *Wings* slightly enfumed, venation black, costa finely yellow; pterostigma deep blackish-brown (3 mm.), nodal index $\frac{10\text{–}15}{13\text{–}11}\Big|\frac{15\text{–}11}{11\text{–}13}$.

Female.—Martin states that the female is very like the male, but the yellow spots on the abdomen are much larger; he gives no detailed description.

Distribution.—Darjeeling district; also recorded by Martin from Tonkin, where it is apparently not uncommon.

Type in the Selys collection; a male in the Indian Museum collection, taken in May 1913, and a male in my own collection. These, so far as I am aware, are the only Indian specimens known.

286. **Lamelligomphus risi** (Fraser). (Fig. 82, **b**.)

Gomphus risi Fraser, Mem. Dept. Agric. India, vol. vii, no. 7, pp. 73, 74 (1922).
Heterogomphus risi Fraser, J. Bombay Nat. Hist. Soc. vol. xxix, pp. 331, 678, 679 (1923).
Lamellogomphus inglisi Fraser, J. Bombay Nat. Hist. Soc. vol. xxix, pp. 984, 985, pl. i, fig. 4 (1924); Needham, Rec. Ind. Mus. vol. xxxiv, p. 224 (1932).
Lamellogomphus drummondi Fraser, J. Bombay Nat. Hist. Soc. vol. xxix, pp. 985, 986 (1924).
Megalogomphus risi Laidlaw, Trans. Ent. Soc. London, vol. lxxviii, p. 197 (1930).

Male.—Abdomen (with appendages) 45 mm. Hind-wing 36 mm.

Head: labium pale yellow, the border of median lobe narrowly brown; base of mandibles broadly greenish-yellow; labrum glossy black, with a large transversely oval greenish-yellow spot on each side; anteclypeus and a large spot on each side of postclypeus against the eye, citron-yellow; frons black, greenish-yellow above, this colour slightly overlapping the fore border, base of frons narrowly black, which colour projects for a short distance into the sulcus; rest of head mat black except for an obsolete spot on occiput, which appears to be due to the translucence of a greenish-yellow spot on its posterior surface; border of occipital scale rather sinuous, fringed with long black hairs. Eyes bottle-green. *Prothorax* black, with a small citron-yellow spot on each side. *Thorax* black, marked with yellow as follows:—A mesothoracic collar narrowly interrupted in the middle line; oblique antehumeral stripes extending from the alar sinus, but not meeting the mesothoracic collar; a vestigial humeral spot above; the whole of the sides except for a very broad median black stripe which bears a small transversely oval yellow spot immediately below the insertions of the wings. *Legs* black, *the posterior pair of femora only* bearing an outer yellow stripe; coxæ and trochanters yellow; hind femora furnished with a row of closely-set, evenly spaced, short but robust spines on the outer side; middle femora with longer, less closely-set spines. *Wings* palely enfumed, costa finely yellow to well beyond the node, nodal index $\frac{12\text{–}17}{13\text{–}12}\Big|\frac{17\text{–}12}{13\text{–}14}$; pterostigma well braced, black, 3 rows of cells between *Rii* and *IRii*. *Abdomen* tumid at base as far as segment 2, very narrow and cylindrical as far as extreme apex of 7, 8 to 10 greatly expanded, black, marked with yellow as follows:—Segment 1 with a large inferior lateral spot and a linear transverse apical dorsal spot; 2 broadly yellow on the sides, including the robust oreillets, and a linear mid-dorsal stripe broadening basally and longitudinally cleft at the base by a brown mark on the dorsal carina; 3 to 6 with basal rings occupying about the basal fourth of segments, all nearly cleft by an invasion of the black along the dorsal carina posteriorly; 7 with rather more than the basal half greenish-yellow, the basal half of this marking on the sides squarely pale brown; 8 to 10 unmarked. *Anal appendages* about the length of segments 8 and 9 together; superiors yellow, the posterior two-thirds straight as seen in profile, the apical third curling strongly downwards and then actually backwards, so that finally its dorsum comes into contact with the dorsal surface of the inferior appendage; the latter black, cleft to its base, the branches curving at first downwards and then bent at a right angle and prolonged to overlap the superiors, a little dilated at base, then thin and cylindrical for as far as middle third, where they dilate rather abruptly and taper to the end, separated at base to enclose a long oval space. *Genitalia*: lamina broad, depressed and obtusely notched; inner hamules converging, long, sinuous, broad at base and tapering to a fine outwardly turned point; outer hamules broad and conical, projecting well beyond the genital sac and bearing a stiff pencil of hairs at the apices; lobe broad and deep, funnel-shaped, but not markedly prominent, surface black and coarsely corrugated.

Female.—Abdomen 45 mm. Hind-wing 38 mm.

Very similar to the male; markings similar in every respect. Wings rather more enfumed, pterostigma well braced, blackish-brown, 2 or 3 rows of cells between *Rii* and *IRii*, nodal index $\frac{13\text{–}19 \mid 18\text{–}12}{12\text{–}14 \mid 13\text{–}13}$. *Anal appendages* yellow, very fine, and rather longer than segment 10. Vulvar scale very short, triangular, deeply encased by the expanded sides of segments 8 and 9. *Legs* coloured as in the male, hind femora with a row of short but robust closely-set spines on the proximal half, and 3 to 6 much longer, more robust and more widely spaced spines on the distal half.

Distribution.—One pair collected by Mr. C. M. Inglis on the Riyang River, Mungpoo, DARJEELING DISTRICT, 1,800 ft., 17. v. 23. A female collected by Mr. H. V. O'Donel at Hasimara tea estate, Duars, BENGAL.

The markings in the Hasimara female, which is distinctly teneral, are distinctly broader, and there is some evidence of a broken yellow stripe traversing the lateral black stripe of the thorax, and there are also small baso-lateral spots on segments 8 and 9.

This species differs from *L. biforceps* in the colour of the labium, the presence of lateral yellow spots on the post-clypeus, the vestigial character of the humeral stripe, the absence of mid-dorsal markings on segments 5 and 6, and greater number of antenodal nervures; from *L. camelus* (Mart.) in the absence of excrescences on segment 8, in its smaller size, etc.; from *L. acinaces* in the shape of the anal appendages; and from *L. nilgiriensis* in its much larger size and markings.

Types of *risi* (a female), *inglisi* (a male), and *drummondi* (a teneral female) all in the Author's collection.

287. **Lamelligomphus cacharicus** Fraser.

Lamellogomphus cacharicus Fraser, Mem. Dept. Agric. India, vol. viii, no. 8, pp. 81–83 (1924).
Lamelligomphus cacharensis Laidlaw, Trans. Ent. Soc. Lond. vol. lxxviii, p. 194 (1930).

Male unknown.

Female.—Abdomen 42 mm. . Hind-wing 38 mm.

Head: labium dirty yellow; labrum black, marked with a small comma-shaped spot at each outer end; bases of mandibles yellow; ante- and postclypeus glossy black, latter marked with a small yellow spot on each side against the eyes; frons black, marked with a broad yellow stripe along the crest; vertex and anterior part of occiput black, posterior part of latter dark brown, simple, straight, no armature, fringed with black hairs. *Prothorax* black, marked with yellow as follows:—A small spot on each side; the posterior lobe narrowly; two small points on mid-dorsum and a narrow anterior collar. *Thorax* black, marked with yellow as follows:—A narrowly interrupted mesothoracic collar; antehumeral stripes confluent with the latter and forming inverted 7's; humeral stripe entirely absent; laterally a broad posthumeral stripe; and the whole of the metepimeron yellow, divided by a broken black stripe. *Legs* black, coxæ and trochanters yellow, anterior femora striped with yellow on the inner side; hind femora with a row of short robust moderately closely-set spines. *Wings* hyaline, pale rays of saffron in subcostal and cubital spaces extending outwards as far as the arc; pterostigma brown, covering 5 to 6 cells, braced robustly; membrane obsolete; 1 to 2 rows of cells between *Rii* and *IRii* at distal end of pterostigma; nodal index $\frac{10\text{–}15 \mid 14\text{–}10}{10\text{–}10 \mid 11\text{–}11}$; 1 cubital nervure in all wings; 1 to 2 rows of postanal cells in fore-wing, 4 rows in hind-wing. *Abdomen* black, marked with yellow as follows:—A narrow mid-dorsal stripe tapering from segment 1 to the transverse suture on 3; the sides of all these three segments broadly yellow, but narrowing progressively from 1 to 3 and not quite extending to apical border of latter; the dorsal and lateral stripes separated by a broad black stripe towards either side; 4 to 7 with broad basal rings incomplete below and covering nearly one-fourth of the segments; 8 with a small baso-dorsal spot; 9 and 10 laterally. *Anal appendages* dark brown, minute, pointed. Vulvar scale short, rounded, not projecting (this organ crushed and not very easily made out in the type; the specimen is slightly teneral, and the black colouring, especially of abdomen, not completely matured).

Distribution.—Dilkhoosh, Cachar, ASSAM.

This insect is distinguished from *L. acinaces* and *L. nilgiriensis* by the mesothoracic collar and antehumeral stripes being confluent. The same character and the absence of a spine on the occiput will distinguish it from *L. malabarensis*. The absence of a humeral stripe will separate it from *L. biforceps* and *L. risi*.

Type in the British Museum.

288. **Lamelligomphus nilgiriensis** (Fraser). (Fig. 83, a, b.)

Onychogomphus biforceps nilgiriensis Fraser, Rec. Ind. Mus. vol. xxiv, pp. 425, 426, pl. xi, fig. 2 (1922).

Lamellogomphus biforceps nilgiriensis Fraser, J. Bombay Nat. Hist. Soc. vol. xxix, pp. 65, 332 (1923).

Lamellogomphus nilgiriensis Fraser, J. Bombay Nat. Hist. Soc. vol. xxix, pp. 986–988, pl. i, fig. 1 (1924); id., Rec. Ind. Mus. vol. xxvi, pp. 427, 477 (1924); Laidlaw, Trans. Ent. Soc. Lond. vol. lxxviii, p. 193 (1930); Fraser, Rec. Ind. Mus. vol. xxxiii, p. 447 (1931); Needham, Rec. Ind. Mus. vol. xxxiv, pp. 223, 224 (1932).

Male.—Abdomen (with appendages) 36–41 mm. Hindwing 30–34 mm.

Head: labium entirely black; labrum black, marked with two oval transverse yellow spots (large or small); bases of mandibles and anteclypeus greenish-yellow (but in one specimen this latter is quite black); postclypeus black; frons greenish-yellow, its base rather broadly black, a prolongation of this colour invading the floor of sulcus; rest of head black, often with a rounded yellow spot behind the occiput, only visible when the head is tilted forwards. Eyes deep bottle-green, or in younger specimens bluish-green. *Prothorax* black, occasionally unmarked but usually bearing a small geminate yellow spot at the centre of posterior lobe, and outside this an even smaller spot. *Thorax* black, marked with greenish-yellow as follows:—Upper part of mid-dorsal carina only occasionally; a mesothoracic collar interrupted in the middle line; oblique antehumeral stripes, generally connected with the mesothoracic collar, but in a small percentage of specimens more or less widely separated therefrom (in such cases the lower end of the stripe is squared off, not pointed as in *L. acinaces*); humeral stripe absent (only in a single male was there a vestigial spot representing the upper part of a humeral stripe); laterally two broad stripes separated by an almost equally broad black stripe, the posterior the broadest and covering the greater part of the metepimeron, the anterior end slightly narrower, the black stripe nearly always with a linear longitudinal yellow spot at its upper part close to the insertions of the wings, but this occasionally absent on one or both sides. *Wings* hyaline; pterostigma black, covering 5-6 cells, not robustly braced; nodal index variable, in two specimens $\frac{10\text{–}15}{12\text{–}10}\Big|\frac{16\text{–}11}{11\text{–}10}$, $\frac{11\text{–}14}{12\text{–}10}\Big|\frac{14\text{–}10}{10\text{–}12}$; 2 to 3 rows of cells between *Rii* and *IRii*. *Legs* black, without markings. *Abdomen* black, marked with bright citron-yellow as follows:—

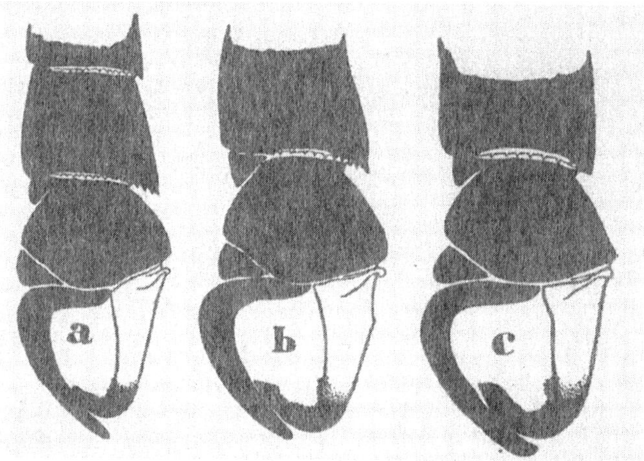

Fig. 83.—Anal appendages of (a) *Lamelligomphus nilgiriensis* (Fraser), male; (b) *Lamelligomphus nilgiriensis annaimallaicus* Fraser, male; (c) *Lamelligomphus biforceps* (Selys), male. All viewed from the left side. (Camera lucida sketches drawn to the same scale.)

Segment 1 with a triangular apical spot on the dorsum and a large apico-lateral spot; 2 with a dorsal longitudinal stripe broadening at the middle, tapering at the apex, the oreillets and an apico-lateral spot varying in size; 3 with a large baso-dorsal spot deeply cleft behind by an invasion of the black on dorsal carina; 4 to 6 with two dorsal triangular spots situated close to the base; 7 with nearly the basal half; 8 with a basal spot low down on the sides, this sometimes very minute or rarely absent; when very large accompanied by a smaller apico-lateral spot; 9 and 10 unmarked. (In one specimen the dorsal spots on 4 to 6 are almost obsolete and that on 7 is cut into two by the black of dorsal carina.) *Anal appendages* (fig. 83, a) black, the outer and upper surfaces of the superiors bright citron-yellow to nearly as far as the apex (in one specimen the yellow is restricted to a short linear streak on the basal half of the outer side only). In shape very similar to those of *L. biforceps* and *L. risi*. *Genitalia* scarcely differing from those of *L. risi*, the outer hamules, however, much longer and tapering, projecting very prominently from the genital sac.

Female.—Abdomen 36 mm. Hind-wing 33 mm.

Very similar to the male, abdomen stout and cylindrical. Markings differing as follows:—Spots on labrum smaller; band on frons cut into two large oval spots by the black at base of frons joining up with that on front; antehumeral stripe separated from the mesothoracic collar (but this may be variable); first lateral yellow stripe on thorax with its upper part isolated; the yellow markings on sides of abdominal segment 2 confluent, basal spots on 3 and 7 separated by the black on dorsal carina, baso-lateral spots on 8 very minute and the apical spot never present. Occiput armed with a pair of very long robust spines at its centre, the apices of which curl outwards. *Wings* enfumed and distinctly saffronated at the base. *Anal appendages* rather longer than segment 10, tapering, black with a bright yellow tip, a long triangular protuberance between them; 7 to 10 progressively shortening. Vulvar scale half the length of segment 9, cleft to its base so as to form two small triangular leaf-like processes which project backwards and somewhat downwards so as to be easily visible in profile.

Larva.—Total length 23 mm. Length of hind femur 6 mm. Greatest breadth (at about centre of abdomen) 9 mm.

Head moderately broad and quadrate, antennæ with basal segment small and cylindrical, third broad, flattened, triangular, fourth rudimentary, exists as a tiny spine at inner lateral angle of third. A duplicated tubercle on the prothorax. Wing-sheaths broad, extending to segment 6. Abdomen depressed, strongly keeled from segments 4 to 8, which bear robust spines on the carina, 7 to 10 with stout lateral spines. Mask very short, extending to base of first pair of legs, middle lobe rounded, fringed with rather long stiff brissæ, mentum angulated, the whole mask nearly quadrate.

Distribution.—SOUTH INDIA: Nilgiri-Wynaad, Coorg, South Kanara, Annaimallai, Palni, and Travancore Hills, from the end of April to the end of June.

The insect, which is very local, frequents shady mountain streams, generally those with clean gravelly bottoms, and is found settled on rocks or twigs in mid-stream. When disturbed it immediately rises perpendicularly to trees overhanging the stream. In Coorg it prefers streams almost entirely hidden and closed in by overhanging cane-brakes, where it may be found settled on rocks or on the gravelly beach, or occasionally hawking to and fro over runnels or rapids, to which places the female usually resorts to lay her eggs. Whilst ovipositing, the female hovers some 2 feet or less over the surface of the water and drops her eggs perpendicularly into the swirling stream.

L. nilgiriensis is easily distinguished from *L. acinaces*, which occurs in the same districts, by the shape of its anal appendages. These two species stand somewhat apart from *L. biforceps* and *risi* by their relatively smaller size, this being due to the shorter abdomen. *L. nilgiriensis* differs from *L. biforceps* in the absence of a humeral stripe and of mid-dorsal spots on segments 3 to 6 of the abdomen; from *L. risi* in its much smaller size, in its confluent antehumeral stripes, and in the absence of a vestigial humeral spot, etc. Its female differs from all others in the presence of two robust occipital spines.

Type in the British Museum.

289. **Lamelligomphus nilgiriensis annamallaicus**, new race.

Specimens of *L. nilgiriensis* found to the south of the Palghat Gap are very much larger than those from Coorg and the Nilgiris; the anal appendages also differ slightly in shape (fig. 83, **b**).

Type in my own collection from Mudis Hills, S. India.

290. **Lamelligomphus malabarensis** (Fraser).

Onychogomphus biforceps Fraser (nec Selys), Rec. Ind. Mus. vol. xxiv, pp. 424, 425, pl. xi, fig. 10 (1922).
Lamellogomphus malabarensis Fraser, J. Bombay Nat. Hist. Soc. vol. xxix, pp. 990, 991 (1924); Laidlaw, Trans. Ent. Soc. Lond. vol. lxxviii, p. 193 (1930); Needham, Rec. Ind. Mus. vol. xxxiv, p. 224 (1932).
Lamellogomphus malabaricus Fraser, Rec. Ind. Mus. vol. xxvi, pp. 427, 477 (1924); id., ibid. vol. xxxiii, p. 448 (1931).

Male unknown.

Female.—Abdomen 42 mm. Hind-wing 35 mm.

Head: labium pale yellow; labrum yellow, broadly edged with black, with a prolongation of this colour from the base joining the anterior black border; anteclypeus yellow, postclypeus black, with a large spot on each side close against the eyes; frons black traversed by a broad yellow stripe on the crest, constricted at its middle by an approximation of the black; vertex and occiput black, the latter with a yellow spot at its middle which is raised into a small tubercle. *Prothorax* black with a large yellow spot on each side. *Thorax* black, marked with greenish-yellow as follows:—A mesothoracic collar narrowly interrupted in the middle line; oblique antehumeral stripes falling well short of the mesothoracic collar; no vestige of a humeral stripe; sides greenish-yellow, traversed by a broad black stripe which includes a yellow stripe interrupted above. *Legs*: femora yellow mottled with black, hind femora with a row of 9 or 10 closely-set very short, very robust black spines. *Wings* hyaline; pterostigma black, covering five cells, well braced; only 1 row of cells between *Rii* and *IRii* at level of outer end of pterostigma; nodal index $\frac{10\text{–}16}{9\text{–}11}\left|\frac{17\text{–}10}{11\text{–}11}\right.$. *Abdomen* black, marked with yellow as follows:—Segment 1 with a dorsal spot and its sides broadly; 2 with a dorsal stripe, bilobed and extending from base to apex and its sides very broadly; 3 with a broad basal ring and a spot situated on the middle of the mid-dorsal carina; 4 to 6 with subdorsal basal spots confluent across the carina, and mid-dorsal spots as on segment 3, but progressively smaller from 4 to 6, on 6 almost obsolete; 7 with the basal half yellow; 8 with a large baso-lateral spot; 9 and 10 unmarked. *Anal appendages* yellow, small and pointed, the intermediate process also yellow. Vulvar scale small, deeply cleft to its base into two small triangular leaf-like processes; a shallow depression on segment 9 beneath the vulvar scale very similar to that in *M. lineatus*.

Distribution.—MALABAR, a single specimen taken at Palghat by Mr. T. N. Hearsey, 16. vi. 21.

From *L. nilgiriensis* this species is distinguished by the yellow occiput without spines and by the mid-dorsal spots on segments 3 to 6; the latter character also serves to separate it from *L. acinaces*, an additional point of difference being the presence of a yellow stripe traversing the medio-lateral black stripe of the thorax; from *L. biforceps* the entire absence of the humeral stripe will at once distinguish it; lastly it differs from *L. risi* in the presence of mid-dorsal yellow spots on segments 3 to 6 and of a lateral spot on segment 8 (unmarked in *L. risi*), and in segments 9 and 10 being without yellow markings.

Type in Author's collection.

291. **Lamelligomphus acinaces** (Laidlaw).

Onychogomphus acinaces Laidlaw, Rec. Ind. Mus. vol. xxiv, pp. 407–408, text-fig. 20 (1922).

Lamellogomphus biforceps acinaces Fraser, J. Bombay Nat. Hist. Soc. vol. xxix, pp. 65, 332 (1923).

Lamellogomphus acinaces Fraser, J. Bombay Nat. Hist. Soc. vol. xxix, pp. 988–990, pl. i, fig. 2 (1924); id., Rec. Ind. Mus. vol. xxvi, pp. 427, 477 (1924); Laidlaw, Trans. Ent. Soc. Lond. vol. lxxviii, p. 193 (1930); Fraser, Rec. Ind. Mus. vol. xxxiii, p. 447 (1931); Needham, Rec. Ind. Mus. vol. xxxiv, p. 224 (1932).

Male.—Abdomen (with appendages) 39 mm. Hind-wing 30 mm.

Head: labium black, yellow at the base; labrum black, marked with a pair of transversely oval greenish-yellow spots; anteclypeus yellow, postclypeus black; frons greenish-yellow above, black in front, the base narrowly black, this colour sending a prolongation forward into the sulcus which meets the black on the front margin and cuts the greenish-yellow area into two oval spots; vertex and occiput black. (In some specimens there is a small yellow spot behind the occiput as in *L. nilgiriensis*.) *Prothorax* entirely black. *Thorax* black, marked with greenish-yellow as follows:—The lower part of the dorsal carina (in about half of the speciments examined); a mesothoracic collar, complete where the middle line exists but broken when this is absent; an oblique antehumeral stripe not connected with the collar, squared above but *tapering to a point below* (no trace of a humeral stripe); sides greenish-yellow with a broad median black stripe, usually marked above near the insertions of the wings by a small linear yellow spot. *Legs* black, unmarked except for a broad greenish-yellow stripe on the flexor surface of the anterior femora; armature similar to that in *L. nilgiriensis*. *Wings* hyaline, rays of yellow tinting in the subcostal and cubital spaces; pterostigma black, covering 5 to 5½ cells, braced; usually 2 but sometimes 3 rows of cells between *Rii* and *IRii*; nodal index $\frac{10\text{–}16}{11\text{–}10}\Big|\frac{15\text{–}11}{11\text{–}11}$, $\frac{12\text{–}15}{13\text{–}10}\Big|\frac{16\text{–}12}{10\text{–}12}$. *Abdomen* black, marked with greenish-yellow as follows:—Segment 1 with an apical triangular spot, its base resting on the apical border and continuous with a mid-dorsal stripe on 2 which expands medially and tapers apically, also with an apical lateral spot; 2 with two large yellow lateral spots, the proximal including the oreillet and its surrounding area; 3 with a large basal spot nearly cut in two by an invasion of black along the dorsal carina; 4 to 6 with smaller spots nearly or quite cut in two; 7 with its basal half yellow; 8 with only a basal spot on the sides very variable in size, usually quadrate and acutely indented on its apical border; 9 and 10 unmarked. *Anal appendages* black; superiors with the upper and outer surfaces bright yellow as far as the apices, in some specimens entirely yellow and in such the upper surface of the inferior is of the same colour. Differing entirely in shape from those of all other species, the superiors truncate, tapering sinuously backwards and sloping slightly downwards towards the apices, equal in length to the two last segments of the abdomen; inferior slightly longer, divided into two closely parallel branches almost as far as its base, the distal halves curved gently upwards. *Genitalia* similar to those of *L. nilgiriensis*, the inner hamules much stouter and more robust, the outer shorter and stouter, the pencils of hairs at their apices very long and prominent.

Female.—Abdomen 39–40 mm. Hind-wing 32–35 mm.

Abdomen tumid at base, stout, parallel-sided, and cylindrical as far as the tip ; black, marked exactly as in the male except that the whole side of the second segment is bright yellow, this in some specimens confluent with the broad mid-dorsal stripe ; segment 8 unmarked. The dorsal carina on segment 7 usually finely black, so that the basal spot is cut in two. *Anal appendages* yellow, very short and conical. Occiput fringed with long black hairs, slightly notched in the middle, without spines. Legs as in male, but the hind femora furnished with a row of very long, very robust, very widely-spaced spines as in the female of *L. nilgiriensis*.

Distribution.—Coorg, North and South Kanara. Dr. S. Kemp took many specimens on streams flowing down the Mangalore Ghat from Coorg to Kanara. Mr. C. Souter, Commissioner of Coorg, found it swarming at Bhagmandala, Coorg, and wrote: "After having taken about 50 males I grew tired of taking more and contented myself with capturing females, of which I was fortunate enough to take six out of 10 specimens seen. All were hovering over a deep pool formed by damming up the river below, and were busy ovipositing by dropping their eggs plumb into the stream."

Found in company with *L. nilgiriensis*. Easily distinguished from all other species of the genus by its abnormally-shaped anal appendages.

The species is very closely related to *L. nilgiriensis*, and apart from the appendages the only reliable character by which to distinguish them is the pointed lower end of the antehumeral stripe in *L. acinaces* and the absence of occipital spines in the female. *L. acinaces* is a far more static insect than *L. nilgiriensis*, which, as has been noted above, exhibits considerable variation.

The *type*, in the Indian Museum, was taken by Dr S. Kemp in North Kanara (probably October 1916). Specimens in the Laidlaw, Morton, Author's, and British Museum collections.

Genus **NEPOGOMPHUS**, gen. nov. (Fig. 84.)

Onychogomphus (pars) Selys, Bull Acad Belg vol xxi, pt 2, p 30 (1854); id., Mon. Gomph. p. 15 (1857); Kirby, Cat. Odon. p. 57 (1890); Williamson, Proc. U.S. Nat. Mus. vol. xxxiii, pp. 310, 311 (1907); Laidlaw, Rec. Ind. Mus. vol. xxiv, pp. 408, 409 (1922); id., Trans. Ent. Soc. Lond. vol. lxxviii, p. 195 (1930).

A genus of very small dragonflies ; colour black, marked sparingly with greenish- or citron-yellow.

Head very large for the size of the insect, frons moderately angulated ; face oblique ; occiput very small, its posterior border simple, concave. Wings: reticulation close; tornus rather strongly angulated; base of hind-wing deeply excavated, very oblique; anal triangle 3-celled; arc situated between the first and second or opposite the second antenodal nervure; 2 transverse nervures between the sectors of arc from arc to bifurcation of *Rs* in fore-wing, only 1 in hind-wing; one row of postanal cells with a few double cells in fore-wings, 3 to 4 rows in hind-wings; a rudimentary anal loop present, made up of the first postanal cell split into two cells; no incomplete basal antenodal nervures present; nodal index moderately high; primary antenodal nervures the first and the fifth; discoidal cells entire, that of fore-wing with basal and distal sides equal and nearly half as long again as the costal, the

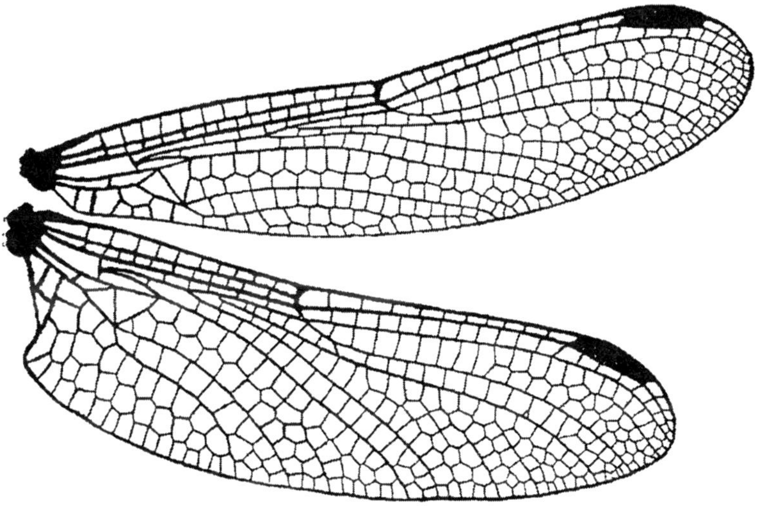

Fig. 84.—Wings of *Nepogomphus modestus* (Selys), male.

distal side angulated, that of hind-wing not elongated, the distal side slightly longer than the costal and the costal a little longer than the basal, this cell sometimes connected to the lower sector of arc by a short but definite stalk as in *Merogomphus*; pterostigma of moderate length, swollen at its middle, nearly as long as one-third the distance from node to proximal end of pterostigma, braced; *IA* in fore-wing not pectinated, only 2 rows of cells between this nervure and the wing-border; *Cuii* and *IA* in hind-wing divergent at wing-border; discoidal field with only 2 rows of cells nearly to wing-border; only a single cubital nervure in all wings; subtrigones and hypertrigones all entire. Legs short, hind femora extending only as far as the apical border of segment 1 and furnished with a group of very small, numerous, and closely-set spines, which near the distal end of limb become arranged into two rows; hind tibial spines moderately long and slim. Abdomen of male very tumid at base and very slim and cylindrical from segment 3 to the base of 7, from which point the abdomen becomes greatly expanded again, especially the apical end of segment 8. Anal appendages of male: superiors forcipate, conical, the apices curling evenly downward, nearly double the length of segment 10; inferior bifid almost to its base, the two branches very closely apposed and extending and curling upwards to beyond the apices of the superiors. Genitalia of male: lamina projecting, strongly and evenly arched; anterior and posterior hamules very similar, rather short, compressed processes ending in an acute hook-like point; the posteriors nearly double the width of the anteriors; lobe scrotal-shaped, of comparatively enormous size, and closely resembling the same structure in *Cyclogomphus*, but deep black in colour.

Genotype, *Onychogomphus modestus* Selys.

Distribution.—ASSAM, BENGAL, and UPPER BURMA. Two species are known from within our limits which differ in markings and the shape of the anal appendages. They breed in montane streams and may be found resting on rocks in mid-stream or resting on foliage in the neighbourhood. Larvæ unknown.

The genotype was considered by Selys to be closely related to *Onychogomphus saundersi*, but it and *N. walli* are only half the size of *O. saundersi* and differ in the venation and genitalia. Thus *IA* in the fore-wing is not pectinate and encloses a maximum of 2 rows of cells between itself and the border of the wing, *Cuii* and *IA* in the hind-wing are markedly divaricate at the border of the wing, the discoidal field is continued as 2 rows of cells almost to the border of the wing, the anal triangle has only 3 cells, the discoidal cell in the hind-wing is not as elongate as in *O. saundersi*, the anal appendages resemble those of *Lamelligomphus* more than *Onychogomphus*, and, lastly, the lobe of the genitalia is quite unlike anything found in the latter genus, being more akin to that found in the genus *Cyclogomphus*. These two species are among the smallest known in the family GOMPHIDÆ.

Key to Species of Nepogomphus.

Branches of inferior anal appendage with a robust superior subapical spine .... *modestus* (Selys), [p. 285.

Branches of inferior anal appendage without a superior subapical spine .... *walli* (Fras.), p. 286.

292. **Nepogomphus modestus** (Selys).

Onychogomphus modestus Selys, Bull. Acad. Belg. (2) vol. xlvi, p. 423 (1878); id., Ann. Soc. Ent. Belg. vol. xxxviii, p. 168 (1894); Williamson, Proc. U.S. Nat. Mus. vol. xxxiii, pp. 310, 311 (1907); Laidlaw, Rec. Ind. Mus. vol. xxiv, pp. 371, 408, 409 (1922); Fraser, J. Bombay Nat. Hist. Soc. vol. xxix, pp. 65, 333 (1923); id., ibid. vol. xxx, pp. 108, 109 (1924); Ris, Zool. Meded. Leiden, vol. x, pp. 30, 31, 46 (1927); Laidlaw, Trans. Ent. Soc. Lond. vol. lxxviii, p. 195, text-figs. 30, 31, 46 (1930); Needham, Rec. Ind. Mus. vol. xxxiv, p. 223 (1932).

Lindenia modesta Kirby, Cat. Odon. p. 60 (1890).

Onychogomphus diminutivus Fraser, J. Bombay Nat. Hist. Soc. vol. xxx, pp. 109, 110, pl. i, fig. 1 (1924); Ris, Zool. Meded. Leiden, vol. x, p. 46 (1927); Laidlaw, Trans. Ent. Soc. Lond. vol. lxxviii, p. 195 (1930); Needham, Rec. Ind. Mus. vol. xxxiv, p. 222 (1932).

Male.—Abdomen 27–29 mm. Hind-wing 23–25 mm.

Head: labium cinereous; labrum black with a yellow basal spot of variable size on each side; bases of mandibles yellow; anteclypeus black; postclypeus black or obscurely brownish, with or without a yellow spot on each side; rest of head black save for a transverse yellow stripe on the crest of frons, which is invaded at the middle by the black at base of frons; eyes green during life. *Prothorax* black. *Thorax* black, with a narrowly interrupted yellow mesothoracic collar, and oblique yellow antehumeral stripes well separated or confluent at a point with the mesothoracic collar; laterally broadly yellow, with the two sutures mapped out in black, these two stripes united in the middle by a short black bridge. *Legs* black, fore femora on the inner side and middle pair at the tips yellow. *Wings* hyaline; pterostigma black or blackish-brown, poorly braced, covering 3 to 4 cells; 1 or 2 rows of postanal cells in fore-wing, 4 rows in hind-wing; anal triangle 3-celled: nodal index of two specimens $\frac{10\text{–}13 \mid 14\text{–}10}{11\text{–}10 \mid 10\text{–}10}$, $\frac{8\text{–}14 \mid 13\text{–}9}{8\text{–}10 \mid 9\text{–}8}$. *Abdomen* black, marked with yellow as follows:—Segment 1 broadly on the sides; 2 on the sides (including the oreillets) but not at the base, and with a lanceolate mid-dorsal stripe which does not quite extend to the apical border; 3 to 6 with narrow basal rings; 7 on its basal half, but not laterally beyond the subdorsum; remaining segments unmarked. *Anal appendages* (fig. 85, **a**): superiors bright yellow, changing to brownish-black towards the apices; inferior dark reddish-brown, cleft deeply into two branches which are closely apposed throughout and which, just before the apex above, present a very robust tooth. *Genitalia*: lamina hood-like, broadly arched, fringed with long black hairs; anterior hamules short, narrow, notched at apex, where they end in an outer spine; posterior hamules longer and broader, ending in a point which is furnished with a pencil of hairs; lobe very large and globular, resembling that of a *Cyclogomphus*, very prominent in profile.

Female.—Abdomen 27 mm. Hind-wing 24 mm.

Closely similar to the male and exactly so as regards the markings, save that the ring on segment 7 covers only about the basal third. Wings palely tinted with yellow at the bases; pterostigma reddish-brown between black nervures, well braced, covering 4 to 5 cells; nodal index $\frac{10\text{–}15}{11\text{–}11}\Big|\frac{15\text{–}9}{12\text{–}11}$; *anal appendages* shortly conical, bright yellow; vulvar scale short, triangular, deeply emarginate.

Distribution.—BENGAL, ASSAM, BURMA, and Sumatra; probably widely but sparsely distributed throughout Malaysia. I possess two males from Shillong, Assam, and also the *type* of *O. diminutivus*, which is undoubtedly synonymous with *N. modestus*, and which comes from the Naga Hills, Assam. The female described above is a specimen in my own collection from Hasimara, Duars, Bengal, about which there can be no doubt of the correct relationship as it is exactly similar to the male.

Selys failed to notice the prominent subapical spines on the inferior appendage; I have seen another male from Bengal in the McLachlan collection, named by Selys as *O. modestus*, which has the spines well developed, and although I have not been able to examine the type I feel sure the omission in his description was accidental. The female described, with some doubt, as that of *O. modestus* by Selys is more probably that of *N. walli*, which also has oval medial spots on the abdominal segments. A teneral female of *N. modestus* which I possess from Hasimara, Bengal, is without these spots, and thus resembles the mature form from the same locality.

Type, a male in the Selys collection from Bengal.

293. **Nepogomphus walli** (Fraser). (Fig. 85, **b**.)

Onychogomphus walli Fraser, J. Bombay Nat. Hist. Soc. vol. xxx, p. 109 (1924); Laidlaw, Trans. Ent. Soc. Lond. vol. lxxviii, p. 196 (1930); Needham, Rec. Ind. Mus. vol. xxxiv, p. 222 (1932).

Male.—Abdomen 27 mm. Hind-wing 23 mm.

Head: labium pale yellow; labrum black, with a large oval citron-yellow spot on each side well separated by the black; anteclypeus black, with a transverse yellow spot at its base; postclypeus black, with a small yellow spot on each side against the eyes; bases of mandibles yellow; frons yellow above, this colour invaded posteriorly by the black; eyes green during life; rest of head black; occiput concave, fringed with very long blackish-brown hairs. *Prothorax* black, the lower part of its sides and a small point on each side of the middle lobe citron-yellow. *Thorax* black on dorsum, marked with yellow as follows:—A narrow uninterrupted mesothoracic collar; oblique antehumeral stripes which taper below and finally become confluent with the mesothoracic collar; laterally broadly citron-yellow, with a narrow black stripe on each suture, stripes not joined at the middle as in *N. modestus*; underside yellow. *Legs* black, the fore-femora striped with yellow within. *Wings* hyaline, bases tinted with yellow; pterostigma dark ochreous, framed in black nervures, well braced, covering 4 cells; distal side of discoidal cells angulated as in *N. modestus*; 4 cells in the anal triangle; one row of cells or a few doubled cells in the anal field of fore-wings; nodal index $\frac{10\text{–}12}{9\text{–}9} \Big| \frac{11\text{–}10}{10\text{–}9}$. *Abdomen* black, marked with citron-yellow as follows:—Segment 1 on the sides broadly, the yellow area extending up along the apical border

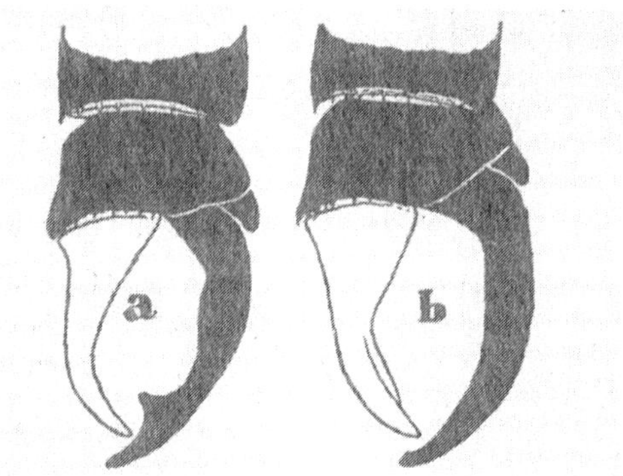

Fig. 85.—Anal appendages of (a) *Nepogomphus modestus* (Selys), male; (b) *Nepogomphus walli* (Fraser), male. Right lateral views.

and becoming nearly confluent over the dorsum; 2 with the sides broadly, including the whole of the oreillets, also a truncate subapical stripe, and a bilobate mid-dorsal stripe, the posterior lobe of which is tapered but does not quite extend to the apical border; 3 with a narrow basal ring, incomplete below, and a long oval mid-dorsal spot situated midway between the ring and the apical end of segment; 4 to 6 marked similarly, but the mid-dorsal spot becoming progressively smaller until nearly obsolete on 6; 7 with a broad basal ring which tapers away to nearly as far as the apical end of segment; remaining segments unmarked. *Anal appendages* (fig. 85, b): superiors yellow; inferior without any vestige of a subapical spine above, deeply cleft for its apical two-thirds. *Genitalia* closely similar to those of *N. modestus*, the lamina fringed with coarse black hairs which show prominently in profile, the anterior hamules not emarginate at apex.

Female.—Abdomen 25 mm. Hind-wing 21 mm.

Closely similar to the male in colour and markings; differs as follows:—The yellow on crest of frons interrupted; antehumeral stripes not confluent with the mesothoracic collar; mid-dorsal oval spots on abdomen only present on segments 3 and 4, and ring on segment 7 of less extent. Anal appendages pale yellow, shortly conical; vulvar scale short, triangular, deeply emarginate at apex.

Distribution.—Maymyo, UPPER BURMA, during June.

As mentioned above, the female described by Selys as that of *O. modestus* is really that of *N. walli*. This species differs from *N. modestus* not only in the shape of the inferior anal appendage of the male, which has no subapical spine above, but in the presence of mid-dorsal oval spots on abdominal segments 3 to 6 in both sexes, these spots being absent in *N. modestus*; also in the separate lateral black stripes on the thorax.

Type, an adult male, and allotype (rather teneral) from the same locality, in my collection.

Genus **STYLOGOMPHUS** Fraser. (Fig. 86.)

Stylogomphus Fraser, Mem. Dept. Agric. India (Ent.), vol. vii, no. 7, pp. 69, 70 (1922); id., J. Bombay Nat. Hist. Soc. vol. xxix, pp. 64, 332 (1923); id., ibid. vol. xxx, p. 397 (1925); Laidlaw, Trans. Ent. Soc. Lond. vol. lxxviii, p. 185 (1930); Needham, Rec. Ind. Mus. vol. xxxiv, pp. 218, 222 (1932).

Size very small; colour glossy black, marked with bright greenish-yellow.

Head comparatively large; frons well angulated; occiput simple, posterior border straight, dorsum deeply grooved. Wings: reticulation close; tornus angulated; base of hind-wing deeply excavate; anal triangle 3-celled; arc situated between the first and second or more commonly opposite the second antenodal nervure; 2 or occasionally 3 nervures between the sectors of arc from arc to bifurcation of *Rs* in fore-wings, only 1 in hind-wings; a single row, with occasional double cells, of postanal cells in fore-wings, 4 to 5 rows in hind-wings; anal loop absent, the first postanal cell in hind-wing extending back only as far as middle of base of subtrigone; no incomplete basal antenodal nervures; nodal index moderately high; primary antenodals the first and the fifth or sixth; discoidal cells all entire, that of fore-wing subequilateral, the distal side slightly the longer and angulated at its middle; that of hind-wing with the distal side twice the length of basal and slightly longer than the costal, elongate in the length of wing ; pterostigma small and swollen, barely or less than one-third the length of distance from node to proximal end of pterostigma, strongly braced ; *IA* in fore-wing only pectinate at its distal end ; *Cuii* and *IA* in the hind-wing divaricate at wing-border ; only a single cubital nervure in all wings ; subtrigones and hypertrigones entire in all wings, the subtrigone of fore-wing occasionally four-sided, its costal and distal sides failing to meet at the discoidal cell. Legs moderately long, hind femora extending as far as the apical border of segment 1 and furnished with 2 rows of short, robust spines which are gradually more widely spaced and of more robust build towards the distal end of limb ; tibial spines slim, moderately numerous and rather widely spaced. Abdomen tumid at base, very narrow and cylindrical from segment 3

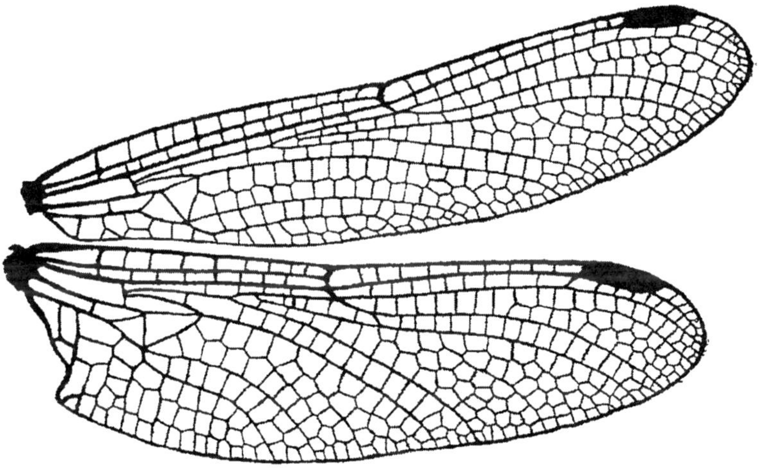

Fig. 86.—Wings of *Stylogomphus inglisi* Fraser, male.

to the middle of 7, thence dilated considerably. Anal appendages : superiors rather longer than segment 10, broad at base, tapering to a fine apex and with two obtuse spines on the outer border near base ; inferior closely similar to that of *Leptogomphus*, triangular, cleft for only half its length, its two branches thick and parallel, the interposing notch very narrow. Genitalia : lamina arched, emarginate, projecting but slightly ; anterior hamules long slender curved processes (like a rodent's incisors) ; posterior hamules broader and spatulate ; lobe purse-shaped, its border finely emarginate, but slightly prominent in profile.

Genotype, *Stylogomphus inglisi* Fraser.

Distribution.—Bengal only.

The writer has observed the single species of this genus settled on rocks in the beds of small mountain streams, but it was very shy and arose swiftly to trees in the neighbourhood, so that it is probably arboreal by nature, frequenting streams only for the purpose of pairing; the larva is unknown.

The genus is probably closely related to the genus *Leptogomphus*. The anal appendages resemble those of *Merogomphus*, but the small size of *S. inglisi* and its venation will prevent any confusion.

294. Stylogomphus inglisi Fraser. (Pl. III, fig. 1; text-fig. 87.)

Stylogomphus inglisi Fraser, Mem. Dept. Agric. India (Ent.), vol. vii, no. 7, pp. 70, 71, pl. vii, figs. 3, 3 *a*, & 3 *b* (1922); id., J. Bombay Nat. Hist. Soc. vol. xxix, pp. 64, 332 (1923); id., ibid. vol. xxx, p. 112, text-fig. 2, viii, ix (1924); id., ibid. vol. xxx, pp. 397, 398 (1925); Laidlaw, Trans Ent. Soc. Lond. vol. lxxviii, p. 185 (1930); Needham, Rec. Ind. Mus. vol. xxxiv, p. 222 (1932).

Male.—Abdomen 25·5 mm. Hind-wing 21 mm.

Head: labium yellowish; labrum black, marked laterally with an oval yellow spot; bases of mandibles yellow; anteclypeus black, yellow medially, postclypeus black; frons black, its crest traversed with yellow; occiput black, simple. *Prothorax* black, posterior lobe margined with yellow. *Thorax* black on dorsum, marked with bright yellow as follows:—A complete mesothoracic collar, oblique antehumeral oval spots well separated from the alar sinus above and mesothoracic collar below. Laterally yellow, the sutures mapped out in black, the two fine stripes connected by a short stripe at their middle. Tergum spotted with yellow. *Legs* black, coxæ and trochanters yellow. *Wings* hyaline, pale saffron at base; pterostigma black, covering $2\frac{1}{2}$ to 4 cells, strongly braced; nodal index in three specimens $\frac{9\text{–}12}{9\text{–}9}\Big|\frac{11\text{–}9}{7\text{–}9}$, $\frac{8\text{–}11}{8\text{–}9}\Big|\frac{11\text{–}8}{8\text{–}7}$, $\frac{8\text{–}12}{8\text{–}8}\Big|\frac{12\text{–}10}{8\text{–}10}$; anal triangle 3-celled. *Abdomen* black, marked with yellow as follows:—Segment 1 with the sides broadly and a broad dorsal spot; 2 with the sides broadly, including the oreillets and a mid-dorsal fusiform stripe; 3 with its base laterally and a mid-dorsal oval basal spot; segments 4 to 6 with narrow basal rings slightly interrupted on the dorsum; remaining segments black. *Anal appendages* (fig. 87): superiors pale yellow; inferior black. *Genitalia*: lamina arched, hood-like; anterior hamules very narrow and very long, strongly curled; posterior hamules equally broad, truncate at end and with a tuft of hairs projecting from the middle of the free border; lobe tumid, globular, emarginate at brim.

Female.—Abdomen 22 mm. Hind-wing 27 mm.

Larger and more robust than the male, and differing as follows :—Fore and middle femora striped with yellow; black stripes on sides of thorax not joined at the middle; abdominal segment 7 with a pair of basal dorsal spots ; wings deeply tinted with yellow at extreme base, this colour paling as far as the level of arc. Occiput broadly and shallowly

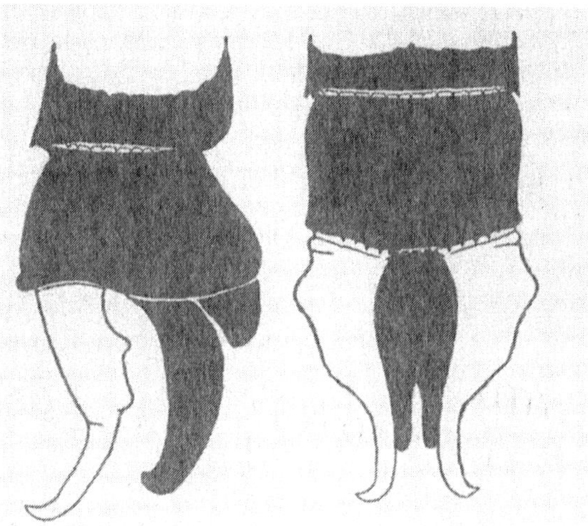

Fig. 87.—Anal appendages of *Stylogomphus inglisi* Fraser, male. Right lateral and dorsal views.

concave. *Anal appendages* pale yellow, shortly conical; vulvar scale short, triangular, emarginate at apex.

Distribution.—The *type* male in the British Museum is from the Teesta Valley, DARJEELING DISTRICT. A pair in the Darjeeling Museum and two other males in my collection are from Mungpoo, in the same district, taken during June.

Genus **MEGALOGOMPHUS** Campion. (Fig. 88.)

Heterogomphus Selys, Bull. Acad. Belg. vol. xxi, pt. 2, p. 27 (1854); id., Mon. Gomph. p. 94 (1857); Kirby, Cat. Odon. p. 57 (1890); Williamson, Proc. U.S. Nat. Mus. vol. xxxiii, pp. 275, 276, 315, 316 (1907); Laidlaw, Rec. Ind. Mus. vol. xxiv, p. 412 (1922); Fraser, J. Bombay Nat. Hist. Soc. vol. xxix, pp. 63, 331, 673, 674 (1923).

Megalogomphus Campion, Ann. Mag. Nat. Hist. (9) vol. xii, p. 668 (1923); Laidlaw, Trans. Ent. Soc. Lond. vol. lxxviii, p. 196 (1930); Needham, Rec. Ind. Mus. vol. xxxiv, pp. 217, 218, 221 (1932).

Allogomphus Needham, Zool. Sinica, ser. A, vol. xi, fasc. 1, pp. 19, 35 (1930).

Very large, robust dragonflies, usually glossy or mat black marked with bright citron-yellow.

Head of great size in proportion to that of the body, triangular; frons angulated; occiput low and simple in the male, sometimes spined in the female. Wings: reticulation very close; tornus markedly angulate; base of hind-wing rather deeply excavate; membrane narrow; anal triangle 4-celled, very rarely 3-celled; arc situated variably, usually between the first and third antenodal nervures; 2 to 3 nervures between the sectors of arc from arc to bifurcation of *Rs* in fore-wings, only 1 in hind-wing; 2 rows of postanal cells in fore-wings, 5 in hind-wings; rudimentary anal loop present, of two cells, which extends widely proximal to base of subtrigone; incomplete basal postcostal nervures very rarely present; nodal index very high; primary antenodal nervures the first and the seventh; discoidal cells of very similar shape in the fore- and hind-wings, but that of fore-wing more oblique,

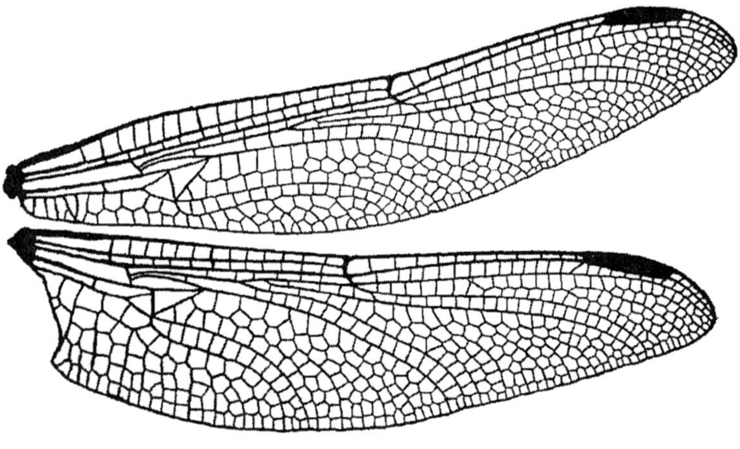

Fig. 88.—Wings of *Megalogomphus superbus* Fraser, male.

costal side much longer than basal, distal side slightly longer than costal especially in the fore-wings, entire (very rarely traversed once); pterostigma equal in length to one-fourth the distance from node to distal end of pterostigma, braced, very slightly expanded at the middle; *IA* in fore-wing markedly pectinate; *Cuii* and *IA* in hind-wing parallel almost as far as the termen; only 1 cubital nervure in all wings; hypertrigones and subtrigones all entire; *Riv*+*v* and *MA* undulated. Legs short, hind femora extending to hinder border of thorax only, furnished with a group of short numerous spines on the flexor surface in both sexes, but rather more robust in the female. Abdomen tumid at base, narrow and cylindrical from segment 3 to the base of 7, thereafter expanded, markedly so on segments 8 and 9 which have narrow lateral wing-like projections. Anal appendages: superiors twice the length of segment 10, projecting straight backwards, tapering finely to the end; inferior deeply bifid into two long narrow branches, which are themselves deeply bifid at the apex and slightly divaricate. Genitalia: lamina depressed, emarginate; anterior hamules narrow, long, their apices ending in a fine hooked spine; posterior hamules very similar, but stouter, and the hooked apices facing those of the anterior pairs; lobe moderately large and tumid.

Genotype, *Heterogomphus smithi* Selys.

Distribution.—The WESTERN GHATS, CEYLON, the HIMALAYAS, ASSAM, Indo-China, Java, Sumatra, and China.

The genus comprises some of the largest and most beautiful species of dragonflies, six occurring within our limits.

The larvæ have been described by Needham; they resemble those of *Gomphus* or *Burmagomphus*, but are, of course, much larger.

Key to Indian Species of Megalogomphus.

1.	Head entirely yellow except at back of eyes; abdomen yellow, with broad medial black rings on segments 2 to 7 and apical rings on 8 to 10	*flavicolor* (Fras.), [p. 303.
	Head black, marked with yellow or grass-green; abdomen black, marked with dorsal and basal yellow spots	2.
2.	Segments 8 to 10 unmarked; occiput with two spines on the posterior border	*bicornutus* (Fras.), [p. 303.
	Segments 8 and 9 with large baso-dorsal yellow spots; occiput without spines....	3.
3.	Abdomen black, with very broad dorsal yellow markings; the lateral black stripe on thorax very narrow, covering less than the posterior half of mesepimeron; mesothoracic collar not interrupted and very broadly confluent with the antehumeral stripes	*smithi* (Selys), [p. 294.
	Abdomen with very narrow dorso-medial markings on segments 1 to 6; the lateral black thoracic stripe very broad, covering the whole of mesepimeron except for some small included spots; mesothoracic collar more or less interrupted and not confluent with the antehumeral stripes or at a point only........................	4.
4.	A narrow humeral stripe present; abdominal segments 3 to 6 with the medial third bright reddish-brown	*superbus* Fras., [p. 300.
	Humeral stripe absent, but a small upper humeral spot often present; ground-colour of all abdominal segments, including 3 to 6, mat black	5.

5.	Occiput yellow; femora striped obliquely with yellow; abdominal segments 3 to 6 with a yellow dorsal stripe tapering from base to apex of segments	[p. 298. *ceylonicus* (Laid.).
	Occiput without yellow marks; femora entirely black; segments 3 to 6 with a chain of yellow dorsal spots	[p. 296. *hannyngtoni* (Fras.).

295. **Megalogomphus smithi** (Selys). (Pl. IV, fig. 3; text-fig. 89, b.)

Heterogomphus smithii Selys, Bull. Acad. Belg. vol. xxi, pt. 2, p. 29 (1854); id., Mon. Gomph. p. 97 (1857); id., Bull. Acad. Belg. (2) vol. xxxvi, p. 495 (1873); Kirby, Cat. Odon. p. 57 (1890); Foerster, Wien. Ent. Zeit. vol. xxi, p. 21 (1905); Williamson, Proc. U.S. Nat. Mus. vol. xxxiii, p. 315 (1907); Laidlaw, Rec. Ind. Mus. vol. xxiv, p. 412 (1922); Fraser, J. Bombay Nat. Hist. Soc. vol. xxix, pp. 63, 331 (1923).

Heterogomphus smithi Fraser, J. Bombay Nat. Hist. Soc. vol. xxix, pp. 677, 678 (1923).

Megalogomphus smithii Fraser, J. Bombay Nat. Hist. Soc. vol. xxx, pp. 50, 51, fig. B (1924); Needham, Rec. Ind. Mus. vol. xxxiv, p. 222 (1932).

Megalogomphus smithi Laidlaw, Trans. Ent. Soc. Lond. vol. lxxviii, p. 196 (1930).

Allogomphus smithii Needham, Zool. Sinica, ser. A, vol. xi, fasc. 1, pp. 35, 36 (1930).

Male.—Abdomen (with appendages) 52–55 mm. Hind-wing 43–44 mm.

Head: labium greenish-yellow, middle lobe bordered with black, lateral lobes black at the extreme base; labrum and face greenish-yellow, the former finely bordered with blackish-brown; postclypeus bordered with ochreous below; lower border of frons narrowly blackish-brown anteriorly; frons above sinuously and narrowly black; vertex black; occiput greenish-yellow above and behind and fringed with long coarse blackish-brown hairs; eyes bottle-green during life. *Prothorax* black, middle lobe with a greenish-yellow geminate spot on the dorsum and a larger pyriform spot on each side. *Thorax* black on dorsum, marked with a very broad unbroken mesothoracic collar which is broadly confluent with short, broad, oblique antehumeral stripes; a small yellow cordate upper humeral spot on the outer side of each of the former stripes; laterally broadly greenish-yellow, with a moderately broad black stripe over the postero-lateral suture which bifurcates below to enclose a small and a large yellow spot above and below the spiracle respectively; underside yellow, with a broad lyrate pattern of dark brown. *Legs* black, femora broadly bright yellow on the external surface. *Wings* hyaline; pterostigma dark brown, with a fine pale yellow border posteriorly and framed in black nervures, irregularly braced, covering 4 to 5 cells; 3 to 4 cells in the anal triangle;

anal loop of hind-wings well formed in some specimens, made up of 2 to 3 cells; nodal index $\dfrac{12\text{–}15 \mid 15\text{–}12}{11\text{–}12 \mid 14\text{–}14}$, $\dfrac{11\text{–}18 \mid 17\text{–}11}{13\text{–}13 \mid 13\text{–}10}$. *Abdomen* black, broadly marked with citron-yellow as follows:—Sides of segments 1 and 2 and the base of 3 broadly so; 1 with a broad dorsal marking; 2 with a broad trilobate mid-dorsal stripe extending from base to apex, the middle lobe the smaller and pointed outwards, the apical lobe scutellate; oreillets very large, yellow, with a few tiny spines on the free border; 3 to 6 broadly yellow on the dorsal basal two-thirds, the jugal suture, which is black, partially dividing the yellow into basal and apical areas; 7 with its basal two-thirds entirely ringed with yellow; 8 and 9 with lateral elongate spots, that

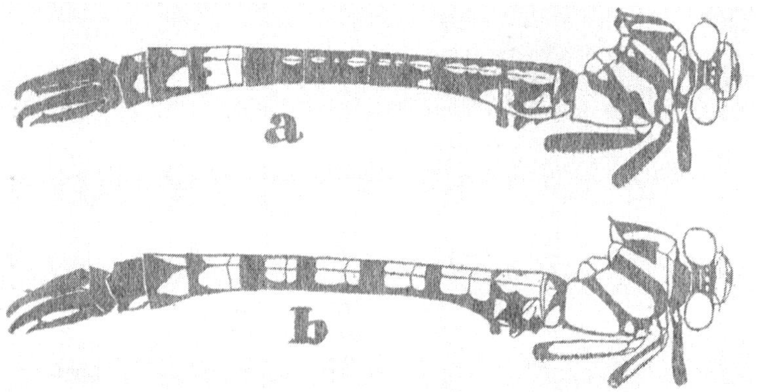

Fig. 89.—Body-markings of (a) *Megalogomphus hannyngtoni* (Fraser), male; (b) *Megalogomphus smithi* (Selys), male.

on the former segment interrupted apically; 10 unmarked. *Anal appendages* (fig. 89, *b*) blackish-brown.

Female.—Abdomen 50 mm. Hind-wing 50 mm.

Closely similar to the male, but more robust, with a shorter, stouter abdomen, which is of the same length as the wings. In the single specimen which I have seen the labium is entirely golden yellow, the occiput is fringed with very short hairs and has on each side a stout conical eminence almost amounting to a spine; the wings are tinted with pale yellow near the base, the anal loop being as highly developed as in the genus *Ophiogomphus* and formed of 3 cells; the pterostigma is longer, covering 6 to 7 cells, and poorly braced, the brace being situated slightly proximal to the proximal end of the pterostigma; nodal index $\dfrac{12\text{–}19 \mid 20\text{–}12}{13\text{–}14 \mid 13\text{–}12}$; the abdominal markings are similar to but broader than in the male, except on segment 7, where the yellow area is deeply invaded by the black on the mid-dorsal and ventral aspects. *Anal appendages* shortly conical. Vulvar scales short, about one-fourth the length of segment 9, triangular, narrowly emarginate at apex, and with two basal black scale-like processes superimposed upon them, basal to which is a transverse conical eminence.

Distribution.—BENGAL and ASSAM.

The species is easily determined by the broad extent of its yellow markings, excelled only by *M. flavicolor*. It is the only species in which the legs are black marked with yellow, *M. flavicolor* having the legs entirely yellow, whilst other species have them entirely black. In its colour and markings it appears most nearly related to *M. superbus*.

The *type* is a male from Sylhet, in the British Museum, and there is another male in the Darjeeling Museum from Sikkim. I possess a pair from Nowgong, Assam, and a male from the Duars, Bengal, all taken in April.

296. **Megalogomphus hannyngtoni** (Fraser). (Pl. IV, fig. 1; text-fig. 89, **a**.)

Heterogomphus hannyngtoni Fraser, J. Bombay Nat. Hist. Soc. vol. xxix, pp. 674–676, text-fig. 4, pl. i, figs. 1 & 1 *a* (1923).

Megalogomphus hannyngtoni Fraser, Rec. Ind. Mus. vol. xxvi, pp. 428, 478, 479 (1924); id., Mem. Dept. Agric. India, vol. viii, no. 8, pp. 79–81, text-figs. 1–3 (1924); id., J. Bombay Nat. Hist. Soc. vol. xxx, pp. 50, 51 (1924); Laidlaw, Trans. Ent. Soc. Lond. vol. lxxviii, p. 197 (1930); Fraser, Rec. Ind. Mus. vol. xxxiii, p. 448 (1931); Needham, Rec. Ind. Mus. vol. xxxiv, p. 222 (1932).

Male.—Abdomen (with appendages) 58 mm. Hind-wing 48 mm.

Head very large and massive, triangular; labium chrome-yellow; labrum bright greenish-yellow margined narrowly with black; face black, the lower epistome citron-yellow and an apple-green spot on either side against the eyes; frons apple-green above and in front with two small black points in the sulcus just in front of the vesicle; vertex and occiput black, the hinder border of latter raised and scale-like, with a small rounded tubercle on the superior surface, fringed with long black hairs. Eyes bottle-green. *Prothorax* black, with a postero-lateral yellow spot and two small spots on mid-dorsum. *Thorax* black; markings bright greenish-yellow (in some specimens bright citron-yellow above, changing to bright apple-green below) as follows:—The outer angles of the alar sinus, a broad subtriangular oblique dorsal stripe more or less separated from a mesothoracic collar which may be either entire or interrupted (in one specimen the separation between these two markings is bright reddish-brown), a rudimentary humeral stripe represented by a small upper spot; laterally two broad apple-green stripes separated by a broad black band, the upper part of which bears a small yellow spot, the posterior stripe covering the whole of metepimeron. *Wings* hyaline, long and rather broad, reticulation close ; pterostigma long, blackish-brown, braced ; *IRii* very indistinct, only 2 rows of cells between it and *Rii*. Only 1 cross-nervure between the sectors of arc in all wings ; nodal index $\frac{13\text{–}19 \mid 19\text{–}15}{14\text{–}14 \mid 14\text{–}14}$. *Legs* short and robust, black ; fore coxæ and trochanters yellow ; hind femora with two rows of short robust closely set black spines. *Abdomen* black, marked with bright citron-yellow as follows :—Segment 1 with a small apical dorsal triangle and the whole of the sides below ; 2 with a dorsal stripe, broad at the base and tapering to, but not reaching, the apex, laterally a broad stripe, broadest at the base and including the oreillet, interrupted just after the latter structure, lastly a narrow stripe bordering the genitalia ; 3 with a trilobed dorsal stripe and a lateral wedge-shaped spot at the base ; 4 and 5 with a chain of three dorsal spots, the apical one cordate and larger than the two basal ; 6 with a single basal dorsal spot ; 7 with rather more than the basal half yellow, the margins of the yellow area concave laterally ; 8 and 9 with small lateral basal triangular spots ; 10 entirely black. *Anal appendages* (fig. 89, a) black. *Genitalia* : lamina scuttle-shaped, its sides straight and at right angles to the dorsum ; internal hamules projecting from under the lamina, long curving robust hooks ; external hamules narrow triangular plates projecting perpendicularly from the genital orifice ; lobe with a broad black base, narrowing to a truncate neck which projects markedly from the genital sac and bifurcates at its apex, the bifurcations curling inwards.

Female.—Abdomen 56 mm. Hind-wing 55 mm.

Very similar to the male, but the ratio of wing-length to that of abdomen strikingly different. The wings are enormously lengthened and broadened (probably to give the insect a mechanical advantage for carrying the heavy weight of the very robust body ; apart from the abdomen being slightly shorter, the female is a much more robust and bulkier insect than the male). Labrum bright citron-yellow instead of apple-green ; frons with a black semicircle at the base of upper surface ; occiput flat, with a robust spine at either end against the eyes. The lateral spot of prothorax much larger than in the male. Metepimeron citron-yellow with a tinge of green at its centre ; markings of abdomen much broader and more crowded than in the male, the dorsal marking on segment 2 being trilobed, its basal part square, the median large and broadly oval, whilst the apical lobe is small and round, the lateral stripe on this segment is unbroken at its middle ; the dorsal marking on 3 is partially constricted to form four lobes, whilst laterally a broad stripe runs its entire length, broken slightly at the transverse suture ; 4 and 5 have an additional lateral basal spot ; 6 has this spot and another still larger following it ; 9 has a minute lateral basal spot. *Anal appendages* very small, conical, tapering black. Vulvar scale very small, deeply cleft, the apices of the two lobes thus formed curling strongly inward towards each other. Wings hyaline, nodal index similar to that of male.

Distribution.—SOUTH INDIA : submontane areas of Coorg and Vayitri, Malabar Wynaad. Moderately common on streams near Bhagmandala, Coorg, but rare in Malabar.

This species is found haunting the banks of jungle mountain streams, perching on twigs or more rarely settling on rocks. The male when settled bears a strong resemblance to *Ictinus*, but may be recognized by its abdomen, which is held stiffly and straight outward instead of curved, scimitar-like, as in *Ictinus*. Like the latter insect it rests with the head lowermost, the body inclined upwards. When disturbed it plunges downwards as it takes flight. It travels long distances up and down stream, but makes long rests, and if followed up is soon met with again. A female was taken in the act of ovipositing in the deep shade of a tree overhanging the stream at a point where the current was very swift and racing over a bed of pebbles. It was performing a series of figure-of-eight evolutions, striking the surface of the water at the waist of each figure-of-eight. This and the first males were taken on the Cauvery near Bhagamandala (where the river rises), two subsequent males being taken at Hallery, near Mercara, on a mountain stream which empties lower down into the Hatty River, a large tributary of the Cauvery. The streams here are not more than 10 feet across, and I have never seen the insect on the main river where it becomes much wider.

Type in the British Museum, specimens in my own collection and several other private collections.

297. **Megalogomphus ceylonicus** (Laidlaw).

Heterogomphus ceylonicus Laidlaw, Rec. Ind. Mus. vol. xxiv, p. 412, text-fig. 21 (1922); Fraser, J. Bombay Nat. Hist. Soc. vol. xxix, pp. 63, 331, 676, 677 (1923).
Heterogomphus sp. Laidlaw, Spolia Zeylanica, vol. xii, p. 342 (1924).
Megalogomphus ceylonicus Laidlaw, Trans. Ent. Soc. Lond. vol. lxxviii, p. 197 (1931); Needham, Rec. Ind. Mus. vol. xxxiv, p. 222 (1932); Fraser, Ceylon J. Sci., B, vol. xviii, p. 22, fig. 1 (1933).

Male.—Abdomen (with appendages) 47–48 mm. Hind-wing 41 mm.

Head: labium brownish-yellow; labrum yellow, with a broad frame of black from the base of which a prolongation nearly cuts the yellow into two spots; bases of mandibles yellow; anteclypeus and a lateral spot on each side of the postclypeus yellow; frons black in front, citron-yellow above where the base is narrowly black; vertex and occiput black, the latter with a median yellow spot; eyes bottle-green during life. *Prothorax* black, with a large median dorsal spot and a smaller geminate spot yellow. *Thorax* black, marked with citron-yellow as follows:—A broad mesothoracic collar narrowly interrupted in the middle line, broad oval oblique antehumeral stripes and a vestigial humeral spot. Laterally broadly yellow, with a broad median oblique black stripe. *Legs* black, coxæ and the extensor surfaces of femora yellow. *Wings* hyaline; pterostigma black, strongly braced, covering 5 to 6 cells; a vestigial anal loop of 2 cells present; nodal index $\frac{12\text{–}18}{11\text{–}13}\left|\frac{17\text{–}14}{13\text{–}12}\right.$. *Abdomen* black, marked with citron-yellow as follows:—Segment 1 broadly on sides and narrowly on apical margin; 2 with a broad trilobate mid-dorsal stripe broadest at the middle, and its sides including the oreillets broadly; 3 with a continuation of the yellow on its sides for about its basal third and a mid-dorsal stripe tapering from base to apex of segment; 4 to 6 with the mid-dorsal stripe only, this diminishing in length from segment to segment; 7 with its basal half yellow; 8 and 9 each with a baso-lateral spot, this largest on the latter segment; 10 unmarked. *Anal appendages* black, long and tapering, parallel or slightly convergent at apices; inferior deeply bifid, each branch nearly as long as the superior appendages and bearing a robust spine on the inner side of its apex.

Female.—Abdomen 49 mm. Hind-wing 46 mm.

Closely similar to the male but more robust, with shorter abdomen and greater expanse of wing. It differs in the following particulars:—The labium is yellow; the citron-yellow of the frons is narrowly interrupted above by a prolongation of the black at base; prothorax with an additional yellow linear spot on each side; the median oblique black stripe on sides of thorax with an upper and a lower yellow spot; all femora broadly yellow on the outer sides; wings broader and with closer reticulation; nodal index $\frac{11\text{–}18}{11\text{–}12}\left|\frac{18\text{–}10}{14\text{–}12}\right.$; abdomen with segments 4 to 6 bearing baso-lateral spots; 7 with its basal two-thirds yellow and rather more extensively so on the sides; 10 with a pair of tiny subdorsal spots; other markings similar to those of the male. Vulvar scales very short, deeply and narrowly cleft, the sides of the fissure prolonged into two long robust points.

Distribution.—Nalande and Balangoda, CEYLON.

M. ceylonicus is nearly related to *M. hannyngtoni* of South India, but is a much smaller insect and is differently marked. An exuvium of this insect in the Colombo Museum greatly resembles that of *M. hannyngtoni* except in its much smaller size.

The *type*, a male collected by Col. Yerbury, is in the British Museum, and masqueraded for long under the name of *Ictinus rapax*, which insect it greatly resembles. The allotype female is in my own collection, and was taken by Col. F. Wall at Nalande, 16. x. 24; there is another female in the Colombo Museum, these three being the only specimens known.

298. **Megalogomphus superbus** Fraser. (Pl. IV, fig. 2; text-figs. 45, 90, & 91, a.)

Megalogomphus superbus Fraser, Rec. Ind. Mus. vol. xxxiii, pp. 448, 460–463, text-figs. 3 & 4 (1931); Needham, Rec. Ind. Mus. vol. xxxiv, p. 222 (1932).

Male.—Abdomen 53–56 mm. Hind-wing 42–43 mm.

Head: labium citron-yellow; labrum, bases of mandibles, clypeus, and frons bright grass-green, paler on labrum and clypeus; labrum and frons above very narrowly black at base, bases of mandibles narrowly black on their free border, the lower part of frons and the adjacent upper part of postclypeus narrowly black, and, on the latter, two black streaks running obliquely on to the anteclypeus; eyes bottle-green above, paling to greenish-yellow below; vertex black; occiput citron-yellow, behind black except at centre, where is a small citron-yellow spot confluent with that on upper surface. *Prothorax* black, middle lobe with a small medial geminate spot on posterior border and a short stripe of the same colour to the outer side of the medial spot; posterior lobe small, tumid, posterior border rounded and covered with long black hairs. *Thorax* velvety black marked with tender foliage green as follows:—Thick inverted figures of 7 formed by confluence of oblique antehumeral stripes and an interrupted mesothoracic collar; narrow humeral stripes, the upper end swollen and often slightly separated from the lower part of stripe; a broad stripe on the mesepimeron, notched posteriorly; the whole of metepimeron; a large triangular upper spot between these two last stripes and a lower yellowish spot; the hinder part of antealar sinus and a stripe crossing the tergum between the origins of wings. *Legs* short, robust, black; coxæ, trochanters, and proximal halves of the two hind pairs of femora, a spot on the coxæ and one on the base of the fore femora bright citron-yellow. *Wings* hyaline;

pterostigma black, braced strongly, covering 6 cells; all triangles entire; anal loop of 2 cells or occasionally of 1 large cell; nodal index $\frac{11\text{–}20}{12\text{–}13}\Big|\frac{18\text{–}11}{13\text{–}12}$. *Abdomen* coloured as follows:—Segment 1 black, lower part of sides and middle of apical border dorsally pale grass-green; 2 black, with a trilobed mid-dorsal stripe bordered with yellow apically and with ferruginous elsewhere, oreillet yellow bordered with grass-green, ventral border citron-yellow and sending an upward prolongation to partially encircle the oreillet; 3 with a narrow trilobed mid-dorsal stripe, green basal to the

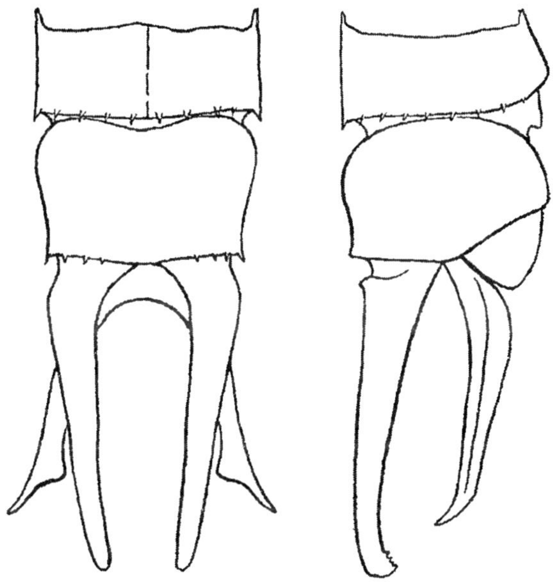

Fig. 90.—Anal appendages of *Megalogomphus superbus* Fraser, male. Dorsal and right lateral views.

jugal suture, yellow thereafter, a triangular basal lateral spot, the portion between it and the dorsal stripe black, the apical third of segment black, the medial third except on mid-dorsum bright reddish-brown; 4 to 6 similar to 3, except that the latero-basal spot is very minute and the dorsal stripe becomes more or less obliterated after the jugal suture, so that the middle third of all segments is entirely reddish-brown; 7 with basal two-thirds citron-yellow, apical third black, this extending slightly into the yellow laterally and still less so along the mid-dorsal carina; 8 dark reddish-brown changing to black on dorsum and with a diffuse latero-basal spot of citron-yellow edged with ferruginous ; 9 similar but with the lateral spot much larger and brighter yellow ; 10 reddish-brown, the base narrowly citron-yellow subdorsally. *Anal appendages* (fig. 90) reddish-brown, paler at base.

Female.—Abdomen 52 mm. Hind-wing 46 mm.

Very similar to the male but much more robust and with a thick, cylindrical abdomen. A tiny black point in centre of labrum ; a small round citron-yellow spot on vertex just behind the ocellar space ; posterior lobe of prothorax shaped similarly to that of the male, but with two small dorsal ochreous spots ; thorax with the antehumeral stripes well separated from the mesothoracic collar, the narrow humeral stripe always complete, the mesepimeron with a fine medial prolongation from the green area, the medial upper spot prolonged downwards and nearly confluent with the prolongation from mesepimeral stripe ; yellow on femora more extensive. *Wings* broader and longer, pterostigma covering from 7 to 9 cells. *Abdomen* similar but the lateral spots on segment 2 fused to form a broad even stripe ; 3 with the basal lateral spot prolonged into a stripe extending two-thirds the length of segment and with little evidence of reddish-brown colour in the middle ; remaining segments similar to those of the male, but the basal spots on 4 to 6 larger and more conspicuous. Vulvar scale short, triangular, with a deep, narrow cleft at its medial border ; *anal appendages* long, reddish-brown, yellow outwardly.

Distribution.—SOUTH INDIA : Waliyar Forest, Malabar ; Kalar, Nilgiris, and Bolovumpatti Hills, South Coimbatore district, where it is very common, its habits being similar to those of *M. hannyngtoni.*

M. superbus differs from *M. hannyngtoni* in the red medial markings of abdominal segments 3 to 6, in the occiput being yellow instead of black and the face less black, and in the presence of a humeral stripe, etc. The same abdominal characters will serve to differentiate it from *M. ceylonicus* (a larger insect), as well as the presence of a humeral stripe and the fusion of the antehumeral stripe with the mesothoracic collar. The narrow mid-dorsal abdominal spots are entirely different from the broad dorsal markings on all segments from 2 to 8 of *M. smithi* Selys. In its size and markings the species falls about midway between the first two with which it is contrasted here. It is quite the most beautiful species of the family GOMPHIDÆ yet discovered.

Type in the British Museum. Specimens in the Author's collection taken at Bolovumpatti, S.E. Coimbatore district, 12th April, 1931.

299. Megalogomphus bicornutus (Fraser).

Gomphus bicornutus Fraser, Mem. Dept. Agric. India (Ent.), vol. vii, no. 7, p. 72 (1922).
Heterogomphus bicornutus Fraser, J. Bombay Nat. Hist. Soc. vol. xxix, p. 679 (1923).
Megalogomphus bicornutus Laidlaw, Trans. Ent. Soc. Lond. vol. lxxviii, p. 197 (1930); Needham, Rec. Ind. Mus. vol. xxxiv, p. 222 (1932).

Male unknown.

Female.—Abdomen 47 mm. Hind-wing 40 mm.

Head: labium black; labrum marked with two transverse oval yellow spots; anteclypeus yellow; postclypeus black; frons black, its crest bright yellow; occiput black, furnished with two long robust medial spines on its posterior border which project somewhat backwards. *Prothorax* black, marked with dorsal and lateral spots. *Thorax* black, marked with bright yellow as follows:—An antehumeral oblique stripe joined to a slightly incomplete mesothoracic collar, so as to form an inverted 7 on each side; a vestigial humeral stripe on each side represented by a small superior spot; laterally yellow, marked with a very broad medial oblique stripe which bears an upper and a lower yellow spot; tergum spotted with yellow. *Legs* short, black, trochanters spotted with yellow; femora bearing short, fine spines. *Wings* tinted with yellow at bases as far as discoidal cells, and for rather more than this extent in the subcostal space; pterostigma black, rather short, braced; 2 rows of cells between *Rii* and *Riii*, beginning nearer the node than the pterostigma; 3 rows of cells between *Rii* and *IRii* at level of distal end of pterostigma; nodal index $\frac{14\text{–}18 \mid 18\text{–}15}{15\text{–}14 \mid 14\text{–}16}$. *Abdomen* black, marked with bright yellow as follows:—Segment 1 with a large lateral spot and a smaller apical dorsal spot; 2 almost entirely yellow, marked only by a fine black apical spot on the dorsum, which extends for a short distance along the mid-dorsal carina; 3 with nearly its basal half yellow dorsally and rather less than this laterally; 4 to 6 with small basal semilunar spots which just meet over the dorsum; 7 with rather more than its basal half yellow; remaining segments black. *Anal appendages* black, shortly conical.

Distribution.—Shillong, Assam, taken in June whilst ovipositing by a roadside stream.

Type, a unique female, in the British Museum.

300. Megalogomphus flavicolor (Fraser). (Fig. 91, **b**.)

Heterogomphus flavicolor Fraser, J. Bombay Nat. Hist. Soc. vol. xxix, p. 678, pl. i, figs. 2, 2 *a* (1923).
Megalogomphus flavicolor Laidlaw, Trans. Ent. Soc. Lond. vol. lxxviii, p. 197 (1930); Needham, Rec. Ind. Mus. vol. xxxiv, p. 222 (1932).

Male unknown.

Female.—Abdomen 60 mm. Hind-wing 51 mm.

Head, *thorax*, and *abdomen* uniformly bright ochreous, marked sparingly with black as follows:—Back of upper part of eyes glossy black; prothorax blackish; thorax with a dorsal stripe running parallel and close to the mid-dorsal suture, tapering above, where it extends to the alar sinus, clubbed below, and not reaching the anterior border of thorax, a humeral stripe tapering below, broadening and bifurcating above to enclose a spot of the ground-colour; laterally a fine black line on the posterior suture. The humeral and lateral markings are connected to their fellows over the tergum. *Legs* short and robust, entirely yellow save for some black spines; hind femora with two rows of short, black, evenly-spaced spines which merge into a group of shorter, stouter

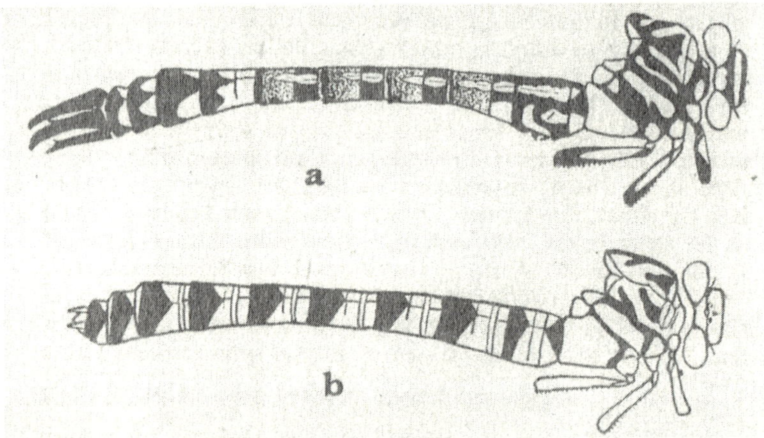

Fig. 91.—Body-markings of (a) *Megalogomphus superbus* Fraser, male; (b) *Megalogomphus flavicolor* (Fraser), female.

spines at the base of limb. *Wings* hyaline, costa bright yellow along its outer border; pterostigma long, black, braced; only a single nervure in all wings between the sectors of arc; 5 to 6 rows of cells in the anal field; the discoidal field very irregular at its beginning, at first a row of 3 cells, which is followed by rows of 2 or 3 cells with no regularity; 3 rows of cells between *Rii* and *IRii*; nodal index $\frac{10\text{–}19 \mid 19\text{–}11}{12\text{–}15 \mid 15\text{–}11}$.

Abdomen yellow, with a narrow black basal ring on segment 1 and broad rather diffuse apical rings on 2 to 9, all extending basally along the mid-dorsal carina for a short distance. Vulvar scale very small, deeply emarginate, and cleft into two small triangular scales.

Distribution.—Duars, Bengal.

M. flavicolor is remarkable for its massive robust build and the great length and breadth of its wings, and also by its almost uniform yellow colour, which is sufficient to distinguish it from all other species of the genus. Martin's *H. unicolor* from Tonkin most nearly approaches it.

Type, a female, in the British Museum ; another female in the Darjeeling Museum.

Genus **OPHIOGOMPHUS** Selys. (Fig 92.)

Ophiogomphus Selys, Bull. Acad. Belg. vol. xxi, pt. 2, p. 39 (1854) ; id., Mon. Gomph. p. 76 (1857) ; Laidlaw, Rec. Ind. Mus. vol. xxiv p. 414 (1922) ; Fraser, J. Bombay Nat. Hist. Soc. vol. xxix, pp. 62, 330 (1923) ; id., ibid. vol. xxx, pp. 398, 399 (1925) ; Laidlaw, Trans. Ent. Soc. Lond. vol. lxxviii, p. 190 (1930) ; Needham, Zool. Sinica, ser. A, vol. xi, fasc. 1, pp. 19, 36 (1930) ; id., Rec. Ind. Mus. vol. xxxiv, pp. 217, 218, 222 (1932).

Diastatomma Kirby, Cat. Odon. p. 61 (1890).

Of rather large size, build robust ; colour pale green sparingly marked with black.

Head large, triangular ; frons angulated ; occiput simple in the male, but sometimes furnished with two or more spines along the posterior border in the female. Wings : reticulation close ; tornus sharply angulated ; base of hind-wing oblique, moderately excavated ; membrane very narrow, almost obsolete ; anal triangle 4-celled ; arc situated between the first and second antenodal nervures or opposite the second ; only 2 nervures between the sectors of arc from arc to bifurcation of *Rs* in the fore-wing, 1 in hind-wing ; 2 rows of postanal cells in fore-wing, 6 in hind-wing ; anal loop well formed, made up of 3 cells and extending proximally nearly to the level of the cubital nervure ; no basal incomplete antenodal nervures ; primary antenodals the first and the fifth ; discoidal cells entire, that of fore-wing with costal and basal sides equal, the distal somewhat longer and a little angulate, that of the hind-wing but slightly elongate, costal and distal sides about equal, the basal distinctly shorter ; pterostigma short and swollen, less than one-third the length of distance from node to proximal end of pterostigma ; *IA* in fore-wing markedly pectinate ; *Cuii* and *IA* in hind-wings markedly divergent ; only 1 cubital nervure in all wings ; subtrigones and hypertrigones entire in all wings. Legs short but robust, hind femora extending only as far as the apical border of abdominal segment 1 and furnished with numerous scattered spines on the flexor surface which merge into two closely-set rows of short spines towards the distal end of limb ; tibial spines moderately long, slim, and numerous. Abdomen tumid at basal segments, moderately narrow and cylindrical from segment 3 to the base of 7 ; the apical end of latter and segments 8 to 10 again somewhat dilated. Anal appendages: superiors rather longer than segment 10, simple, slightly forcipate, or shortly conical; inferior bifid for its apical half or three-fourths, the two branches closely apposed and shorter than the superiors. Genitalia: lamina projecting, arched and emarginate; anterior hamules short slim processes; posterior hamules robust flattened structures with obtuse apex; lobe small, purse-shaped, its edges greatly elevated.

Genotype, *Libellula cecilia* Fourcroy.

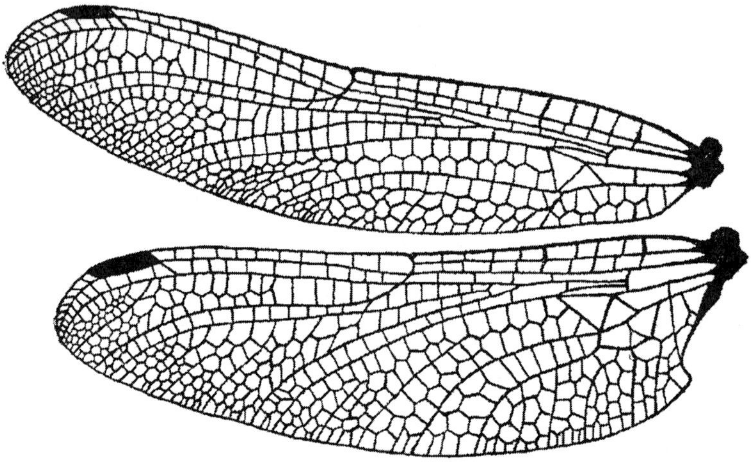

Fig. 92.—Wings of *Ophiogomphus reductus* Calvert, male.

Distribution.—Holarctic (Europe, Northern Asia, and North America); one species is found in Kashmir.

In general facies the species of the genus are remarkably like those of *Gomphus*, but they are more modern insects, to judge from the fusing of the branches of the inferior anal appendix and the well-formed anal loop in the hind-wing.

The larvæ breed in clean gravelly-bottomed streams running through meadows or open lands; some, however breed, in lakes.

301. **Ophiogomphus reductus** Calvert. (Pl. III, fig. 2; text-figs. 92 & 93.)

Ophiogomphus reductus Calvert, Proc. Acad. Sci. Phil. pp. 150–152 (1898); Laidlaw, Rec. Ind. Mus. vol. xxiv, p. 414 (1922); Fraser, J. Bombay Nat. Hist. Soc. vol. xxix, pp. 62, 330 (1923); id., ibid. vol. xxx, pp. 399–401, text-fig. 1, pl. i, fig. 2 (1925); Laidlaw, Trans. Ent. Soc. Lond. vol. lxxviii, p. 190 (1930); Needham, Rec. Ind. Mus. vol. xxxiv, p. 222 (1932).

Male.—Abdomen 37–38 mm. Hind-wing 33–34 mm.

Head : labium and labrum yellowish-green ; face and frons grass-green, the latter narrowly black at its base ; vertex black, with an oval yellow spot just posterior to the ocelli ; occiput pale green, separated from the previous spot by a narrow black streak, fringed with black hairs, simple. *Prothorax* yellow, marked with black as follows :—A transverse stripe just behind its anterior border, a small spot at either end of the posterior lobe, and a larger spot on each side. *Thorax* grass-green, marked sparingly with black as follows :—The alar sinus and upper part of mid-dorsal carina finely, a narrow wedge-shaped humeral spot situated in the length of and about the middle of the humeral region, the humeral and postero-lateral sutures finely black. *Legs* yellow marked with black, tarsi black, spotted with yellow on the extensor surface, tibiæ with a longitudinal black stripe on either side, femora with an outer distal stripe, tapering and fading away about the middle of the femora ; hind and middle femora furnished with a group of very short minute black spines. *Wings* hyaline, costa greenish-yellow as far as pterostigma, which is green, framed in black nervures, covering 3 cells, braced ; anal loop of 2 to 3 cells, wide but shallow ; anal triangle of 4 cells ; 5 rows of postanal cells in hind-wing, 2 in the fore-wing ; 2 rows of cells in discoidal field as far as level of node ; nodal index $\frac{10\text{–}12}{9\text{–}9}\Big|\frac{13\text{–}10}{9\text{–}10}$; 2 to 3 rows of cells between *Rii* and *IRii* at distal end of pterostigma. *Abdomen* yellow, marked with black as follows :—Segment 1 with a subdorsal subbasal streak and two small spots near the apical margin, from the upper one of which springs a peculiar tuft of long hairs ; 2 with the apical border ringed with black, but incomplete over the dorsum, a large spot on the sides running from the jugal suture backwards, but not reaching the apical border, and sending a prolongation upwards which does not quite meet its fellow from the other side ; 3 to 6 each with a thick apical ring and an irregular black stripe on each side which sends up prolongations dorsally at the jugal suture and subapically, these on 5 and 6 fusing at the apex of segments with the apical black rings (the extent of these markings is very variable, and the prolongations may meet over the dorsum, cutting up the ground-colour into yellow dorsal oval spots) ; 7 with the lateral stripe much thicker and the apical ring very fine and followed by a yellow ring ; on 8 the lateral stripe extends the whole length of segment, fusing broadly with the apical ring, and is not deficient at the base as in the other segments ; 9 similar to 6, but the ends of the lateral stripe curling up to meet over the dorsum so as to enclose an oval spot of the ground-colour ; 10 with a broad lateral spot on the basal two-thirds of the segment, meeting its fellow only at a point on the dorsum and tapered posteriorly, its apical border finely black. *Anal appendages* (fig. 93) yellow, superiors tipped with black, inferior brown at the sides. *Genitalia*: lamina short and narrow, arched, folded laterally on itself and deeply notched; inner hamules black, the apex deeply bifid, chelate, long, narrow, and projecting; outer hamules greenish, digitate as seen in profile, the apex bevelled and pointed; lobe tumid at base and then constricted like the neck of a bottle, the apex expanded in two leaf-like processes, yellow margined with black. Oreillets moderately large, greenish-yellow.

Female.—Abdomen 37–39 mm. Hind-wing 35–37 mm.

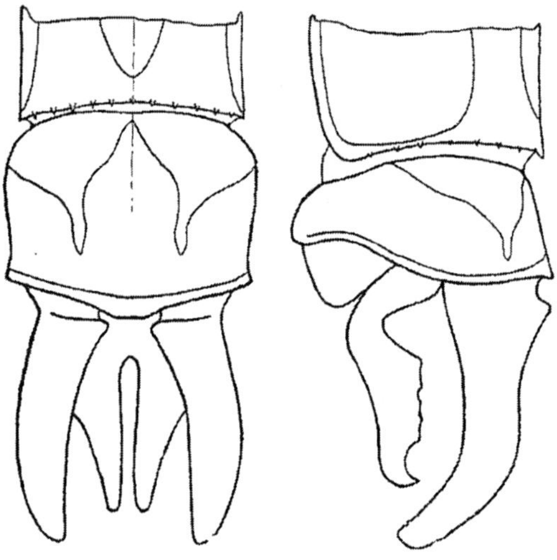

Fig. 93.—Anal appendages of *Ophiogomphus reductus* Calvert, male. Dorsal and left lateral views.

Almost exactly similar to the male, the markings somewhat more restricted. The humeral linear spot almost obsolete (entirely so in some of Mr. Bainbrigge Fletcher's specimens). Anal loop larger and made up of 3 cells, as deep as wide; discoidal field with 2 rows of cells to well short of level of node. Armature of hind femora very specialized; following a short group of small spines there is a row of moderately widely spaced, robust, and rather long spines; armature of middle femora similar to that of male. *Anal appendages* short, yellow, conical, and tapering, separated by a large yellow conical process. Vulvar scale very short, deeply bifid into two small triangular processes which are markedly divergent.

Distribution.—KASHMIR only. The above descriptions were made from a series of specimens collected by Mr. T. Bainbrigge Fletcher at Gulmarg, 8,500 ft., 23. viii. 23, and Yusimarg, 7,500 ft., 15. viii. 23.

The delicate grass-green colour of the head and thorax will distinguish this species from all other Indian Gomphines.

Type in the United States National Museum.

Genus **MEROGOMPHUS** Martin. (Fig. 94.)

Merogomphus Martin, Mission Pavie, Neuropt. vol. iii, p. 214 (1904); Williamson, Proc. U.S. Nat. Mus. vol. xxxiii, pp. 316, 317 (1907); Laidlaw, Trans. Ent. Soc. Lond. vol. lxxviii, p. 184 (1930); Needham, Zool. Sinica, ser. A, vol. xi, fasc. 1, pp. 19, 20, 68 (1930).

Indogomphus Fraser, Rec. Ind. Mus. vol. xxiv, p. 422 (1922); id., J. Bombay Nat. Hist. Soc. vol. xxix, pp. 64, 332 (1923); Needham, Rec. Ind. Mus. vol. xxxiv, pp. 217, 218 (1932).

Size large to medium, colour mat black marked with bright greenish-yellow.

Head rather small and wide, frons markedly angulate, shallowly grooved above, occiput simple, straight. Wings long and broad; reticulation very close; tornus angulate; base of hind-wing excavated, oblique; anal triangle 3-celled; arc situated at the second antenodal nervure or between the second and third; 3 transverse nervures between the sectors of arc from arc to bifurcation of *Rs* in fore-wings, only 1 in hind-wings; 1 or 2 rows of postanal cells in fore-wing, 4 to 5 in hind-wing; anal loop absent; the first postanal cell of hind-wing not extending proximally as far as the proximal end of base of subtrigone; a basal incomplete antenodal usually present, but occasionally absent in one or more wings; nodal index high; primary antenodal nervures the first and the fifth, sixth, or seventh; discoidal cell of fore-wing subequilateral, but the distal side usually distinctly longer than the costal or basal, entire; that of hind-wing entire, elongate in length of wing, costal and distal sides subequal, the latter the longest and nearly double the length of basal; pterostigma braced, rather short, slightly longer than distance between node and proximal end of pterostigma; *IA* in fore-wings markedly pectinate; *Cuii* and *IA* in hind-wing divaricate near border of wing; only 1 cubital nervure in all wings; subtrigones and hypertrigones entire in all wings; no supplementary nervure arising from the distal side of discoidal cells, the latter, in the hind-wings especially, often being separated from the lower sector of arc by a short stalk; subtrigones quite occasionally four-sided. Legs very long; hind femora extending to the apical border of abdominal segment 2 or even on to the base of segment 3, furnished in the male with several pairs of very long, very widely spaced, robust spines, between which are others much shorter and more closely-set; in the female similar but longer spines, 5 pairs of which are usually present. Abdomen broadened at base, then narrow or excessively narrow and cylindrical as far as the base of segment 7, from the apical end of which the abdomen again expands broadly, especially at segment 8; segment 9 nearly as long as 8, segment 10 very small and short. Anal appendages: superiors lyrate or curled like the horns of a bull, or in some species simple short divaricate structures similar to those found in genus *Gomphus*; inferior deeply and broadly notched, the two branches usually widely divaricate. Genitalia variable; lamina usually depressed and emarginate; anterior hamules rather short and slim; posterior hamules

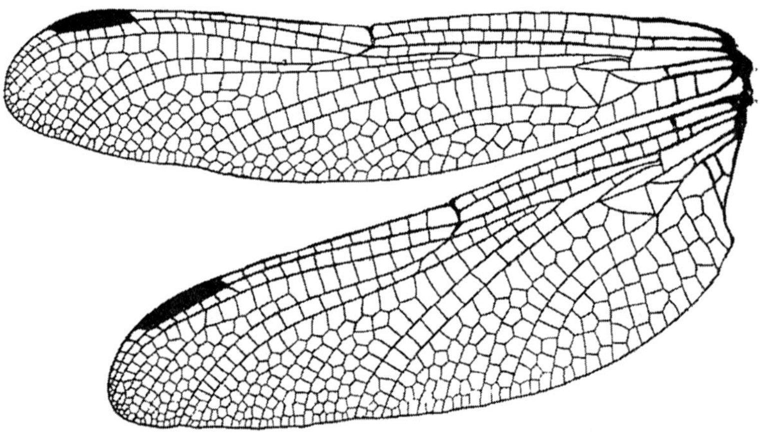

Fig. 94.—Wings of *Merogomphus longistigma longistigma* (Fraser), male.

very robust, sinuous, projecting, and ending in a recurved spine; lobe shaped like the spout of a jug, projecting markedly.

Genotype, *Merogomphus paviei* Martin.

Distribution.—WESTERN GHATS, COORG, ASSAM, BENGAL, BURMA, and Indo-China; probably throughout Malaysia also. Two species only are found within our limits.

The genus is characterized by the long narrow thorax and abdomen, the latter with the end segments curiously elongate, and also by the very distinctive armature of the legs; the stalked discoidal cell of the hind-wing is also a characteristic although not a constant feature in the genus. The larvæ breed in submontane streams; they resemble those of *Burmagomphus* and related genera, except that the terminal segments are elongate, though not nearly to the same extent as in *Macrogomphus*.

Key to Indian Species of Merogomphus.

Superior anal appendages strongly curled, lyrate when seen together .... *longistigma* (Fras.), [p. 311.
Superior anal appendages simple, short, and tapered, strongly divaricate .... *martini* (Fras.), p. 313.

302. **Merogomphus longistigma longistigma** (Fraser). (Fig. 95, a.)

Indogomphus longistigma Fraser, Rec. Ind. Mus. vol. xxiv, pp. 422–424, pl. xi, fig. 8 (1922); id., J. Bombay Nat. Hist. Soc. vol. xxix, pp. 64, 332 (1923); id., Rec. Ind. Mus. vol. xxvi, pp. 428, 478 (1924); id., J. Bombay Nat. Hist. Soc. vol. xxx, pp. 402, 403, pl. i, fig. 6 (1925); Needham, Rec. Ind. Mus. vol. xxxiv, p. 224 (1932).
Merogomphus longistigma Laidlaw, Trans. Ent. Soc. Lond. vol. lxxviii, p. 185 (1930); Fraser, Rec. Ind. Mus. vol. xxxiii, pp. 448, 460 (1931).

Male.—Abdomen 44 mm. Hind-wing 35 mm.

Head : eyes bottle-green ; labium, labrum, and face glossy black ; frons black, with a broad greenish-yellow stripe above and a fine black line at its base. Occiput simple, fringed with long black hairs, greenish-yellow, black at either end. *Prothorax* black ; posterior lobe, a small oval spot just in front of it, and a transverse anterior band yellow. *Thorax* black, marked with bright yellow as follows :—A complete mesothoracic collar which sends a prolongation along the mid-dorsal carina as far as the alar sinus, a narrow dorsal stripe running alongside and parallel with the mid-dorsal carina, reaching the alar sinus above but not the mesothoracic collar below, a vestigial humeral stripe represented by an upper spot and a more or less evident fine line below (sometimes quite obsolete). Laterally greenish-yellow, traversed by two closely parallel medial black stripes. Underside black, marked by a fine V-shaped spot. *Legs* long and slim. Anterior femora greenish-yellow on flexor surface, otherwise all femora entirely black. *Wings* hyaline, long and narrow ; pterostigma pale brownish-yellow, that of hind-wing considerably larger than that of fore-wing, 3·5 mm. to 5 mm. ; nodal index of two specimens $\frac{11\text{–}16}{12\text{–}11} \Big| \frac{15\text{–}12}{10\text{–}12}, \frac{13\text{–}17}{12\text{–}14} \Big| \frac{17\text{–}12}{11\text{–}12}$; an incomplete basal antenodal usually present in one or all wings, rarely absent in all wings. *Abdomen* black, marked with bright yellow as follows :—Segment 1 with a broad stripe on dorsum and a large quadrate spot on each side ; 2 with an L-shaped spot on each side, the underside of the oreillets, a short stripe on their upper surface, and a trilobed mid-dorsal band ; 3 with a baso-lateral triangular spot, a latero-ventral stripe tapering from the basal end and a narrow mid-dorsal stripe which may be cut in two by the finely black mid-dorsal carina (the jugal suture is also occasionally finely black, cutting this stripe transversely); 4 to 6 with similar markings, but the lateral stripe absent and the mid-dorsal stripe well separated from the basal yellow area; 7 with its basal half or rather less broadly yellow, with a prolongation of this colour apically along the dorsal carina; 8 to 10 with only the mid-dorsal carina finely yellow. *Anal appendages* (fig. 95, a): superiors yellow, lyrate; inferior black. *Genitalia*: lamina rather depressed, only slightly visible in profile, rather broadly arched, black; inner hamules aborted, outer very robust, with a prominent mid-rib, which is continued on as a robust forwardly curled spine, black; lobe prominent, deeply cleft, its mouth expansive, lips markedly everted.

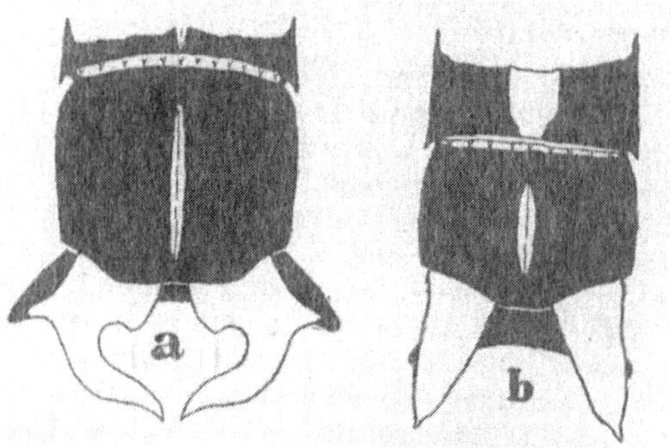

Fig. 95.—Anal appendages of (a) *Merogomphus longistigma longistigma* (Fraser), male; (b) *Merogomphus martini* (Fraser), male. Dorsal views.

Female.—Abdomen 44–46 mm. Hind-wing 37 mm.

Very similar to the male, differing as follows:—Occiput simple, with short hairs fringing it, no medial spines as in the two other members of the genus (these are probably absent on account of the highly specialized appendages of the male); bases of mandibles yellow, abdomen with segment 9 about the same length as 8, which is slightly dilated, as is also the apical half of 7, 9 dilated but tapering rapidly to segment 10, which is very short and narrow. The long spines on the hind femora are more numerous than in the male. *Anal appendages* small, conical, pale yellow, as is also a conical protuberance between them. Wings slightly enfumed, pterostigma light brown, the difference in size more marked than in male. (One female has two incomplete basal antenodals in one fore-wing.)

Distribution.—SOUTH INDIA: Nilgiri Wynaad, Malabar Wynaad, Coorg, and Chanar, Travancore.

I have found *M. longistigma* on the southern slopes of the Nilgiris in August; it haunts montane streams at altitudes of about 3000 ft., but is an uncommon insect. It is the largest species of the genus in India.

Type (from Gudalur, Nilgiris) and allotype in the British Museum; specimens in my own collection.

303. Merogomphus longistigma tamaracherriensis Fraser.

Merogomphus longistigma tamaracherriensis race or sp. ? Fraser, Rec. Ind. Mus. vol. xxxiii, p. 460 (1931).

Male.—Abdomen 40 mm. Hind-wing 30 mm.

Differs from typical *M. longistigma* as follows:—Occiput entirely black; mid-dorsal spot on segment 3 of abdomen isolated, on 4–6 entirely absent; segment 8 variable, in some specimens with the basal half yellow, the extreme base of segment finely black, and the apical border of the yellow area markedly crenate, in others the yellow area is reduced to a mid-dorsal basal tiny diamond-shaped spot; segments 9 and 10 usually unmarked, but occasionally a fine mid-dorsal streak on 9 and a tiny mid-dorsal apical point of yellow on 10.

Distribution.—Tamaracherri, South Malabar, inhabiting marsh-lands or bogs at the foot of hills.

Type in the Author's collection.

304. Merogomphus martini (Fraser). (Fig. 95, b.)

Platygomphus martini Fraser, Mem. Dept. Agric. India (Ent.), vol. vii, no. 7, pp. 68, 69 (1922); id., J. Bombay Nat. Hist. Soc. vol. xxix, pp. 62, 330 (1923).

Burmagomphus duarensis Fraser, J. Bombay Nat. Hist. Soc. vol. xxiv, pp. 421, 422 (1922); id., J. Bombay Nat. Hist. Soc. vol. xxix, pp. 62, 330 (1923).

Indogomphus duarensis Fraser, J. Bombay Nat. Hist. Soc. vol. xxx, p. 112, fig. 2, xiii (1924); id., ibid. vol. xxx, p. 405, pl. i, fig. 4 (1925).

Indogomphus martini Fraser, J. Bombay Nat. Hist. Soc. vol. xxx, pp. 403, 404 (1925); Needham, Rec. Ind. Mus. vol. xxxiv, p. 224 (1932).

Merogomphus martini Laidlaw, Trans. Ent. Soc. Lond. vol. lxxviii, p. 185 (1930).

Male.—Abdomen 34 mm. Hind-wing 26 mm.

Head entirely black save for bases of mandibles and a broad bright yellow stripe on upper surface of frons; occiput straight, simple, fringed with black hairs. *Prothorax* black, with a large citron-yellow spot on each side. *Thorax* black on dorsum marked with yellow as follows:—An oblique ante-humeral stripe, confluent below with a slightly interrupted mesothoracic collar; a vestigial humeral stripe represented by a small upper spot. Laterally greenish-yellow, marked by two fine black lines which map out the sutures. *Wings* hyaline, slightly saffronated at the extreme base; pterostigma pale brown, braced, short; incomplete basal antenodal nervures entirely absent or irregularly present in one or more wings; nodal index of two specimens $\frac{11\text{–}15}{11\text{–}10}\Big|\frac{16\text{–}11}{10\text{–}11}$, $\frac{12\text{–}17}{11\text{–}12}\Big|\frac{15\text{–}11}{12\text{–}11}$. *Legs* entirely black save for the anterior femora, which are pale whitish-green inwardly; hind legs of great length. *Abdomen* black, marked with citron-yellow as follows:—Segment 1 entirely, save for two black dorsal spots; 2 with a trilobed mid-dorsal stripe, the oreillets and a lateral spot in continuation of the last; 3 and 4 with the dorsal carina finely yellow and a lateral stripe, broad at base of segments, tapering thence to nearly the apical border; 5 to 7 with the baso-lateral spots meeting over the dorsum, the ring thus formed much wider on 6 and occupying the basal half of 7. Remaining segments entirely black. The markings on segments 5 and 6 are subject to some variation; in some specimens from Shillong the basal ring on 6 is almost as broad as that on 7. *Anal appendages* (fig. 95, **b**): superiors pale yellow, inferior black. *Genitalia*: lamina arched, border emarginate, distinctly visible from the side; inner hamules narrow, long, projecting, ending in a sharply turned-back point; outer hamules much stouter, projecting beyond apices of former, bevelled near the apex, which ends in a blunt forwardly directed point; lobe cupped, projecting nearly as far as outer hamules.

Female.—Abdomen 35 mm. Hind-wing 29 mm.

Very similar to the male. Occiput slightly concave, armed in some specimens with medial spines similar in character to those of *O. cerastes*; in one female a long robust spine is present slightly to the left of the middle line, in a second female there is a spine on both sides of the middle line, but one is longer than the other. *Prothorax* finely bordered with yellow along the posterior border of posterior lobe, in addition to the lateral spot. *Wings* pale yellow at base; pterostigma pale brown; 2 rows of cells between *Rii* and *IRii*. *Legs*: armature similar to that of the male, but the spines more robust. *Abdomen*: segment 1 entirely yellow, 2 with the whole of the sides yellow as well as the dorsal trilobed stripe; 4 to 6 with the basal rings very narrow and not extended at all along the sides. *Anal appendages* pale yellow, conical, pointed, short. Vulvar scale rather hidden by the overlapping of the dilated sides, rather long, triangular, very slightly bifid at apex.

Distribution.—Duars, BENGAL. The species is also moderately common at Shillong, ASSAM, and Maymyo, BURMA, during June and July.

The basal incomplete nervure is more commonly found in the fore-wing, and is never present in the hind-wing of females;

occasionally it is entirely absent in both sexes. The form of the superior anal appendages will serve to distinguish this species from *M. longistigma*, whose appendages are more like those found in genus *Heliogomphus*.

Type (from Hasimara, Duars) in the British Museum; type of *I. duarensis* in my own collection.

Subfamily EPIGOMPHINÆ.

The members of this subfamily are characterized by having more than two transverse nervures between the sectors of the arc in the hind-wing and by a much larger number in the fore-wing. The forking or bifurcation of *Rs* is usually asymmetrical—pronouncedly so in some of the genera (*Leptogomphus, Heliogomphus,* and *Microgomphus*). Generic characters are the shape of the anal appendages of the male, the genitalia, the presence or absence of a basal incomplete antenodal nervure and of braces to the pterostigma, and the length and character of the armature of the hind femora in both sexes.

Distribution.—Cosmopolitan. Seven Indian genera are here included in this subfamily, but Laidlaw places one of these (*Sieboldius*) in a separate subfamily or series which he has named *Hagenius*.

Key to Indian Genera of Epigomphinæ.

1.	Discoidal cells of fore- and hind-wings of the same shape and size, both elongate in the length of wing; more than four cells in the anal triangle; a supplementary nervure springing from the distal side of discoidal cell	SIEBOLDIUS Selys, [p. 316.
	Discoidal cell of fore-wing shorter than that of hind-wing and of a different shape; never more than 4 cells in the anal triangle; no supplementary nervure springing from the distal side of discoidal cell	2.
2.	An incomplete basal antenodal nervure present in most wings	3.
	Incomplete basal antenodal nervures always absent	4.
3.	Three or more rows of cells between *IA* and the posterior margin in fore-wings; the forking of *Rs* symmetrical; superior anal appendages of male with an inner medial branch	MACROGOMPHUS Selys, [p. 339.
	Only two rows of cells between *IA* and the margin in fore-wings; the forking of *Rs* asymmetrical; superior anal appendages without branches	LEPTOGOMPHUS Selys, [p. 362.

4.	Superior anal appendages with an inner branch, inferior but slightly notched at apex	MICROGOMPHUS Selys, p. 351.
	Superior anal appendages without an inner branch, inferior deeply notched and branched	5.
5.	Superior anal appendages lyrate and with an outer robust spine................	HELIOGOMPHUS Laidlaw, p. 322.
	Superior anal appendages not lyrate, shortly conical or simply curved and without an outer spine	6.
6.	Discoidal cell of hind-wing traversed by a nervure; more than one cubital nervure in all wings	PERISSOGOMPHUS Laidlaw, p. 319.
	Discoidal cell of hind-wing not traversed; only one cubital nervure present in all wings	ACROGOMPHUS Laidlaw, p. 335.

Genus **SIEBOLDIUS** Selys. (Figs. 45 *a* (D) & 96.)

Sieboldius Selys, Bull. Acad. Belg. vol. xxi, pt. 2, p. 83 (1854); id., Mon. Gomph. p. 243 (1857); Kirby, Cat. Odon. p. 76 (1890); Williamson, Proc. U.S. Nat. Mus. vol. xxxiii, pp. 272, 285 (1907); Laidlaw, Rec. Ind. Mus. vol. xxiv, p. 372 (1922); Fraser, J. Bombay Nat. Hist. Soc. vol. xxxi, pp. 887, 888 (1927); Laidlaw, Trans. Ent. Soc. Lond. vol. lxxviii, p. 175 (1930); Needham, Zool. Sinica, ser. A, vol. xi, fasc. 1, pp. 19, 29 (1930); id., Rec. Ind. Mus. vol. xxxiv, pp. 217, 218 (1932).

Size large; glossy or mat black marked sparingly with bright citron-yellow.

Head relatively small to the size of the insects; frons not at all prominent, rather flattened; occiput projecting, laminate, notched at middle or fringed with coarse hairs. Thorax very robust, dwarfing the head. Wings long and rather narrow, reticulation close, tornus but slightly angulate, base of hind wing very slightly excavate; membrane short and narrow; anal triangle 3- to 6-celled; arc between the first and second or second and third antenodal nervures; 4 or 5 nervures between the sectors of arc from arc to bifurcation of *Rs* in fore wings, only 2 in hind wings; 2 rows of postanal cells in fore wings, 4 or 5 in the hind; anal loop present, made up of 3 or 4 cells and extending proximal to base of subtrigone; a basal incomplete antenodal nervure present in all wings; nodal index high; primary antenodals the first and the seventh, eighth, or ninth; discoidal cells of similar shape in fore and hind-wings, very elongate, costal side at least twice as long as the base, the distal side strongly curved, traversed by a nervure in all wings; pterostigma long and narrow, braced; *Cuii* and *IA* in fore-wing nearly parallel as far as border of wing, the latter markedly pectinate; *IA* in hind-wing with four or five sectors running parallel to wing-border; 2 to 3 cubital nervures in all wings; subtrigones and hypertrigones all entire; a supplementary nervure running outwards from the distal side of discoidal cells. All main nervures, especially *Riii*, *IRiii*, *Riv+v*, and *MA*, curving strongly down towards the termen of wings. Legs of great length, the hind-femora extending to the apical end of abdominal segment 2 and furnished with a row of very closely set spines, these gradually lengthening and becoming wider spaced towards the distal end of the limb; hind tibial spines very closely set, robust, and relatively short. Abdomen long and cylindrical, tumid at base; segment 10 markedly rounded, dome-shaped. Anal appendages: superiors short, thick, furnished with robust ventral spines; inferior broad and quadrate, shallowly and broadly bifid. Genitalia:

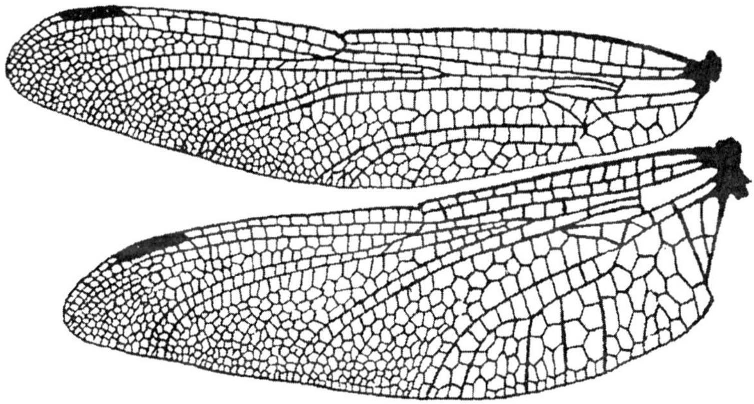

Fig. 96.—Wings of *Sieboldius nigricolor* (Fraser), male.

lamina prominent, sloping anteriorly; anterior hamules short thin stylets; posterior hamules broad, robust, and furnished with an apical recurved hook; lobe very large, scrotal-shaped.

The larva is blattiform in shape, the abdomen being extremely depressed and enormously broadened; these characters, together with its black colour, make it resemble one of the dead leaves amongst which it is found. The larva bears much resemblance to those of *Lamelligomphus*, although on adult characters the genera appear to be unrelated (fig. 45*a*, D).

Genotype, *Sieboldius japponicus* Selys.

Distribution.—BURMA; Malaysia, Indo-China, Sumatra, Japan, and China.

305. Sieboldius nigricolor (Fraser). (Fig. 97.)

Hagenius nigricolor Fraser, Mem. Dept. Agric. India (Ent.), vol. viii, no. 8, pp. 76–79, pl. x, figs. 1, 2, 5–7 (1924).
Sieboldius nigricolor Fraser, J. Bombay Nat. Hist. Soc. vol. xxxi, pp. 888, 889, text-figs. 2, 3 (1927); Laidlaw, Trans. Ent. Soc. Lond. vol. lxxviii, p. 176 (1930); Needham, Zool. Sinica, ser. A, vol. xi, fasc. 1, p. 30, pl. v, fig. 2 (1930).

Male.—Abdomen 54 mm. Hind-wing 49 mm.

Head glossy black, upper surface of frons bright greenish-yellow, its base very narrowly black; eyes probably bottle-green during life. Occiput projecting squarely back as a thick plate which overhangs the prothorax, its hinder border slightly notched at its middle and densely fringed with short stiff black hairs, its upper surface coarsely corrugated and a little hollowed out; vesicle raised into two robust points very much as in the LIBELLULINÆ. *Prothorax* mat black, with two small yellow points on mid-dorsum. *Thorax* black, marked with yellow as follows:—A complete mesothoracic

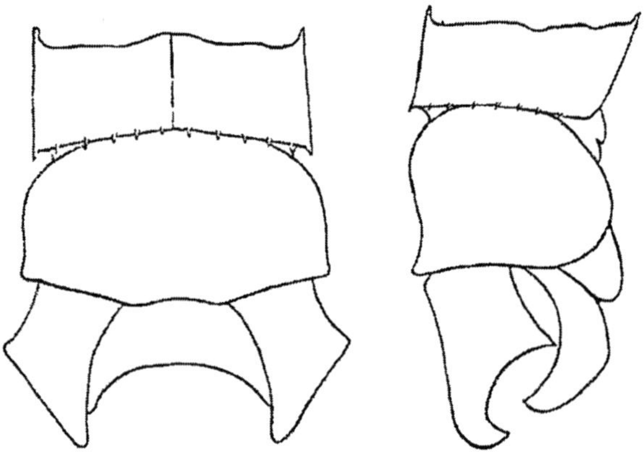

Fig. 97.—Anal appendages of *Sieboldius nigricolor* (Fraser), male. Dorsal and right lateral views.

collar, confluent with a median spot on lower half of mid-dorsal carina; narrow antehumeral stripes not confluent with the mesothoracic collar, divergent in front, the upper ends turning out at a little more than a right angle. No humeral stripe. Laterally two oblique yellow stripes, a narrow one bordering the hinder border of the humeral suture and a broad one covering the greater part of the metepimeron. *Legs* black. *Wings* hyaline, pterostigma black, braced, 6 mm. in length; costa finely yellow; membrane nearly obsolete; 3 rows of cells between *Rii* and *IRii* at level of distal end of pterostigma; 2 cubital nervures in all wings; 2 rows of postanal cells in fore-wing, 5 in hind-wing; nodal index $\frac{13\text{–}18}{14\text{–}14}\Big|\frac{20\text{–}12}{13\text{–}15}$; anal triangle 3-celled. *Abdomen* black, marked with yellow as follows:—Sides of segments 1 and 2 broadly, the colour extending up along apical margin as far as dorsum on 2; a fine obscure mid-dorsal carinal line on 2 and 3, the latter with the basal part of sides yellow; 4 to 7 with basal paired lunules separated by the dorsal crest; 8 with its basal half yellow and rather more than that of its sides; 9 with a small baso-lateral spot; 10 unmarked. *Anal appendages* (fig. 97) black. *Genitalia*: lamina depressed, deeply and very narrowly cleft at the middle; anterior hamules thin stylets directed backwards and downwards, converging towards each other, their apices curled a little outwards; posterior hamules very broad, long, robust, the apices quadrate, a robust hook, shaped like a tiger's claw, springing from the outer corner; lobe enormously inflated, rounded, glossy black.

Female unknown.

Distribution.—Southern Shan States, BURMA, taken in September.

Type, a unique male in my own collection.

Genus **PERISSOGOMPHUS** Laidlaw. (Fig. 98.)

Perissogomphus Laidlaw, Rec. Ind. Mus. vol. xxiv, pp. 383, 384 (1922); Fraser, J. Bombay Nat. Hist. Soc. vol. xxix, pp. 61, 329 (1923); id., ibid. vol. xxxi, pp. 742, 743 (1926); Laidlaw, Trans. Ent. Soc. Lond. vol. lxxviii, p. 187 (1930); Needham, Rec. Ind. Mus. vol. xxxiv, pp. 217, 218, 225 (1932).

A monotypic genus of medium sized dragonflies; colour bright grass-green, marked with black.

Head robustly built, frons rounded, face rather oblique; occiput simple, hinder border straight and fringed with long black hairs, its upper surface bearing a sulcus and a small median tubercle. Wings: reticulation close; tornus well angulated; base of hind-wing oblique, moderately excavated; anal triangle 4-celled; arc situated opposite the second or between the second and third antenodal nervures; 4 transverse nervures between the sectors of arc from arc to bifurcation of *Rs* in fore-wings, 2 in hind-wings; 2 rows of postanal cells in the fore-wings, 5 in hind-wings; a distinct anal loop present and made up of 2 to 3 cells by a division of the first postanal cell, which extends well proximal to base of subtrigone; no basal incomplete antenodal nervures present; nodal index high; primary antenodals the first and the sixth or seventh; discoidal cells usually traversed (this feature very variable; occasionally all are entire, most usually only those of the hind-wings are traversed, but in one specimen only the cells of the fore-wings are traversed), that of fore-wing with the sides subequal, the distal slightly angulated, that of the hind-wing elongate in length of wing, the distal side longer than the costal and the costal longer than the basal, the traversing nervures running from the basal to the distal sides; pterostigma short, only slightly more than one-third the distance from node to proximal end of pterostigma, swollen at its middle, braced; *IA* in fore-wing markedly pectinate; *Cuii* and *IA* in hind-wing parallel to border of wing; 2 to 3 cubital nervures in fore-wings, 2 in hind-wings; all subtrigones and hypertrigones entire. Legs moderately long, hind femora extending to apical border of abdominal segment 1 and furnished with a group of short, numerous,

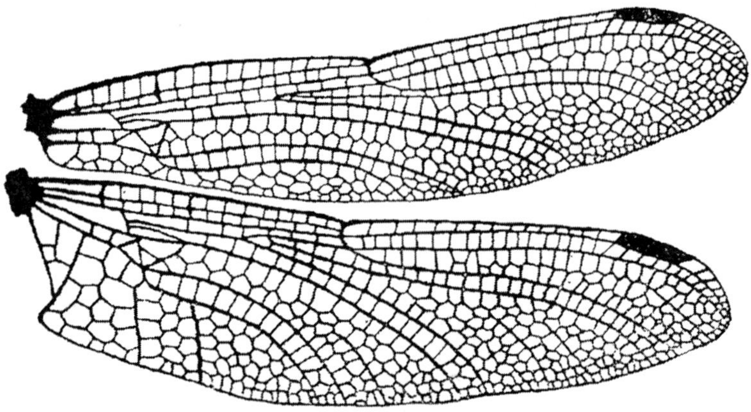

Fig. 98.—Wings of *Perissogomphus stevensi* Laidlaw, male.

closely-set spines on the proximal half which are gradually resolved into two rows of rather more robust and slightly longer spines at the distal end; in the female these spines longer, less numerous, and more widely spaced. Abdomen tumid at base, thin and cylindrical from segment 3 to 7, thereafter gradually dilating again as far as segment 10; segments 8 and 9 and the apical half of 7 with slight lateral dilatations. Anal appendages: superiors closely similar to those of genus *Gomphus*, short, about as long as segment 10, simple, conical, and ending in an acute apex, rather widely divaricate; inferior deeply cleft into two equally divaricate branches. Genitalia: lamina narrowly and deeply emarginate, projecting prominently; anterior hamules long and narrow, terminating in long, fine, backwardly curved spines; posterior hamules much more robust, stout at base, terminating in a robust forwardly curved spine ; lobe-shaped like a pitcher, with its spout strongly emarginate.

Genotype, *Perissogomphus stevensi* Laidlaw.

Distribution.—Assam and Bengal.

The genus is probably closely allied to *Heliogomphus*, although the species differ markedly in their habits, which in *Perissogomphus* are arboreal. The larvæ are unknown, but most probably breed in mountain streams, the adults being found resting on foliage in adjacent evergreen jungle.

306. **Perissogomphus stevensi** Laidlaw. (Pl. IV, fig. 5 ; text-fig. 98.)

Perissogomphus stevensi Laidlaw, Rec. Ind. Mus. vol. xxiv, pp. 371, 384–387, text-figs. 6–8 (1922); Fraser, J. Bombay Nat. Hist. Soc. vol. xxix, pp. 61, 329 (1923); id., ibid. vol. xxxi, pp. 743, 744, text-figs. 2–4, pl. i, fig. 4 (1926); Laidlaw, Trans. Ent. Soc. Lond. vol. lxxviii, p. 187 (1930); Needham, Rec. Ind. Mus. vol. xxxiv, p. 225 (1932).

Male.—Abdomen 36–38 mm. Hind-wing 33–34 mm.

Head: labium olivaceous; labrum olivaceous-brown, its anterior border narrowly black, the outer corners and a pair of small median spots near middle of anterior border yellow ; anteclypeus and adjacent portion of postclypeus olivaceous ; postclypeus dark olivaceous-brown with an ill-defined large area against the eyes yellow. Frons deeply and widely notched at its centre, a beautiful grass-green above and in front, its extreme base above narrowly black ; vertex and occiput black. *Prothorax* black, the posterior lobe, a spot almost confluent with it, and a narrow anterior collar greenish-yellow. *Thorax* black on dorsum, grass-green laterally. The dorsum with a broad grass-green, M-shaped marking, formed by the confluence of broad antehumeral stripes which extend from the alar sinus, with a complete mesothoracic collar below ; the middle limb of the M is formed by the mid-dorsal carina, which is green in its lower two-thirds, and confluent with the mesothoracic collar below. Laterally the vestiges of an upper dark brown narrow stripe, and the hinder suture finely mapped out in green. *Legs* black, inner surfaces of fore femora and distal half of same area on the mid-femora greenish-yellow. *Wings* hyaline, a faint tinge of yellow at the extreme base in the cubital area. Pterostigma blackish-brown between thick black nervures ; nodal index of two specimens $\frac{12\text{–}16 \mid 17\text{–}14}{14\text{–}13 \mid 12\text{–}13}$, $\frac{13\text{–}18 \mid 17\text{–}11}{13\text{–}12 \mid 12\text{–}13}$. *Abdomen* black, marked as follows :—Segments 1 to 7 with a mid-dorsal citron-yellow stripe, broad on 1, trilobed on 2 and 3, broad at base on 4 to 6, on which segments it is confluent at the base with a narrow complete ring, and thereafter tapers away to a fine point; on 7 similar but much broader at the base, tapering less abruptly, and confluent at the base with a broad ventro-lateral stripe which runs the full length of the segment; sides of segments 1 and 2 broadly grass-green; a triangular baso-lateral spot on 3 yellow; sides of 8 to 10 broadly yellow-ochre, as also the apical half of dorsum of 10, where are seen two small round subdorsal subapical black spots. *Anal appendages* yellow; superiors as long as segment 10, a little divaricate, conical, tapering, acuminate, the apex slightly upturned and bearing beneath 4 or 5 small blackish-brown spines; inferior about two-thirds the length of superiors, curling slightly up, deeply cleft into two short concial bluntly pointed divaricate branches. Genitalia (see under genus).

Female.—Abdomen 38–41 mm. Hind-wing 35–37·5 mm.

Closely similar to the male, the face, however, more yellowish, the labium paler and the markings on abdomen more extensive, the whole of segment 10 being yellow. Wings much more broadly tinted with yellow at base, as far out as end of cubital space and along costa as far as node. (In one female the whole of the wings are deeply and evenly enfumed, probably due to old age.) The anal loop is as highly developed as in the male. Hind femora, as well as the other pairs, broadly yellow internally, and a fine diffuse stripe of the same colour on the outer surface. *Anal appendages* very short, conical, pointed, yellow. Vulvar scale (see under genus).

Distribution.—Gopaldhara, ASSAM (*H. Stevens*) and Pashok, DARJEELING DISTRICT (*C. M. Inglis*) during May and June. I took a male at Mungpoo, Darjeeling District, which was settled on an evergreen bush.

The generic name implies "redundancy," and is well suited to this insect, in which the venation is extraordinarily close considering the size of the insect and by comparison with other Gomphines.

Type in the Indian Museum.

Genus **HELIOGOMPHUS** Laidlaw. (Fig. 99.)

Heliogomphus Laidlaw, Rec. Ind. Mus. vol. xxiv, pp. 378, 379 (1922); Fraser, J. Bombay Nat. Hist. Soc. vol. xxix, pp. 63, 331 (1923); id., ibid. vol. xxx, pp. 846, 847 (1925); Laidlaw, Trans. Ent. Soc. Lond. vol. lxxviii, p. 180 (1930); Needham, Rec. Ind. Mus. vol. xxxiv, pp. 217, 218, 220 (1932).

Size medium; colour black, marked with greenish-yellow; superior anal appendages lyrate in shape and pale in colour.

Head broad, frons rather rounded, occiput small, concave or straight at the posterior border. Wings: reticulation close; tornus rather rounded, base of hind-wing very oblique and shallowly excavate; membrane obsolete; anal triangle

3-celled; arc situated opposite the second or between the second and third antenodal nervures; 6 or 7 transverse nervures between the sectors of arc from arc to bifurcation of *Rs* in fore-wing, 4 or 5 in hind-wing; a single row of postanal cells in fore-wing, 3 to 4 rows in hind-wing; first postanal cell in hind-wing not extending proximal to base of subtrigone; anal loop absent; no basal incomplete antenodal nervures present; nodal index high; primary antenodals the first and the fifth; discoidal cell entire in all wings, that of fore-wing nearly equilateral, that of hind-wing slightly elongate in length of wing, costal and distal sides equal and nearly half as long again as the basal, the cell in the hind-wing quite occasionally connected to the lower sector of arc by a short stalk as in *Merogomphus*; pterostigma very short and swollen, only about one-fifth the length of distance from node to proximal end of pterostigma, without a brace or with the brace

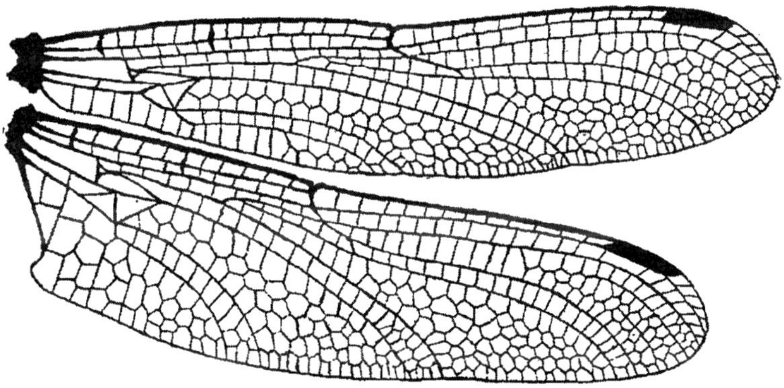

Fig. 99.—Wings of *Heliogomphus promelas* (Selys), male.

poorly developed; *IA* in fore-wing not pectinate and with only 1 or 2 rows of cells between it and the margin of wing; *Cuii* and *IA* in hind-wing variable, entirely parallel or widely divergent at wing-border; 1 or 2 cubital nervures in fore-wing, only 1 in hind-wing; all subtrigones and hypertrigones entire. Legs variable in length, hind femora usually extending to about the middle of abdominal segment 2, but occasionally only to the base of segment 1, furnished with a group of numerous, short, closely-set spines on the proximal half, which become arranged into rows on the distal half; tibial spines rather widely spaced, moderately long, and very slim. Abdomen tumid at base, narrow and cylindrical from segment 3 to 7, the apical end of the latter and segments 8 to 10 again dilated, but only moderately so. Anal appendages: superiors lyrate, forcipate, curled strongly like the horns of a cow and with a more or less robust outer spine near the base; inferior with the two branches very widely divaricate and almost in line, so that they project outwardly from beneath the superiors. Genitalia: lamina depressed, deeply and narrowly emarginate; anterior hamules short, slim processes pointed at apex; posterior hamules but slightly longer but more robust and with a short recurved spine at apex; lobe projecting slightly, flask-shaped, emarginate at lip.

Genotype, *Heliogomphus selysi* Fraser (as *Leptogomphus nietneri* Selys).

Distribution.—INDIA, BURMA, and CEYLON; Malaysia, Java, Sumatra, Borneo, China, Indo-China, and the Philippines. Eight species are known from within our limits, of which four occur in Ceylon, two in the Western Ghats, and two in north-east India and Burma.

The species are characterized by the zygopterous shape of the head, which is strongly reminiscent of that of *Pseudophœa*, and also by the small size of the pterostigma, which is unbraced, and by the almost uniform lyrate shape of the superior anal appendages.

The larvæ have not been described; they breed in montane and submontane streams at altitudes of 1,500 ft. and upwards, and the adult insect is to be found settled on foliage or ferns alongside the parent stream. In the Annaimallai Hills I found *H. promelas* (Selys) breeding in mere seepages on the hill-sides and restricted to very small areas. They are never found away from the vicinity of heavy jungle, and prefer districts where the rainfall is of the heaviest.

Key to Indian Species of Heliogomphus.

1.	Segment 7 of abdomen with a broad basal yellow ring	*promelas* (Selys), [p. 327.
	Segment 7 without a basal ring	2.
2.	A superior humeral yellow spot present on thorax	*nietneri* (Selys), [p. 326.
	No superior humeral spot present	3.
3.	A yellow spot on each side of the postclypeus; mesothoracic collar broadly interrupted....	*ceylonicus* (Selys), [p. 330
	Postclypeus without a yellow spot laterally .	4.
4.	Yellow stripe on crest of frons interrupted at middle; mesothoracic collar very broad and uninterrupted; hind femora dark yellow on the inner side	*walli* Fras., p. 331.
	Yellow stripe on crest of frons uninterrupted at middle; mesothoracic collar narrower, usually uninterrupted;' hind femora black	5.
5.	Lateral spine of superior anal appendages very long and robust	6.
	Lateral spine of superior anal appendages shorter and obtuse	7.

6. Lateral spine of superior anal appendages directed obliquely backwards, so that a deep notch occurs on the appendage immediately after the spine .... *spirillus* (Fras.), [p. 334.
Lateral spine of superior anal appendages directed horizontally outwards, so that no notch occurs on the side of appendage after it. *lyratus* Fras., [p. 333.

7. Yellow markings on sides of abdomen extending as far as segment 6 .... *kalarensis*, sp. n., [p. 329.
Yellow markings on sides of abdomen extending only as far as segment 4 .... *selysi* Fras., p. 325.

307. **Heliogomphus selysi** Fraser. (Fig. 100, a.)

Leptogomphus nietneri (pars) Selys, Ann. Mus. Civ. Genova, (2) vol. x, pp. 474, 476 (1891); Martin, Mission Pavie Indo-Chine, vol. iii. p. 214 (1904).
Leptogomphus ? nietneri Williamson, Proc. U.S. Nat. Mus. vol. xxxiii, pp. 292, 293, 295 (1907).
Heliogomphus nietneri (pars) Laidlaw, Rec. Ind. Mus. vol. xxiv, pp. 378–380, text-fig. 3 (1922); id., Philip. J. Sci. vol. xxviii, p. 560 (1925).
Heliogomphus selysi Fraser, J. Bombay Nat. Hist. Soc. vol. xxx, p. 850 (1925); Laidlaw, Trans. Ent. Soc. Lond. vol. lxxviii, p. 180 (1930); Needham, Rec. Ind. Mus. vol. xxxiv, p. 220 (1932).

Male.—Abdomen 32–35 mm. Hind-wing 28–30 mm.

Head : labium dark yellow ; labrum glossy black, with two small triangular basal whitish-green spots ; bases of mandibles greenish-yellow ; rest of face, frons, and upper surface of head black, save for a transverse stripe of yellow on crest of frons

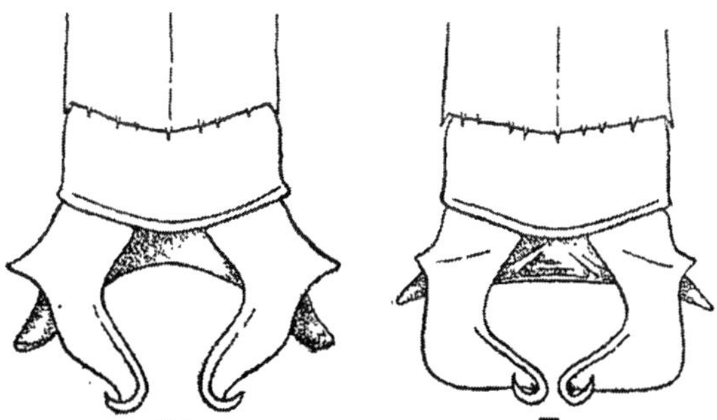

a b

Fig. 100.—Anal appendages of (a) *Heliogomphus selysi* Fraser, male ; (b) *Heliogomphus nietneri* (Selys), male. Dorsal views.

constricted at its middle. Occiput simple, low, slightly convex, fringed with short hairs. *Prothorax* black, a geminate spot in front of the posterior lobe, a small lateral spot and a narrow anterior collar of yellow. *Thorax* black, marked as in *H. promelas*, the mesothoracic collar usually finely interrupted. *Legs* black. *Wings* hyaline; nodal index 11–16/15–12; 2 cubital nervures in all wings; pterostigma short, covering 4 to 5 cells, dark brown. *Abdomen* black, with a fine dorsal greenish-yellow stripe on segments 1 to 3, becoming very obscure on 4, and variably present or absent on 5 and 6; laterally segments 1 and 2, including the oreillets, broadly yellow; a baso-lateral spot on segment 3; remaining segments unmarked. *Anal appendages* (fig. 100, **a**): superiors yellow or in some specimens blackish-brown at the base, pale yellow thereafter, the outer tooth at the base more robust than in *H. promelas*, otherwise of the same shape as in that species; inferior black, as in *H. promelas*.

Female unknown.

Distribution.—The *type* is a male from Leito, Burma, collected by Mr. Fea and now in the Genoa Museum. A pair in the Indian Museum from the Garo Hills, Tura, Assam, 1500 ft. (*Kemp*) was described by Laidlaw, and Mr. T. Bainbrigge Fletcher has also taken this insect on the Gauhati road near Shillong, Assam, 1000 ft., 2. viii. 19.

This species is easily distinguished from all others of the genus by the total absence of abdominal markings after segment 4.

308. **Heliogomphus nietneri** (Selys). (Fig. 100, **b**.)

Gomphus ? nietneri Selys, Bull. Acad. Belg. (2) vol. xlvi, p. 449 (1878).
Anisogomphus nietneri Kirby, Cat. Odon. p. 69 (1890); id., J. Linn. Soc. Lond. (Zool.) vol. xxiv, p. 557 (1893).
Leptogomphus nietneri (pars) Selys, Ann. Mus. Civ. Genova, (2) vol. x, pp. 474–476 (1891).
Leptogomphus ? nietneri Williamson, Proc. U.S. Nat. Mus. vol. xxxiii, p. 295 (1907); Ris, Suppl. Ent. no. 1, p. 69 (1912).
Heliogomphus nietneri Fraser, Rec. Ind. Mus. vol. xxiv, p. 416 (1922); id., J. Bombay Nat. Hist. Soc. vol. xxix, pp. 63, 331 (1923); Laidlaw, Spolia Zeylanica, vol. xii, p. 339 (1924); Fraser, J. Bombay Nat. Hist. Soc. vol. xxx, pp. 847, 850 (1925); Laidlaw, Philip. J. Sci. vol. xxviii, pp. 559, 560 (1925); id., Trans. Ent. Soc. Lond. vol. lxxviii, p. 181 (1930); Needham, Rec. Ind. Mus. vol. xxxiv, p. 220 (1932); Fraser, Ceylon J. Sci., B, vol. xviii, p. 27 (1933).

Male.—Abdomen 35 mm. Hind-wing 30 mm.

Head: labium pale yellow, middle lobe bordered with brown; labrum glossy black, marked with two moderately broad, closely apposed, bright citron-yellow spots; face black, frons traversed by a broad stripe of greenish-yellow which slightly overlaps the fore border; vertex black, occiput yellow in its anterior third, black posteriorly, depressed, its crest slightly elevated, fringed with short black hairs. *Prothorax* black, with a broad anterior collar of greenish-yellow. *Thorax* coal-black, marked with greenish-yellow as follows :—A slender mesothoracic collar, slightly interrupted at its middle ; a narrow oblique antehumeral stripe not confluent below with the collar nor extending up as far as the alar sinus, pointed below, squared above ; a small upper humeral spot. Laterally greenish-yellow, marked with two broad black stripes, the anterior one the broader, the posterior nearly terminal. Under surface pale yellow. (The position of the hinder stripe as given by Selys seems to me rather doubtful; it is more probably situated along the lateral suture.) *Legs* black, femora brown within and outwardly at the base. *Wings* hyaline, costal border finely yellow ; 2 cubital nervures in the fore-wing ; anal border of hind-wing very oblique ; pterostigma yellow, 2·5–3 mm., slender, covering 3½ cells, not braced ; membrane absent ; 16/17 antenodals, 14 postnodals. *Abdomen* with large rounded oreillets, bearing minute black denticles along the hinder border. Colour black, marked with bright yellow as follows :—Segment 1 with a triangular dorsal spot ; 2 with a bilobed dorsal spot ; 3 to 7 with only the dorsal carina finely marked but 6 and 7 with larger basal spots ; sides of 1 and 2 broadly yellow ; 3 to 8 with baso-lateral spots tapering apically ; 8 and 9 with apico-lateral spots ; 10 unmarked. *Anal appendages* (fig. 100, *b*) : superiors white, turning to brown at the tips.

Distribution.—CEYLON, in montane or submontane areas such as Ramboddah.

Type, a unique male, in the Hagen collection, Museum of Comparative Zoology, Cambridge, Mass., U.S.A.

309. **Heliogomphus promelas** (Selys). (Fig. 101, a.)

Gomphus promelas Selys, Bull. Acad. Belg. (2) vol. xxxvi, p. 498 (1873); Fraser, J. Bombay Nat. Hist. Soc. vol. xxix, pp. 62, 330 (1923).

Æshna promelas Kirby, Cat. Odon. p. 68 (1890).

Gomphus ? promelas Williamson, Proc. U.S. Nat. Mus. vol. xxxiii, p. 305 (1907); Laidlaw, Rec. Ind. Mus. vol. xxiv, pp. 371, 398 (1922).

Heliogomphus pruinans Fraser, Rec. Ind. Mus. vol. xxiv, pp. 416, 417, pl. xi, fig. 3 (1922); id., J. Bombay Nat. Hist. Soc. vol. xxix, pp. 63, 331 (1923); id., Rec. Ind. Mus. vol. xxvi, pp. 427, 473 (1924); Laidlaw, Philip. J. Sci. vol. xxviii, p. 560 (1925).

Heliogomphus promelas Fraser, J. Bombay Nat. Hist. Soc. vol. xxx, pp. 848, 849, text-fig. 1, pl. i, fig. 6 (1925); Laidlaw, Trans. Ent. Soc. Lond. vol. lxxviii, p. 182 (1930); Fraser, Rec. Ind. Mus. vol. xxxiii, pp. 444, 447 (1931); Needham, Rec. Ind. Mus. vol. xxxiv, p. 220 (1932).

Male.—Abdomen 42 mm. Hind-wing 32 mm.

Head black, marked with yellow ; labium yellowish-brown ; labrum black, marked with two basal greenish-yellow spots whose opposing borders are concave ; mandibles greenish-white ; face black ; frons marked above with a greenish-white transverse stripe; occiput black, depressed; eyes bottle-green. *Prothorax* black, the posterior lobe, a geminate spot in front of it, and an anterior collar yellowish-green; under surface pruinosed white. *Thorax* black on dorsum, greenish-yellow on sides; a complete yellow mesothoracic collar, short yellow antehumeral stripes not quite reaching the alar sinus above and not nearly reaching the mesothoracic collar below, parallel with the mid-dorsal carina. Laterally the two sutures mapped out in black. Under surface pruinosed white, especially in fully mature specimens. *Legs* black; hind femora armed with a row of very closely set, very small spines. *Wings* hyaline, occasionally slightly enfumed; pterostigma blackish-brown, very rarely braced, and then only poorly so and not usually in more than one or two of the wings; only a single cubital nervure to all wings; costal side of

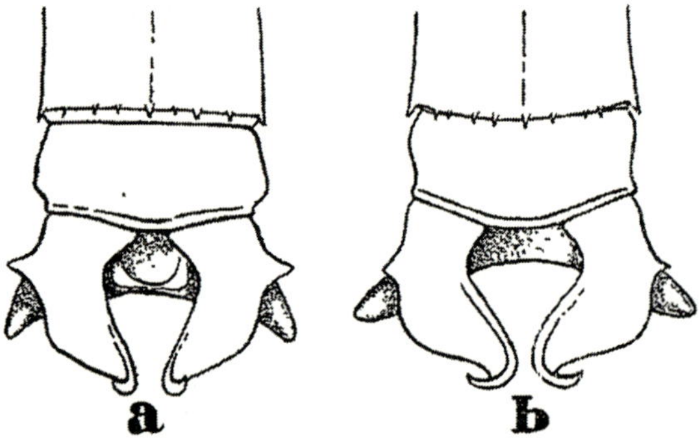

Fig. 101.—Anal appendages of (a) *Heliogomphus promelas* (Selys), male; (b) *Heliogomphus kalarensis* Fraser, male. Dorsal views.

discoidal cell of hind-wing twice as long as the basal; nodal index $\frac{11\text{–}15 \mid 15\text{–}12}{12 \;\; 11 \mid 12 \;\; 12}$; other points as for genus. *Abdomen* black, marked with pale greenish-yellow as follows:—A narrow mid-dorsal stripe extending from segment 1 to 5, thickest on 2 where it may be lobulated; a large lateral spot on segment 1 and a similar but larger one on segment 2 which envelops the oreillet; small baso-lateral spots on 3 to 6, progressively smaller on each segment; 7 with a ring occupying about its basal third; lastly an occasional dorsal spot on segment 8; remaining segments entirely black. *Anal appendages* (fig. 101, *a*) similar to the genotype; superiors black at base, pale green to yellow at apices; inferior black. *Genitalia*: lamina very depressed, deeply bifid, hood-shaped; inner hamules narrow, stylet-shaped, pointed; outer much more robust, elongate, the inner side of apex curling slightly up ; lobe black, globular, rather prominent, its upper border narrowly but rather deeply notched.

Female.—Abdomen 39–42 mm. Hind-wing 35–38 mm.

Very similar to the male, but more robust. Occiput simple, depressed. The antehumeral stripes rather longer and sometimes confluent at a point with the mesothoracic collar. The yellow spots on the sides of abdomen more extensive, those on 3 to 6 extending from the base to the transverse suture and on 3 to 5 continued after a short interruption as an elongate spot which does not quite reach the apical border of segments. On segment 6 there may be a vestigial mediolateral spot. *Legs* with longer spines on the femora. Vulvar scale triangular, extending for about one-third the length of segment 9, glossy black. *Wings* usually distinctly tinted with yellow at the extreme base, broader. *Anal appendages* black, with an inner yellow stripe.

Distribution.—SOUTH INDIA : Mettupalayam and Kotagiri Ghats, Nilgiri Hills ; Annaimallai Hills, Coimbatore District, and in Coorg and Travancore montane areas.

This species is found on small mountain streams or occasionally restricted to tiny brooks or seepages on the hill-sides. It occurs in colonies, and, where found, is usually moderately common. It is easily distinguished from other species by the broad basal ring on segment 7.

Type, a female in the Selys collection. Allotype male in the British Museum ; type of *H. pruinans* in my own collection.

310. **Heliogomphus kalarensis**, sp. n. (Fig. 101, **b**.)

Heliogomphus kalarensis Fraser (nom. nud.), Ceylon J. Sci., B, vol. xviii, p. 29, fig. 4, *d* (1933).

Male.—Abdomen 37 mm. Hind-wing 32 mm.

Head : labium brownish-yellow ; labrum glossy black, with a pair of bright citron-yellow semilunar spots at base ; clypeus blackish-brown, the postclypeus with a small yellow spot on each side ; frons broadly citron-yellow in front and above ; vertex black ; occiput black, with a quadrate yellow spot on its posterior half and behind. *Prothorax* and *thorax* black, marked with citron-yellow as follows :—Prothorax with the posterior and anterior lobes and a small geminate spot on the posterior part of the mid-dorsum of middle lobe ; thorax with a complete mesothoracic collar, straight antehumeral stripes, almost parallel and nearly confluent with the mesothoracic collar at their lower pointed ends, a tiny vestigial upper humeral spot ; the sides very broadly yellow, marked with very narrow oblique black stripes on the two sutures. *Legs* black, coxæ and trochanters yellow. *Wings* hyaline, tinted with golden yellow, this colour fading but distinct as far out as the level of the node; pterostigma bright yellow, framed narrowly in black, not braced, covering 4 cells; a single cubital nervure in all wings; 3 cells in the anal triangle; anal field 3 cells deep; nodal index $\frac{12\text{–}14}{10\text{–}11}\left|\frac{14\text{–}12}{11\text{–}12}\right.$. *Abdomen* black, marked with yellow as follows:—Segment 1 with a small triangular basal dorsal spot, and its sides broadly; 2 with a basal spot which includes the oreillet, an apical lateral spot, and a trilobate mid-dorsal narrow stripe extending from end to end of segment; 3 to 6 with small basal lateral spots and the mid-dorsal carina very finely; 7 with a ring covering its basal third; remaining segments unmarked. *Anal appendages* (fig. 101, **b**): superiors closely similar to those of *H. nietneri* and *promelas*, but the spine is directed more backwards than in either and the appendage is more curled and rounded than in *H. promelas* and less angulated outwardly than in *H. nietneri*; inferior similar to that of *H. nietneri*.

Distribution.—Nilgiris, SOUTH INDIA. A single male, the *type*, is in my own collection, and was taken near Kalar at the foot of the Mettupalayam Ghat during March.

H. kalarensis differs from *H. nietneri* in having only a single cubital nervure in all wings, in the absence of lateral stripes on the abdomen, and in the occiput being yellow posteriorly. From *H. promelas*, from the same locality, the yellow occiput will at once distinguish it, as well as the presence of an upper humeral spot and the shape of the superior anal appendages.

311. **Heliogomphus ceylonicus** (Selys).

Gomphus ? ceylonicus Selys, Bull. Acad. Belg. (2) vol. xlvi, p. 455 (1878); Williamson, Proc. U.S. Nat. Mus. vol. xxxiii, p. 305 (1907); Laidlaw, Rec. Ind. Mus. vol. xxiv, p. 398 (1922).
Æshna ceylonica Kirby, Cat. Odon. p. 68 (1890).
Gomphus ceylonicus Fraser, J. Bombay Nat. Hist. Soc. vol. xxix, pp. 62, 330 (1923).
Gomphus ceylanicus Laidlaw, Spolia Zeylanica, vol. xii, p. 342 (1924).
Heliogomphus ceylonicus Fraser, J. Bombay Nat. Hist. Soc. vol. xxx, pp. 849, 850, pl. i, fig. 5 (1925); Laidlaw, Trans. Ent. Soc. Lond. vol. lxxviii, p. 181 (1930); Needham, Rec. Ind. Mus. vol. xxxiv, p. 220 (1932); Fraser, Ceylon J. Sci., B, vol. xviii, p. 32 (1933).

Male unknown.

Female.—Abdomen 41 mm. Hind-wing 39 mm.

Head: labium pale brown; labrum black, marked with two basal yellow spots; base of mandibles yellow; ante- and postclypeus black, the latter with a medial yellow spot and another on each side against the eyes; frons black with its crest broadly yellow; upper surface of head black, including the occiput, which is fringed with long black hairs, is slightly raised in the middle in a small triangular obtuse point, has six spines near the eyes and a rough tubercle behind. Back of eyes glossy black, with a spot of bright yellow at the middle. *Thorax* black, marked with yellow as follows :—A broadly interrupted mesothoracic collar, narrow oblique antehumeral stripes not extending as far as the alar sinus above or to the mesothoracic collar below, a small upper humeral spot. Laterally greenish-yellow marked by two narrow black stripes on the sutures. All these stripes confluent below at the trochanters. *Wings* hyaline ; pterostigma large, broad, pale yellow (possibly teneral), 3–4 mm. long, covering 3–4 cells, unbraced ; nodal index 14–16/16–14. *Abdomen* black, marked with yellow as follows :—A fine mid-dorsal line on segments 1 to 6, tapering apically on 1 and 2, finer on 3–6, and not extending quite to the apex of 6 ; 6 with a broader basal marking ; sides of 1 and 3 broadly yellow, 4 to 9 with small baso-lateral spots, and a larger one on each side of 7. *Legs* black. *Anal appendages* and the conical protuberance between them black.

Distribution.—CEYLON, Ramboddah Pass (*Nietner*).

The close resemblance of this species to the female of *H. promelas*, remarked upon by Selys, and the unbraced character of the pterostigma, places it without doubt in the genus *Heliogomphus*. Although clearly very closely related to other species, it differs in so many respects that it deserves to retain its specific rank.

Type, a unique female in the Hagen collection, Museum of Comparative Zoology, Cambridge, Mass., U.S.A.

312. **Heliogomphus walli** Fraser. (Fig. 102, a.)

> *Heliogomphus walli* Fraser, J. Bombay Nat. Hist. Soc. vol. xxx, pp. 849–851 (1925) ; Laidlaw, Trans. Ent. Soc. Lond. vol. lxxviii, p. 182 (1930) ; Needham, Rec. Ind. Mus. vol. xxxiv, p. 220 (1932) ; Fraser, Ceylon J. Sci., B, vol. xviii, p. 33 (1933).

Male.—Abdomen 36 mm. Hind-wing 31 mm.

Head : labium with the lateral lobes yellow, middle lobe black ; labrum black, marked with two large triangular citron-yellow spots at the base ; rest of head black except for a moderately broad citron-yellow stripe on crest of frons which is nearly cut in two by a prolongation of the black at base of frons into the median sulcus ; occiput naked, with a narrow transverse sulcus at its posterior part, an anterior ridge, and a short medial ridge with a depression on each side running straight back from the anterior ridge. *Prothorax* black, with a narrow anterior collar. *Thorax* black, marked with greenish-yellow or greenish-white as follows :—A very narrow slightly interrupted mesothoracic collar, very short antehumeral stripes strongly divergent below but not extending the whole length of dorsum either above or below; laterally broadly greenish-yellow, with a narrow oblique black stripe on each suture, that on the anterior suture the broader; underside yellow, pruinosed with white when mature. *Wings* hyaline; pterostigma dark ochreous heavily bordered with black nervures, covering 3 to 4 cells, unbraced; 2 cubital nervures in all wings; anal triangle 3-celled; nodal index $\frac{15\text{–}16 \mid 16\text{–}14}{13\text{–}12 \mid 11\text{–}13}$. *Legs* black. *Abdomen* black, marked with citron-yellow as follows:—Segment 1 with a small triangular mid-dorsal basal spot and its sides broadly; 2 with a latero-basal spot which includes the oreillets, a latero-apical spot on each side, and a short, very narrow mid-dorsal stripe; 3 to 8 with small latero-basal spots on each side, large and

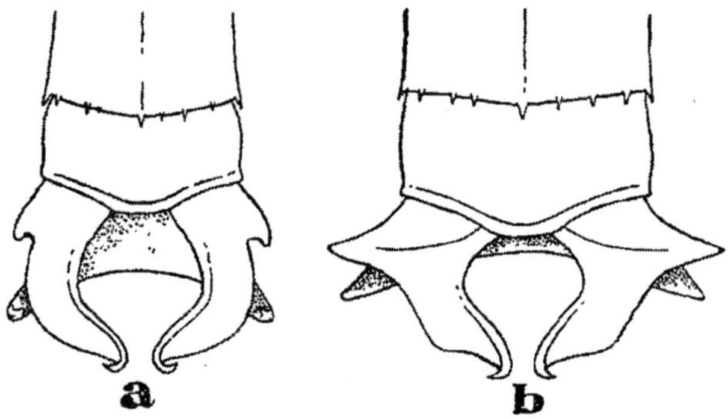

Fig. 102.—Anal appendages of (a) *Heliogomphus walli* Fraser, male; (b) *Heliogomphus lyratus* Fraser, male. Dorsal views.

triangular on 8; 7 and 8 with a latero-apical spot on each side also; mid-dorsal carina finely yellow from segment 3 to 7. *Anal appendages* (fig. 102, a): superiors black for the basal third, yellow thereafter; inferior black.

Female.—Abdomen 38 mm. Hind-wing 36 mm.

Closely similar to the male, but the yellow on crest of frons slightly interrupted at the middle; occiput sculptured exactly as in the male. Wings similar, but one or more occasionally with only a single cubital nervure, and the pterostigma may be poorly braced in one or more wings. Legs with the hind femora yellow on the inner side. Segment 8 of abdomen unmarked. *Anal appendages* pale yellow, shortly conical; vulvar scale about one-third as long as segment 9, slightly bifid at apex, triangular, black.

Distribution.—Ceylon: Balangoda, Rakvana, and Nalande, from May to September.

Differs from *H. nietneri* in the colour of the labium, in the occiput being entirely black and the costal borders of the wings black, and in the absence of an upper humeral stripe. The shape of the anal appendages and the character of the sculpturing of the occiput will serve to distinguish this species from others of the genus.

Type female and a paratype male in my collection; allotype male in the Colombo Museum.

313. **Heliogomphus lyratus** Fraser. (Fig. 102, **b**.)

Heliogomphus lyratus Fraser, Ceylon J. Sci., B, vol. xviii, p. 31, fig. 4, *a* (1933).

Male.—Abdomen 34 mm. Hind-wing 30 mm.

Head: labium with the lateral lobes yellow, tips of these and middle lobe black; labrum black, with two large triangular greenish-yellow spots at base; rest of head black save for a broad greenish-yellow stripe on crest of frons which extends slightly on to the anterior surface and is nearly cut in two by the prolongation of the basal black area medially above the frons; occiput naked, with a narrow posterior sulcus and a transverse anterior ridge, which, however, has no ridge running back from its middle as in *H. walli*. *Prothorax* black, with a narrow anterior yellow collar. *Thorax* black, marked with greenish-yellow or greenish-white as follows:—A very narrow, slightly interrupted mesothoracic collar, very short antehumeral stripes, strongly divergent below and not extending the whole length of dorsum; sides broadly greenish-yellow marked with two narrow black stripes, one on each lateral suture. Underside yellow, pruinosed white when mature. *Legs* black, coxæ yellow externally. *Wings* hyaline; pterostigma dark ochreous, heavily bordered with black, covering 3 to 4 cells, not braced; 2 cubital nervures in fore-wings, or rarely but 1, and always only 1 in hind-wings; anal triangle 3-celled; membrane obsolete; nodal index $\frac{15\text{–}16 \mid 16\text{–}14}{13\text{–}12 \mid 11\text{–}13}$. *Abdomen* black, marked with citron-yellow as follows:—Segment 1 with a minute mid-dorsal triangular basal spot and its sides broadly; 2 with a latero-basal spot which includes the oreillets, a latero-apical spot on each side, and a short very narrow mid-dorsal stripe; 3 to 7 with small triangular baso-lateral spots on each side; the mid-dorsal carina finely yellow on segments 3 and 4; remaining segments unmarked. *Anal appendages* (fig. 102, **b**): superiors black for the basal two-thirds, apical third yellow tipped with black; inferior black.

Female.—Abdomen 33 mm. Hind-wing 30 mm.

Differs in a few respects from the male:—Stripe on crest of frons broader and not constricted at its middle; a small superior humeral spot on each side of thorax; abdomen marked with a greenish-yellow mid-dorsal stripe on segments 1 to 7, triangular on 1, narrow and linear on the remaining segments, and not quite extending to the apical ends of 5 to 7, a linear greenish-yellow spot on sides of 1 to 8, very broad on 1 and 2 and base of 3, broken into an apical and a basal linear spot on 4–8. *Anal appendages* shortly conical, yellow; vulvar scale nearly half as long as segment 8, broadly triangular, and narrowly cleft for about half its length. Wings palely enfumed; only one cubital nervure in all wings.

Distribution.—Haldumulla, CEYLON, during June.

This species resembles *H. walli* in colour, both differing from *H. nietneri* in the colour of the labium, in the occiput being entirely black, the upper humeral spot absent (at least in the male), and in the costa being black instead of yellow. The enormous size of the lateral spine of the superior anal appendages will serve to distinguish *H. lyratus* from all other species of the genus. The sculpturing of the upper surface of the occiput will distinguish the female from that of *H. walli*.

Type male and allotype female in the Colombo Museum.

314. **Heliogomphus spirillus** (Fraser). (Fig. 103.)

Leptogomphus spirillus Fraser, Mem. Dept. Agric. India (Ent.), vol. vii, no. 7, pp. 71, 72, pl. vii, figs. 2 & 2 *a* (1922).

Heliogomphus spirillus Fraser, J. Bombay Nat. Hist. Soc. vol. xxix, pp. 63, 331 (1923); Laidlaw, Trans. Ent. Soc. Lond. vol. lxxviii, p. 182 (1930).

Male.—Abdomen 32 mm. Hind-wing 28 mm.

Head: labium with middle lobe black, lateral lobes brownish-yellow; rest of head black, save for a broad bright citron-yellow stripe on crest of frons; occiput naked, simple. *Prothorax* black, with a narrow anterior yellow collar, a large yellow spot on each side and a small geminate spot on dorsum of posterior lobe. *Thorax* black, with a broad, citron-yellow, uninterrupted mesothoracic collar, and elongate-oval, citron-yellow antehumeral spots which do not extend the whole length of dorsum and are strongly divergent below; sides broadly yellow, with two narrow black oblique stripes, one on each lateral suture; under surface yellow, pruinosed with white. *Legs* black, coxæ yellow externally. *Wings* hyaline; pterostigma black, poorly braced; only a single cubital nervure in all wings; anal triangle 3-celled; nodal index $\frac{13\text{–}16}{11\text{–}10}\Big|\frac{16\text{–}13}{10\text{–}11}$.

Abdomen black, marked with citron-yellow as follows:—Segment 1 with a small triangular baso-dorsal spot and its sides broadly ; 2 with a bilobate mid-dorsal stripe, broadest near the base, and a baso-lateral spot which includes the oreillets ; 3 with its mid-dorsal carina finely yellow nearly as far as the apical end ; 4 to 7 similar, but the marking becoming progressively shorter and, in some specimens, entirely absent. *Anal appendages* (fig. 103) : superiors bright yellow, inferior black.

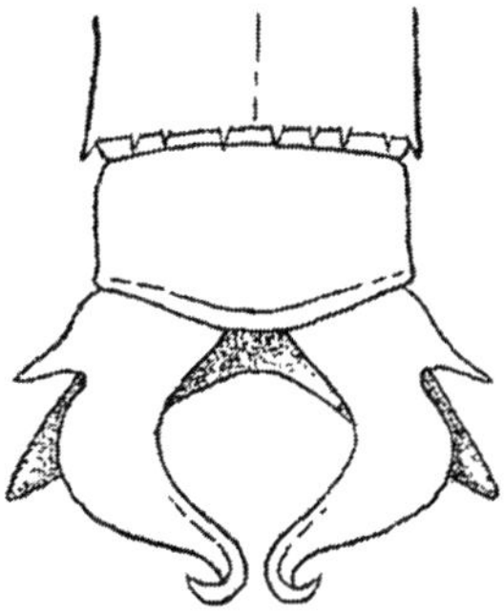

Fig. 103.—Anal appendages of *Heliogomphus spirillus* (Fraser), male. Dorsal view.

Distribution.—ASSAM : Garo and Khasi Hills.

The species may be distinguished from others of the genus by its unmarked labrum, by the braced pterostigma (unusual in the genus), and by the shape of the superior anal appendages.

Type male in the Indian Museum ; paratype male in the Pusa collection.

Genus **ACROGOMPHUS** Laidlaw. (Fig. 104.)

Acrogomphus Laidlaw, Proc. Zool. Soc. Lond. pp. 439, 440 (1925); Fraser, J. Bombay Nat. Hist. Soc. vol. xxxi, pp. 739, 740 (1926); Laidlaw, Trans. Ent. Soc. Lond. vol. lxxviii, p. 191 (1930); Needham, Rec. Ind. Mus. vol. xxxiv, pp. 217, 218, 222 (1932).

Size rather large, build robust ; colour glossy black, sparingly marked with bright greenish-yellow.

Head large, triangular, frons well angulated, shallowly grooved above ; occiput simple. Wings palely tinted at base ; reticulation very close ; tornus angulated, prominent, base of wing oblique, moderately excavated ; membrane narrow, almost obsolete ; anal triangle 4-celled ; arc situated between the first and second or second and third antenodal nervures ; 3 to 5 transverse nervures between the sectors of arc from arc to bifurcation of *Rs* in fore-wings, 2 to 3 in hind-wings ; 1 to 2 rows of postanal cells in fore-wings, 3 to 4 in hind-wing ; anal loop present ; first postanal cell in hind-wing divided and extending inwards well proximal to the proximal end of base of subtrigone; basal incomplete antenodal nervures absent; nodal index high; primary antenodals the first and the fifth or sixth; discoidal cells entire, that of fore-wing with distal side longer than costal and the latter slightly longer than basal, that of hind-wing with costal and distal sides subequal and both nearly twice the length of basal, this cell very elongate in the length of wing; pterostigma rather short and stout, braced, equal in length to about one-third the distance from node to proximal end of pterostigma; *IA* in fore-wing markedly pectinate; *Cuii* and *IA* in hind-wing parallel as far as border of wings; only 1 cubital nervure in fore-wing, 1 to 2 in hind-wing; all subtrigones and hypertrigones entire. Legs rather short, hind femora extending only as far as base of first abdominal segment,

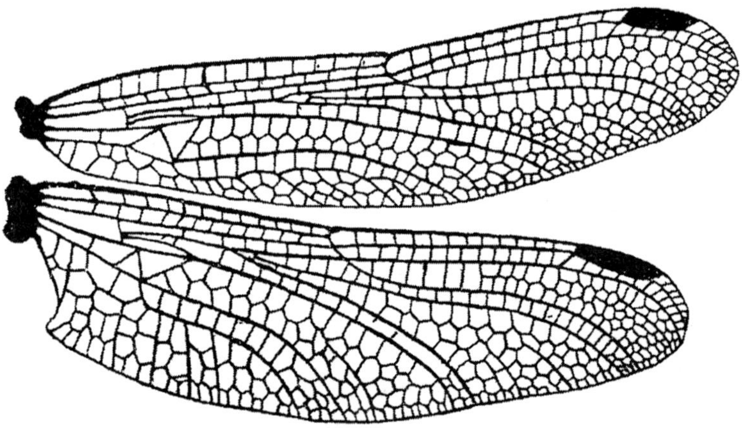

Fig. 104.—Wings of *Acrogomphus fraseri* Laidlaw, male.

furnished with a group of short, numerous, closely-set spines on the proximal half of the flexor surface, which merge into two rows of slightly longer and more robust spines at the distal end; tibial spines numerous, closely-set and rather short. Abdomen tumid at base, narrow and cylindrical from segment 3 as far as the middle of segment 7, from which point it gradually widens again, segments 8 and 9 having short but distinct wing-like prolongations laterally. Anal appendages: superiors simple, curved, and slightly forcipate, tapering to the end, rather longer than segment 10; inferior bifid, its long branches slightly divaricate. Genitalia: lamina narrow, arched, and prominent; anterior hamules long narrow processes ending in a small recurved hook-like spine; posterior hamules broad, flattened, sinuous processes, recurved back and tapering to a fine point; lobe moderately large, scoop-shaped.

Genotype, *Acrogomphus fraseri* Laidlaw.

Distribution.—WESTERN INDIA, from the montane areas of Coorg, Annaimallai, and Mudis Hills, and the High Range of Travancore, ASSAM, BURMA, Malaysia, and Indo-China.

Species of the genus are arboreal by nature, often resting at great heights on trees, only occasionally coming down to the beds of turbulent mountain streams, in which they breed in wild areas. Only one species occurs within our limits. (*O. circularis* Selys is not an *Acrogomphus*, as formerly surmised by Laidlaw; a specimen of this insect which I possess has the venation typical of that of *Onychogomphus.*)

The larvæ resemble those of *Burmagomphus* (p. 212).

315. **Acrogomphus fraseri** Laidlaw (Figs. 104 & 105.)

Acrogomphus fraseri Laidlaw, Proc. Zool. Soc. Lond. pp. 441–443 (1925); Fraser, J. Bombay Nat. Hist. Soc. vol. xxxi, pp. 740, 741, text-fig. 3, pl. i, fig. 6 (1926); id., Rec. Ind. Mus. vol. xxxiii, pp. 459, 460 (1931); Needham, Rec. Ind. Mus. vol. xxxiv, pp. 217, 218 (1932).

Male.—Abdomen 44 mm. Hind-wing 38 mm.

Head: labium variegated with shades of brown; labrum black, with a greenish-yellow semilunar spot on either side of the middle line; bases of mandibles, anteclypeus, the upper part of frons, and its anterior border yellow. Base of frons above, vertex, and occiput black, the latter simple, emarginate, and with a minute notch at its centre. *Prothorax* black, unmarked. *Thorax* black, marked with bright citron-yellow as follows:—A narrow interrupted mesothoracic collar, a pair of wedge-shaped antehumeral stripes not extending as far up as the alar sinus, and falling well short of the mesothoracic collar below, squared above, pointed and divergent below. Laterally a broad stripe on the mesepimeron, the posterior three-fourths of the metepimeron, and a small spot of yellow above and between them. *Legs* black, inner surfaces of anterior femora bright yellow, a small spot of the same colour at the middle of the outer side of mid-femora and a larger spot on the hind femora. *Wings* pale saffron throughout, this colour deepening gradually to a golden yellow near the bases. Nodal index variable, in three specimens $\dfrac{11\text{–}18 \mid 18\text{–}11}{12\text{–}12 \mid 12\text{–}11}$, $\dfrac{11\text{–}16 \mid 16\text{–}12}{12\text{–}12 \mid 12\text{–}12}$, $\dfrac{9\text{–}15 \mid 16\text{–}8}{10\text{–}11 \mid 11\text{–}9}$; pterostigma black, braced, covering 5 cells; 1 cubital nervure in all wings; anal loop well developed, formed of 2 cells, and extending proximally nearly to level of cubital nervure; discoidal field in fore-wing of 2 rows of cells to beyond level of node; membrane black.

Abdomen black, marked with bright citron-yellow as follows :—Segment 1 with a triangular dorsal spot, its base resting on the apical border of segment, laterally a large spot ; 2 with a linear mid-dorsal stripe slightly expanded basally, and two spots on each side, one of which includes the large oreillet, the other apical ; 3 to 6 each with a pair of subdorsal basal spots decreasing in size from 3 to 6 ; 3 also with a small mid-dorsal subbasal spot ; 7 with the basal half yellow, the yellow invaded by the black dorsal carina behind and in front, and separated narrowly from the base of segment by a narrow basal black ring ; 8 with a subbasal spot on each side, and 9 with a similar but much smaller spot (absent in some specimens). *Anal appendages* (fig. 105) : superiors dark reddish-brown, almost black ; inferior black. *Genitalia* : lamina

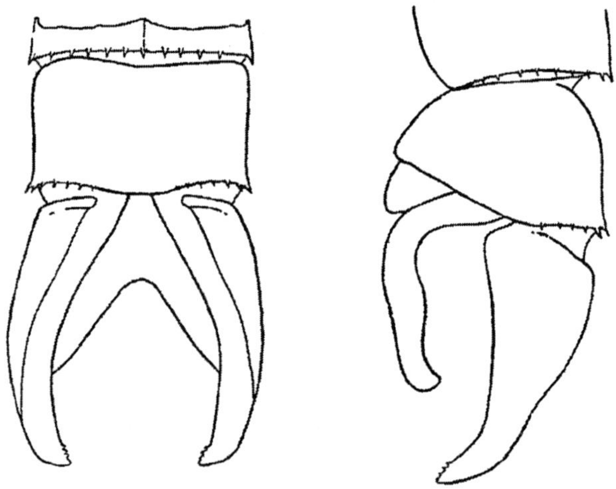

Fig. 105.—Anal appendages of *Acrogomphus fraseri* Laidlaw, male. Dorsal and left lateral views.

small, prominent, narrowly arched ; anterior hamules thin, moderately long, directed backwards, the apices ending in a small needle-like point which turns abruptly outwards ; posterior hamules somewhat similar in shape. but larger and more robust, sinuous, ending in a point which is directed backwards and inwards ; lobe scoop-shaped, prominent, black.

Female.—Abdomen 43 mm. Hind-wing 40 mm.

Very similar to the male, the wings broader, rather less deeply tinted than in the male ; abdomen broader, stouter, cylindrical throughout. Markings a little broader ; the two spots on sides of segment 2 confluent to form a complete lateral stripe ; the carinal spot on 3 larger ; lateral spots on 8 and 9 absent ; labium pale brown ; occiput similar to that of male, but with a large conical spine at each end, situated well behind inner corner of eyes. Spines on femora fewer, more widely spaced, and more robust. Nodal index rather higher, in three specimens $\frac{12\text{–}21}{14\text{–}15}\Big|\frac{18\text{–}12}{15\text{–}13}$, $\frac{13\text{–}18}{14\text{–}14}\Big|\frac{19\text{–}14}{14\text{–}14}$, $\frac{12\text{–}21}{13\text{–}15}\Big|\frac{20\text{–}12}{15\text{–}14}$; 1 cubital nervure in fore-wing, 1 to 2 in hind-wing. Vulvar scale less than half the length of segment 9, bifid almost to its base, its two branches divaricate, separated by a broad rounded notch, the branches broadly and bluntly triangular.

Distribution.—SOUTH INDIA: Coorg, at altitudes of 3,000 ft., Mudis Hills, and the montane areas of Travancore.

Only a few pairs of this rare and beautiful insect are known, but possibly its rareness is due to its habit of roosting at great heights, often as much as 100 feet or more above the ground. I have watched females ovipositing in the Sampaji stream, Coorg, generally in tunnels formed by cane-brakes arching and covering over the stream. They drop their eggs in clean water where the current is swift and the bottom sandy or gravelly. At such spots they hover about 1 inch above the surface of the stream, glide forward for a few feet, reverse rapidly in their own length, and return. This manœuvre is repeated again and again, a hovering pause taking place before each turn to give time for a fresh batch of eggs to be extruded. With each glide the insect dips and strokes the eggs off on the water's surface. Females are occasionally seen crossing open glades, flying low, and trailing the abdomen as if it were fractured. With the aid of field-glasses males could be seen at times perched on prominent dead twigs on the tops of trees, or soaring at great heights in company with *Chlorogomphus campioni* and *Zygonyx iris*.

Type in the Laidlaw collection; specimens in the British Museum, Morton, and my own collections.

Genus **MACROGOMPHUS** Selys. (Figs. 45 *a* (B) & 106.)

Heterogomphus Selys (pars), Bull. Acad. Belg. vol. xxi, pt. 2, p. 27 (1854); id., ibid. (2) vol. vii, p. 334 (1859).

Macrogomphus Selys, Mon. Gomph. pp. 87, 428 (1857); Kirby, Cat. Odon. p. 63 (1890); Williamson, Proc. U.S. Nat. Mus. vol. xxxiii, pp. 287–290 (1907); Laidlaw, Rec. Ind. Mus. vol. xxiv, p. 375 (1922); Fraser, J. Bombay Nat. Hist. Soc. vol. xxix, pp. 60, 329 (1923); id., ibid. vol. xxxi, pp. 733, 734 (1926); Laidlaw, Trans. Ent. Soc. Lond. vol. lxxviii, p. 183 (1930); Needham, Zool. Sinica, ser. A, vol. xi, fasc. 1, pp. 19, 43, 44 (1930); id., Rec. Ind. Mus. vol. xxxiv, pp. 217, 218 (1932).

Size large and build robust; colour black, marked with bright citron-yellow; eighth and ninth abdominal segments of great length.

Head moderately large, transversely narrow ; frons strongly angulate ; occiput simple, straight. Wings : reticulation very close ; tornus strongly angulate ; base of hind-wing deeply excavate ; anal triangle 3-celled ; arc opposite the second antenodal nervure or between the second and third ; 5 to 6 transverse nervures between the sectors of arc in fore-wing from arc to bifurcation of *Rs*, 3 to 4 in hind-wing ; 2 rows of postanal cells in fore-wing, 4 to 5 in hind-wing ; anal loop absent, the first postanal cell not extending proximal to the base of subtrigone ; a basal incomplete antenodal nervure present in all wings ; nodal index high ; primary antenodals the first and the fifth, sixth or seventh ; discoidal cells entire, that of the fore-wings with costal and basal sides equal, the distal but slightly longer, that of hind-wing elongate in length of wing, the distal side slightly longer than

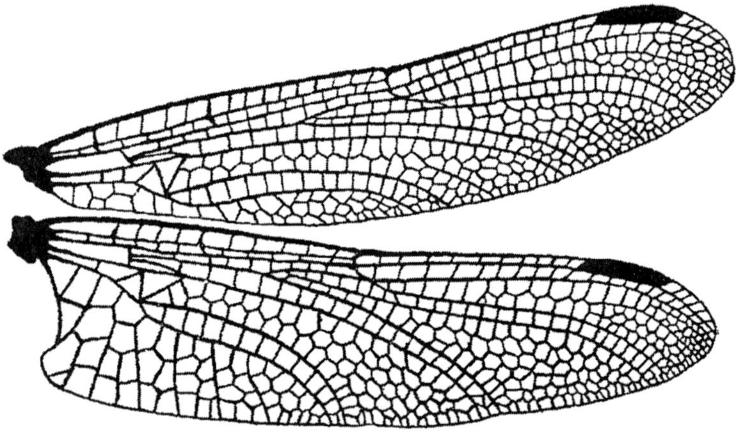

Fig. 106.—Wings of *Macrogomphus seductus* Fraser, male.

costal, the costal nearly half as long again as the basal ; pterostigma long and narrow, more than one-third as long as distance from proximal end of pterostigma to node, usually but poorly braced ; *IA* in fore-wing markedly pectinate ; *Cuii* and *IA* in hind-wing parallel to wing-border ; usually 2 cubital nervures in the fore-wings, only 1 in hind-wing ; subtrigones and hypertrigones entire in all wings. Legs robust but short, hind femora not extending beyond the posterior border of thorax, and furnished with a group of numerous short but stout spines on the flexor surface, which form two rows towards the distal end of limb ; hind tibial spines of same limb slim, closely-set, and short. Abdomen tumid at base, then slim and cylindrical from segment 3 to base of segment 7, the latter dilating apically, 8 very broad, 9 tapering from base to apex, and *nearly as long as segments 7 and 8 together*, 10 very small and very short. Anal appendages: superiors rather longer than segment 10, conical and acutely pointed at apex, furnished with a long medial inner spine, strongly divaricate; inferior broadly and deeply bifid, its two branches sinuous, long and tapering. Genitalia: lamina arched, depressed; anterior hamules short, narrowly spatulate; posterior hamules very broad and foliate, ending in a fine curled spine; lobe broadly emarginate at lip, scrotum-shaped.

Genotype, *Heterogomphus robustus* Selys.

Distribution.—WESTERN INDIA, the DECCAN, CEYLON, BENGAL, ASSAM, BURMA, Malaysia, Indo-China, Java, Sumatra, Borneo, and Tibet. Six species are known to occur within out limits, one of which is confined to Ceylon, one to the whole of Western India south of Bombay, and the others to N.E. India and Burma.

The larvæ (fig. 45 *a*, B), which breed in submontane streams, are curious insects, the terminal segments being prolonged into a narrow, tubular, syphon-like structure to enable the insect to breathe whilst the rest of its body is immersed in mud and sand. At the time of emergence, the rocks, in the bed of the streams in which they breed, are often encrusted with nymphal exuviæ of this species, and yet not a single imago will be found on the wing in the neighbourhood. In two years spent in Coorg only a single imago was seen, and yet thousands of exuviæ were observed during the same time. In the Deccan at this season great numbers of adults were observed perched on the twigs of Babul trees about a mile away from the parent river, so that it is evident that, as soon as they emerge, they betake themselves to the jungle, where copulation takes place, the female only returning to the streams to oviposit. In heavy evergreen jungle such as as found in Coorg, the adults probably perch high in trees, and so are rarely seen by the collector.

Key to Indian Species of Macrogomphus.

1.	Two narrow black stripes on sides of thorax..	*montanus* (Selys), [p. 345.
	Only a single broad black stripe on sides of thorax	2.
2.	Legs black, anterior femora on inner side yellow	3.
	Legs black, anterior femora unmarked	5.
3.	Abdomen with yellow basal rings on segments 3 to 6; vertex with a median yellow spot	*annulatus* (Selys), [p. 342.
	Abdomen with paired yellow basal spots on segments 3 to 6; vertex without a median yellow spot	4.

4.	Labrum black, with two large yellow spots at base; segment 8 with a yellow subdorsal spot on each side; prothorax marked with yellow laterally and on dorsum	*lankanensis* Fras., [p. 349.
	Labrum black, the yellow spots almost obsolete; segment 8 and prothorax unmarked	*wynaadicus* Fras., [p. 344.
5.	Labrum black, unmarked; occiput with a bifid tubercle above; at least 19 antenodal nervures in fore-wings	*robustus* (Selys), [p. 346.
	Labrum with two large yellow spots at base; occiput simple, without a bifid tubercle; not more than 15 antenodal nervures in fore-wings	*seductus* Fras., [p. 347.

316. **Macrogomphus annulatus** (Selys). (Figs. 107 & 108, **a**.)

Heterogomphus annulatus Selys, Bull. Acad. Belg. vol. xxi, pt. 2, p. 28 (1854).

Macrogomphus annulatus Selys, Mon. Gomph. pp. 92, 405 (1857); id., Bull. Acad. Belg. (2) vol. xxviii, p. 170 (1869); Kirby, Cat. Odon. p. 63 (1890); Martin, Mission Pavie Indo-Chine, vol. iii, p. 213 (1904); Williamson, Proc. U.S. Nat. Mus. vol. xxxiii, pp. 288, 290 (1907); Fraser, Rec. Ind. Mus. vol. xvi, pp. 461, 462, pl. xxxiii, fig. 3, pl. xxxiv, figs. 4, 4 *a* (Larva) (1919); id., J. Nat. Hist. Soc. Siam, vol. iii, p. 457 (1919); Laidlaw, Rec. Ind. Mus. vol. xxiv, pp. 371, 376 (1922); Fraser, J. Bombay Nat. Hist. Soc. vol. xxix, pp. 61, 329 (1923); id., Rec. Ind. Mus. vol. xxvi, pp. 427, 470, 471 (1924); id., J. Bombay Nat. Hist. Soc. vol. xxxi, pp. 734–736, text-fig. 2, i, ii, v, pl., fig. 1 (1926); Laidlaw, Trans. Ent. Soc. Lond. vol. lxxviii, p. 183 (1930); Fraser, Rec. Ind. Mus. vol. xxxiii, p. 447 (1931); Needham, Rec. Ind. Mus. vol. xxxiv, p. 220 (1932); Fraser, J. Siam. Soc., Nat. Hist. Suppl. vol. ix, pp. 137, 139, text-fig. viii (1933); id., Ceylon J. Sci., B, vol. xviii, pt. i, pp. 19, 20, 25, text-fig. 2, *a*, *b* (1933).

Male.—Abdomen 45 mm. Hind-wing 35 mm.

Head black, marked with citron-yellow as follows :—Lateral lobes of labium ; bases of mandibles ; two very large transversely oval spots on labrum ; a large spot on each side of postclypeus against the eye, connected narrowly along the lower border of postclypeus ; upper surface of frons except at its extreme base, where a small median black point projects into the yellow ; an obscure spot on the vertex. Occiput straight, slightly raised, its border fringed with black hairs. *Prothorax* black, marked with a small yellow spot on either side and a narrow anterior yellow collar. *Thorax* black, with two thick wedge-shaped citron-yellow antehumeral dorsal stripes, tapering to a point above, broadly confluent with a narrow interrupted mesothoracic yellow collar below, the inner border of these stripes parallel, the outer divergent below. Middle of ante-alar sinus narrowly yellow, sides very broadly so, and marked by a broad medial oblique black stripe which is bifid below and encloses a yellow spot. *Legs* black, coxæ and trochanters with a large yellow spot. *Wings* hyaline, with a pale yellow ray in the cubital space. Nodal index $\frac{9\text{–}16}{10\text{–}11}\Big|\frac{16\text{–}11}{11\text{–}10}$; 2 *cubital nervures in all wings* ; pterostigma dark brownish-black. *Abdomen* black, marked with citron-yellow as follows:—Segment 1 with the sides and dorsum broadly; 2 with a lanceolate dorsal bilobed stripe, the oreillets broadly and a latero-basal stripe; 3 to 6 with broad basal complete rings equal to nearly one-fourth the length of segments; 7 with its basal half yellow on dorsum, rather more than this laterally; 8 and 9 with a baso-lateral triangular spot. *Anal appendages* (fig. 107): superiors yellow, the inner border, branch, and extreme apex rusty brown; inferior black. *Genitalia*: lamina forming a gothic arch, prominent, coated with straight hairs; anterior hamules moderately

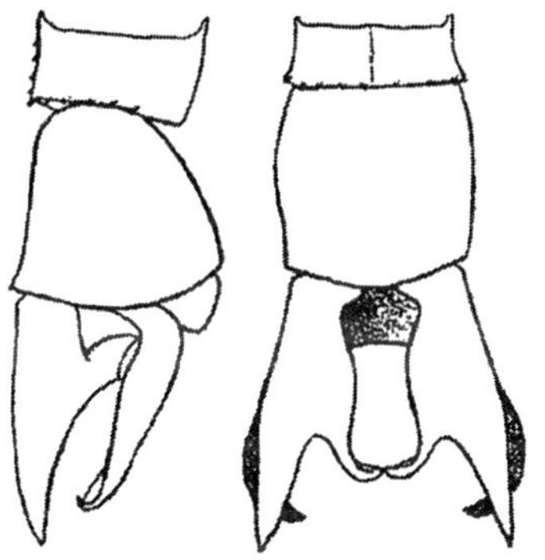

Fig. 107.—Anal appendages of *Macrogomphus annulatus* (Selys), male. Right lateral and dorsal views.

long, sinuous, pointed at apex as seen in profile, spatulate as seen from below; posterior hamules very robust, broad, sloping backwards and markedly projecting, the extreme apex ending in a fine curled spine directed forwards and inwards; lobe scrotum-shaped, black, corrugated, deeply bifid into two conical nipple-like processes.

Female.—Abdomen 49 mm. Hind-wing 38 mm.

Differs in a few respects from the male, as follows:—Yellow markings more extensive; stripe on frons broader; the spot on vertex large and conspicuous, lying just in front of occiput; thorax nearly always with an upper humeral spot (rarely present in the male), and nearly always with two spots on the lateral black band; pterostigma slightly longer, nodal index higher, $\frac{11\text{–}17}{11\text{–}14}\Big|\frac{18\text{–}9}{12\text{–}10}$; occiput with a bifid or trifid tubercle on its hinder border, occasionally separated as 2 or 3 robust spines; abdominal segment 9 longer and 10 shorter than in the male. Vulvar scale glossy black, short, broad at base, tapering to a blunt point, bifid at apex, but the two branches closely apposed.

Distribution.—The DECCAN bordering on the Western Ghats. The *type*, in the British Museum, probably came from Poona or Satara, where I found the species quite common during September and October. It breeds in the Moolah River and from thence flies inland for long distances, nearly every Babul tree harbouring one or two specimens.

317. **Macrogomphus wynaadicus** Fraser.

Macrogomphus wynaadicus Fraser, Rec. Ind. Mus. vol. xxvi, pp. 427, 471, 472 (1924); id., J. Bombay Nat. Hist. Soc. vol. xxxi, pp. 736, 737 (1926); Laidlaw, Trans. Ent. Soc. Lond. vol. lxxviii, p. 183 (1930); Fraser, Rec. Ind. Mus. vol. xxxiii, p. 447 (1931); Needham, Rec. Ind. Mus. vol. xxxiv, p. 220 (1932); Fraser, J. Siam. Soc., Nat. Hist. Suppl. vol. ix, p. 139 (1933).

Male unknown.

Female.—Abdomen 53 mm. Hind-wing 45 mm.

Closely related to *M. annulatus*, of which it may be a subspecies, but larger, and differs as follows:—Yellow spots on labrum almost obsolete; face entirely black save for a small spot at each end of postclypeus; occiput a little emarginate and without any vestige of the spines found in *M. annulatus*, a dark brown line anteriorly replaces the conspicuous yellow spot seen in that species; prothorax entirely black; thorax without any humeral marking, the lateral black line, on the contrary, is marked with a medial yellow interrupted line. *Wings*, in the type, rather darkly and evenly enfumed, but this may be mere evidence of old age; reticulation closer than in *M. annulatus*, the pterostigma covers 6–7 cells instead of only 4, and the nodal index is higher, $\frac{14\text{–}20 \mid 20\text{–}14}{13\text{–}15 \mid 14\text{–}13}$; 2 cubital nervures in all wings. *Abdomen* differing more markedly in having paired spots instead of complete rings; segment 2 with the dorsal stripe broadly broken at its centre; 3 with a large baso-lateral dorsal spot widely separated from its fellow by the black dorsal carina; 4 to 6 similar, but the spots smaller; 7 with the yellow basal half split as far as its base by a black dorsal carina; 10 longer and narrower than in *M. annulatus*. Vulvar scale shorter, with convex hinder border, not bifid.

Distribution.—SOUTH-WEST INDIA: Nilgiri Wynaad, Coorg, and the Western Ghats from North Kanara to Malabar.

The adult is rarely seen, although the larvæ and exuvia may be found abundantly in any stream throughout the area.

I have seen vast numbers of exuvia on rocks in the Kibribetta stream, North Coorg, but not one imago was to be seen in the neighbourhood.

Type in my own collection.

318. **Macrogomphus montanus** Selys. (Fig. 108, **b**.)

Macrogomphus montanus Selys, Bull. Acad. Belg. (2) vol. xxviii, p. 171 (1869); Kirby, Cat. Odon. p. 63 (1890); Williamson, Proc. U.S. Nat. Mus. vol. xxxiii, pp. 289, 290 (1907); Laidlaw, Rec. Ind. Mus. vol. xxiv, pp. 371, 377, 378, text-fig. 2 (1922); Fraser, J. Bombay Nat. Hist. Soc. vol. xxix, pp. 61, 329 (1923); id., Mem. Dept. Agric. India (Ent.), vol. viii, p. 81, pl. ix, figs. 5, 7 (1924); id., J. Bombay Nat. Hist. Soc. vol. xxxi, pp. 738, 739, pl., fig. 3 (1926); Laidlaw, Trans. Ent. Soc. Lond. vol. lxxviii, p. 183 (1930); Needham, Zool. Sinica, ser. A, vol. xi, fasc. 1, p. 44 (1930); id., Rec. Ind. vol. xxxiv, p. 220 (1932); Fraser, J. Siam. Soc., Nat. Hist. Suppl. vol. ix, p. 139 (1933).

Male.—Abdomen 50 mm. Hind-wing 38–40 mm.

Head black, marked with yellow as follows :—Two rather large oval basal spots on labrum ; whole of labium ; lower border of postclypeus ; a stripe on upper border of frons. Occiput brown, its border raised into a small conical tubercle at the middle. *Prothorax* black, unmarked. *Thorax* black, marked with antehumeral stripes similar to those of *M. seductus*, but rather longer and wider (roughly these marks may be compared to the outline of a tintack with its head below) ; in

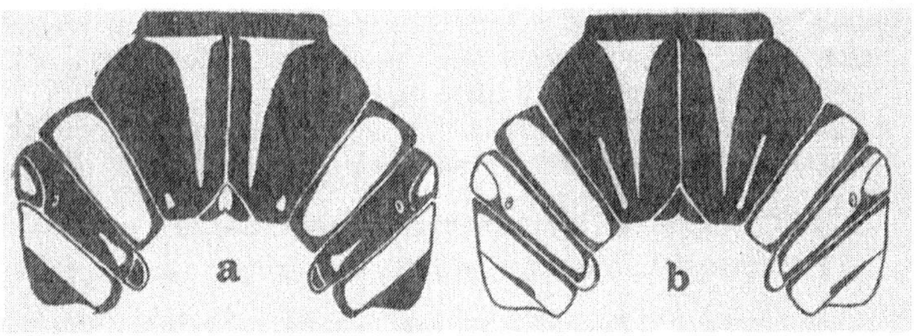

Fig. 108.—Thoracic markings of (**a**) *Macrogomphus annulatus* (Selys), male ; (**b**) *Macrogomphus montanus* (Selys), male.

addition, there is a narrow humeral stripe gradually tapering away to a fine point below. Sides broadly yellow, the two lateral sutures finely mapped out in black. *Wings* hyaline, palely enfumed, reticulation close, brown ; pterostigma dark brown, covering 5 to 6 cells ; nodal index $\frac{12\text{–}18}{10\text{–}11} \Big| \frac{20\text{–}11}{12\text{–}11}$. *Legs* dark blackish-brown. *Abdomen* black, marked with yellow as follows :—Segment 1 almost entirely yellow ; 2 with a basal ring confluent with a dorsal stripe and a broad spot on the oreillets ; **3** to **6** with basal rings as in *M. annulatus*, but longer and occupying about one-third of the segments ;

3 and 4 with these rings prolonged along the dorsum shortly; 7 with the basal half yellow and also prolonged along the dorsum; 9 with a small baso-lateral spot. *Anal appendages* very similar to those of *M. annulatus*, thick at base, divaricate, tapering to a point in the apical half; the inner branch springing from the middle of appendage, sloping downwards and inwards, black in colour, its apex bevelled to a point and extending a little beyond the apices of superior appendages; inferior appendage black, nearly as long as superiors, widely and deeply forked, branches more divaricate than superiors.

Female.—Abdomen 47 mm. Hind-wing 40 mm.

Very similar to male, differs in having the yellow markings rather more extensive. Wings tinted with yellow at base and along costal margin; pterostigma pale brown; nodal index slightly higher. Occiput yellow, its border bicrenulate and with a tubercle situated behind, similar to but larger than that of male. (This tubercle is not visible from the front.) Segment 2 of abdomen has the dorsal stripe broader and bilobate, the rings on the other segments distinctly broader. Anal appendages yellow, conical, very tiny and inconspicuous. Vulvar scale similar to that of *M. seductus*. Legs dark reddish-brown; tibiæ on the outer side, and also the hind femora, reddish-brown.

Distribution.—Sylhet, ASSAM.

This species is at once distinguished from all other Indian members of the genus by the presence of a humeral stripe and the sides of the thorax being marked with two fine black lines instead of the thick medial black stripe.

Type in the Selys collection. A pair in the Indian Museum determined by Selys, but without data as to locality.

Macrogomphus robustus (Selys). (Fig. 109, a.)

Heterogomphus robustus Selys, Bull. Acad. Belg. vol. xxi, pt. 2, p. 27 (1854).

Macrogomphus robustus Selys, Mon. Gomph. p. 88 (1857); Kirby, Cat. Odon. p. 63 (1890); Williamson, Proc. U.S. Nat. Mus. vol. xxxiii, pp. 289, 290 (1907); Laidlaw, Rec. Ind. Mus. vol. xxiv, pp. 371, 375, 376, text-fig. 1 (1922); Fraser, J. Bombay Nat. Hist. Soc. vol. xxix, pp. 61, 329 (1923); id., ibid. vol. xxxi, p. 737 (1926); Laidlaw, Trans. Ent. Soc. Lond. vol. lxxviii, p. 183 (1930); Needham, Zool. Sinica, ser. A, vol. xi, fasc. 1, p. 44 (1930); id., Rec. Ind. Mus. vol. xxxiv, p. 220 (1932); Fraser, J. Siam. Soc., Nat. Hist. Suppl. vol. ix, p. 139 (1933).

Male.—Abdomen missing in the type, but by analogy with other species about 45 mm. Hind-wing 42 mm.

Head black, frons depressed, not notched, forming a very obtuse angle in front, marked with an orange stripe in front of ocelli; a blackish spot at side of mandibles; ocelli and lower lip partly brown; an obscure brown spot on clypeus; occiput not scale-like, but with a large raised bifid tubercle at its middle, hairy in front and behind. *Prothorax* black, with a basal ring and a lateral spot yellow. *Thorax* black, with 6 orange stripes, the two in front cuneiform, separated from each other, broader towards the mesothoracic notch, which they do not quite reach, their apex not quite reaching the alar sinus; two lateral, slightly oval stripes, situated under the wings. A yellow spot at middle of antealar sinus and a large dorsal band of same on interalar space. *Legs* entirely black. *Wings* moderately broad, tinted with yellow ochre, especially towards the base; reticulation black; pterostigma brown between black nervures, a little yellowish, moderately

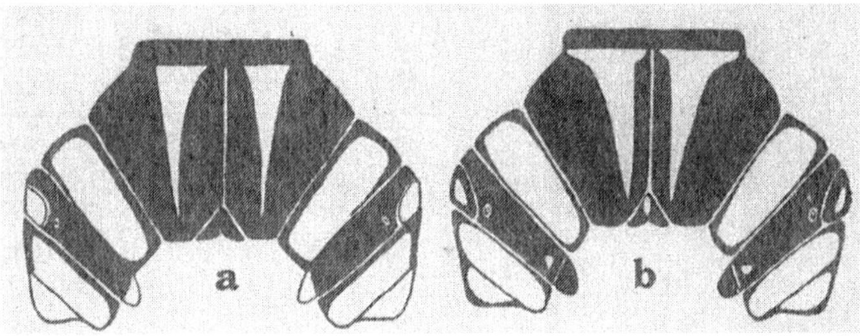

Fig. 109.—Thoracic markings of (**a**) *Macrogomphus robustus* Selys, male; (**b**) *Macrogomphus seductus* Fraser, male.

broad, 4 to 5 mm. long, covering 6 cells; anal border very excavate; membrane black, very slender, but extending as far as anal angle; 19 antenodal nervures in fore-wings, 14 in hind-wings, 13 postnodal nervures in all wings.

Female.—Unknown.

Distribution.—Tibet.

This species is distinguished by its entirely black labium, by the bifid tubercle on the occiput, and by its high nodal index. The latter character, and the simple occiput, have induced me to separate from it *M. seductus*, to which species belong the two males and the single female in the Indian Museum placed by Laidlaw and Selys as *M. robustus*.

Type, a unique male in the Selys collection.

319. **Macrogomphus seductus** Fraser. (Figs. 109, **b** & 110.)

Macrogomphus robustus Laidlaw (*M. seductus* Fraser, nec *M. robustus* Selys), Rec. Ind. Mus. vol. xxiv, pp. 375, 376, text-fig. 1 (1922).

Macrogomphus seductus Fraser, J. Bombay Nat. Hist. Soc. vol. xxxi, pp. 737, 738, pl., fig. 2, text-figs. 1 *a*, 2, iii, iv, v (1926); Laidlaw, Trans. Ent. Soc. Lond. vol. lxxviii, p. 184 (1930); Needham, Rec. Ind. Mus. vol. xxxiv, p. 220 (1932); Fraser, J. Siam. Soc., Nat. Hist. Suppl. vol. ix, p. 139 (1933).

Male.—Abdomen 50 mm. Hind-wing 40 mm.

Head: labium black; labrum glossy black, with two transversely-elongate greenish-yellow spots at base; anteclypeus black; postclypeus black, but its lower border and a large spot on either side against the eyes greenish-yellow; frons black in front, broadly yellow above, where its base only is finely black; vertex and occiput black, the latter simple, neither spined nor tuberculated, fringed with long black hairs behind. *Prothorax* black, marked with a narrow anterior yellow collar. *Thorax* black, marked with citron-yellow as follows:—Antehumeral stripes tapering to a fine point above at the alar sinus, very broadly confluent below with a slightly interrupted mesothoracic collar. No humeral marking. Sides broadly yellow, marked with a broad medial oblique black band, which is itself marked with an upper small spot and a lower large spot. *Legs* entirely black, but trochanters and coxæ marked with yellow. *Wings* tinted with pale yellow, more

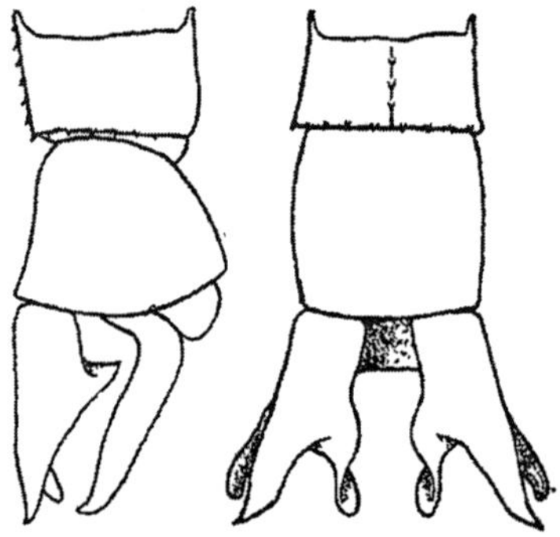

Fig. 110.—Anal appendages of *Macrogomphus seductus* Fraser, male. Right lateral and dorsal views.

deeply so at bases, especially in the cubital space; pterostigma black (dark brown by transmitted light), covering 5 to 6 cells, braced; nodal index $\frac{12\text{–}15}{10\text{–}10}\Big|\frac{14\text{–}10}{11\text{–}10}$; membrane brown; anal triangle 3-celled. *Abdomen* black; segment 1 with its apical half citron-yellow, the basal black area encroaching on this subdorsally, where is a well-marked ridge resembling a second rudimentary oreillet bearing a fringe of long hairs; 2 with a broad basal ring, the oreillets except for a fine blackish-brown border, and its sides broadly yellow, the black on this segment restricted to a broad subdorsal band ending basally abruptly at the jugal suture; 3 to 6 with broad basal yellow rings slightly notched by the black dorsal carina and covering nearly one-third of segment 3, one-fourth of 4 and rather less of 5 and 6; 7 with the basal half yellow; 8 with an irregular baso-lateral yellow stripe; 9 and 10 unmarked. *Anal appendages* (fig. 110): superiors pale yellow, the apices carneous, as also the inner branch; inferior glossy black. *Genitalia*: lamina narrow, more prominent than in *M. annulatus*, deeply and roundly arched; anterior hamules spatulate, narrow at base, broadening apically, convergent; posterior hamules very robust, flattened, bluntly pointed at the apex, where there is a pencil of long fine hairs; subapically is a small recurved spine; lobe similar to that of *M. annulatus*.

Female.—Abdomen 56 mm. Hind-wing 44 mm.

Very similar to male. Occiput slightly concave, simple. *Prothorax* black, with three tiny obscure spots on the posterior lobe in addition to the anterior collar. *Thorax* and *legs* not differing from those of the male. *Wings* enfumed, the tinting only evident at the bases; nodal index $\frac{12\text{–}19}{11\text{–}12}\Big|\frac{18\text{–}12}{13\text{–}11}$; 2 cubital nervures in all wings. *Abdomen* similar to that of the male but the subdorsal black fascia on segment 2 confluent over the dorsum, and the dorsal carina on segments 4 to 7, especially 6 and 7, black, the basal rings being split into basal paired spots. *Anal appendages* creamy-yellow, very short, conical. Vulvar scale very short, glossy black, split as far as its base into two small but broad triangular processes.

Distribution.—A pair in my collection (the male the *type*) from Hasimara, Duars, BENGAL (*H. V. O'Donel*). Two males and a female in the Indian Museum from Sibsagar, ASSAM, the males doubtfully referred to *M. robustus* by Laidlaw, the female labelled *M. robustus* by De Selys himself, although he never described the female of *M. robustus*, and mentions it as unknown.

The differences between *M. robustus* and *M. seductus* have already been pointed out under the description of the former. From *M. montanus* the present species is easily distinguished by the total absence of a humeral stripe, etc.

320. **Macrogomphus lankanensis** Fraser. (Fig. 111.)

Macrogomphus annulatus Selys (*M. lankanensis*, nec *annulatus*) Laidlaw, Spolia Zeylanica, vol. xii, p. 340 (1912).
Macrogomphus annulatus ? Laidlaw (*M. lankanensis* Fraser, nec *M. annulatus* Selys), Spolia Zeylanica, vol. xii, p. 340 (1924).
Macrogomphus lankanensis Fraser, Ceylon J. Sci., B, vol. xviii, pp. 20, 24–26, text-fig. 2, *c* & *d* (1933).

Male.—Abdomen 45 mm. Hind-wing 33 mm.

Head: labium black, lateral lobes yellowish; labrum black, marked with two large triangular citron-yellow spots; bases of mandibles citron-yellow; ante- and postclypeus black, the latter with two small yellow spots below; frons black, the crest broadly yellow, this colour overlapping on to upper half of anterior surface, the base of frons above, vertex, and occiput black, the latter fringed with short black hairs. *Prothorax* black, with a moderately large rounded spot on each side of the middle lobe and a small oval one on middle of posterior lobe. *Thorax* black, marked with citron-yellow as follows :—An antehumeral stripe parallel with the mid-dorsal carina, thickened below and tapered above as far as the ante-alar sinus, this stripe shaped like a tintack with the head below ; a small superior humeral spot ; two broad lateral stripes, the posterior of which covers the anterior half of the metepimeron, and between these stripes an upper and a lower spot. *Legs* black, fore femora with an oval yellow spot on the inner side. *Wings* hyaline ; pterostigma blackish-brown, braced, covering five cells : anal triangle 3-celled ; 2 cubital nervures in fore-wings, 1 in hind-wing ; nodal index $\frac{10\text{–}16}{10\text{–}11}\left|\frac{15\text{–}11}{11\text{–}10}\right.$. *Abdomen* black, marked with

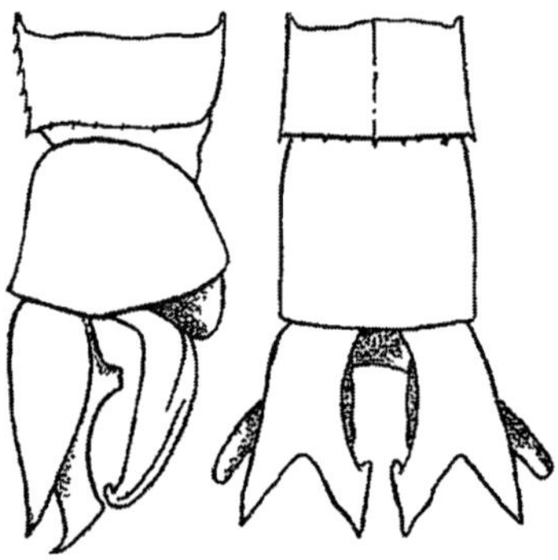

Fig. 111.—Anal appendages of *Macrogomphus lankanensis* Fraser, male. Dorsal and right lateral views.

citron-yellow as follows :—Segment 1 with a mid-dorsal apical triangular spot and another large triangular apical spot on each side ; 2 with a trilobate mid-dorsal stripe in continuation of the dorsal spot on segment 1, a very large spot on each side which includes the oreillets, and a small triangular apical lateral spot ; 3 to 6 with large baso-lateral spots ; 7 with its basal half yellow, this area with a small black triangular baso-dorsal point ; 8 with a linear subdorsal spot on each side ; remaining segments unmarked. *Anal appendages* (fig. 111) : superiors creamy yellow except the tip of inner branch, which is black ; inferior black.

Distribution.—CEYLON only. Murunkhan, N.P., and Haragama, from June to August ; probably widely distributed in various parts of the island in submontane areas.

Differs from *M. annulatus*, its nearest relative, in the shape of the anal appendages and in the less extensive yellow markings. I have not had an opportunity of studying the female, but according to Laidlaw the yellow spots on the labrum are smaller, the postclypeus is almost entirely black, with two very small yellow points, the occiput is entirely black, the antehumeral stripes are narrower, and the humeral spot is obsolete.

Type male in the British Museum, formerly in the Colombo Museum.

Genus **MICROGOMPHUS** Selys. (Figs. 45 *a* (A) & 112.)

Microgomphus Selys, Mon. Gomph. p. 100 (1857); id., Bull. Acad. Belg. (2) vol. vii, p. 533 (1859); Kirby, Cat. Odon. p. 63 (1890); Williamson, Proc. U.S. Nat. Mus. vol. xxxiii, pp. 295, 296 (1907); Laidlaw, Rec. Ind. Mus. vol. xxiv, p. 380 (1922); Fraser, J. Bombay Nat. Hist. Soc. vol. xxix, pp. 60, 329 (1923); id., ibid. vol. xxx, pp. 851, 852 (1925); Laidlaw, Trans. Ent. Soc. Lond. vol. lxxviii, p. 182 (1930); Needham, Zool. Sinica, ser. A, vol. xi, fasc. 1, p. 19 (1930); id., Rec. Ind. Mus. vol. xxxiv, pp. 217, 218, 221 (1932).

Size very small, the genus containing the smallest species of the family; colour black, marked with greenish-yellow; superior anal appendages curiously branched.

Head comparatively large for the size of the insects, broad, frons angulated, occiput flat, ridged, concave posteriorly. Wings: reticulation moderately close; tornus angulated; base of wing oblique; slightly excavated; membrane obsolete; anal triangle 3-celled; arc situated opposite the second antenodal nervure; 5 to 6 transverse nervures between the sectors of arc from arc to bifurcation of *Rs* in fore-wings, 3 to 4 in hind-wings; only 1 row of postanal cells in fore-wings, 2 to 3 rows in hind-wings; anal loop absent, the first postanal cell in hind-wing extending proximal to the middle of subtrigone only; no incomplete basal antenodal nervures present; nodal index high; primary antenodal nervures the first and the fifth; discoidal cells entire, that of fore-wing an equilateral triangle, that of hind-wing with the distal side longer than the costal and the latter longer than the basal; pterostigma short and stout, equal to about one-third the distance from node to proximal end of pterostigma, usually braced; *IA* in fore-wing not pectinate, very flatly curved, and with only a single row of cells between it and the border of wing; *IA* and *Cuii* in the hind-wing divergent only at the wing-border; only 1 cubital nervure in all wings; subtrigones and hypertrigones entire in all wings. Legs short, the hind femora extending back as far as the apical border of abdominal segment 1 and furnished with a group of very short, numerous, and closely-set spines which merge into two rows at the extreme distal end of limb; hind tibial spine fine and moderately long and widely spaced. Abdomen short, tumid at base, slim and cylindrical from segment 3 to base of 7, then markedly expanded again, especially segments 8 and 9. Anal appendages: superiors rather longer than segment 10, conical, pointed at apex, and with a slim medial spine or hook; inferior triangular, bifid only at apex, the two short branches curled upwards. Genitalia: lamina depressed, deeply emarginate; anterior hamules slim and rather short processes; posterior hamules robust, flattened, ending in a short recurved hook; lobe very prominent, purse-shaped, emarginate at lip.

Genotype, *Microgomphus chelifer* Selys.

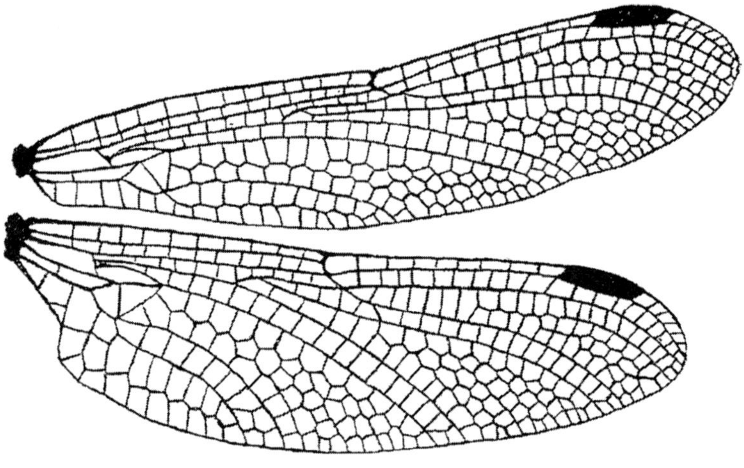

Fig. 112.—Wings of *Microgomphus torquatus* (Selys), male.

Distribution.—Western India and the more humid parts of the Deccan at altitudes of 2000 and upwards, Burma, Malaysia, Java, Sumatra, Indo-China, and Borneo. Six species found within our limits.

Species of the genus are arboreal by nature, but quite occasionally the males descend and settle on rocks in mid-stream; they do not appear to wander far from their parent streams, and may be found settled on evergreens, usually beside the water. The larvæ breed in deep pools in sub-montane streams flowing through heavy primary jungle; they are broad and flat, resembling those of *Lamelligomphus*, but much smaller, and without the flattened antennæ of that genus.

Closely related by venation to the genus *Heliogomphus*.

Key to Indian Species of Microgomphus.

1.	Antehumeral stripes not confluent with the mesothoracic collar	2.
	Antehumeral stripes confluent with the mesothoracic collar	5.
2.	Segments 4 and 5 of abdomen without mid-dorsal oval yellow spots; 8 to 10 without yellow markings	3.
	Segments 4 and 5 with mid-dorsal oval yellow spots; 8 and 10 with yellow markings	*torquatus* (Selys), [p. 353.
3.	Vertex with a small yellow spot	*verticalis* (Selys), [p. 357.
	Vertex unmarked	4.
4.	Labrum black, marked with two large triangular greenish spots	*souteri* Fras., p. 355.
	Labrum unmarked	*minusculus* (Selys), [p. 360.
5.	Antehumeral stripes joining the middle of mesothoracic collar each side; stripes on sides of thorax vestigial, only a short inferior remnant present; an upper humeral spot present	*loogali* Fras., p. 358.
	Antehumeral stripes joining the ends of the mesothoracic collar so as to form inverted figure 7's; stripes on sides of thorax present and meeting to form a black Y; no upper humeral spot present	*lilliputians* Fras., [p. 357.

321. **Microgomphus torquatus** (Selys). (Figs. 112 & 113, a.)

Cyclogomphus torquatus Selys, Bull. Acad. Belg. vol. xxi, pt. 2, p. 63 (1854); id., Mon. Gomph. p. 108 (1857); Kirby, Cat. Odon. p. 69 (1890); Williamson, Proc. U.S. Nat. Mus. vol. xxxiii, p. 297 (1907).

Microgomphus torquatus Laidlaw, Rec. Ind. Mus. vol. xxiv, pp. 371, 380, 383, text-figs. 4, *a* & *b*, and 5 (1922); Fraser, J. Bombay Nat. Hist. Soc. vol. xxix, pp. 60, 329 (1923); id., ibid. vol. xxx, pp. 852, 853, pl. i, fig. 4, text-fig. 3 (1925); Laidlaw, Trans. Ent. Soc. Lond. vol. lxxviii, p. 182 (1930); Needham, Rec. Ind. Mus. vol. xxxiv, p. 221 (1932).

Microgomphus torquatus torquatus Fraser, Rec. Ind. Mus. vol. xxvi, pp. 427, 474 (1924); id., ibid. vol. xxxiii, p. 447 (1931).

Male.—Abdomen 22–24 mm. Hind-wing 21 mm.

Head: labium yellow; labrum greenish-yellow, the base and front border narrowly black, the basal marking produced as a triangular mark in the middle line; frons, bases of mandibles, ante- and post-clypeus greenish-yellow, the frons with a broad black transverse stripe where it joins the post-clypeus, and sending downwards a short oblique submedial prolongation on the postclypeus at either side; vertex and occiput black, the latter with a low, slightly concave crest which is bordered with about ten minute spines. *Prothorax* black, with an anterior collar, a small medial spot on the posterior lobe and a larger spot on each side yellow. *Thorax* black, marked with yellow as follows:—A complete mesothoracic collar which is confluent with a fine yellow line on the lower part of the mid-dorsal carina; a very oblique fusiform antehumeral stripe, not confluent with the mesothoracic collar, nor reaching the alar sinus above; an upper humeral spot which splits the humeral black above so as to form an upright black Y. Laterally greenish-yellow, with a black stripe on the posterior suture, which is joined at its middle by a narrow oblique black line descending from beneath the fore-wing and is thus converted into a second black Y. *Legs* black, hind femora on the extensor surface with a greenish-yellow stripe which tapers apically. *Wings* hyaline; pterostigma pale brown between black nervures, unbraced, covering 4 cells; 3 rows of postanal cells in fore-wing; nodal index $\frac{9\text{–}12}{9\text{–}10} \Big| \frac{12\text{–}8}{9\text{–}8}$. *Abdomen* black, marked with bright greenish-yellow as follows:—Segment 1 with an apical spot on mid-dorsum, narrowly confluent with a broad lateral fascia along the apical border of segment; 2 with a mid-dorsal bilobed spot not quite reaching the apical border, the sides broadly yellow, including the large oreillets, the subdorsal black stripes curving down behind the oreillets; 3 similarly marked, but the lateral yellow border usually finely divided by the black transverse suture; 4 and 5 with broad basal rings occupying about one-third the length of segments, a longitudinal oval spot on the mid-dorsal carina at the middle of each segment; 6 similar, but the dorsal spot absent, the basal ring on the sides overlapping the transverse suture for a short distance; 7 with its basal half or more yellow, this colour extending nearer the apical border on the sides than on the dorsum; 8 very similar to 7; 9 with only a small ventro-lateral medial spot; 10 unmarked. *Anal appendages* (fig. 113, *a*): superiors yellow, tipped on the outer side and bordered beneath with black. (In many specimens the superior appendages have a minute but distinct spine on the outer side near the apex, but it does not appear to be at all constant.) *Genitalia*: lamina black, depressed, broadly emarginate; inner hamules slim, spine-like, shorter than the outer, which are robust, parallel-sided, with the outer corner acutely spined and projecting backwards and downwards from the genital sac, yellow at base, black at apex; lobe of penis bulbous, black, deeply notched, very prominent.

Female.—Abdomen 27 mm. Hind-wing 23 mm.

Closely resembling male, but more robust and larger. Occiput shallowly concave, with 4 to 5 minute black spines bordering it on either side of the middle line. *Wings* tinted at the bases; two rows of cells between *IA* and the hinder margin of fore-wing. Lateral markings of *abdomen* more extensive; segments 4 to 6 with a long oval spot on the middle of each side; the mid-dorsal oval spot on segment 5 absent; 7 with the lateral yellow finely divided by the black transverse suture; 8 with the basal ring interrupted on the mid-dorsal carina. *Anal appendages* very small, yellow, conical, pointed. Vulvar scale bright yellow, triangular, half the length of segment 9, bifid for half its length, the branches closely apposed.

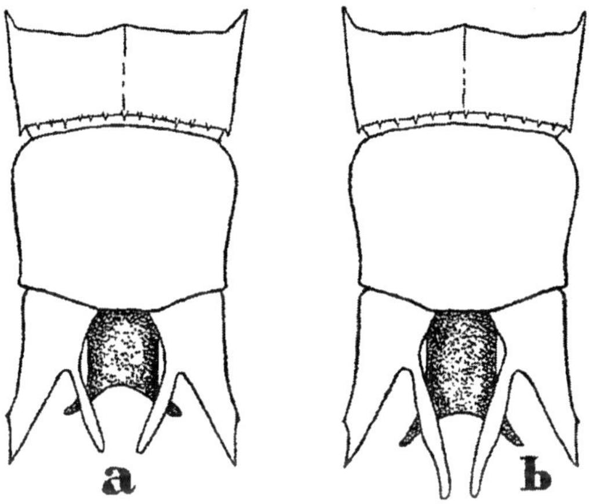

Fig. 113.—Anal appendages of (**a**) *Microgomphus torquatus* (Selys), male; (**b**) *Microgomphus souteri* Fraser, male. Dorsal views.

Distribution.—The western parts of the Deccan, Poona and Satara. The *type*, a female in the Selys collection, is probably from the former locality, where I have found it rather common during the wetter months of the year. Males may be seen resting on stones or rocks in the bed of the Byrobah Nullah and also in the neighbouring canal and Mullah River, Poona, or they may be beaten up from evergreens bordering these streams.

322. **Microgomphus souteri** Fraser. (Fig. 113, **b**.)

Microgomphus torquatus souteri Fraser, Rec. Ind. Mus. vol. xxvi, pp. 427, 474 (1924); id., ibid. vol. xxxiii, p. 447 (1931).
Microgomphus souteri Fraser, J. Bombay Nat. Hist. Soc. vol. xxx, pp. 853, 854, pl. i, fig. 5, text-fig. 3 (1925); Laidlaw, Trans. Ent. Soc. Lond. vol. lxxviii, p. 182 (1930); Needham, Rec. Ind. Mus. vol. xxxiv, p. 221 (1932).

Male.—Abdomen 29 mm. Hind-wing 24 mm.

Head: labium pale yellow; labrum glossy black, with two large triangular greenish-white spots nearly confluent at the middle and separated from the anterior border by a narrow black stripe; anteclypeus greenish-white; postclypeus and front of frons glossy black, the lower outer part of the former greenish-white, continuing the colour of the anteclypeus as a transverse stripe, which traverses the whole face at that level; upper surface of frons greenish-white; vertex and occiput black, latter slightly concave, fringed with yellow hairs, but no minute spines visible; eyes bottle-green. *Prothorax* black, with a broad greenish-yellow anterior collar. *Thorax* black, markings differing from those of *M. torquatus* as follows:—Mesothoracic collar slightly interrupted in the middle; mid-dorsal carina not marked with yellow below; no upper humeral spot; antehumeral stripes narrower and shorter; black stripe on postero-lateral suture broader, usually so broadly confluent with the shorter oblique stripe that the enclosed yellow ground-colour is blotted out. *Legs* entirely black. *Wings* with reticulation closer and the nodal index higher, $\frac{9\text{–}15}{12\text{–}11}\Big|\frac{16\text{–}11}{11\text{–}11}$, otherwise resembling those of *M. torquatus*. *Abdomen* black, marked with greenish-yellow as follows:—Segment 1 with a narrow apical border; 2 with a broad basal ring, which includes the oreillets and a short apico-lateral spot (not always present); 3 to 6 with narrow basal rings occupying about one-sixth of the segments; 7 similar, but the ring occupying the basal third and laterally overlapping the transverse suture; 8 to 10 unmarked. *Anal appendages* (fig. 113, b) differing markedly from those of *M. torquatus* as follows:— Superiors with the inner branches much longer, springing from the appendages much nearer the base, and extending beyond their apices; the small outer spine near the apex much more noticeable; inferior more robust, slightly longer than in *M. torquatus*. *Genitalia* similar to those of *M. torquatus*, but the apex of outer hamules curling forwards.

Female.—Abdomen 28 mm. Hind-wing 25 mm.

Differs from the male in markings in the same respects as does the female from the male of *M. torquatus*; thus there are lateral oval spots on abdominal segments 4 to 6 (but much smaller than in *M. torquatus*), and segments 8 to 10 are without markings; the basal marking on segment 7 is strictly limited behind by the transverse suture.

Distribution.—Confined to COORG and SOUTH KANARA.

Habits similar to the last. Breeds in the Sampaji and Hallery Rivers and the Kibribetta stream, 2,000–3,500 ft., Coorg.

Type in the British Museum; paratypes in my own collection.

323. **Microgomphus ? verticalis** (Selys).

Cyclogomphus verticalis Selys, Bull. Acad. Belg. (2) vol. xxxvi, p. 501 (1873); Kirby, Cat. Odon. p. 70 (1890); Williamson, Proc. U.S. Nat. Mus. vol. xxxiii, pp. 296, 297 (1907); Fraser, J. Bombay Nat. Hist. Soc. vol. xxix, pp. 64, 332 (1923).
Microgomphus verticalis Fraser, J. Bombay Nat. Hist. Soc. vol. xxx, p. 857 (1925); Needham, Rec. Ind. Mus. vol. xxxiv, p. 221 (1932).
Cyclogomphus ? verticalis Laidlaw, Trans. Ent. Soc. Lond. vol. lxxviii, p. 187 (1930).

Male unknown.

Female.—Abdomen 27 mm. Hind-wing 25 mm.

Similar to *M. torquatus*, differing only in the markings being less extensive, the face largely black. Labrum with the yellow entirely enclosed as two large spots; ante- and postclypeus black, with a medial yellow spot and another on each side against the eyes, frons similar, but the vertex with a small yellow spot between the eyes.

The short description given by Selys, meagre as it is, clearly shows that this species, which he classed as a *Cyclogomphus*, is really a *Microgomphus*, and that it is closely related to, if not conspecific with, *M. souteri*. The location of the *type* is unknown, but it was said to be in the Moore collection, and is given as from "India."

324. **Microgomphus lilliputians** Fraser.

Microgomphus lilliputians Fraser, J. Bombay Nat. Hist. Soc. vol. xxix, pp. 60, 329 (1923); id., ibid. vol. xxx, pp. 855, 856, pl. i, fig. 7 (1925); Laidlaw, Trans. Ent. Soc. Lond. vol. lxxviii, p. 182 (1930); Needham, Rec. Ind. Mus. vol. xxxiv, p. 221 (1932).

Male unknown.

Female.—Abdomen 20 mm. Hind-wing 18 mm.

Head: labium pale yellow; labrum black, with two yellow spots; bases of mandibles yellow; ante- and postclypeus black, former with a central oval spot, latter with a yellow spot on each side against the eyes; frons with the crest traversed with yellow; rest of head, including the occiput, black, latter simple, sinuous, a little notched at its middle. *Prothorax* with a small spot on the posterior lobe and a larger spot at the sides, otherwise black. *Thorax* black, marked with bright yellow as follows:—An interrupted mesothoracic collar; oblique antehumeral stripes which are broadly confluent with the mesothoracic collar and form with it inverted 7's; laterally greenish-yellow, marked on the postero-lateral suture by a thick black stripe, which, by joining an oblique black shorter stripe running from beneath the fore-wing, forms a black Y. *Legs* entirely black, the spines on hind femora rather more widely spaced and robust than is usual in the genus. *Wings* hyaline, tinted with yellow at the bases, reticulation rather close for so small an insect, discoidal cells and subtrigones all angulated where joined by secondary nervures, sectors of arc apposed for a long distance, especially in the hind-wings, discoidal field in the fore-wings beginning with a single row of cells and continued as such for a distance of 3 rows, *IA* in fore-wings with only a single row of cells between it and hinder border (and only 6 cells in length), only 2 rows of cells in the postanal area of hind-wing; pterostigma brown, unbraced in two of the wings, poorly so in the others, covering 3 cells; nodal index $\frac{7\text{–}12}{8\text{–}8}\Big|\frac{11\text{–}7}{9\text{–}7}$. *Abdomen* black, marked with yellow as follows:—Segment 1 wholly black; 2 with broad subdorsal longitudinal stripes confluent over the base of segment and finely divided transversely by the transverse suture; 3 with a continuation of these stripes, which do not, however, extend as far as the apical border; 4 to 7 with broad basal rings extending as far as the transverse sutures; remaining segments unmarked. *Anal appendages* small, conical, yellow tipped with black. Vulvar scale rudimentary.

Distribution.—A single female from Tenasserim, LOWER BURMA, 18. iv. 23.

This species, which is the smallest known Gomphine, closely resembles *Tetrathemis platyptera* in size, general appearance, and in other respects, for example the discoidal field of the wing begins with a single row of cells. The only species likely to be confounded with it is *Microgomphus chelifer* from the Malay Peninsula, which is somewhat larger, has a higher nodal index, the lateral thoracic stripe simple and not Y-shaped. The two are closely related, and may even be conspecific. *M. lilliputians* is the only species known to me in which the discoidal field of the fore-wing begins with a single cell and in which the female has only one row of cells between the hinder margin of fore-wing and *IA*.

Type in the Author's collection.

325. **Microgomphus loogali** Fraser. (Fig. 114.)

Microgomphus loogali Fraser, J. Bombay Nat. Hist. Soc. vol. xxix, pp. 60, 329 (1923); Laidlaw, Trans. Ent. Soc. Lond. vol. lxxviii, p. 182 (1930).

Microgomphus burmicus Fraser, J. Bombay Nat. Hist. Soc. vol. xxx, pp. 854, 855, pl. i, fig. 6, text-fig. 3, iii (1925); Needham, Rec. Ind. Mus. vol. xxxiv, p. 221 (1932).

Male.—Abdomen 31 mm. Hind-wing 28 mm.

Head: labium pale whitish-green, the middle lobe with its base and free border blackish-brown; labrum black, with a large rounded whitish-green spot at each outer corner (a very unusual situation for this marking); bases of mandibles greenish-white, and a transverse stripe of the same colour at the middle of anteclypeus; rest of face and upper surface of head glossy black except for the crest of frons, which is greenish-white, finely divided at its centre by a narrow black isthmus. Occiput concave at its centre, with a slight angular projection at its outer ends. *Prothorax* black. *Thorax* black marked with greenish-yellow as follows:—Very oblique broad antehumeral stripes which run from near the alar sinus and are broadly confluent with an interrupted mesothoracic collar below; a small upper humeral spot; sides broadly greenish-yellow, with a broad black stripe on the first lateral suture and a short remnant of a similar stripe on the

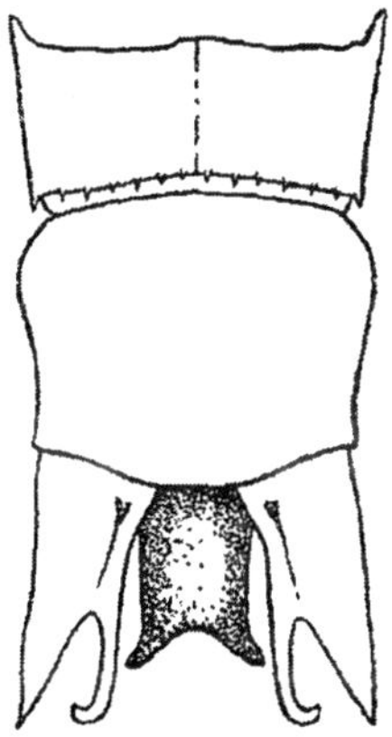

Fig. 114.—Anal appendages of *Microgomphus loogali* Fraser, male. Dorsal view.

lower part of the second suture (another unusual type of marking). *Legs* entirely black. *Wings* hyaline, very palely enfumed; pterostigma dark brown between black nervures, covering 4 cells, short, unbraced, or the brace poorly developed and arising from the costa a short distance from the pterostigma; nodal index $\frac{12\text{–}16 \mid 17\text{–}11}{12\text{–}12 \mid 12\text{–}11}$; 3 rows of postanal cells in hind-wing, 2 rows of cells between *IA* and the hinder margin in fore-wing. *Abdomen* black, marked with greenish-yellow as follows:—Segment 1 with the sides broadly and a stripe on the mid-dorsal carina, which is flanked by a broad black stripe subdorsally; 2 similar, the lateral yellow area including the oreillet, the stripe on the carina tapering apically and expanded at the level of the transverse suture; 3 with the mid-dorsal carina finely yellow and a large ventro-lateral basal spot on each side; 4 to 6 with small triangular basal spots on each side; 7 with a broad basal yellow ring occupying the basal fifth; remaining segments unmarked. *Anal appendages*: superiors very robust, widely divaricate, tapering to a point, greenish-white, tipped with black; from the middle of the inner side a slim pale yellow branch springs and inclines inwards to meet its fellow from the opposite side, the extreme apex of each branch strongly recurved outwards, the fusion of these branches with the main appendage demarcated by a fine groove; inferior triangular, not bifid, as long as superiors, black. *Genitalia*: hamules broad, short, tipped by a small spine; vesicle of penis bulbous, globular, black.

Female.—Very similar to the male, but markings on abdomen more extensive; on segments 3 to 6 a large triangular baso-lateral spot which extends to the transverse suture, beyond which, and only narrowly separated from the basal spot, is a narrow stripe extending almost to the apex of each segment; in one specimen there is also a minute basal lateral spot on segment 8. *Thorax* and *prothorax* pruinosed white beneath when mature. *Anal appendages* short, conical, pale greenish-white.

Distribution.—One male and two females collected by Col. F. Wall, I.M.S., at Maymyo, Northern Shan States, UPPER BURMA, 19–26. vi. 24 and 10. vii. 24.

This species is remarkable for its large size (it is the largest known species of the genus) and for the unusual markings on the labrum and on the sides of thorax. The two rows of cells between *IA* and the hinder margin of the fore-wing is clearly a response to the need of a greater wing area to support a weightier insect.

Types of both *M. loogali* and *M. burmicus* in the Author's collection.

326. **Microgomphus ? minusculus** (Selys).

Cyclogomphus ? minusculus Selys, Bull. Acad. Belg. (2) vol. xlvi, p. 468 (1878); Kirby, Cat. Odon. p. 70 (1890); Laidlaw, Trans. Ent. Soc. Lond. vol. lxxviii, p. 186 (1930).

Cyclogomphus minusculus Williamson, Proc. U.S. Nat. Mus. vol. xxxiii, pp. 296–298 (1907); Laidlaw, Rec. Ind. Mus. vol. xxiv, p. 390 (1922); Fraser, J. Bombay Nat. Hist. Soc. vol. xxix, pp. 64, 332 (1923); id., ibid. vol. xxxi, p. 163 (1926).

Male unknown.

Female.—Abdomen 22 mm. Hind-wing 21 mm.

Head: labium yellow; labrum black; rest of face black except for a small spot on each side of the postclypeus;

frons yellow along its crest, but its base and a prolongation forwards (which meets the black on face and front of frons) black; occiput simple, straight, black. *Prothorax* black, its posterior lobe yellow. *Thorax* black in front and on dorsum, yellow at the sides, marked as follows:—Short, oblique, isolated antehumeral yellow stripes, approximating above, divaricate below, not confluent with the mesothoracic collar of the same colour; a black stripe on the first lateral suture, which is confluent at its middle with a similar stripe on the second lateral suture, both stripes moderately thick, the anterior one confluent below with the humeral black stripe. *Wings* hyaline, reticulation black, costa finely yellow; pterostigma moderately long, thick, pale between black nervures, covering 3 cells (3 mm.); nodal index 9–12 | 11–10 in fore-wings. *Legs* short, black, inner sides of femora yellow, armed with very short spines. *Abdomen* equal in length to the wings, black, brownish beneath, marked and ringed with yellow as follows:—Segment 1 yellow, with its base narrowly black and interrupted mid-dorsally; 2 with a mid-dorsal trilobed stripe enclosed by broad subdorsal black stripes, the ventro-lateral border yellow, marked with a small black spot, the basal articulation finely black; 3 black, with an interrupted dorsal stripe and a ventro-baso-lateral spot interrupted by the transverse suture; 4 to 7 black, marked with complete basal rings which occupy about one-sixth the length of segments; remaining segments black, unmarked; 8 and 9 not dilated; 10 very short. *Anal appendages* conical, pointed, very small, pale yellow, with a small similarly coloured protuberance between them. Vulvar scale very short.

Distribution.—Between Moolai and Moorlut, 4,000–6,000 ft., TENASSERIM.

Apparently the *type* was in the McLachlan collection, as its description was communicated to De Selys by Mr. McLachlan, but I have been unable to find it in that collection, so presume that it has been lost. As no details of its venation are known, it is impossible to place the species with any accuracy. De Selys gives the following note:—"It is the smallest species of the subgenus [Genus *Cyclogomphus*]. It appears to be related to *torquatus* [*Microgomphus*] by the black dorsal carina of the thorax (yellow in all other species). It differs by its smaller size, the absence of a yellow humeral stripe, the narrower mesothoracic collar, which is not prolonged as far as the yellow trochanters of the anterior legs. So long as we do not know the anal appendages of the males of *torquatus* and *minusculus*, there will remain some doubt as to the correct place of these species."

Genus **LEPTOGOMPHUS** Selys. (Fig. 115.)

Leptogomphus Selys, Bull. Acad. Belg. (2) vol. xlvi, p. 442 (1878); Kirby, Cat. Odon. p. 70 (1890); Williamson, Proc. U.S. Nat. Mus. vol. xxxiii, pp. 291–295 (1907); Laidlaw, Rec. Ind. Mus. vol. xxiv, p. 378 (1922); Fraser, J. Bombay Nat. Hist. Soc. vol. xxix, pp. 63, 331 (1923); id., ibid. vol. xxxi, pp. 882, 883 (1927); Laidlaw, Trans. Ent. Soc. Lond. vol. lxxviii, p. 179 (1930); Needham, Zool. Sinica, ser. A, vol. xi, fasc. 1, pp. 19, 45 (1930); id., Rec. Ind. Mus. vol. xxxiv, pp. 217, 218, 220 (1932).

Size medium or rather robust; colour mat black, marked with bright citron-yellow.

Head triangular, rather wide, frons rounded or moderately angulated, face very oblique, occiput usually simple, slightly concave. Wings: reticulation close; tornus angulated; base of hind-wing very oblique and rather short, rather deeply excavated; membrane obsolete; anal triangle 3-celled; arc situated opposite the second antenodal nervure or between the second and third, the latter situation the more common; 3 to 5 transverse nervures between the sectors of arc from the

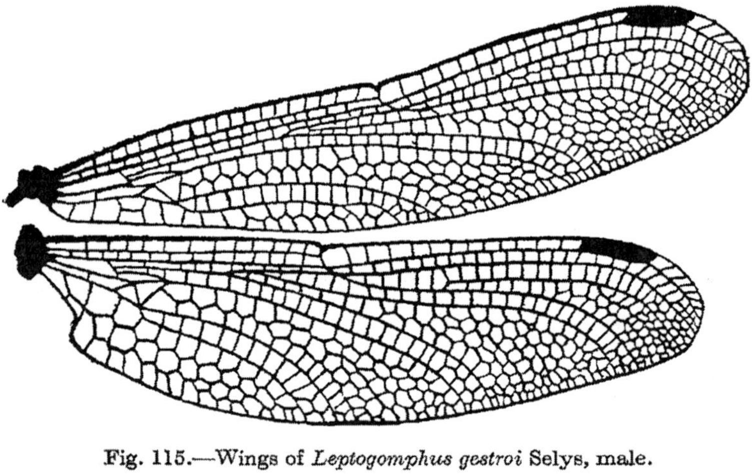

Fig. 115.—Wings of *Leptogomphus gestroi* Selys, male.

arc to bifurcation of *Rs* in fore-wing, 3 to 4 in hind-wing; 1 or 2 rows of postanal cells in fore-wing, 3 to 4 in hind-wings; the first postanal cell in hind-wing not extending proximal to the base of subtrigone; anal loop absent; a basal incomplete antenodal nearly always present in all wings; nodal index high; primary antenodals the first and the sixth, seventh or eighth; discoidal cells entire, that of fore-wing subequilateral, the distal side but slightly longer than the two others, that of the hind-wing a little elongate in the length of wing, distal side slightly longer than the costal and half as long again as the basal; pterostigma very short, equal to about one-fourth or less of the distance from node to proximal end of pterostigma, unbraced; *IA* in fore-wing not pectinated or but slightly so at its distal end, only 2 rows of cells between it and the margin of wing; *Cuii* and *IA* in hind-wing parallel or more or less widely divergent at wing margin; 1 or 2 cubital nervures; all subtrigones and hypertrigones entire. Legs variable in length, hind femora extending to the apical border of segment 1 or middle of segment 2 of abdomen and armed with 2 rows of spines, very closely set at proximal end of limb, then gradually lengthening and with a single long robust spine at the extreme distal end; hind tibial spines moderately long, numerous and closely-set. Abdomen rather dilated at base, then slim and cylindrical or robust, and cylindrical from segment 3 to segment 7, from which point it gradually dilates as far as the end. Anal appendages: superiors conical, acute at apex and spined beneath; inferior usually more or less triangular, bifid for about half its length, the two branches more or less closely apposed. Genitalia: lamina depressed or more or less projecting and deeply and angularly arched; anterior and posterior hamules variable in the species; lobe scrotum-shaped, rather hidden, lip narrow and shallowly emarginate.

Genotype, *Leptogomphus semperi* Selys.

Distribution.—ASSAM, BURMA, Java, Sumatra, Indo-China, China, Borneo, and Formosa. Four species only found within our limits, one of which is found in Assam, the others in Burma.

Species of the genus are extremely rare, the Javan being the only one which has been found in any numbers. Nothing is known of their ecology, except that Bainbrigge Fletcher found the larva of *L. bidentatus* in a small cement tank, surely the most exiguous place to find a Gomphine larva!

The genus is closely related to *Heliogomphus* by its very short and unbraced pterostigma, narrow zygopterous head, etc.

Key to Indian Species of Leptogomphus.

1.	Antehumeral stripe confluent with the mesothoracic collar, thus forming yellow, inverted, 7-like markings	2.
	Antehumeral stripe well separated from the mesothoracic collar	3.
2.	Humeral stripe complete; only a single black stripe on sides of thorax, situated on the postero-lateral suture	*inclitus* Selys, [p. 364.
	Humeral stripe interrupted above; two black stripes on sides of thorax, the anterior interrupted below	*maculivertex* Selys, [p. 369.

3. Humeral stripe absent; only a single broad black stripe on sides of thorax .... *bidentatus* Fras., [p. 367.
A humeral stripe present; two black stripes on the sides of thorax, the posterior incomplete below .... *gestroi* Selys, [p. 365.

327. **Leptogomphus inclitus** Selys.

Leptogomphus inclitus Selys, Bull. Acad. Belg. (2) vol. xlvi, p. 444 (1878); Kirby, Cat. Odon. p. 70 (1890); Williamson, Proc. U.S. Nat. Mus. vol. xxxiii, pp. 291–294, text-figs. 19, 20 (1907); Laidlaw, Rec. Ind. Mus. vol. xxiv, pp. 371, 378 (1922); Fraser, J. Bombay Nat. Hist. Soc. vol. xxix, pp. 63, 331 (1923); id., ibid. vol. xxxi, p. 883 (1927); Laidlaw, Trans. Ent. Soc. Lond. vol. lxxviii, p. 179 (1930); Needham, Rec. Ind. Mus. vol. xxxiv, pp. 220, 221 (1932).

Male unknown.

Female.—Abdomen 36 mm. Hind-wing 32–35 mm.

Head: labium yellow; labrum greenish-yellow, bordered anteriorly with black; face black, the centre of postclypeus yellow; crest of frons yellow, invaded at its middle from behind by black, or the yellow area cut in two by the median black mark meeting the black on front of frons; vertex and occiput black, latter concave, a ridge on each side near the eyes, the area between these depressed. Area behind eyes tumid, black above, yellow below. *Prothorax* black, with a mid-dorsal geminate yellow spot and a larger spot on each side. *Thorax* black in front as far as first lateral suture, with a narrow antehumeral stripe which may or may not be confluent with a narrow mesothoracic collar, in the former case forming an inverted 7; external to these an equally narrow humeral stripe. Laterally yellow, the second suture mapped out in black and confluent with an upper black border below the wings. *Wings* hyaline, reticulation black, close; pterostigma brown between black nervures, covering 4–5 cells, 3·5 mm. in length; 14 to 19 antenodal nervures in fore-wing, 9 to 11 postnodals. *Legs* short, hind femora 6·5 mm. in length, black; the inner sides of anterior two pairs and the greater part of the inner sides of hind pair yellow. *Abdomen* black, marked with yellow as follows:—A narrow yellow stripe running along the mid-dorsal carina from segments 1 to 7, broad on 1 and 2, tapering on 3; a lateral stripe on the same segments, broad from segment 1 to base of 3, broken into an elongate basal spot and an elongate subapical rounded spot on 3 to 7, vestigial in character on segments 5 to 7; remaining segments unmarked. *Anal appendages* short, conical, pale yellow, the apices ferruginous, separated by a conical structure as long as themselves. Vulvar scale moderately narrow, extending to the middle of segment 9, bifid at apex.

Distribution.—BURMA. The *type* (in the McLachlan collection) from Moolai, Lower Burma, (*Prof. Wood-Mason*), is much darker than two other females from Upper Burma, which may possibly belong to a distinct species.

This species bears a close resemblance to *L. semperi* from the Philippines, a comparison with which serves to place it in the genus *Leptogomphus*, in spite of the fact that the male is unknown.

328. **Leptogomphus gestroi** Selys. (Fig. 116.)

Leptogomphus gestroi Selys, Ann. Mus. Civ. Genova, (2) vol. x, pp. 476, 477 (1891); Martin, Mission Pavie Indo-Chine, vol. iii, p. 214 (1904); Williamson, Proc. U.S. Nat. Mus. vol. xxxiii, pp. 291–294 (1907); Laidlaw, Rec. Ind. Mus. vol. xxiv, pp. 371, 378 (1922); Fraser, J. Bombay Nat. Hist. Soc. vol. xxix, pp. 63, 331 (1923); id., ibid. vol. xxxi, pp. 884, 885, text-fig. 2, i, ii (1927); Laidlaw, Trans. Ent. Soc. Lond. vol. lxxviii, p. 180 (1930); Needham, Rec. Ind. Mus. vol. xxxiv, pp. 220, 221 (1932).

Male.—Abdomen 40–42 mm. Hind-wing 32–35 mm.

Head: labium pale yellow, the middle lobe bordered with black; labrum palest yellow, almost white, its base very narrowly, its anterior border broadly black, the yellow area almost bisected by a fine black medial line and a lower brownish fusiform spot; bases of mandibles yellowish-white;

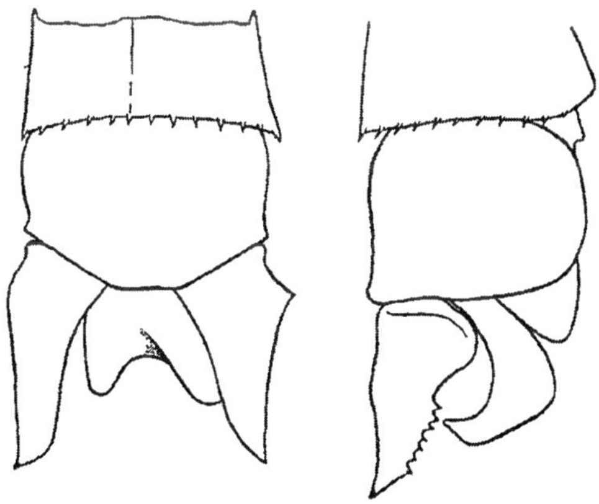

Fig. 116.—Anal appendages of *Leptogomphus gestroi* Selys, male. Dorsal and right lateral views.

ante- and postclypeus black; frons pale greenish-yellow above and over the crest, its base very finely black; vertex and occiput black, but the latter with a small median yellow point, emarginate behind, especially at its middle. Area behind eyes yellow except the upper border. *Prothorax* black, a medial yellow spot on the posterior lobe confluent with a larger paler yellow spot on the dorsum of median lobe, and a large outer lateral spot on each side. *Thorax* black, marked with yellow as follows :—A pair of antehumeral stripes running parallel with the mid-dorsal carina, not confluent with the mesothoracic collar below, but extending nearly to the alar sinus above ; a long, curved, very narrow humeral stripe, its upper end slightly expanded ; the mesothoracic collar very narrowly interrupted. Laterally yellow, the humeral black slightly overlapping the humeral suture, the first lateral suture and the upper part of the second mapped out in black. *Legs* black, anterior pair of femora yellow on the inner sides ; hind femora with an outer yellow stripe, which fades away distally, armed with two rows of robust short spines, the inner row moderately widely spaced, more robust and less numerous than the outer row, which ends in a single larger distal spine ; middle and anterior pairs with fewer but more robust spines. *Wings* hyaline ; pterostigma dark brown between black nervures, covering 3 to 4 cells, unbraced ; anal triangle 3-celled ; 3 rows of postanal cells in hind-wings ; a basal incomplete antenodal nervure present in all wings ; nodal index $\frac{11\text{–}14}{10\text{–}10}\Big|\frac{16\text{–}10}{11\text{–}11}$; 5 cross-nervures in the fore-wings between the sectors of arc. *Abdomen* black, marked with greenish-yellow as follows :—A small dorsal spot and the sides broadly of segment 1 ; a trilobed dorsal stripe and the sides broadly on 2 ; the mid-dorsal carina of 3 to 7 finely, a baso-lateral spot followed by a medio-lateral spot on 3, smaller baso-lateral spots on 4 and 5 ; remaining segments unmarked. The abdomen from the apex of segment 7 gradually and progressively broadens as far as the apical border of 10, which latter is rather massively square, its dorsum dome-shaped, its apical border prolonged squarely between the anal appendages. *Anal appendages* (fig. 116) : superiors bright yellow, darker below and at the inner and outer ends of base ; inferior black. *Genitalia* prominent ; lamina arched, very depressed ; anterior hamules broad at base and nearly as far as apex, directed straight outwards, the borders inwardly curled, the apex ungulate and recurved at an obtuse angle ; posterior hamules much larger, also directed straight outwards, tongue-shaped, sinuous, the apex ending in a forwardly directed, very robust, curled thick spine ; lobe tumid, prominent, its anterior border deeply but narrowly emarginate.

Female.—Very similar to male, but slightly larger and more robust. Occiput with a broad rounded notch at its centre, on each side of which is a robust spine. *Anal appendages* short, conical, pointed, yellow, separated by a conical yellow protuberance. Vulvar scale broad at base, tapering as far as the apex of ninth segment, its apical half split into two slightly separated lamellæ.

Distribution.—BURMA.

The *type,* a male in the Selys collection, is from Leito, and was taken, towards the end of May, by Mr. Fea. There is a female in the same collection, the only specimen of that sex known. Col. F. Wall, I.M.S., has taken a second male at Maymyo, Upper Burma, 1. vii. 25, which is now in the author's collection; it differs from the type in possessing an occiput almost exactly similar to that of the female as described by Selys. The Selysian male appears to be more teneral than the Maymyo one, the markings of which are pale greenish-white to greenish-yellow on the thorax. This species is closely related to *L. lansbergei* Selys, from Java.

329. **Leptogomphus bidentatus** Fraser. (Fig. 117.)

Leptogomphus bidentatus Fraser, J. Bombay Nat. Hist. Soc. vol. xxxiv, pp. 752, 753, text-figs. 1, 2, *a*, *b* (1930); Needham, Rec. Ind. Mus. vol. xxxiv, pp. 220, 221 (1932).

Male.—Abdomen 38 mm. Hind-wing 33 mm.

Head black, marked with yellow; labium with lateral lobes yellow; labrum black, with a small yellow spot on each side; anteclypeus with a small median yellow spot; post-clypeus and frons broadly yellow; vertex black, but with a small rounded spot behind the ocelli, and the whole of the occiput yellow. *Prothorax* black, with a large yellow spot on the median and posterior lobes and a small spot on each side of the former. *Thorax* black, marked with yellow as follows:—A mesothoracic collar interrupted in the middle; an oblique antehumeral stripe, the upper ends of which are squared and closely opposed to the antealar sinus, the lower ends pointed and divergent and not meeting the meso-thoracic collar; laterally marked with two broad yellow stripes, the posterior of which covers about four-fifths of the metepimeron; between these two stripes a small upper spot. *Legs* black, spines of hind femora gradually lengthening towards the distal end, very robust and rather short; hind femora extending to distal end of segment 1. *Wings* hyaline, tinted with yellow at the base; pterostigma black or dark yellow framed in black, braced, covering 3 to 4 cells; nodal index $\frac{14\text{–}15}{12\text{–}10}\Big|\frac{14\text{–}13}{10\text{–}13}$; discoidal cells traversed once in the hind-wings; 5 rows of cells in the anal field; anal loop absent. *Abdomen* black, marked with yellow as follows:—Segment 1 with a quadrate mid-dorsal and a large lateral spot, its apical border narrowly yellow; 2 with a trilobate mid-dorsal stripe and two large spots on each side, one of which includes the oreillet and extends to the base; 3 to 9 with paired subdorsal basal spots, which on 7 and 8 extend distally to the jugal suture; 10 with a small lateral spot. *Anal appendages* (fig. 117) black.

Female.—Abdomen 38 mm. Hind-wing 35 mm.

Markings very similar to those of male, abdominal markings differing as follows:—Segment 2 with the lateral spots confluent with one another to form a broad stripe; 3 to 6 with additional lateral spots in line with the basal, elongate on 3 and 4, very small and short on 5 and 6; 10 unmarked. Head with the vertex raised and sloping back as a thick quadrate plate, the hind corners of which are prolonged as robust spines. *Anal appendages* short, black, simple.

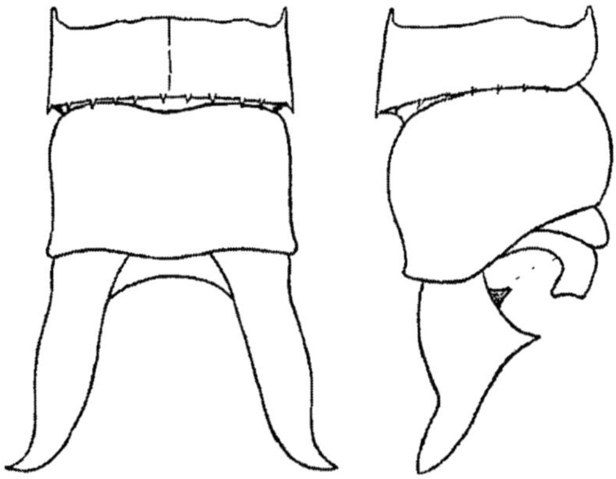

Fig. 117.—Anal appendages of *Leptogomphus bidentatus* Fraser, male. Dorsal and right lateral views.

Vulvar scale one-fourth the length of segment 9, triangular, narrowly bifid so as to form two closely apposed scales.

Distribution.—Shillong, ASSAM. Two females and one male taken by Mr. T. Bainbrigge Fletcher, 19. vii. 28. One female was found emerging on the side of a small cement tank.

This species is interesting from its extraordinary breeding place and also from the triangles of the hind-wings being traversed, a characteristic shared by *L. retroflexus* Ris and *L. scorpio* Ris, from both of which it differs in the shape of the anal appendages and other characters. The shape of the superior anal appendages will also serve to differentiate it from *L. semperi* Selys, *L. assimilis* Krug., *L. gracilis* Krug., *L. gestroi* Selys, and *L. lansbergei* Selys. From *L. inclitus* Selys it is to be distinguished by the yellow frons and by the absence of a humeral stripe. The genitalia appear to ally it closely to *L. sauteri* Ris, *L. perforatus* Ris, and *L. scorpio*

Ris; the tapering ends of the superior appendages distinguishing it from the two former species.

Type and paratype in my own collection.

330. **Leptogomphus (?) maculivertex** Selys.

Leptogomphus ? maculivertex Selys, Ann. Mus. Civ. Genova, (2) vol. x, pp. 478, 479 (1891); Williamson, Proc. U.S. Nat. Mus. vol. xxxiii, pp. 292–295 (1907); Laidlaw, Rec. Ind. Mus. vol. xxix, pp. 371, 378 (1922); Fraser, J. Bombay Nat. Hist. Soc. vol. xxix, pp. 63, 331 (1923); id., ibid. vol. xxxi, pp. 886-887 (1927); Laidlaw, Trans. Ent. Soc. Lond. vol. lxxviii, p. 180 (1930); Needham, Rec. Ind. Mus. vol. xxxiv, pp. 220, 221 (1932).

Male unknown.

Female.—Abdomen 33 mm. Hind-wing 31 mm.

Head: labium black, its base broadly yellow; anteclypeus black, postclypeus black, with a rounded spot on each side bordering the eyes; crest of frons yellow, this colour slightly overlapping the anterior surface and invaded behind by the black of base of frons; occiput and a medial spot on vertex yellow, former black behind, slightly emarginate and with a convex scallop at its middle. *Prothorax* black, its dorsum and the posterior lobe yellow. *Thorax* black as far as the humeral suture, marked with yellow as follows:—A slightly interrupted mesothoracic collar, narrow antehumeral stripes, which form inverted figure 7's by confluence with the collar below, a fine humeral stripe on each side, broken up into an upper spot and a lower fine line; mid-dorsal carina with a small medial spot; laterally yellow, with the upper part of the first lateral suture and the whole of the second mapped out in black. *Wings* hyaline, reticulation black, close; pterostigma dark brown, covering 4 to 5 cells, 3 mm. in length; 15 antenodal nervures and 12 postnodals in fore-wings; *basal incomplete antenodal nervure absent. Legs* black, femora yellow, marked with an outer black stripe and armed with black spines, which are moderately robust, short, and numerous; hind femur 5 mm. in length. *Abdomen* slender, the borders of segments 8 and 9 slightly dilated; black, marked with yellow as follows:—A triangular dorsal spot on segment 1; a trilobed mid-dorsal stripe on 2; sides of these two segments broadly yellow; 3 to 7 with the mid-dorsal carina finely but not quite extending to the apical borders of segments 3 to 6, and occupying only the basal two-thirds of 7; the dorsal stripes on segments 3 to 6 indented laterally by an encroachment of the black at the jugal sutures; remainder unmarked. *Anal appendages* slender, cylindrical, pointed, rather longer than segment 10, separated by a conical yellow protuberance. Vulvar scale robust, yellow at base, tapering, narrow, extending to the apical border of segment 10, deeply cleft into two closely apposed lamellæ.

Distribution.—Meteleo, BURMA, collected by Mr. Fea, September 10.

The absence of the incomplete basal nervure suggests that it is rather a *Heliogomphus* than a *Leptogomphus*. The very long vulvar scale is analogous to that of *Gomphus nilgiricus* and *Onychogomphus M-flavum*. The discovery of the male is necessary to settle its identity.

Type female, unique, in the Selys collection. I have not seen this specimen, a better knowledge of the venation of which is needed to place it correctly.

Subfamily ICTININÆ.

A small subfamily which contains many of the largest species of the suborder ANISOPTERA. Species of this group are characterized by the close reticulation of their wings, and by the discoidal cells being of different shape in the fore- and hind-wings, these cells being traversed by one or more nervures. As in the last subfamily, there are more than two transverse nervures between the sectors of the arc, usually a great many more so in the fore-wing; *IRiii* and *MA* each possess a well-marked supplementary nervure, whilst the subtrigones of the fore-wings are divided up into 2 or 3 cells. Generic characters are the presence or absence of lateral leaf-like expansions on segments 8 and 9 of the abdomen and the shape of the anal appendages.

Distribution.—With the exception of one genus, which is from South America, all are confined to the Oriental and Ethiopian regions. Two genera only are found within Indian limits.

Key to Indian Genera of the Subfamily Ictininæ.

Segment 8 of abdomen widely dilated and with wing-like lateral projections; superior anal appendages acute at apex	ICTINUS Rambur, [p. 370.
Segment 8 of abdomen not dilated; superior anal appendages obtuse at apex	GOMPHIDIA Selys, [p. 381.

Genus ICTINUS Rambur. (Fig. 118.)

Ictinus Rambur, Ins. Névrop. p. 171 (1842); Selys, Bull. Acad. Belg. (2) vol. xxi, p. 86 (1854); id., Mon. Gomph. p. 263 (1857); Kirby, Cat. Odon. p. 77 (1890); Williamson, Proc. U.S. Nat. Mus. vol. xxxiii, pp. 272, 278–280 (1907); Laidlaw, Rec. Ind. Mus. vol. xxiv, p. 373 (1922); Fraser, J. Bombay Nat. Hist. Soc. vol. xxix, pp. 60, 328, 661 (1923); Laidlaw, Trans. Ent. Soc. Lond. vol. lxxviii, p. 176 (1930); Needham, Zool. Sinica, ser. A, vol. xi, fasc. 1, pp. 19, 20 (1930); id., Rec. Ind. Mus. vol. xxxiv, pp. 217, 219 (1932).

Size large and build robust; colour mat black, marked with bright citron-yellow or greenish-yellow.

Head large and triangular, frons strongly angulated, occiput simple or slightly concave, and fringed with short hairs. Wings: reticulation very close: tornus strongly angulated; base of hind-wing deeply excavated; membrane long and narrow; anal triangle 5-celled; arc situated between the first and second antenodal nervures; 10 to 12 transverse nervures between the sectors of arc from the arc to bifurcation of *Rs* in fore-wings, 6 in hind-wings; 2 rows of postanal cells in fore-wings, 5 to 6 in hind-wings; a distinct anal loop present, made up of 4 to 5 cells and extending proximal to base of subtrigone; no basal incomplete postcostal nervures present; nodal index very high; primary antenodals the first and seventh; discoidal cell in fore-wings subequilateral, the distal side slightly longer than the two others which

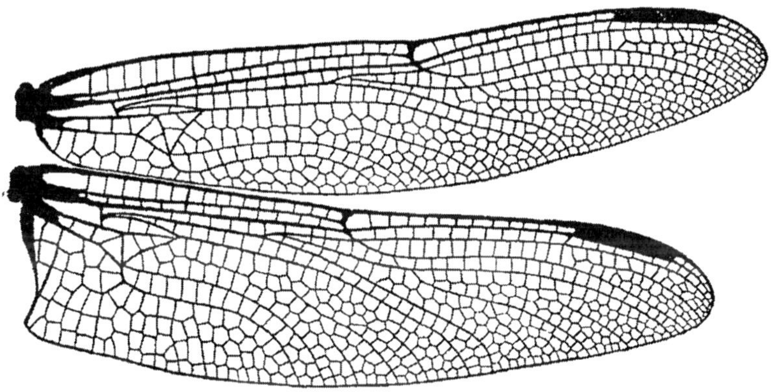

Fig. 118.—Wings of *Ictinus rapax* (Rambur), male.

are equal, made up of 2 to 4 cells by traversing nervures; that of hind-wing 3-celled, narrow and elongate in the length of wing, costal side twice the length of basal and about equal to the distal which is sinuous; pterostigma elongate and narrow, about twice the length of the distance between the node and proximal end of pterostigma, braced; *IA* in fore-wing but slightly or not pectinate; *Cuii* and *IA* in hind-wing parallel to wing-border; 2 to 4 cubital nervures in fore-wings, 2 in hind-wings; subtrigone in fore-wings 2-celled, in hind-wings entire or rarely 2-celled; hypertrigones with 2 traversing nervures; a supplementary nervure running outwards from distal side of discoidal cells; several strong intercalated nervures present. Legs robust, hind femora extending slightly beyond the end of posterior border of thorax and furnished with two rows of spines, which are numerous and closely set on the proximal half of limb, longer, more robust, and much more widely spaced on the distal half; hind tibial spines very closely set and very short and numerous. Abdomen dilated at basal segment, narrow and cylindrical from segment 3 to the base of 7, dilated from 7 to 9, segment 8 markedly expanded and bearing lateral leaf-like projections on each side, segment 10 being short and small. Anal appendages: superiors half as long again as segment 10, narrow and pointed at apex, the inferior very much shorter, deeply bifid, the two branches rather widely divaricate. Genitalia: lamina projecting slightly, depressed at centre; anterior hamules long robust, hook-like, organs; posterior hamules short, flattened, and tongue-like; lobe short, scoop-shaped; a fringe of short yellow hairs covering the genital sac and obscuring the hamules.

Genotype, *Diastatomma rapax* Rambur.

Distribution.—Tropical Africa, INDIA, BURMA, CEYLON, Southern Asia, Java, Sumatra, China, the Philippines, and Australia.

Included in this genus are some of the largest and most striking insects of the Order, four species occurring within our limits. *I. rapax*, the most common species of the genus, has a number of varieties which were formerly classified as species; abundant material from many sources has convinced me that their specific value can no longer be upheld.

The larvæ usually live in stagnant water, but occasionally appear to be compelled by the absence of the latter to breed in streams; when this is the case, they take to deep pools in the course of sluggish rivers, such spots conforming closely to their natural habitats. The limpet-like shape of the larva, with its broad flat abdomen, suggests that at one time it was an inhabitant of swift torrential streams, adapted to clinging to rocky surfaces.

Key to Indian Species of Ictinus.

1.	Face largely black; posterior border of thorax margined with black; femora largely black	2.
	Face largely yellow; posterior border of thorax not margined with black; femora largely yellow	3.
2.	Segment 8 without a basal yellow ring	*pertinax* Selys, [p. 377.
	Segment 8 with a broad yellow ring	*rapax* (Ramb.), [p. 373.
3.	Labrum not bordered with black; back of head black; lateral expansion of segment 8 margined with yellow	*atrox* Selys, p. 379.
	Labrum bordered with black; back of head marked with yellow; lateral expansions of egment 8 unmarked with yellow	*angulosus* Selys, [p. 378.

331. **Ictinus rapax** (Rambur). (Fig. 119, **b**.)

Diastatomma rapax Rambur, Ins. Névrop. p. 169 (1842).
Ictinus vorax Rambur (female) Rambur, Ins. Névrop. p. 171 (1842).
Ictinus rapax Selys, Bull. Acad. Belg. vol. xxi, pt. 2, p. 90 (1854); id., Mon. Gomph. p. 276 (1857); Kirby, Cat. Odon. p. 77 (1890); Martin, Mission Pavie Indo-Chine, p. 217 (1904); Williamson, Proc. U.S. Nat. Mus. vol. xxxiii, p. 279 (1907); Fraser, J. Nat. Hist. Soc. Siam, vol. iii, p. 457 (1919); id., J. Bombay Nat. Hist. Soc. vol. xvii, p. 541 (1921); Laidlaw, Rec. Ind. Mus. vol. xxiv, pp. 370, 373, 374 (1922); id., Spolia Zeylanica, vol. xii, p. 339 (1924); id., Trans. Ent. Soc. Lond. vol. lxxviii, p. 176 (1930); Needham, Zool. Sinica, ser. A, vol. xi, fasc. 1, pp. 23–25 (1930); Fraser, Rec. Ind. Mus. vol. xxxiii, p. 447 (1931); Needham, op. cit. vol. xxxiv, p. 219 (1932); Fraser Ceylon J. Sci., B, vol. xviii, pt. i, pp. 19, 20 (1933).
Ictinus præcox Selys, Bull. Acad. Belg. vol. xxi, pt. 2, p. 89 (1854); id., Mon. Gomph. p. 275 (1857); id., Bull. Acad. Belg, vol. xlvi, p. 677 (1878); Kirby, Cat. Odon. p. 77 (1890); Williamson, Proc. U.S. Nat. Mus. vol. xxxiii, p. 279 (1907); Laidlaw, Rec. Ind. Mus. vol. xxiv, p. 373 (1922); id., Trans. Ent. Soc. Lond. vol. lxxviii, p. 176 (1930).
Ictinus mordax Selys, Mon. Gomph. p. 433 (1857); Kirby, Cat. Odon. p. 77 (1890); Williamson, Proc. U.S. Nat. Mus. vol. xxxiii, p. 279 (1907); Laidlaw, Rec. Ind. Mus. vol. xxiv, p. 373 (1922); id., Trans. Ent. Soc. Lond. vol. lxxviii, p. 176 (1930).
Ictinus rapax race ? *mordax* Selys, Bull. Acad. Belg. vol. xxxv, pt. 2, p. 768 (1873).
Ictinus rapax rapax Fraser, J. Bombay Nat. Hist. Soc. vol. xxix, pp. 60, 328, 661–663, pl. ii, figs. 1, 1 *a* (1923); id., Rec. Ind. Mus. vol. xxvi, pp. 427, 470 (1924); Laidlaw, Trans. Ent. Soc. Lond. vol. lxxviii, p. 176 (1930).
Ictinus rapax prœcox Fraser, J. Bombay Nat. Hist. Soc. vol. xxix, pp. 60, 328, 663, 664 (1923); id., Rec. Ind. Mus. vol. xxvi, pp. 427, 470 (1924); Laidlaw, Trans. Ent. Soc. Lond. vol. lxxviii, p. 176 (1930).
Ictinus rapax mordax Fraser, J. Bombay Nat. Hist. Soc. vol. xxix, pp. 60, 328, 663 (1923); id., Rec. Ind. Mus. vol. xxvi, pp. 427, 470 (1924); Laidlaw, Trans. Ent. Soc. Lond. vol. lxxviii, p. 176 (1930).

Male.—Abdomen (with appendages) 52 mm. Hind-wing 40 mm.

Head: eyes bluish-grey; labium yellow, bordered with brownish; labrum yellow, with a heavy black border and a median prolongation of black from the base which may or may not meet the black anterior border; face and frons greenish-yellow, with a black stripe traversing lower part of front of frons and expanding upwards at its middle to cut the yellow area into two large lateral triangular spots; anteclypeus yellow; postclypeus black, with a large lateral yellow spot on either side; a black spot at base of frons above in the middle line which does not extend as far forwards as the crest; vertex black; occiput greenish-yellow, fringed with short yellow hairs. *Prothorax* black, marked with yellow. *Thorax* black, marked with yellow or greenish-yellow follows :—A complete mesothoracic collar; oblique dorsal spots, short and broad above, pointed below, where they are widely separated from the mesothoracic collar; a large central spot in alar sinus; spots on tergum; a humeral stripe represented by an upper triangular spot and often a lower broad streak. Laterally yellowish-green, with a median broad black stripe and the posterior border of the metepimeron narrowly yellow; vestiges of a third yellow stripe on the median lateral black band, sometimes an upper yellow spot, sometimes an upper and a lower, or again a row of three yellow spots (I have never seen a complete yellow band). *Wings* clear or, when fully mature, slightly enfumed. Pterostigma black, braced, long, covering 5 to 6 cells; discoidal cell of fore-wings with 4, of hind-wings with 3 cells; sub-trigone of fore-wing with 2 cells, of hind-wing with 1 cell;

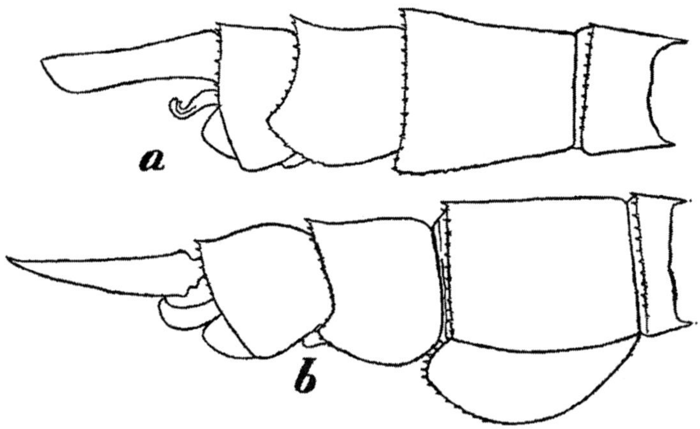

Fig. 119.—Terminal abdominal segments and anal appendages of (**a**) *Gomphidia kodaguensis* Fraser, male; (**b**) *Ictinus rapax* (Rambur), male. Seen from the right side.

hypertrigones traversed once or twice; membrane whitish; nodal index $\frac{23\text{–}13}{14\text{–}16} \Big| \frac{21\text{–}12}{15\text{–}13}$; 3 cubital cells in fore-wing, 2 in hind-wing; 5 cells in anal triangle. Discoidal field begins with a row of 4 cells and is continued as rows of 2. *Legs* black, coxæ and trochanters yellow, and a stripe of the same colour on flexor surface of anterior femora; hind femora extending just beyond hind margin of thorax, furnished with an inner and outer row of spines, those of the inner row closely set and numerous at the base, gradually lengthening and more widely spaced towards the apex, those of the outer row closely set and small near the base, followed by

5 or 6 robust, widely spaced spines, the longest in the middle of the series. *Abdomen* black, marked with bright yellow as follows :—Segment 1 with an apical dorsal stripe, confluent with a large triangular spot on dorsum of 2, and a narrow lateral apical stripe, confluent with a broad lateral spot on 2, which involves but does not pass beyond the oreillet ; 3 with nearly the basal half yellow, this turning to pure white low down on sides and beneath ; 4 to 6 with large dorsal basal spots confluent across the middle line except at apices ; 7 with basal half, 8 with rather more than basal half, yellow, the black here indenting the yellow on the dorsal carina ; the leaf-like expansions all black ; 9 with a lateral basal stripe and a small apical lateral spot ; 10 either unmarked or with a small subbasal, subdorsal spot on either side and a dorsal subapical pair of small spots, the basal spot sometimes much lower down on the sides (in one specimen all four spots are joined to form a bow-like stripe on the sides and dorsum). *Anal appendages* (fig. 119, b) black, as long as the two last segments, cylindrical and tapering. Inferior much shorter, deeply bifid, black. *Genitalia* : lamina tumid, broad, and deeply cupped along the free border ; internal hamules robust hooks ; external short flat and tongue-like, all hidden beneath a fringe of stout bright yellow hairs directed inwards and covering the genital orifice ; lobe short, trowel-shaped.

Female.—Abdomen 50 mm. Hind-wing 42–44 mm.

Very similar to the male ; the yellow markings more extensive ; the abdomen much stouter, laterally compressed and shorter. The humeral stripe is nearly always almost complete and the lateral black band is always spotted, and there may be an almost complete yellow stripe here. Segment 10 is usually entirely black. *Anal appendages* short, conical, black. Occiput raised, a robust spine situated at its middle, which in some specimens is minutely bifid, black, with the floor of occiput yellow. Wings with a dark brown basal marking extending as far as the first antenodal nervure. (This is often present also in the male, but never so well defined as in the female). Vulvar scale black, deeply cleft into two narrow tongue-like processes which extend nearly to the base of segment 10.

Var. *mordax* Selys.

Differs from typical *I. rapax* in the greater extent of yellow on the face ; the pterostigma rather longer, segment 10 entirely black, whilst 9 has only small lateral spots. Malabar specimens, in my collection, have the whole of the ante- and postclypeus yellow save for two minute black points at the middle of the latter ; the occiput is almost entirely yellow and has a much longer and better developed central spine. The lateral black stripes on the sides of the thorax are very variable, being sometimes entirely divided by yellow, whilst in others, especially in the males, there are only small upper and lower spots.

Var. *præcox* Selys.

This is said to differ from typical *I. rapax* in its slightly longer abdomen and slightly shorter wings. The superior anal appendages are slightly longer and less abruptly truncate; the wing-like borders on segment 8 have the denticles more pronounced. The face is usually darker, the yellow on frons being divided into two spots by the confluence of the black in front with the black above; segments 3 to 6 have shorter basal rings, whilst segment 10 has rather extensive yellow markings.

Distribution.—Throughout INDIA, BURMA, CEYLON, and Malaysia, except in desert areas.

A rather variable insect, but the same varieties crop up everywhere in the same localities, so that it is evident that *mordax* and *præcox* cannot be accepted as subspecies or geographical forms.

The insect, which is a very common one, breeds in both running and still waters, but principally in the latter. In Malabar and Coorg it abounds on every large tank, but during the dry months takes to the beds of rivers. Here it may be seen perched on a prominent twig facing the water, head inclined downwards and abdomen held well up. Should it be disturbed either by a rival, a passing female, or by the collector, it dives gracefully towards the surface of the water and then banks and turns, flying swiftly off along the borders of the pond or stream. If disturbed it usually returns to its resting place again and again, or settles close by until the danger is past, when it again returns to its first resting place. Females are rarely seen, and then only when coming to oviposit; their stay then is of very brief duration. Pairing takes place over water, and is not of more than a minute or two's duration, after which a few eggs are deposited by swift dips over the water, the insect then rising and disappearing high over the tops of neighbouring trees. Unlike most Gomphines, the males frequently engage in fierce combat, especially if females are frequenting their locality.

The *type* of *I. rapax* is in the Serville collection, that of *mordax*, from Assam, is in the Selys collection, whilst that of *præcox*, from Pondicherry, S. India, is in the Copenhagen Museum.

332. **Ictinus pertinax** Selys.

Ictinus pertinax Selys, Bull. Acad. Belg. vol. xxi, pt. 2, p. 88 (1854); id., Mon. Gomph. p. 270 (1857); Kirby, Cat. Odon. p. 77 (1890); Williamson, Proc. U.S. Nat. Mus. vol. xxxiii, pp. 279, 281 (1907); Fraser, J. Bombay Nat. Hist. Soc. vol. xxix, pp. 60, 328 (1923); Laidlaw, Trans. Ent. Soc. Lond. vol. lxxviii, p. 177 (1930); Needham, Rec. Ind. Mus. vol. xxxiv, p. 219 (1932).

Male.—Abdomen 56 mm. Hind-wing 38 mm.

Head: labium yellowish; labrum black, with two small oval spots at base or the two spots larger and slightly confluent at the centre of lip; ante- and postclypeus black, the latter with a small rounded spot on each side; frons black, its crest and upper part of front citron-yellow, divided by a prolongation of black from the base; rest of head black; eyes greenish during life. *Prothorax* black, unmarked. *Thorax* black, marked with yellow as follows:—A narrow complete mesothoracic collar; antehumeral stripes very short and divergent below; narrow humeral stripes broadly interrupted, consisting of an upper spot and a short stripe below it. Laterally yellow, with two median oblique black stripes more or less confluent at their middles; a very narrow border of black at the posterior part of thorax. *Legs* black, fore femora with a broad yellow stripe on the inner side and the middle pair with a vestige of the same marking. *Wings* hyaline, extreme bases slightly tinted with brown; pterostigma black, narrow, covering 6 cells; costa finely yellow; discoidal cell of fore-wing made up of 3 cells by a confluence of 3 nervures which meet in the centre of the cell, that of hind-wing with 2 or 3 cells; nodal index $\frac{11\text{–}22}{13\text{–}12}\Big|\frac{21\text{–}12}{14\text{–}17}$; anal triangle 4-celled. *Abdomen* black, marked with citron-yellow as follows:—Segment 1 unmarked; 2 with a large lateral spot which includes the oreillet, and, on the mid-dorsal carina, a lanceolate stripe which does not extend to the end of segment; 3 to 6 with dorso-basal spots deeply indented behind by the black mid-dorsal carina and occupying about one-third the length of each segment (slightly less on 6); 7 with its basal half yellow on the dorsum and subdorsum; 8 and 9 each with a broad, rounded, basal lateral spot on each side; 10 unmarked. Segment 8 is bordered with long, very narrow wings which overlap the apex of segment and are bordered with denticles only at the extreme end. *Anal appendages* black, shaped similarly to those of *I. rapax*.

Female.—Abdomen 51 mm. Hind-wing 44 mm.

Differs only from the male in sexual characters, and in its more robust abdomen. *Anal appendages* black, shortly conical, well separated; vulvar scale as long as segment 9, deeply bifid, each division shaped like an arrow-head.

Distribution.—BURMA, Malaysia, Indo-China and China. So far, the allotype, which was taken in Burma, is the only record from within Indian limits.

Differs from *I. rapax* in the black anteclypeus, and abdominal segment 8 with lateral spots instead of a basal ring and segment 10 entirely black.

Type a female in the Berlin Museum; allotype and three other males in the Williamson collection, Michigan University.

333. **Ictinus angulosus** Selys.

Ictinus angulosus Selys, Bull. Acad. Belg. vol. xxi, pt. 2, p. 92 (1854); id., Mon. Gomph. p. 281 (1857); Kirby, Cat. Odon. p. 77 (1890); Williamson, Proc. U.S. Nat. Mus. vol. xxxiii, p. 279 (1907); Laidlaw, Rec. Ind. Mus. vol. xxiv, pp. 370, 374 (1922); Fraser, J. Bombay Nat. Hist. Soc. vol. xxix, pp. 60, 329, 665, 666 (1923); Laidlaw, Trans. Ent. Soc. Lond. vol. lxxviii, p. 177 (1930); Needham, Rec. Ind. Mus. vol. xxxiv, p. 219 (1932).

Male.—Abdomen 53 mm. Hind-wing 43 mm.

Head: labium and labrum yellow, with the borders finely black; rest of face and frons yellow, with a vestige of black on anteclypeus, 2 small points on postclypeus and a large spot on front of frons, the latter with a broad black basal line above; vertex and occiput both yellow, the former black in front, the latter narrowly bordered with black. Back of eyes black, with a large yellow spot above. *Prothorax* almost entirely black. *Thorax* black, marked with yellow as follows:—A complete mesothoracic collar, broad dorsal oblique stripes converging above and resting on the borders of the alar sinus, pointed below, but not meeting the mesothoracic collar. Laterally broadly yellow, the sutures outlined in black. Tergum spotted with yellow. *Legs* black, femora broadly yellow on the outer sides, a fine yellow line on the outer sides of tibiæ; hind femora with 2 rows of spines, the distal 7 or 8 being much longer than the others. *Wings* hyaline, bases clouded with dark brown as far as the first antenodal nervure; costa finely yellow; pterostigma reddish-brown; discoidal cells with 3 cells; 19 antenodal and 11 postnodal nervures in fore-wings; membrane blackish-brown, especially at base. Base of wing deeply excavate, the tornal angle very prominent. *Abdomen* black, marked with yellow as follows:—Segment 1 broadly on dorsum, the marking constricted at its middle; 2 with a broad even dorsal stripe not extending as far as the apex, the sides including the oreillets broadly; 3 to 6 with large lanceolate basal spots, extending nearly to the apex on segment 3 and for about three-fourths the length of the others; 7 with a ring occupying the basal two-thirds (rather less on the sides); 8 with the basal half and its sides yellow, the moderately large leaf-like expansions entirely black and not visibly denticulate along borders; 9 black above, yellow laterally; 10 similarly coloured. *Anal appendages* black, superiors a little longer than segment 10, subcylindrical, excavate within, pointed, moderately divergent; inferior about half the length of anteriors, forked, branches straight but separated.

Female.—Unknown.

Distribution.—INDIA (locality unknown).

The abdominal markings are sufficiently distinctive to separate this from other Indian species.

Type in the Saunders collection, British Museum.

334. **Ictinus atrox** Selys.

Ictinus atrox Selys, Bull. Acad. Belg. vol. xxi, pt. 2, p. 92 (1854); id., Mon. Gomph. p. 282 (1857); id., Bull. Acad. Belg. (2) vol. xlvi, p. 677 (1878); Kirby, Cat. Odon. p. 77 (1890); Williamson, Proc. U.S. Nat. Mus. vol. xxxiii, p. 279 (1907); Laidlaw, Rec. Ind. Mus. vol. xxiv, pp. 370, 374 (1922); Fraser, J. Bombay Nat. Hist. Soc. vol. xxix, pp. 60, 329, 664, 665 (1923); Laidlaw, Trans. Ent. Soc. Lond. vol. lxxviii, p. 177 (1930); Needham, Rec. Ind. Mus. vol. xxxiv, p. 219 (1932).

Male.—Abdomen 51 mm. Hind-wing 41·5 mm.

Head: labium yellowish; labrum, face, and frons yellow, the labrum very finely bordered with black, a fine transverse line across lower part of frons and two tiny spots of black below this; upper part of frons black at base; vertex and occiput yellow, a black stripe crossing the former just behind the ocelli. *Prothorax* black, marked with yellow laterally. *Thorax* black in front, marked with yellow as follows:—A complete mesothoracic collar, broad oblique dorsal stripes not meeting the alar sinus above nor the collar below; the alar sinus, and a complete humeral stripe slightly constricted about its middle. Laterally broadly yellow, marked with fine black lines on the anterior and posterior lateral sutures. *Legs* entirely yellow, with black spines. *Wings* hyaline, tinted with yellow, a basal dark brown mark extending as far as the first antenodal nervure; pterostigma yellow, 7 mm.; membrane ashy grey. *Abdomen* black, marked with yellow as follows:—Segment 1 entirely yellow; 2 with a broad triangular spot, its apex just reaching the apical border of segment and its basal angles fusing with a broad lateral stripe which involves the oreillets and extends from base to apical border of segment; 3 to 7 with the basal half of each yellow; 8 with very large wing-like lateral processes, of which the base is broadly yellow, this continuous with a broad yellow ring covering rather more than the basal half of the segment; 9 and 10 with narrow basal rings which extend apically on the sides of each segment. *Anal appendages* very similar to those of *rapax*, black.

Female.—Abdomen 54 mm. Hind-wing 44 mm.

Head: lips and face reddish-yellow, changing to citron-yellow on the frons, which has a fine sinuous basal black band slightly notched before the ocelli; vertex and occiput yellow, the former margined with black in front, the latter margined with brown, its border concave, with a small spine on either side of the concavity, fringed on the outer side of the spines with fine yellow hairs; back of eyes glossy black, with a lateral vestigial yellow spot. *Prothorax* black, broadly bordered with dark yellow on each side. *Thorax* black, marked with yellow as follows:—A complete mesothoracic collar; moderately narrow oblique dorsal bands pointed below and diverging widely, but not meeting the collar below; a humeral band a little tapered above, broader below and uninterrupted. The sides broadly yellow, the sutures only marked finely with black. *Legs* yellow, marked with black; femora reddish-yellow, fore and middle pairs with an external black stripe broadening apically, hind pair with only a vestige of this and bearing two rows of black spines, the distal five of which are much longer than the others. Tibiæ and tarsi black. *Wings* hyaline, with a slight yellow tint at the bases; costa yellow; pterostigma long, dark yellow between black nervures, covering about 8 cells; membrane brownish; discoidal cell of fore-wing with 3 cells formed by the confluence of 3 nervures in centre of cell, that of hind-wing with only 2 cells; discoidal field begins with a row of 3 cells in fore-wing, followed by rows of 2, in hind-wing beginning with 4 cells; nodal index $\frac{13\text{–}20}{16\text{–}18} \Big| \frac{22\text{–}13}{16\text{–}15}$; subtrigone of fore-wing divided into 2 cells, that of hind-wing entire, small. *Abdomen* tumid at base, a little compressed, black, marked as follows:—Segment 1 with a transverse dorsal stripe and the sides yellow; 2 with a dorsal stripe extending the full length, very broad at base, tapering apically, the sides broadly yellow; 3 to 6 with the basal half yellow, the apical half black, extending basally for some distance along the sides; 7 with basal half yellow, the black encroaching slightly on the dorsum and more so on the sides; 8 with a fine basal black ring, followed by a narrow yellow ring and then black as far as the apical border, the yellow ring extending a little apically on the sides, the wing-like processes black, narrowly yellow at the base, rounded, strongly denticulate except at bases; 9 with a large dorsal black spot, its base at the apical border of segments tapering basally, the rest of segment yellow; 10 black, rather less than the apical half yellow, the apical border finely black. *Anal appendages* black, longer than segment 10, fusiform, pointed. Vulvar scale dark yellow, deeply cleft to its base into two contiguous lamellæ, pointed at apex.

Distribution.—The *type* (a female), in the Selys collection, is from either India or China. The male described above, which I think is undoubtedly conspecific with the female, is in the Pusa Museum, and was probably taken in BIHAR.

I. atrox is distinguished by the large amount of yellow both on the face and body. Selys remarks that it is nearly related to *I. angulosus*, but the description of the latter insect, of which the male alone is known, reads very differently from that of *I. atrox*. The armature of the female occiput differs from other species of *Ictinus*.

Genus **GOMPHIDIA** Selys. (Fig. 120.)

Gomphidia Selys, Bull. Acad. Belg. vol. xxi, pt. 2, p. 86 (1854); id., Mon. Gomph. p. 259 (1857); Kirby, Cat. Odon. p. 76 (1890); Williamson, Proc. U.S. Nat. Mus. vol. xxxiii, pp. 272, 281, 282 (1907); Laidlaw, Rec. Ind. Mus. vol. xxiv, p. 374 (1922); Fraser, J. Bombay Nat. Hist. Soc. vol. xxix, pp. 60, 328, 666, 667 (1923); Laidlaw, Trans. Ent. Soc. Lond. vol. lxxviii, p. 178 (1930); Needham, Zool. Sinica, ser. A, vol. xi, fasc. 1, pp. 19, 25, 26 (1930); id., Rec. Ind. Mus. vol. xxxiv, pp. 217–219 (1932).

Size large, build robust; colour mat black, marked with bright citron-yellow, closely resembling species of the last genus, to which *Gomphidia* is nearly related. Some of the species are the largest insects known in the Order.

Head large and triangular, frons strongly angulated, occiput simple or concave and fringed with short hairs. Wings: reticulation very close; tornus strongly angulated; base of hind-wing excavate; membrane long and narrow; anal triangle 5-celled; arc situated between the second and third antenodal nervures or opposite the second; 8 transverse nervures between the sectors of arc from the arc to bifurcation of *Rs* in fore-wing, 4 to 6 in hind-wing; 2 rows of postanal cells in fore-wing, 4 to 6 in hind-wing; anal loop very irregular, and blending on the proximal side with the common reticulation of the wing; basal incomplete antenodal nervures absent; nodal index very high; primary antenodal nervures the first and the sixth or seventh; discoidal cell in fore-wings with costal and distal sides about equal and both half as long again as the basal, made up of 3 to 4 cells, that of the hind-wings 3-celled, very elongate in length of wing, costal and distal sides approximately equal and double the length of basal, the distal side markedly sinuous; pterostigma elongate and narrow, braced, about twice the length of the distance from node to proximal end of the pterostigma; *IA* in fore-wing but slightly pectinate; *Cuii* and *IA* in hind-wing parallel to the wing-border; 3 to 4 cubital nervures in fore-wing, 2 in hind-wing; subtrigone of fore-wing 2-celled, that of hind-wing entire and sometimes incomplete, its proximal side joining the base of the discoidal cell; hypertrigones traversed twice; a supplementary nervure running outwards from the distal side of discoidal cells; several strong intercalated nervures always present. Legs robust, hind femora extending to posterior end of thorax and furnished with a group of small, closely-set, numerous spines at the proximal half and two rows of longer, more robust spines at the distal half; hind tibial spines short, numerous and very closely-set. Abdomen dilated at base, then narrow and cylindrical as far as segment 7, terminal segments again dilated, but not nearly to the same extent as in genus *Ictinus*, and segment 8 without lateral wing-like prolongations. Anal appendages; superiors nearly twice the length of segment 10, flattened laterally, forcipate and blunt at apex; inferior very short, very deeply bifid, the two triangular branches but slightly

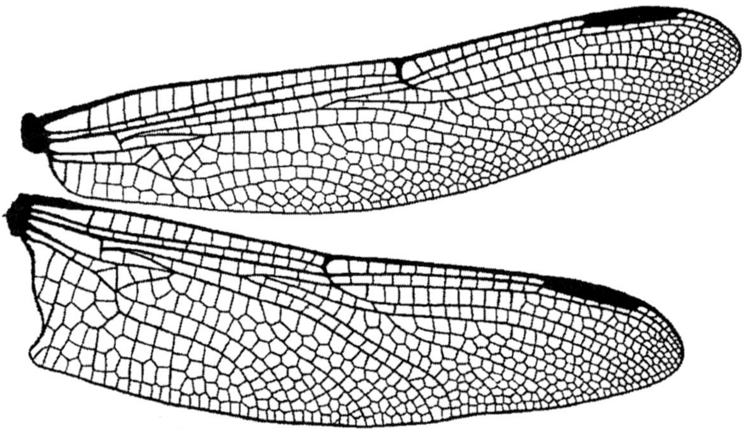

Fig. 120.—Wings of *Gomphidia kodaguensis* Fraser, male.

divaricate. Genitalia: lamina very sinuous, cupped above, and with its centre slightly prolonged; anterior hamules flattened, long, narrow processes minutely spined at apex; posterior hamules much broader and a little shorter, obtuse at apex; lobe shaped like an inverted helmet, broad and moderately deep.

Genotype, *Gomphidia T-nigrum* Selys.

Distribution.—Unlike the last genus, this is purely Oriental in its range, but occurs throughout the region. Five species are found within our limits, of which *G. fletcheri* Fras. is one of the largest dragonflies known. The larvæ, which are similar to those of *Ictinus*, breed in clean submontane streams; this habitat, and certain features of the venation, point to the genus being a more primitive one than *Ictinus*.

Key to Indian Species of Gomphidia.

1.	Face yellow, with T-shaped black mark on upper surface of frons	*T-nigrum* Selys, [p. 383.
	Face black, marked with yellow	2.
2.	Very large species (abdomen more than 60 mm. in length)	*fletcheri* Fras., [p. 386.
	Smaller species (abdomen less than 60 mm. in length)	3.
3.	A superior humeral thoracic spot present	4.
	No superior humeral thoracic spot present ..	5.
4.	Subtrigone of fore-wings made up of 3 cells; abdominal segment 10 entirely black	*pearsoni* Fras., [p. 391.
	Subtrigone of fore-wings made up of 2 cells only; abdominal segment 10 broadly yellow dorsally	*kodaguensis* Fras., [p. 389.
5.	No yellow spots on the medial lateral stripe of thorax; frons entirely yellow above	*williamsoni* [Fras., p. 387.
	An upper and a lower yellow spot on the medial lateral stripe of thorax; frons black, marked with yellow above	*abbotti* Will., [p. 385.

335. **Gomphidia T-nigrum** Selys.

Gomphidia T-nigrum Selys, Bull. Acad. Belg. vol. xxi, pt. 2, p. 86 (1854); id., Mon. Gomph. p. 260 (1857); Kirby, Cat. Odon. p. 76 (1890); Williamson, Proc. U.S. Nat. Mus. vol. xxxiii, p. 282 (1907); Laidlaw, Rec. Ind. Mus. vol. xxiv, pp. 370, 374 (1922); Fraser, J. Bombay Nat. Hist. Soc. vol. xxix, pp. 60, 329, 668, 669, pl. ii, figs. 3, 3 *a* (1923); id., Rec. Ind. Mus. vol. xxvi, pp. 427, 470 (1924); id., ibid. vol. xxxiii, p. 447 (1931); Needham, op. cit. vol. xxxiv, p. 219 (1932).

Male.—Abdomen 53 mm. Hind-wing 38 mm.

Head: eyes bluish-grey; lips, face, and frons bright citron-yellow, upper surface of frons marked with a black line in floor of sulcus, which forms a T by meeting a short medial transverse black line on front of frons; vertex black, the two points of vesicle yellow, occiput largely yellow, its hind border raised, its floor filled by a pyramidal eminence. *Prothorax* brownish-black. *Thorax* black; a broad yellow mesothoracic collar barely interrupted at its middle; two broad, short, dorsal oblique yellow stripes, pointed below and not meeting the mesothoracic collar; the sides broadly yellow, with a broad median black stripe marked above and below with a small upper and a large inferior spot of yellow. Humeral spot vestigial, represented only by a small upper spot, the alar sinus black, the tergum spotted with yellow. *Legs* black, coxæ and trochanters spotted with yellow. Hind femora with two rows of very robust, evenly and widely spaced spines, longest at the middle of femora, crowded at the extreme base, where they are very minute. *Wings* hyaline, costa yellow as far as the pterostigma, which is yellow bordered with black; discoidal cell in fore-wing 4-celled, the basal cell divided into 2 cells, hind-wing 3 cells long; nodal index $\frac{10\text{–}16}{9\text{–}12}\Big|\frac{16\text{–}9}{12\text{–}10}$; 3 cubital cells in fore-wing, 2 in hind-wing; subtrigone in fore-wing 2-celled, that of hind-wing entire; anal triangle 5-celled. *Abdomen* black, broadly marked with yellow as follows:—Segment 1 diffusely yellow on dorsum; 2 yellow except for an irregular black ring at the apex, which extends forwards on either side of dorsum and also below along the ventral border; 3 to 6 with the basal half yellow (rather less on 6); 7 and 8 with narrow black apical rings, broadest on 8, extending forwards along the ventral border on both segments; 9 with a fine lateral stripe at the base and the basal part of dorsal carina finely yellow; 10 with a small dorsal spot. *Anal appendages* brownish; superiors longer than segment 9, broad at base, compressed and of even width thereafter, bevelled at the apex, curving at first outwards and then inwards, the apices meeting; inferior one-third the length of superiors, seen in profile undulated, from below its branches diverging and broadly triangular. *Genitalia* very similar to those of *G. fletcheri* (to which this insect is more closely related than to other species, both by its anal appendages and genitalia); lamina short and broad, bluntly pointed, somewhat excavate; internal hamules long tapering hooks, the apices curling a little outwards; external hamules long narrow and acute tongue-like processes, bright yellow in colour, projecting markedly from the genital sac and hugging the lobe, which is also yellow and broadly funnel-shaped.

Female.—Abdomen 53 mm. Hind-wing 43 mm.

Very similar to the male. Mandibles marked with black at the base; labrum all yellow; occiput concave at its middle; dorsum of segment 2 bearing a triangular yellow spot; the other black markings of abdomen of greater extent; segment 9 without any dorsal stripe; 10 entirely black. Anal appendages short, conical, brownish. Vulvar scale cleft for about three-fourths of its length, the divisions slightly divaricate, extending nearly to base of segment 10.

Distribution.—The *type*, in the Selys collection, is from NORTHERN INDIA (locality unknown). I have never received specimens from that region, but have seen the species in considerable numbers on the Katraj Lake, Poona, DECCAN. In flight it looks very like an *Ictinus*, but its bright yellow colour is sufficient to distinguish it even on the wing. It patrols the borders of the lake for long distances, the males engaging each other in combat as they pass.

336. **Gomphidia abbotti** Williamson.

Gomphidia abbotti Williamson, Proc. U.S. Nat. Mus. vol. xxxiii, pp. 282–285, text-figs. 9, 10 (1907); Fraser, J. Bombay Nat. Hist. Soc. vol. xxix, pp. 60, 329, 672, 673, pl. ii, fig. 4, 4 *a* (1923); Laidlaw, Trans. Ent. Soc. Lond. vol. lxxviii, p. 178 (1930); Needham, Rec. Ind. Mus. vol. xxxiv, p. 219 (1932).

Male.—Abdomen 53 mm. Hind-wing 4 mm.

Head: eyes bluish-green; labium brownish; labrum black, marked with two large transversely-oval lateral yellow spots; anteclypeus yellow, postclypeus black, with a small lateral yellow spot; frons black, its crest narrowly in front and the upper surface yellow, the sulcus black, this colour extending forward nearly to the crest (in Burmese specimens this black area is much restricted); occiput raised, black, fringed with short hairs. *Prothorax* brown, yellowish laterally. *Thorax* black, with a yellow mesothoracic collar slightly interrupted in the middle line; oblique yellow dorsal spots resting on the alar sinus above, widely divaricate below, extending about half-way to the mesothoracic collar; humeral stripe entirely absent; laterally black, marked with a narrow anterior stripe of yellow on the mesepimeron and another, wider, on the metepimeron, between which there is a small upper spot on the black area between the two yellow stripes; tergum spotted with yellow. *Legs* black, armature as in *G. williamsoni*. *Wings* hyaline, evenly and palely enfumed; membrane white; pterostigma dark brown, over 4 to 5 cells, braced; discoidal cell of fore-wing with 4 cells, 3 cells long, the basal divided into 2, that of hind-wing 3 cells long; subtrigones on fore-wing traversed once, that of hind-wing entire; 1 to 2 nervures in hypertrigones; 3 cubital nervures in fore-wing, 2 in hind-wing; nodal index $\frac{9\text{–}18}{10\text{–}13}\Big|\frac{19\text{–}11}{12\text{–}11}$; 5 cells in anal triangle. *Abdomen* black, marked with yellow as follows:—Segment 1 with a dorsal basal spot and a narrow apical one low down on the sides; 2 with a dorsal median spot and a small lateral which involves the oreillets; 3 to 6 with dorsal spots confluent over the dorsal carina except at the extreme apex of spots; 7 with nearly the basal half yellow; 8 with a small basal lateral spot and a smaller one on 9; 10 with a median dorsal spot (this in the Burmese specimens covers the basal half of the dorsum). *Anal appendages*: superiors considerably longer than segment 10, compressed, of about even width to the apex which is squarish, separated in their entire length but converging at the apices; the inferior much shorter, a little sinuous and directed up as seen in profile, deeply cleft, the branches widely divaricate and triangular seen from beneath. *Genitalia* black, tipped with yellow, similar to those of *G. williamsoni*.

Distribution.—The *type* male, in the United States National Museum, is from Trong, Lower Siam. I possess a specimen from Maymyo, BURMA which answers to the above description, the slight variations between it and Mr. Williamson's description being probably due to changes taking place through decomposition.

The entire absence of the humeral stripe separates this species from *G. kodaguensis.*

337. **Gomphidia fletcheri** Fraser.

Gomphidia fletcheri Fraser, J. Bombay Nat. Hist. Soc. vol. xxix, pp. 669, 670, pl. ii, figs. 2, 2 *a*, text-fig. 2 *a* (1923); id., Rec. Ind. Mus. vol. xxvi, pp. 427, 470 (1924); Laidlaw, Trans. Ent. Soc. vol. lxxviii, p. 179 (1930); Fraser, Rec. Ind. Mus. vol. xxxiii, p. 447 (1931); Needham, op. cit. vol. xxxiv, p. 219 (1932).

Male.—Abdomen (with appendages) 63 mm. Hind-wing 42 mm.

Head: eyes bottle-green; middle lobe of labium brownish, lateral lobes yellow; labrum black, marked with two greenish-yellow spots at base; face and frons greenish-yellow, a black stripe across lower part of front of frons which sends a prolongation up to meet a medial black marking on upper surface of frons occupying the floor of sulcus and expanding towards the vesicle; vertex and occiput black, the former raised into two prominent points, the latter raised laterally and medially and fringed with pale-coloured hairs. *Prothorax* black, marked with a yellow anterior collar. *Thorax* black, marked with greenish-yellow as follows:—A broad mesothoracic collar interrupted in the middle line; very short and broad, oblique dorsal stripes rather widely separated from the mesothoracic collar. Laterally broadly yellowish-green, with a medial broad black band which is marked above and below by large yellow spots. Tergum spotted with yellow. *Legs* black, coxæ and trochanters yellow, hind femora with two rows of robust spines, short, numerous and crowded at the base, longest and widely-spaced at the middle, and with 5 or 6 short, closely spaced ones at the apex. *Wings* hyaline, enfumed somewhat patchily with warm brown; pterostigma black, very long, covering 6 to 7 cells, braced; discoidal cell in fore-wing with 3 to 4 cells, that of hind-wing with 3 cells in a line; hypertrigones traversed twice in all wings; subtrigone in hind-wing entire, in fore-wing formed of 3 cells by conjunction of 3 nervures at centre of cell; nodal index $\frac{14\text{–}20}{14\text{–}14}\Big|\frac{20\text{–}12}{16\text{–}14}$; $\frac{15\text{–}18}{14\text{–}14}\Big|\frac{19\text{–}15}{14\text{–}14}$; 2 cubital nervures in hind-wing, 3 to 4 in fore-wing; 6 to 7 cells in anal triangle. (In one specimen the discoidal cells in fore-wings have only

3 cells and the subtrigones have only 2 cells.) *Abdomen* black, marked with yellow as follows :—Segment 1 with a broad dorsal spot and a narrow apical streak low down on the sides confluent with a lateral spot on segment 2 which involves the oreillet and is limited apically by this structure ; 3 to 6 with elongate dorsal spots confluent over the dorsal ridge except at the extreme apices, and gradually diminishing in size from 3 to 6 ; 3 with also the ventro-lateral border narrowly yellow ; segment 7 with the basal half yellow ; 8 with a complete narrow basal ring ; 9 with a mere vestige of this, and 10 entirely unmarked. *Anal appendages* black. Superiors broad at base, compressed in the apical half, apices bevelled off and turning in to meet each other. The inferior only one-third the length of superiors, undulated in profile, flat and triangular as seen from below. *Genitalia* : lobe rather flat, the border emarginate ; internal hamules very long, robust hooks ; external hamules long, flat, narrow, tongue-like lobes projecting well out from the genital sac ; lobe funnel-shaped, prominent, embraced on either side by the outer hamules.

Distribution.—COORG. Two males taken at Hallery, near Mercara, on the borders of a rocky mountain stream, 3,800 ft. altitude.

The insects were very shy and unapproachable, so that I finally had to bring them down with a charge of dust-shot. They bore a close resemblance, when on the wing or resting, to *Ictinus* or to *Megalogomphus hannyngtoni*, for which latter they were mistaken until secured and examined. When settled they rested with the head inclined somewhat downwards and the abdomen held stiffly and straight out, this latter feature distinguishing them from the curved scimitar-like abdomen of *Ictinus*.

The large size of *G. fletcheri* will distinguish it from any other Indian species of *Gomphidia* ; other Oriental species approaching it in size are *G. krugeri* Martin, which has the oblique dorsal stripes joined to an upper humeral spot, and *G. perakensis* Laid., in which the nodal index is much higher than in *G. fletcheri*.

Type in the Author's collection.

338. **Gomphidia williamsoni** Fraser.

Gomphidia williamsoni Fraser, J. Bombay Nat. Hist. Soc. vol. xxix, pp. 670, 671, pl. ii, figs. 5, 5 *a* (1923); Laidlaw, Trans. Ent. Soc. Lond. vol. lxxviii, p. 179 (1930); Needham, Rec. Ind. Mus. vol. xxxiv, p. 219 (1932).

Male.—Abdomen (with appendages) 54 mm. Hind-wing 43 mm.

Head: eyes bottle-green; labium yellow; labrum black, enclosing two moderately large yellow spots; frons greenish-yellow above and in front, a medial basal black spot sometimes present above but never extending as far forwards as the crest; face greenish-yellow below, black above, this colour invading the yellow on the front of frons so as to cut it into two lateral spots; vertex and occiput black. *Prothorax* black, with an anterior yellow collar. *Thorax* black, marked with greenish-yellow as follows:—A mesothoracic collar slightly interrupted in the middle line; broad, oval, oblique, dorsal stripes narrowing below, where they may be connected to the mesothoracic collar or widely separated from it (separated in one male and in the single female examined; connected in two males); laterally two very broad greenish-yellow stripes separated by a broad black stripe which is quite unmarked; the anterior yellow stripe sending a tongue-like process back above, which may have been formed by confluence with an upper yellow spot, the posterior stripe covering the whole of metepimeron; tergum spotted with yellow. *Wings* hyaline, rather deeply enfumed with warm brown; costa black; pterostigma blackish-brown, very long, braced; all discoidal cells 4-celled; anal triangle with 5 cells; nodal index $\frac{10\text{–}20}{12\text{–}14}\left|\frac{18\text{–}11}{14\text{–}11}\right.$. *Legs* entirely black. *Abdomen* black, marked with yellow as follows:—Segment 1 with a dorsal apical triangle and an apical lateral spot low down on the sides; 2 with a broad dorsal stripe not quite reaching the apex, laterally a broad spot of yellow which just meets the dorsal spot at its base and extends to the ventro-lateral border below, involving the oreillet; 3 to 6 with elongate dorsal spots confluent over the dorsal crest save for the extreme apex, these spots gradually decreasing in size from 3 to 6, 3 with a long basal streak low down on the sides; 7 with the basal half yellow, 8 with a tiny lateral basal streak, 9 unmarked; 10 with a small rounded spot on the centre of the dorsum. *Anal appendages* black; superiors long and sinuous, laterally compressed, blunt and bevelled at the apex; inferior very much shorter, leaf-like and curling upwards as seen in profile, flat and triangular as seen from below. *Genitalia*: lamina deeply excavate, more pointed and longer than in *G. fletcheri*; internal hamules short, robust hooks, much shorter than in *G. fletcheri*; external hamules broadly triangular, flat and not markedly projecting from the genital sac; lobe funnel-shaped; external hamules and lobe tipped with yellow.

Female.—Abdomen 54 mm. Hind-wing 45 mm.

Very similar to the male, but the yellow markings broader and better defined. Wings rather more deeply enfumed;

nodal index $\frac{13\text{–}20}{12\text{–}15}\Big|\frac{20\text{–}13}{14\text{–}12}$. Occiput raised, a small point at its centre, finer than that found in *Ictinus*; vesicle black, marked with two small yellow spots. The dorsal oblique stripes widely separated from the mesothoracic collar; the black stripe, traversing the sides of thorax, marked with a large lower spot. Segment 2 of abdomen almost entirely yellow, the lateral band very broad and extending the entire length of the segment, the dorsal stripe also very broad in its basal two-thirds, the apical third connected to it by a narrow neck only; segment 9 with a fine lateral basal streak similar to that on 8; segment 10 unmarked. *Anal appendages* very short, conical, black. Vulvar scale very characteristic, deeply cleft at its base into two long, narrow, tongue-like, foliate processes, the apices of which extend as far as the base of segment 10.

Distribution.—Duars, BENGAL (*H. V. O'Donel*).

The species is closely allied to *G. abbotti* and *G. kodaguensis*, especially to the latter. It differs from *G. kodaguensis* in the individual greenish-yellow area on the upper surface of the frons, with at most a very small black spot at the base, also in the absence of an upper humeral spot and of spots on the lateral black band on the thorax. From *G. abbotti* it differs in the absence of spots on the lateral black band of the thorax and of black on the upper surface of the frons; also in the much greater extent of yellow on segment 2, etc.

Type in the British Museum.

339. **Gomphidia kodaguensis** Fraser. (Fig. 119, a.)

Gomphidia kodaguensis Fraser, J. Bombay Nat. Hist. Soc. vol. xxix, pp. 671, 672 (1923); id., Rec. Ind. Mus. vol. xxvi, pp. 427, 470 (1924); Laidlaw, Trans. Ent. Soc. Lond. vol. lxxviii, p. 179 (1930); Fraser, Rec. Ind. Mus. vol. xxxiii, p. 447 (1931); Needham, op. cit. vol. xxxiv, p. 219 (1932).

Male.—Abdomen (with appendages) 53 mm. Hind-wing 42 mm.

Head: eyes bottle-green; face and frons bright citron-yellow, the latter with black marks in floor of sulcus and on upper part and front, and a transverse black line on the lower part of front, all these black areas confluent; labium yellow; labrum yellow, narrowly bordered with black, the ground-colour split into two large spots by a medial vertical black streak running from the base; occiput black, raised, fringed with very short pale yellow hairs; vertex black, the vesicle elevated into two very acute prominent points. *Prothorax* black, with a narrow anterior yellow collar. *Thorax* black, marked with yellow as follows:—

A mesothoracic collar broadly broken in the middle line; two broad, oval, greenish-yellow dorsal stripes, pointed below and not nearly meeting the mesothoracic collar; a small upper spot representing the rudimentary humeral stripe. Laterally two broad yellow stripes separated by a broad black stripe, which includes an upper and a lower yellow spot; the posterior yellow stripe covering the whole of metepimeron. Tergum spotted with yellow. *Legs* black; coxæ, trochanters, and fore femora yellow. *Wings* hyaline, costa black, the apices and posterior borders palely and evenly enfumed with warm brown; pterostigma very long, black; nodal index $\frac{13\text{–}22}{14\text{–}15}\Big|\frac{22\text{–}15}{15\text{–}13}$; discoidal cell of fore-wings with 3 to 4 cells, of hind-wings with 3 cells (traversed twice); 5 cells in the anal triangle; 5 cubital nervures in fore-wing, 3 in hind-wing; subtrigone in fore-wing traversed once, in hind-wing free; hypertrigones traversed once or twice. *Abdomen* black, marked with yellow as follows:—Segment 1 with a fine apical dorsal ring; 2 with a small oval dorsal spot on the basal two-thirds, a small lateral spot which involves the oreillet, and subdorsally and subapically a very tiny yellow spot; 3 to 6 with dorsal basal elongate spots confluent over the dorsal crest except at the extreme apices; 3 also with a narrow basal streak along the ventro-lateral border; apices of genital lobe and external hamules also tipped with yellow; 7 with the basal half yellow; 8 with an elongate transverse basal spot; 9 unmarked; 10 with a dorsal spot expanding apically and covering almost the entire dorsum. *Anal appendages* (fig. 119, **a**) black. *Genitalia* similar to those of *G. williamsoni*, to which this insect is closely related, as also to those of *G. abbotti*.

Female.—Undescribed. One which I saw but failed to capture appeared to have markings similar to those of the male, but the abdominal rings broader and more conspicuous.

Distribution.—South India: Coorg (in montane areas), Vayitri, Malabar Wynaad, Bolovumpatti Forest, and the Mudis Hills in the Coimbatore district. This species is common in Coorg, especially on the Sampaji and Bhagmandala Rivers, from where the type comes. It is a scarce insect elsewhere.

G. kodaguensis differs from both *G. williamsoni* and *G. abbotti* in having the face and lips almost entirely yellow and by the black on frons joining up with that on the upper surface. From *G. williamsoni* it again differs in the more restricted yellow markings on segments 1 to 3 and in the broader markings on segments 8 to 10 of the abdomen; from *G. abbotti* in the greater number of antenodal nervures and in the vestigial humeral spot on the thorax, which is absent in *G. williamsoni*. *Type* in the British Museum; paratypes in my own, Morton, and Laidlaw collections.

340. **Gomphidia pearsoni** Fraser.

Gomphidia pearsoni Fraser, Ceylon J. Sci., B, vol. xviii, pt. i, pp. 20–22 (1933).

Male.—Abdomen (with appendages) 60 mm. Hind-wing 42 mm.

Head: labrum bright yellow; labium black, with two small citron-yellow spots at base; bases of mandibles yellow at base; anteclypeus brown; postclypeus bright citron-yellow on the lower half, black on the upper half; frons black, with a large citron-yellow triangular spot on each side above, which overlaps slightly on to the anterior surface; vertex and occiput black, the latter fringed with short black hairs; eyes bottle-green during life. *Prothorax* black. *Thorax* black, marked with bright citron-yellow as follows:—A broad, slightly interrupted mesothoracic collar; short pyriform antehumeral stripes, the lower ends pointed and strongly divergent; a small upper humeral spot; two broad stripes on the sides, an anterior at the level of the spiracle, and a much broader stripe covering the posterior two-thirds of the metepimeron; lastly a small upper and a lower spot on the black band dividing these two yellow stripes. *Legs* black. *Wings* hyaline, palely enfumed, especially at the apices; pterostigma long, braced, covering 5 to 7 cells, black; 4 cells in the discoidal cell of fore-wing, 3 in hind-wing; 2 to 3 cells in subtrigone of fore-wing, entire in hind-wing; 3 cubital nervures in fore-wing, 2 in hind-wing; nodal index $\frac{12\text{–}22 \mid 23\text{–}13}{12\text{–}15 \mid 14\text{–}13}$. *Abdomen* black, marked with citron-yellow as follows:—Segment 1 with a vestigial apical dorsal spot and a fine linear stripe on the apical border each side; 2 with the oreillets, a small lateral spot on the apical side of these, and a mid-dorsal triangular basal stripe extending for two-thirds the length of segment; 3 to 6 with small triangular basal spots, emarginate apically and becoming gradually smaller, on 6 almost obsolete; 7 with the basal half yellow, partially bisected on the sides by the black jugal suture; 8 with a narrow linear baso-lateral stripe on each side; remaining segments unmarked. *Anal appendages* black; superiors half as long again as segment 10, compressed, narrow and of even width throughout, bevelled to a point at apex; inferior only one-third as long, deeply emarginate, forming two triangular lobes, the apices of which are slightly turned up and end in three minute teeth.

Female unknown.

Distribution.—CEYLON : Belihul Oya, and Bibile, Gal Oya, from May to July.

This species is closely related to *G. kodaguensis*, but is distinguished by its larger size, the less extent of the yellow markings, segment 10 being quite unmarked. The sub-trigone of the fore-wings is made up of 3 cells instead of 2, as in *G. kodaguensis*. The species also shows a strong resemblance to *G. fletcheri*, but it is smaller and less robust than the latter, and the yellow markings, especially on the face are less extensive.

Type in the Colombo Museum; paratype in my own collection.

ALPHABETICAL INDEX.

[All names printed in *italics* are synonyms. When a species is mentioned on more than one page, the page on which the description occurs is indicated by thickened numerals.]

PRINTED BY TAYLOR AND FRANCIS, RED LION COURT, FLEET STREET.

PLATE I.

Wings of *Rhinocypha* (male).

Fig. 1. *Rhinocypha cuneata* Selys.

2. *Rhinocypha bisignata* Selys.

3. *Rhinocypha iridea* Selys.

Note :—The colours shown are due to iridescence, and are only displayed when the wings are held in certain positions.

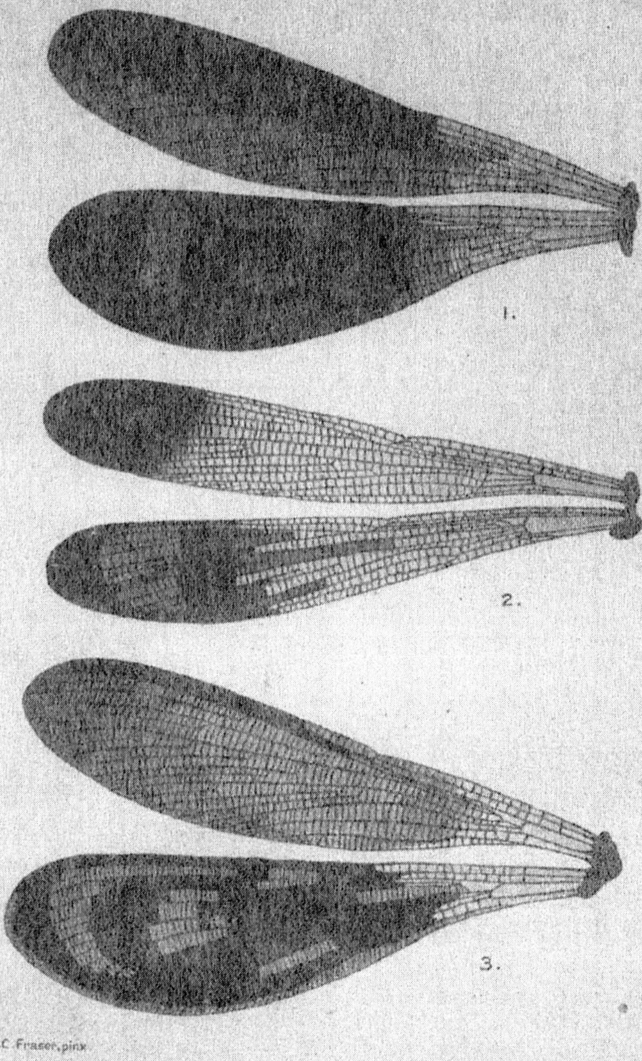

F.C. Fraser, pinx.

RHINOCYPHA

PLATE II.

Wings of *Rhinocypha* (male).

Fig. 1. *Rhinocypha trimaculata* Selys.

2. *Rhinocypha biforata delimbata* Selys.

3. *Rhinocypha trifasciata* Selys.

Note :—The colours shown are due to iridescence, and are only displayed when the wings are held in certain positions.

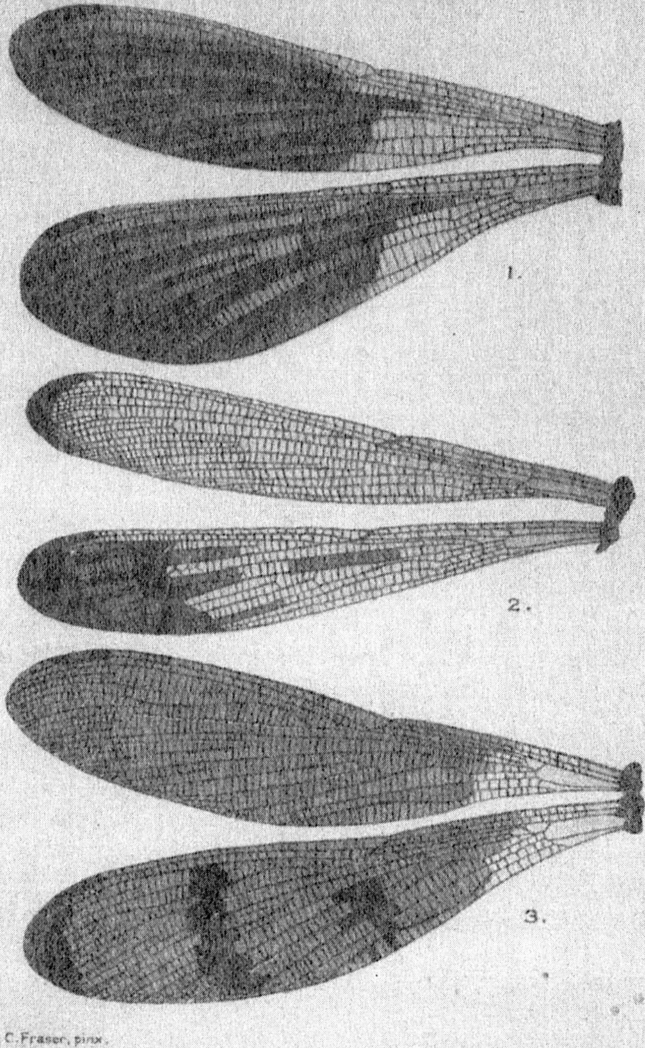

F. C. Fraser, pinx.

RHINOCYPHA

PLATE III.

Body (side view) of Indian Gomphinæ (male).

Fig. 1. *Stylogomphus inglisi* Fras.

2. *Ophiogomphus reductus* Calv.

3. *Onychogomphus saundersi* Selys.

4. *Onychogomphus pulcherrimus* Fras.

5. *Onychogomphus aureus* Laid.

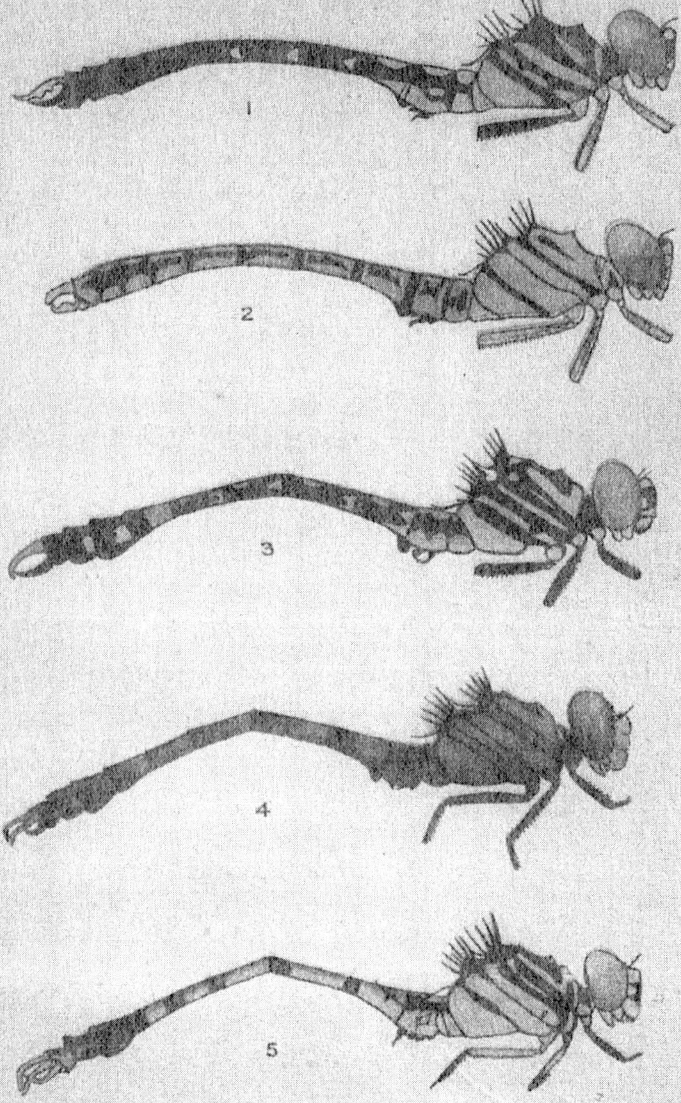

F.C. Fraser. pinx.

GOMPHIDÆ

PLATE IV.

Body (side view) of Indian Gomphinæ (male).

Fig. 1. *Megalogomphus hannyngtoni* (Fras.).

2. *Megalogomphus superbus* Fras.

3. *Megalogomphus smithi* (Selys).

4. *Davidioides martini* Fras.

5. *Perissogomphus stevensi* Laid.

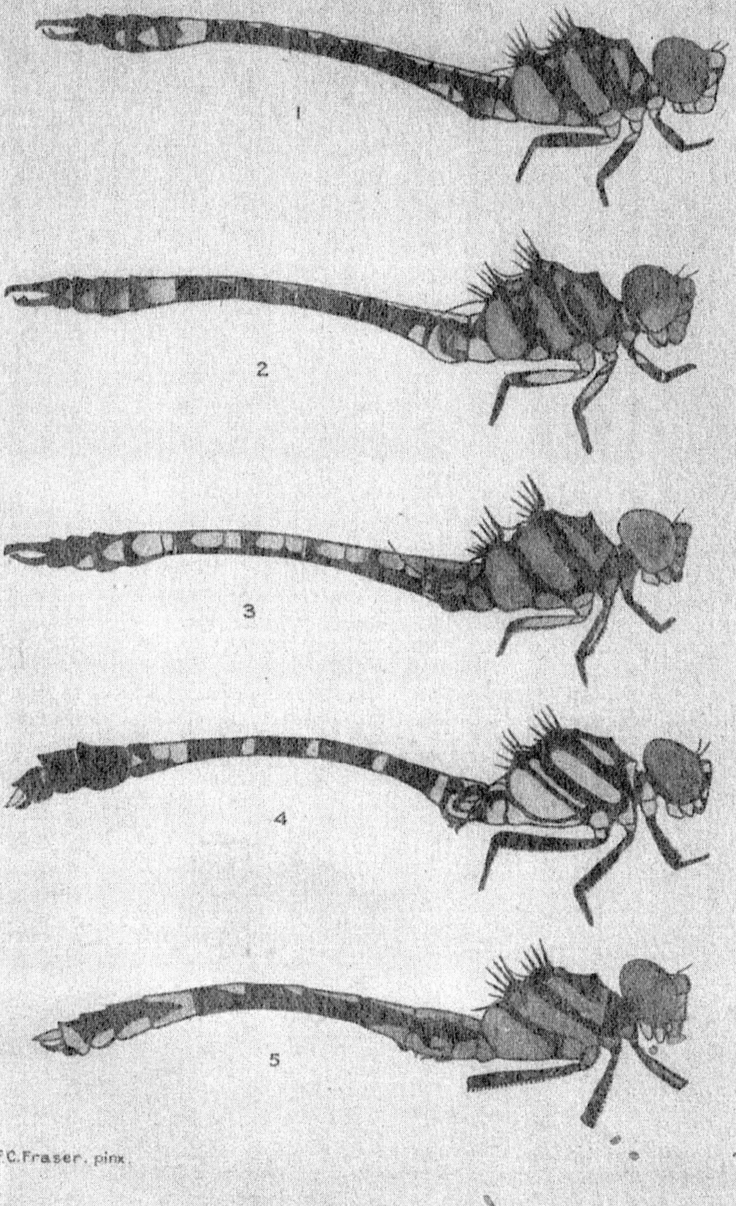

F.C.Fraser, pinx.

GOMPHIDÆ